Pore Scale Phenomena

Frontiers in Energy and Environment

World Scientific Series in Nanoscience and Nanotechnology*

ISSN: 2301-301X

Series Editor-in-Chief: Frans Spaepen (*Harvard University, USA*)

Published

Vol. 4 Plasmonics and Plasmonic Metamaterials: Analysis and Applications
edited by Gennady Shvets (The University of Texas, Austin, USA) and Igor Tsukerman (The University of Akron, USA)

Vol. 5 Inorganic Nanomaterials from Nanotubes to Fullerene-Like Nanoparticles: Fundamentals and Applications
Reshef Tenne (Weizmann Institute of Science, Israel)

Vol. 6 Plasmon Resonances in Nanoparticles
Isaak D. Mayergoyz (University of Maryland, USA)

Vol. 7 Scanning Probe Microscopy for Energy Research: Materials, Devices, and Applications
edited by Dawn A. Bonnell (The University of Pennsylvania, USA) and Sergei V. Kalinin (Oak Ridge National Laboratory, USA)

Vol. 8 Polyoxometalate Chemistry: Some Recent Trends
edited by Francis Secheresse (Université de Versailles-St Quentin, France)

Vol. 9 Handbook of Biomimetics and Bioinspiration: Biologically-Driven Engineering of Materials, Processes, Devices, and Systems
edited by Esmaiel Jabbari (University of South carolina, USA), Deok-Ho Kim (University of Washington, USA), Luke P. Lee (University of California, Berkeley, USA), Amir Ghaemmaghami (University of Nottingham, UK) and Ali Khademhosseini (Harvard University, USA & Massachusetts Institute of Technology, USA)

Vol. 10 Pore Scale Phenomena: Frontiers in Energy and Environment
edited by John Poate (Colorado School of Mines, USA), Tissa Illangasekare (Colorado School of Mines, USA), Hossein Kazemi (Colorado School of Mines, USA) and Robert Kee (Colorado School of Mines, USA)

*For the complete list of volumes in this series, please visit
www.worldscientific.com/series/wssnn

Volume 10

World Scientific Series in **Nanoscience and Nanotechnology**

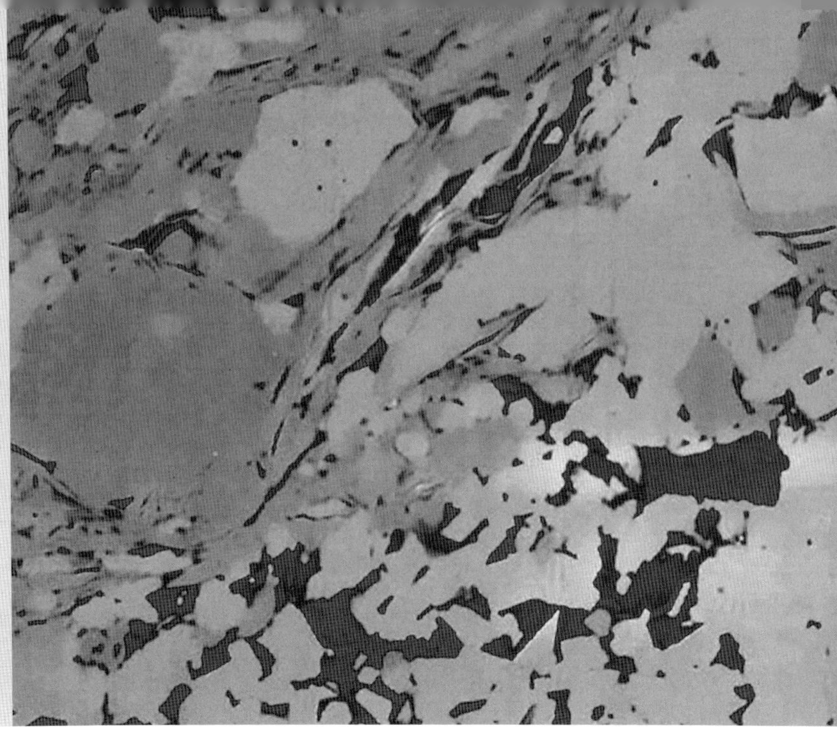

Pore Scale Phenomena

Frontiers in Energy and Environment

Editors

John Poate

Tissa Illangasekare

Hossein Kazemi

Robert Kee

Colorado School of Mines, USA

World Scientific

NEW JERSEY • LONDON • SINGAPORE • BEIJING • SHANGHAI • HONG KONG • TAIPEI • CHENNAI

Published by

World Scientific Publishing Co. Pte. Ltd.

5 Toh Tuck Link, Singapore 596224

USA office: 27 Warren Street, Suite 401-402, Hackensack, NJ 07601

UK office: 57 Shelton Street, Covent Garden, London WC2H 9HE

Library of Congress Cataloging-in-Publication Data

Poate, J. M., author.

Pore scale phenomena : frontiers in energy and environment / John Poate, Tissa Illangasekare, Hossein Kazemi, Robert Kee (Colorado School of Mines, USA).

pages cm -- (World Scientific series in nanoscience and nanotechnology : volume10)

Includes index.

ISBN 978-9814623056 (alk. paper)

1. Porous materials--Mechanical properties. 2. Materials science. 3. Engineering. I. Illangasekare, T. H., author. II. Kazemi, Hossein, 1938– author. III. Kee, R. J., author. IV. Title. V. Series: World Scientific series in nanoscience and nanotechnology ; v. 10.

TA418.9.P6P567 2015

620.1'16--dc23

2015005053

British Library Cataloguing-in-Publication Data

A catalogue record for this book is available from the British Library.

Printed in Singapore

Preface

This monograph consists of a series of chapters devoted to understanding the physical and chemical properties of pore scale phenomena primarily in the energy and environmental worlds. The properties of these phenomena range across length scales from the atomic to the geographic and have large economic and societal implications. The genesis of this endeavor rose through a series of meetings at the Colorado School of Mines where we realized that the school had a critical mass of faculty addressing key pore scale research issues. We have tried to capture the essence of this research in this monograph. The chapters reflect the heterogeneity of the research. Mines has a proud record in applied research and engineering from its founding days and its impact on the mining industry in the US and the world. We are now contributing in many areas of energy and environmental research and trust that this monograph demonstrates this diversity.

We would like to thank and acknowledge our colleagues at Mines and the authors from our collaborating universities, national laboratories, and companies who tackled this endeavor with real enthusiasm. In particular we could not have pulled this off in a timely fashion without the expert assistance of Luis Zerpa, Giovanny Grasso, Jeff Brown and Lisa Kinzel.

John Poate, Hossein Kazemi, Tissa Illangasekare, Robert Kee
Colorado School of Mines
Golden, Colorado
September 2014

Contents

Chapter 1

Overview

John M. Poate
Research and Technology Transfer,
Guggenheim Hall,
Colorado School of Mines, Golden, CO, USA
jpoate@mines.edu

Pore scale phenomena and their connected networks are pervasive in many areas of science. This monograph primarily focuses on those areas pertinent to energy and the environment in terms of oil and gas reservoirs, geohydrology and energy conversion. Transport properties are considered in detail. This chapter introduces the contents of the monograph.

1.1. Introduction

What are pore scale phenomena and why is there so much current interest? The answer is convoluted with a fascinating history of the earth, materials and biological sciences. Natural porous materials abound in the earth, such as expansive bentonite clays and porous limestone and in living organisms, such as the mastoid bone with its porous air cavities. Moreover many fabricated materials and structures now have complicated porous structures. Understanding the properties of these pores in terms of their structure and transport properties is now emerging as a grand challenge in terms of the scaling from the microscopic or nanoscale regime to the macroscopic world. This transition from the nanoscale to a macroscopic world is known as mesoscale science.

Undoubtedly the driver for much of this interest in the earth sciences has been the shale gas revolution. Extraction of natural gas from nanoscale pores in shale or rocks by hydraulic fracturing has transformed the energy agenda of the US. Many fundamental questions remain regarding the fluid flow mechanisms and opportunities for improving flow. There is now a burgeoning interest in similar processes with regard to water extraction and conservation. The fundamental challenges of understanding atomic and molecular transport in the porous structures of energy-conversion devices are profound. The performance of such devices is critically dependent on the transport phenomena.

This is a very large field of science and technology. This monograph primarily concentrates on three areas of the earth and materials sciences that are driving the research and engineering agendas. The opportunities for cutting edge and interdisciplinary research are large. Take, for example, the fields of physical and chemical analysis of surface and bulk properties of materials. Analyzing pores and their characteristics is opening up a new field of physical and chemical analytical techniques. The possibilities of collaboration and synergies between the traditional materials, earth and biological science disciplines could lead to new discoveries and understanding.

1.2. Oil and Gas Reservoirs

The first section of this monograph is introduced by H. Kazemi with a discussion of the physics of flow in petroleum reservoirs. Figure 1.1 shows the dramatic increase in US shale gas production in recent years. As discussed by H. Kazemi, this increase would not have occurred without seminal engineering and scientific advances. The societal and economic impact of this energy source is profound. To quote Daniel Yergin [1], one of the world's leading energy economists and analysts, "Shale gas, the biggest energy innovation since the start of the new century, has turned what was an imminent shortage in the United States into what may be a hundred year supply and may do the same elsewhere in the world. It is dramatically changing the competitive position for everything from nuclear energy to wind power. It has also stoked, in a remarkably short time, a new environmental debate".

This remarkable US revolution was charted and predicted by the Potential Gas Committee [2] which is comprised of 100 volunteer geoscientists and petroleum engineers from industry, academia and state and federal government organizations. The committee functions independently but with technical and administrative assistance from the Potential Gas Agency of the Colorado School of Mines. The committee has been assessing the technically recoverable US natural gas resource since 1964. Prior to 2008 the committee assessed relatively small volumes of gas in shale formations. However, with the increase in shale drilling and production from 2006-2008

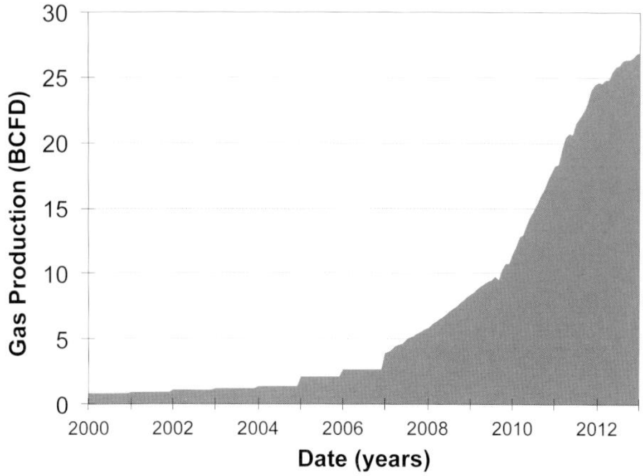

Fig. 1.1. U.S. total shale gas production comprises about 40 percent of total U.S. dry gas production (www.eia.gov/energy in brief/article/about shale gas.cfm).

due to horizontal drilling and multistage hydraulic fracturing, the committee members used private and public data to expand their evaluation of shale reservoirs. While some skepticism was voiced at the time regarding this substantial increase in recoverable resources, shale gas production since 2008 has validated the committees assessments. Indeed the work of US scientists has led to an assessment of the worlds oil and shale gas formation. The technically recoverable resources are ubiquitous throughout the world as shown in Fig. 1.2.

The flow of oil, water or gas through a petroleum reservoir is described by Darcy's Law. The Darcy (D) is a measure of the discharge rate or fluid flow corrected for the viscosity of the fluid and driving pressure gradient. Sandstone and carbonate reservoirs typically have permeabilities in range 10^{-1}–10^{3} mD. These are, for example, the easily accessible and high production reservoirs of the Middle East. Tight shale, for example the Marcellus basin, of the US, is the rock equivalent of mud with permeabilities in the range 10^{-6}–10^{-3} mD. Figure 1.3 shows the inherent permeabilities of accessible and inaccessible reservoirs as a function of rock type. Lateral drilling and hydraulic producing can increase the permeabilities of the tight sandstones and shales by many orders of magnitude for oil and gas production. Numerous examples are shown in this monograph of such shales where the carbonaceous material (oil, gas) and water is contained in pores embedded in the shale matrix.

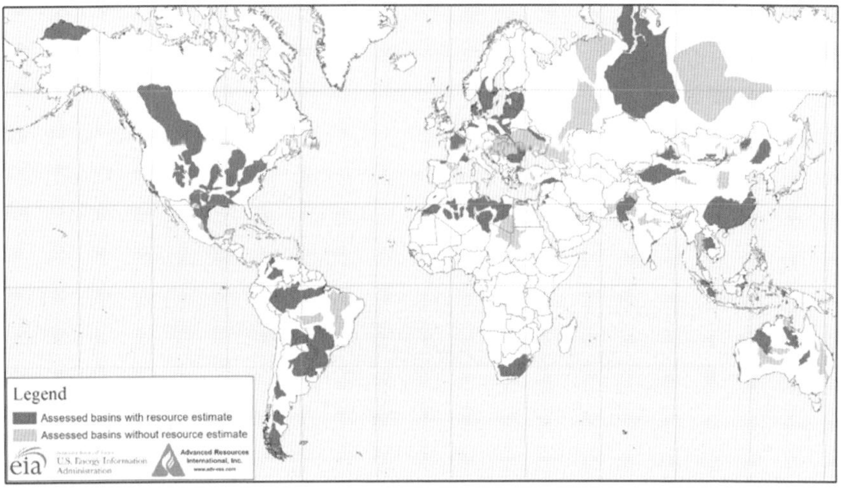

Fig. 1.2. Map of basins with assessed shale oil and shale gas formations (www.eia.gov/analysis/studies/worldshalegas/).

The challenges in simulating, modelling and understanding the behavior of the shales before and after fracturing are daunting. The chapters in this section deal with virtually all aspects of this science and engineering. Figure 1.4 shows an example of the convergence of theory and experiment in modelling fluid flow in fractured networks [3, 4] that are constructed in a two dimensional polymer matrix. These transparent micromodels give beautiful visual progressions of fluid flow in networks that can be artificially configured. In the field, additives are added to hydraulic fracturing fluids to decrease viscosity and increase permeability. In this experiment water flooding is substantially increased by additions of surfactants that decrease viscosity. Such work is allowing the development of models to predict multiphase flow in the field.

A fascinating aspect of the gas reservoir and energy story is that of the methane hydrates. These crystalline solids of water and gas were first documented by Sir Humphrey Davy in 1811 [5]. They were first regarded as a scientific curiosity. Then industry recognized their importance when natural gas pipelines, at high pressures and low temperatures, were blocked by solidifying gas and water. Global studies now show enormous quantities of naturally occurring gas hydrates in porous media. The question is whether these gas resources can be recovered in an efficient and economic fashion.

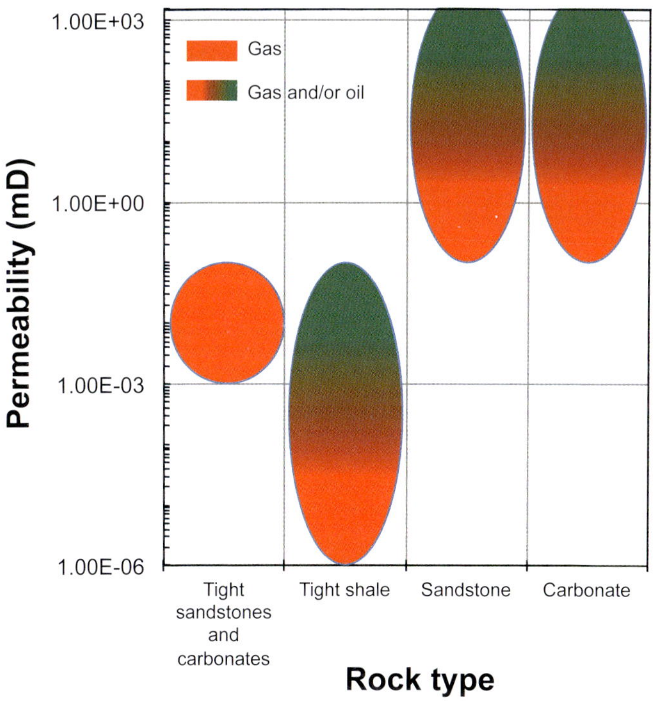

Fig. 1.3. Permeability vs. rock type.

These issues are addressed in the chapters which concern the dissociation and transport mechanisms. These studies bear directly on the environmental concerns of the release of methane under global warming and other geologic events. An environmental consequence of the increase in natural gas production from shale reservoirs has been the substantial decrease in CO_2 production. Gas-fired power stations produce 50% of the total produced CO_2 from coal-fired stations. Whatever the source, there is a need to sequester CO_2. One very novel suggestion [6] is illustrated in Fig. 1.5 showing that CO_2 can form clathrate hydrates, which thermodynamically is favored to displace methane from the hydrate cage. Experiments are underway to inject CO_2 to displace methane from hydrates; thus producing fuel and sequestering the CO_2.

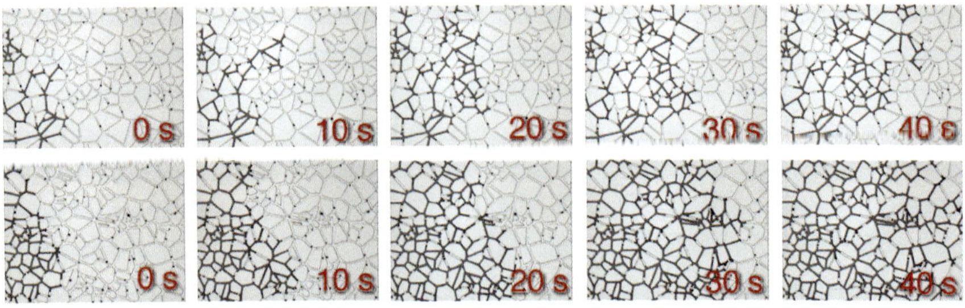

Fig. 1.4. Progression of water flooding (top) and surfactant flooding (bottom) in polymeric organosilicon compounds (PDMS) microfluidic channels. The network contains microchannels that are 4-8 μm wide and 15 μm deep. From Xu et al. [3, 4].

1.3. Hydrology

This section of the monograph is introduced by T. Illangasekare, K. Smits, R. Fucik and H. Davarzani. The first section of the monograph dealt with the accessibility of oil and gas reservoirs and their production which drives the worlds energy economy. An area of equal societal concern in terms, for example, of sustainability and food production is that of water. While the economic driver there is small, in comparison, to the oil and gas industry, the situation is changing. Erratic weather patterns are becoming evident with increased periods of drought. As Ben Franklin [7] remarked in 1746,"When the well is dry, we know the worth of water". Governments and agencies are now addressing with increasing urgency, the research challenges of coupling the earth and meteorological systems for the hydrological cycle.

The hydrological cycle is shown (Fig. 1.6) in the classic figure of Trenberth et al. [8]. The driving forces are radiation and energy. The atmospheric processes cause precipitation for example and the surface and sub-surface processes complete the water cycle. The chapters in this section deal with the many issues of scaling water transport from the nano or small scale to the field scale in the sub surface. The latest modelling and experimental techniques are discussed. An area of particular interest to pore scale studies is that of geochemical reactivity. Injection of CO_2 for sequestration purposes or injection of a working geothermal fluid can be severely impacted by the chemical reactivity of the pores.

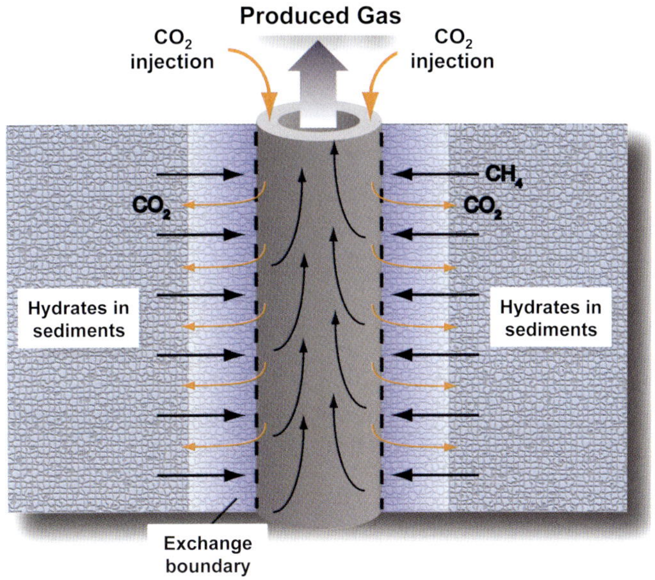

Fig. 1.5. Schematic of a process for CH_4 production from a hydrate reservoir with the injection of CO_2 using a single-well system. From Koh et al. [6].

As presented in the section on oil and gas, horizontal drilling and hydraulic fracturing technologies have reached considerable levels of sophistication. There is now considerable interest in applying those technologies for energy production by enhanced geothermal systems. As an example, two wells, injection and production, are drilled. Fluid is injected into fractured hot rock. The fluid will then gain heat, flow to the fractured zone of the production well and then to the surface. At the surface the heat will be used to generate electricity. For such a system to be successful, knowledge and control of the transport in these fractured, porous zones will be essential to gaining access, in this way, to this clean, inexhaustible, energy source. If this technology succeeds, it will provide a remarkable transition from a fossil fuel energy source to the ultimate terrestrial renewable energy source.

1.4. Materials Science, Energy, Devices & Biology

Materials science, solid state physics and chemistry have undergone a fascinating path from understanding bulk crystals, and their properties to the

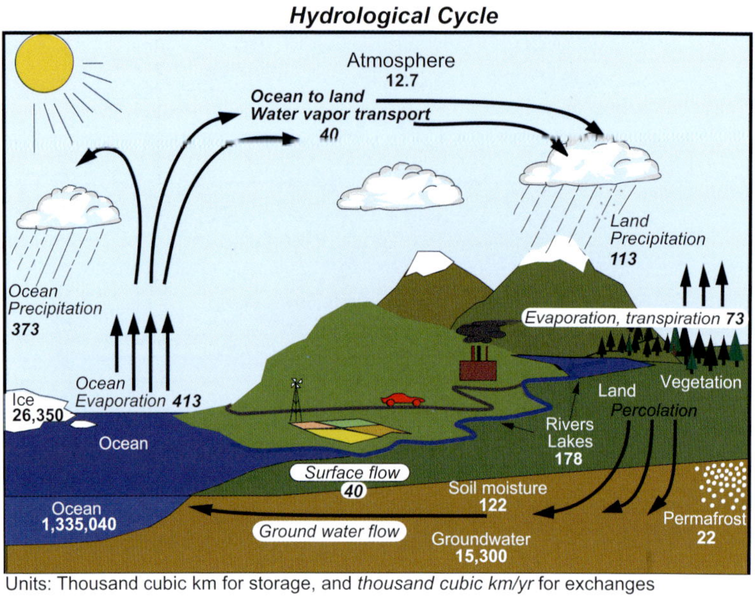

Fig. 1.6. The hydrological cycle. Estimate the main water reservoirs. Plain font for storage (103 km^3) and slanted font for flow of moisture (10^3 km^3 yr^{-1}). From Trenberth et al. [8].

properties of nanoparticles. Understanding the properties of porous solids presents a new challenge. In this section, R. Kee, H. Zhu, G. Goldin and S. Barnett review the subject in terms of porous media transport and chemistry in energy conversion devices. This is a frontier area in pore scale studies and much progress is being made.

Many examples are given throughout this monograph of analytical techniques for analyzing the microstructures of porous materials. This section contains a chapter reviewing and examining the whole portfolio of analytical techniques. Determining pore structures and their physical and chemical properties at the bulk and atomic scales represents, in many ways, a grand challenge for analytical science. Much progress has been made in recent years as demonstrated in this chapter.

A particularly promising area of research and development for porous materials is the fabrication of membranes for separation of gases and liquids. Reviews are given of applications in the energy and desalination fields. Fabrication of nanoporous materials, in particular, mesoporous sil-

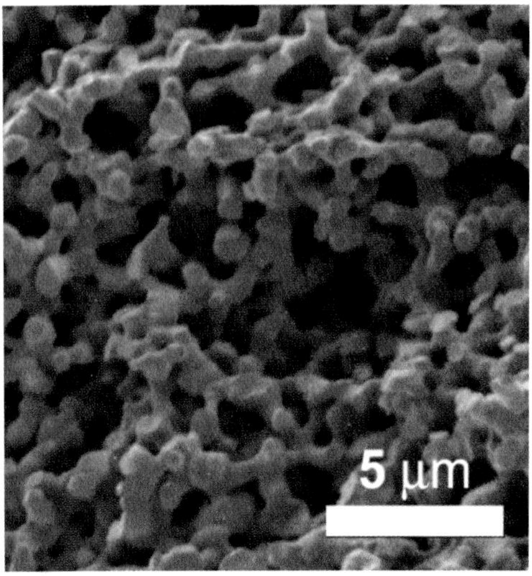

Fig. 1.7. Scanning electron microscope image of a carbon aerogel with an activated surface area of 3100 $m^2 g^{-1}$. From Biener et al. [9].

ica are reviewed in terms of chemical functionality for the biofuel and water purification industries. A burgeoning area of research and development is that of fabricating foams or aerogels. These foams are in many ways the classic example of a pore dominated material with very low densities and high interior surface areas. Figure 1.7 [9] shows a carbon aerogel with an enormous interior surface area of 3100 $m^2 g^{-1}$. Such structures could be used in diverse energy applications such as supercapacitors or hydrogen storage.

The final chapter is in some ways the most fascinating. It reviews the transport of fluids and solutes through porous biological media. Hydraulic permeability and diffusivity determine the rate of fluid transport. This data provides important information on the formation of blood clots. This monograph ranges over many areas of pore scale phenomena in the energy and environmental domain. In this blood transport chapter many disciplines and concepts come together.

Acknowledgements

Hossein Kazemi, Luis Zerpa and John Curtis have been a great help assembling this preamble.

References

[1] D. Yergin, *The Quest: Energy, Security, and the Remaking of the Modern World*. Penguin Group US (2011).
[2] P. G. Committee. Potential gas committee reports significant increase in magnitude of u.s. natural gas resource base. Technical Report, Potential Gas Committee (2013). URL `http://potentialgas.org/press-release`.
[3] W. Xu. Effect of pore geometry on water and surfactant flooding by pdms microfluidic micromodels. Master's thesis, Colorado School of Mines, Golden, CO, USA (2012).
[4] W. Xu, J. T. Ok, F. Xiao, K. B. Neeves, and X. Yin, Effect of pore geometry and interfacial tension on water-oil displacement efficiency in oil-wet microfluidic porous media analogs, *Physics of Fluids*. 26(9):093102 (2014).
[5] H. Davy, On some of the combinations of oxy-muriatic gas and oxygen, and on the chemical relations of the principles to inflamable bodies, *Phil. Trans. Roy. Soc. Lond.* **101**, 1 (1811).
[6] C. A. Koh, A. K. Sum, and E. D. Sloan, Gas hydrates: Unlocking the energy from icy cages, *Journal of Applied Physics*. 106(6):061101 (2009).
[7] B. Franklin. Poor Richard's Almanac (1746).
[8] K. E. Trenberth, L. Smith, T. Qian, A. Dai, and J. Fasullo, Estimates of the global water budget and its annual cycle using observational and model data, *J. Hydrometeor.* **8**(4), 758–769 (2007).
[9] J. Biener, M. Stadermann, M. Suss, M. A. Worsley, M. M. Biener, K. A. Rose, and T. F. Baumann, Advanced carbon aerogels for energy applications, *Energy Environ. Sci.* **4**, 656–667 (2011).

PART 1

Oil & Gas Reservoirs

Chapter 2

Subsurface Pores: Physics of Flow in Petroleum Reservoirs

Hossein Kazemi
Department of Petroleum Engineering,
Colorado School of Mines, Golden, CO, 80401, USA
hkazemi@mines.edu

Pore is a minute opening or space in material, often giving passage to fluids. Pores are abundant in many common materials such as soil, coal, and shale; in leaf, skin and biological tissues; in subsurface aquifers, petroleum reservoirs, and crystalline granite and gneiss deposits; in organic and inorganic membranes, and in synthetic and natural materials. The subsurface pores vary greatly in morphology, size and connectivity. In petroleum applications, the nanometer and micrometer void space between rock grains and rock fragments constitute the reservoir pores. Subsurface pores also include microfractures in the rock matrix and nanometer-size voids in organic matter within the reservoir sediments.

Because of the growing demand for clean energy and water, a new area of science and technology, **the nanoscale science of flow and transport**, has emerged which entails understanding physics of flow at the molecular, nano and macro scales (**Chapter 1**, *Introduction to Pore Scale Science & Technology*, J. Poate). In addition to energy and hydrology, nanoscale science cuts across the materials and biosciences disciplines. This chapter presents a pragmatic overview of flow and transport in petroleum-containing subsurface pores. Supporting material includes explanation of oil production by fluid expansion, water injection, gas injection, enhanced oil recovery (EOR) in conventional reservoirs and unconventional shale, and numerical modeling of flow and transport in reservoirs.

2.1. Introduction

In petroleum reservoirs, pores constitute the storage space for oil, gas, and water. **Pore size** generally means **pore throat diameter or the average**

width of the fracture aperture – both varying from one **nanometer** to a few **micrometers**. Reservoirs also contain larger solution cavities (or vugs) imbedded within the large cluster of smaller pores. The pore size in the **conventional** (or traditional) oil and gas reservoirs is in the micrometer range. In the **unconventional** reservoirs (i.e., shale or very low-permeability clastic sediments) the dominant pore size is in the nanometer range.

Flow and transport in petroleum reservoirs is complex and affected by pore size, morphology, tortuosity, mineralogy, presence of multiple phases, phase appearance or disappearance, interfacial and capillary force, gravity, fluid rheology, and rock-fluid interactions including ion exchange at the pore-fluid interface.

At the end of the reservoirs economic life, on the average, two-thirds of the oil stays behind. This is consistent with the observations that excellent reservoirs produce around fifty percent while poor reservoirs produce around ten percent of the original oil in place. Nonetheless, all reservoirs can produce at very low flow rates for decades beyond their economic life. The low oil recovery results from flow channeling, phase trapping and poor displacement efficiency.

Water injection and infill drilling are the most common improved oil recovery (IOR) techniques used worldwide. Steam injection is used to mobilize viscous oils while miscible hydrocarbon gas injection, including CO_2, is used to enhance production of lighter oils. Continuous injection of CO_2, or water-alternating-CO_2 injection, is effective in enhancing oil production after water flooding. CO_2 is also an effective solvent for heavy viscous oils. Finally, aqueous solutions of low-concentration alkaline-surfactant-polymer mixture are used to produce additional oil after water flooding. These complex techniques are enhanced oil recovery (EOR) methods. EOR techniques are generally capital and labor-intensive, and steam generation requires large energy input. Research to develop EOR techniques for unconventional shale reservoirs is ongoing. Nonetheless, it appears that cyclic injection of CO_2, or a mixture of CO_2 and natural gas liquids, in a group of wells, followed by soak and production periods, can enhance oil recovery from shale reservoirs.

Flow and transport in petroleum reservoirs is modeled using continuum-based numerical simulation. In the continuum formulation, fluids fill the pore space completely, conservation of mass and energy is honored, and conservation of momentum is replaced by Darcy velocity and molecular diffusion flux. While this approach has been extremely successful for solving engineering problems, oil and gas production from very low-permeability shale requires a deeper understanding of flow and transport in micro, meso and macro-scale pores and channels. To achieve this, many investigators have resorted to the lattice-Boltzmann discrete-particle modeling. Lattice-Boltzmann modeling (LBM) is computationally efficient and provides a bridge between the molecular dynamic (MD) modeling and the continuum modeling [1, 2]. Up to now, the LBM formulation in porous media has been used to decipher major issues regarding phase distribution, phase trapping, wall and interphase effects at the pore level. We envision that LBM would bridge the discrete-particle physics of flow in small channels to the Darcy flow in porous media.

While this chapter is an overview of flow and transport in petroleum reservoirs, the next seven chapters, in the oil and gas section, present leading research topics related to oil and gas production. A summary follows:

Chapter 3, *Quatitative Nano and Microscale Rock Characterization (M. Prasad, S. Zargari, and M. Saidian)* addresses porosity, pore-size distribution, and mechanical property measurement of shale using nuclear magnetic resonance (NMR), mercury intrusion porosimetry (MIP), scanning acoustic microscopy (SAM), atomic force acoustic microscopy (AFAM), and nano dynamic mechanical analysis (NanoDMA). This chapter also provides a brief description of scanning electron microscopy (SEM) and micro X-ray computed-tomography (MXCT).

Chapter 4, *Impact of Surface Forces on Coupled Flow and Failure Characteristics in Shale Reservoirs (A. Tutuncu)* presents a non-Hertzian, grain-contact model for determining the impact of surface forces on rock deformation and acoustic properties of shale reservoirs. These results affect both rock failure calculations and seismic image interpretations.

Chapter 5, *Microporosity Characterization of Mud-rich Carbonate Rocks (T.D. Jobe and J.R. Sarg)* describes five distinct, low-permeability, Cretaceous carbonate mudrock facies. A quantitative SEM instrument determined mineralogy, while pore-size distribution was calculated from mercury intrusion porosimetry (MIP) measurements. The results are valuable for assessing flow and transport of hydrocarbons in very low-permeability carbonate mudrocks.

Chapter 6, *Hydrates in Porous Media: New Frontier (L.E. Zerpa and C.A. Koh)* presents the current state of knowledge regarding gas hydrates in oceans and arctic sediments. Gas hydrate production is of great interest as a source of clean energy, but it faces significant challenges.

Chapter 7, *Pore-scale Imaging and Ultrasonic Velocity Measurements* *of Gas Hydrate (M. Schindler and M. Batzle)* presents pore-scale imaging and ultrasonic velocity measurements in gas hydrate sediments. The article concludes that, based on micro X-ray computed-tomography and ultrasonic measurements, methane hydrates formed from gas dissolved in water form in the pore body, mainly away from the grain surfaces, while a small amount of hydrate contacts the grain surfaces. This information is important in assessing the load-bearing character of gas hydrate sediments and decisions regarding production of methane from gas hydrates.

Chapter 8, *Micro and Nafluidic Porous Media Models of Oil and Gas Applications (J.T. Ok, K. Neeves, W. Wu, and X. Yin)* presents construction and use of micro and nanofluidic porous media models as analogs of the subsurface pores. The micromodels provide a new capability to conduct physical measurements in nano and macro-scale pores and channels. **Chapter 9**, *Direct Numerical Simulation of Pore-Scale Fluid Flow and Transport (M.S. Newman, F. Xiao, and X. Yin)* presents numerical simulation of pore-scale flow and transport using lattice Boltzmann modeling (LBM). LBM is a mesoscopic, discrete-particle, numerical modeling method for determining flow and transport behavior at the pore level. LBM could provide a deeper understanding of flow and transport in micro, meso and macro-scale pores.

The remaining chapters in this book address pore-scale physics in aquifers, geothermal reservoirs, material, and bioscience.

2.2. Conventional Reservoirs

Understanding pore-scale physics and chemistry is the key to the development of new technologies for increasing oil and gas production to meet world's growing energy demand. Pores are tiny openings or space in sedimentary rocks giving passage to fluids. Large collections of such tiny pores constitute the main storage space for accumulation of oil and gas in petroleum reservoirs. Similarly, pores provide the storage space for fresh water or brine in aquifers. Figure 2.1 presents a comparison of pores in a prolific carbonate reservoir, a conventional sandstone reservoir, and an unconventional shale reservoir. In the carbonate core (left), we have two kinds of pores the invisible small pores in the rock matrix and the large solution cavities (or vugs) which are in the centimeter range. In the sandstone sample (middle), the pore body diameter is about 60 micrometers, which, depending on porosity and pore connectivity, could amount to millions of equivalent pores in a one-inch long by one-inch diameter core plug.

Fig. 2.1. Images of pores and their dimensions in two conventional reservoirs and an unconventional shale reservoir: (a) low permeability ($k_m = 2.0$ mD), oil-wet Cretaceous vuggy limestone core [3], (b) SEM image of a fine-grained sandstone [4], (c) ion-milled SEM image of a Barnett organic rich shale [4]. Note that the sandstone pore is at least two orders of magnitude larger than the shale pores, while the vugs (solution cavities) in the carbonate core are in the centimeter range.

In the shale sample (right), the pore diameters range from about 10 to 100 nanometers.

We use the following equations to estimate matrix permeability k_m and matrix tortuosity τ_m if we know the modal pore-throat diameter d (from pore-size distribution measurement), matrix porosity ϕ_m, and cementation exponent c. Chapter 5 presents several examples of pore-size distribution for a mud-rich carbonate rock in the Middle East. The cementation exponent for highly cemented material is 2.2 and for unconsolidated sand sediments is 1.3 [5]. We have used these equations to obtain reliable estimates of permeability for engineering applications.

$$k_m = \frac{d^2}{32}\frac{\phi_m}{\tau_m} \tag{2.1}$$

$$\tau_m = \phi_m^{1-c} \tag{2.2}$$

All equations in this chapter are in consistent units. For instance, for Eq. (2.1) and 2.2, if the average pore-throat diameter in a sandstone core plug is 20×10^{-6} meter (20 micrometer), porosity 0.20, and cementation exponent 2.0, tortuosity τ_m yields a value of 5.0 and matrix permeability k_m becomes 5.0×10^{-13} m^2 (or 500 mD). Similarly, assuming 100 nanometers for the pore throat diameter in a shale sample and porosity of 0.05, we calculate tortuosity to be 20 and permeability 781×10^{-21} m^2 (or 0.000781 mD).

2.3. Primary Production and Field Development to Improve Production

To acquaint the reader with field development processes, we present the reservoir management history of Weyburn field, shown in Fig. 2.2. Weyburn is a typical conventional oil field, whose development started with primary production using wide-spaced vertical wells. Primary production relies on the reservoirs natural energy (pressure and fluid compressibility) to produce oil and gas. With the decline of reservoir natural energy, oil production declines. Typically, water flooding follows primary production to add energy to the reservoir and to displace the remaining oil toward the production wells. Figure 2.2 nicely depicts primary production and water flooding. When water breaks through production wells swiftly, it is indicative of flow channeling, which probably was the case in Weyburn, because the reservoir is highly fractured. To alleviate this problem, numerous vertical infill wells were drilled in the unswept reservoir compartments existing between large flow channels. Eventually, horizontal wells were added to improve reservoir sweep further. The improved oil recovery (IOR) process started with water flooding and terminated with horizontal infill wells. At this point in the reservoir's life, nearly thirty percent of the initial oil in place had been produced. To produce additional oil, CO_2 injection was instigated to enhance oil recovery (EOR), which is currently ongoing. With CO_2 injection, a fraction of the CO_2 remains in the reservoir as sequestered CO_2. Weyburn is one of the largest CO_2 sequestration demonstration sites in the world.

The Weyburn history demonstrates utilization of well-known IOR and EOR processes. Interestingly, the estimated ultimate oil production from this field is around forty percent of the initial oil in place in spite of insightful reservoir development and management by the field's operators [6]. This project is a testimony to the fact that flow and transport in porous rocks is

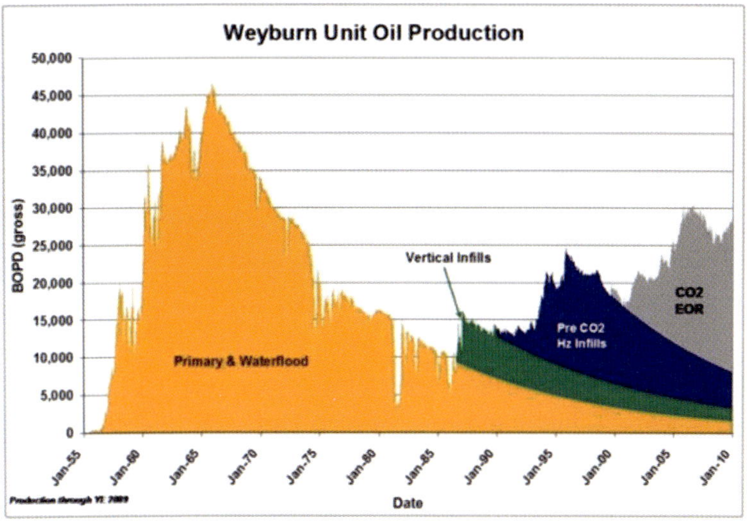

Fig. 2.2. Weyburn field history: Sixty years of insightful reservoir development and management increased ultimate oil recovery [6].

extremely complex. Similarly, in conventional reservoirs, in spite of many engineering innovations to improve oil recovery, two-third of the discovered oil will remain in the reservoir.

2.4. Improved and Enhanced Oil Recovery

In the previous section on Weyburn history, we gave an example of primary production, IOR and EOR in that field. However, in a broader sense, **improved oil recovery (IOR)** includes processes involving injection of water or dry gas to displace oil after primary production, while **enhanced oil recovery (EOR)** includes injecting hydrocarbon-rich gas, CO_2, dilute aqueous chemicals, and steam. Based on current statistics [7] and the adjustments for the unreported data, we conclude that contribution of EOR to the world daily oil production is about 3.1 percent (that is, 2.8 million barrels/day of EOR oil compared to 90 million barrels/day of total oil). Specifically, steam injection yields 1.0 million barrels of oil/day; CO_2, nitrogen, air, hydrocarbon-rich gas and lean gas produce 1.5 million barrels of oil/day; and water-augmented polymer and surfactant produce 0.3 million barrels of oil/day.

A great deal of insight and knowledge has resulted from the analysis of EOR projects in conventional reservoirs, especially in the United States. We believe this knowledge is a wonderful guiding light for developing EOR techniques in unconventional reservoirs. For instance, we conclude that fluid **displacement** is the major EOR mechanism in conventional reservoirs; however, in unconventional reservoirs **fluid extraction** at the fracture-matrix interface is the main EOR mechanism. Thus, EOR applications in unconventional shale could comprise of injection of CO_2, natural gas liquids (NGL), or a mixture of CO_2 and NGL.

We generally assess the quality of subsurface flow in terms of porosity, permeability, relative permeability, and capillary pressure. However, these assessments must also include additional measures such as pore-to-pore connectivity, rock-fluid interactions within the pores, pore surface wettability, and nano-scale pore thermodynamics in unconventional reservoirs when pore confinement is severe.

Wettability and wettability alteration are two important parameters, which affect IOR. Here is our view on how wettability affects oil production: The pore surfaces in sandstone rocks are typically negatively charged because pore walls include clays and silicates. Thus sandstones become hydrophilic (water wet) because polar water molecules become attracted to the negatively charged pore surfaces via hydrogen bonding. In carbonates, the surface charge is positive, which may be the reason for carbonates becoming oleophylic (oil wet) when the negative carboxylate ions, from ionization of the acidic components of oil, bond to the positively charged surfaces. When pores are oil-wet, water flooding produces less oil and when pores are water-wet, water flooding produces more oil. It is interesting that altering formation water composition by mixing with seawater, containing sulfate ions, could improve oil recovery in carbonate reservoirs. In seawater flooding, the wettability alteration from oil-wet to water-wet is the result of favorable ion exchange at the pore walls [8].

We chose Fig. 2.3 to demostrate the diversity of mechanisms for oil production. Figure 2.3(a) shows a high-permeability **water-wet Berea** outcrop core, saturated with irreducible brine and mobile oil, then surrounded by brine in a beaker. In this water-wet core, brine imbibed into the core to displace oil which is apprearing as droplets. In addition to capillary imbibition, gravity head and bouyancy created a gradually decreasing distribution of oil droplets along the core length. Figure 2.3(b) shows a medium-permeability **oil-wet San Anderes carbonate** core from Yates field, USA. The core contains irreducible brine and mobile oil, then sur-

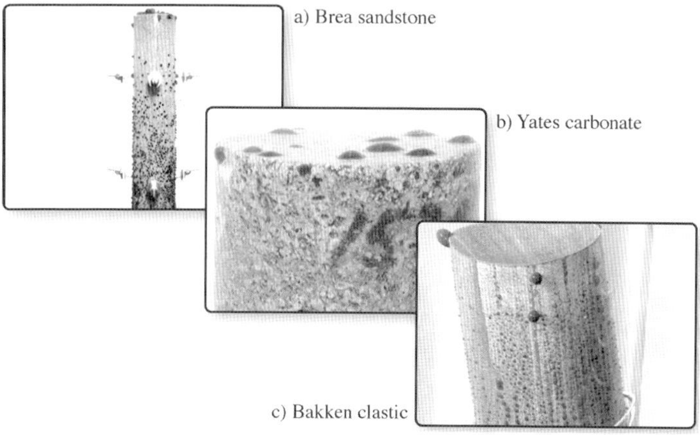

Fig. 2.3. Wettability affects oil production process: (a) a water-wet Berea sandstone core, (b) an oil-wet carbonate core from Yates field, and (c) a core from Bakken unconventional formation.

rounded by brine in a beaker. In this oil-wet rock, water does not imbibe easily; while gravity head and bouyancy cause migration of oil as droplets to the top surface of the core. This image indicates that the oil-rock contact angle is less than 90 degrees, which confirms the oil-wetting condition. Figure 2.3(c) is a Bakken, high-salinity, **unconventional reservoir core** surrounded by low-salinity brine. The low-salinity brine entered the pores by osmotic pressure or capillary imbibition to displace oil [9].

2.5. Unconventional Reservoirs

Oil and gas production from unconventional reservoirs, including shale oil and shale gas, has contributed significantly to the US domestic energy supply in the last ten years, and is rapidly expanding with continued improvements in horizontal well drilling and hydraulic fracturing techniques. These reservoirs are classified as to whether the hydrocarbon source is an integral part of the reservoir rock fabric (self-sourced, as in Eagle Ford), or is adjacent to the reservoir (locally sourced, as in Bakken), or is located at large distances from the reservoir and require significant hydrocarbon migration (externally sourced, as in Austin Chalk) [10]. In unconventional reservoirs, we use multi-stage hydraulic fractures, emanating from a horizontal well, to create large surface areas to improve flow from the reservoir pores by increasing accessibility to the reservoir pores (Fig. 2.4).

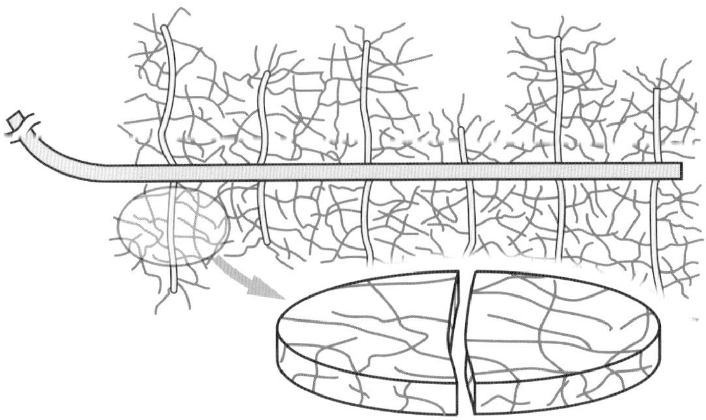

Fig. 2.4. Idealization of a modern horizontal well including several hydraulic fracture stages. The multi-stage hydraulic fractures increase accessibility to the reservoir pores.

In general, pore morphology, internal surface characteristics, and pore-to-pore connectivity are complex issues affecting subsurface fluid dynamics both in conventional and unconventional reservoirs. Specifically, **it is difficult to access or effectively drain the bulk of the subsurface pores because** of phase trapping, flow channeling, reservoir compartmentalization, and poor sweep efficiency. This is why we drill a large number of vertical **infill wells** in the field to access undrained compartments and to contact more reservoir pores. In unconventional reservoirs, these issues are even more impactful. This is why we resort to horizontal wells to contact more of the reservoir pores for production or injection operations, and why we stimulate the unconventional reservoirs by placing large planar fractures perpendicular to the horizontal wells to create large surface areas for pores to flow easier.

Multi-stage hydraulic fracturing increases the effective permeability of shale reservoirs, which we designate $k_{f,\text{eff}}$ and estimate its magnitude using Eq. (2.3).

$$k_{f,\text{eff}} = k_m + k_f \phi_f \tag{2.3}$$

where,

$$k_f = \frac{w_f^2}{12} \tag{2.4}$$

$$\phi_f = \left(\frac{w_f}{L_x} + \frac{w_f}{L_y}\right) \tag{2.5}$$

In the above equations k_f and k_m are fracture and matrix permeability and ϕ_f is the fracture porosity calculated from macrofracture average width w_f and fracture spacing L_x and L_y.

For shale reservoir applications, if the average macrofracture width w_f is 3.0×10^{-6} meters (or 3.0 μm) and fracture spacing is one meter, then the fracture permeability k_f becomes 0.75×10^{-12} m^2 (or 750 mD) and ϕ_f becomes 0.6×10^{-5}. We assume that matrix permeability k_m is the same as the core permeability. The permeability calculated from Eq. (2.3) should be compared with the effective reservoir permeability $k_{f,\text{eff}}$ from transient flow tests in the field to determine the effectiveness of fracture stimulation.

During reservoir depletion, flow rate declines in continuously when the well and reservoir operating conditions remain relatively constant. Equation (2.6) is an empirical equation, which provides a good engineering approximation of the reservoir flow rate as a function of time for conventional reservoirs (Arps, 1944; Garb and Smith, 1987). Engineers use the same equation to forecast the future performance of individual wells in unconventional reservoirs.

$$q(t) = q_i \left(1 + bD_i t\right)^{-1/b} \tag{2.6}$$

where,

D_i = **nominal decline rate** at time zero, T^{-1}

b = rate exponent

In Eq. (2.6), when b is zero, the rate decline is **exponential**, when b is 1.0 the rate decline is **harmonic**, and when b is between zero and one, the rate decline is **hyperbolic**. However, in unconventional shale reservoirs the term $bD_i t$ is generally much larger than one in Eq. (2.6). Thus, shortly after production, we can re-write Eq. (2.6) as shown below:

$$q(t) = q_i \left(bD_i\right)^{-1/b} t^{-1/b} \tag{2.7}$$

When b is 4.0, Eq. (2.7) has the same functional form as the mathematical solution for the **bilinear flow** in an ideal finite-conductivity vertical fracture during early flow times, and when b is 2.0, Eq. (2.7) fits the solution of the **linear flow** in the formation at the intermediate times. Eventually b approaches zero when the outer boundary effects prevails and the flow rate decline becomes **exponential decline**. We have substantiated this simple analysis using field data. The analysis is consistent with the analysis of field data by Refs. [11] and [12].

For engineering applications, we re-write Eq. (2.7) to convert surface flow rates to the bottom-hole conditions and to account for the influence of the bottom-hole pressure on the flow rate behavior:

$$\frac{\Delta p(t)}{q(t)B} = \frac{(bD_i)^{1/b}\,\Delta p(t)}{q_i B_i} t^{1/b} \tag{2.8}$$

Equation (2.8) is merely an extension of the empirical rate-decline equation. The flow-based counterpart of Eq. (2.8) for bilinear flow ($b = 4$) is Eq. (2.9) and for linear flow ($b = 2$) is Eq. 11. Torcuk et al. (2013) derived these equations in an earlier paper [13].

$$\begin{aligned}\frac{\Delta p(t)}{(qB)_t} &= \frac{\sqrt{2\pi}\lambda_t^{-1}}{2\pi h n_{hf}\sqrt{w_{hf}k_{hf}}}\left(\frac{\lambda_t}{(\phi c_t)_{f+m}\,k_{f,\mathrm{eff}}}\right)^{1/4} t^{1/4}\\ &\quad + \frac{\lambda_t^{-1}}{2\pi k_{f,\mathrm{eff}} h n_{hf}} s_{hf}^{\mathrm{well}}\end{aligned} \tag{2.9}$$

Equation (2.9) has the following form for plotting and engineering analysis.

$$\frac{\Delta p(t)}{(qB)_t} = m_{bl}\sqrt[4]{t} + b_{bl} \tag{2.10}$$

$$\begin{aligned}\frac{\Delta p(t)}{(qB)_t} &= \frac{(\pi/2)\,\lambda_t^{-1}}{2\pi\sqrt{k_{f,\mathrm{eff}}}\,(h n_{hf} y_{hf})}\left(\frac{\lambda_t}{(\phi c_t)_{f+m}}\right)^{1/2} t^{1/2}\\ &\quad + \frac{\lambda_t^{-1}}{2\pi k_{f,\mathrm{eff}} h n_{hf}} s_{hf}^{\mathrm{face}}\end{aligned} \tag{2.11}$$

Equation (2.11) also has in the following simple form:

$$\frac{\Delta p(t)}{(qB)_t} = m_l\sqrt{t} + b_l \tag{2.12}$$

We use Eq. (2.10) to determine $k_{f,\text{eff}}$ when flow test data are available. For example, Fig. 2.5 presents the rate-normalized pressure data for a well in Eagle Ford formation. The plot includes simultaneous production of oil, gas, and water.

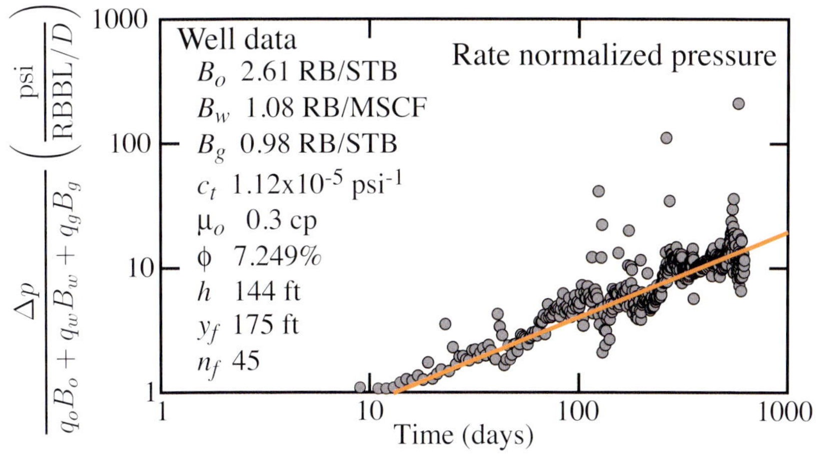

Fig. 2.5. Flow rate normalize field data from an Eagle Ford shale reservoir.

The $k_{f,\text{eff}}$ permeability calculated from this plot is about 0.0007 mD while the matrix or core permeability k_m is 0.00002 mD. The large contrast between these two permeabilities is a strong indication of the presence of macrofractures in the formation, created by the hydraulic fracturing stress distribution effects. Both permeabilities are critical input data for reservoir modeling and engineering analysis.

Comprehensive evaluation of field data, core experiments, and reservoir modeling point to the following reasons for the success of oil and gas production from unconventional reservoirs:

(a) low viscosity of hydrocarbon fluids–both gas and light oils (Fig. 2.6),
(b) high compressibility of fluids,
(c) dominance of very low molecular weight components in the hydrocarbon fluids (Table 2.1),
(d) abnormal high initial reservoir pressure which might have created microfractures,
(e) enhancement of natural micro and macro fractures as a consequence of

the multi-stage hydraulic fracturing (Fig. 2.4),

(f) favorable phase envelop shift of the reservoir hydrocarbon mixture, and favorable gas-oil separation ratio in nanopores in the hydrocarbon two-phase region (Fig. 2.7).

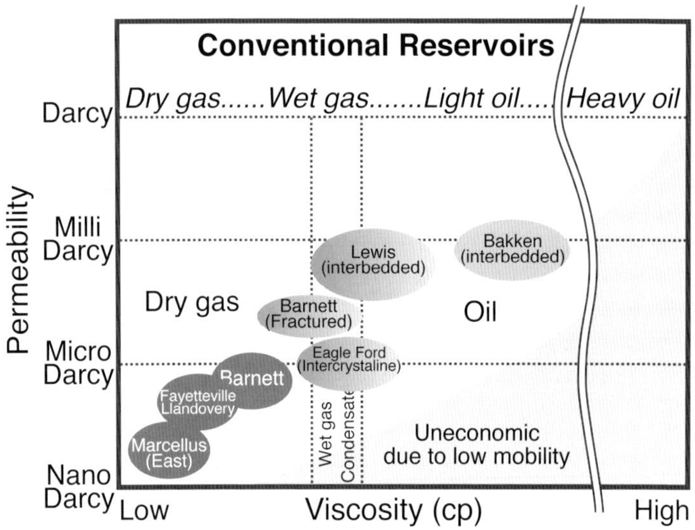

Fig. 2.6. Relative position of shale reservoirs as a function of rock permeability and fluid viscosity (modified from Ref. [14]).

2.6. Numerical Modeling

The fluid **continuum** assumption underlies mathematical formulation of flow and transport equations for traditional engineering applications. These equations are complex and generally require numerical solutions. The inception of **numerical reservoir modeling** in the US petroleum industry dates back to mid-1950s [16], and the commercial use of such models began in mid-1960s. By early 1970, numerical models became the main computational tool for assessing petroleum reservoir performance for the short and long-term needs. The numerical models consist of **grid cells** (volume subdomains) encompassing the entire reservoir, **time steps** (discrete steps in the total time), and a discretized set of flow and transport equations generated by volume integration of the governing differential equations on

Table 2.1. Two fluid samples from the Eagle Ford unconventional formation. Fluid 1 is very light oil and has API gravity of 51.4. Fluid 2 is heavier oil and has API gravity of 47.5 [15].

Components	Fluid 1 Composition (%)	Fluid 2 Composition (%)
H2S	0.00	0.00
N2	0.19	0.18
CO2	2.89	2.78
C1	72.42	69.77
C2	9.26	8.93
C3	5.18	4.99
IC4	1.19	1.15
NC4	2.04	1.96
IC5	0.93	0.90
NC5	1.00	0.96
FC6	1.42	1.37
FC7	1.09	1.42
FC8	0.792	1.21
FC9	0.517	0.93
FC10	0.347	0.73
FC11	0.233	0.58
FC12	0.157	0.45
FC13	0.106	0.36
FC14	0.072	0.28
FC15	0.049	0.22
FC16	0.033	0.18
FC17	0.023	0.14
FC18	0.016	0.11
FC19	0.011	0.09
FC20	0.0075	0.07
FC21	0.0052	0.06
FC22	0.0036	0.05
FC23	0.0025	0.04
FC24	0.0018	0.03
FC25	0.0012	0.02
+C26	0.003	0.02

Properties	Fluid 1	Fluid 2
STO API	51.4	47.5
OGR (STB/MMSCF)	30	100
GOR (SCF/STB)	33,333	10,000

each subdomain. The numerical models generate discrete solutions for each subdomain for each time step. Aquifer and hydrothermal reservoir models also use similar modeling techniques (Chap. 15 of this book, by Wu et al.).

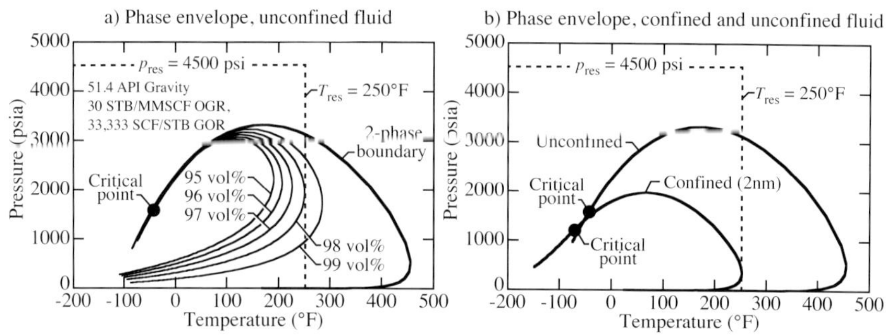

Fig. 2.7. Unshifted and shifted phase envelopes for a fluid system from the Eagle Ford unconventional shale formation: (a) phase envelope in a PVT cell, which is believed to be about the same in macropores (diameter larger than 50 nanometers) of conventional reservoirs, (b) probable phase envelop shift in very small micropores (diameter less than 2 nanometers). These illustrations clearly indicate that crossing the phase envelope takes place at significantly lower dew-point pressures in nanopores, thus delaying liquid condensation in nanopores.

Ever since mid-1980s, new mathematical models, such as the lattice Boltzmann models (LBM), which focus on the pore-scale flow and transport, have become increasingly popular [17]. LBM occupies an important niche between molecular dynamics (MD) and the traditional continuum approach [2]. The LBM is a mesoscopic (2 nm to 50 nm scale) modeling method, which solves the discretized Boltzmann equation of fluid particle distribution to obtain macroscopic solutions for incompressible hydrodynamics problems. Reference [18] provides a clear explanation of the LBM technique and gives solutions for capillary pressure in angular-shaped pores. Furthermore, they applied the LBM technique to simulate two-phase flow in Bentheimer sandstone using its **microstructure images**. Chapter 9 in this book (Newman, et al.) also describes LBM simulation of pore-scale flow and transport in micro and nano models and compares simulation results with measurements. Chapter 8 in this book presents fabrication of nano and micro scale porous media as analogs for oil and gas reservoir porous media to observe capillary retention of water in water-gas experiments. Finally, in Fig. 2.8, we present a schematic of mathematical scaling and averaging in describing the progression from stochastic to deterministic modeling approaches used in engineering applications.

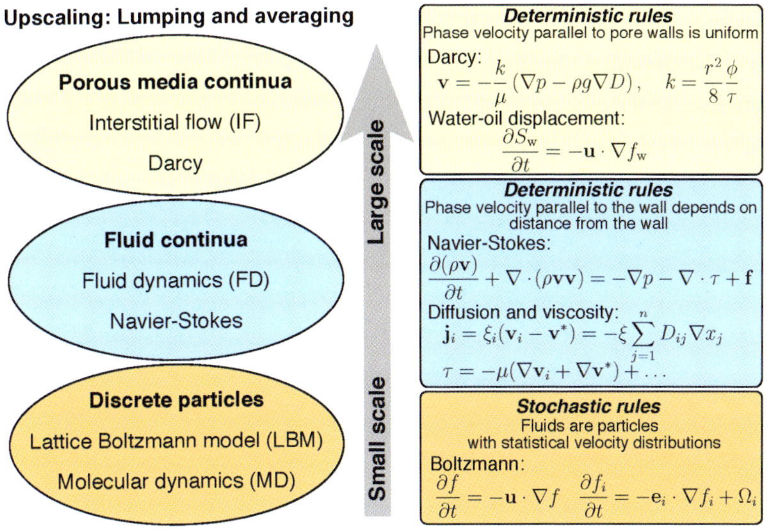

Fig. 2.8. Schematic of mathematical scaling and averaging to describe the progression from stochastic to deterministic formulation of isothermal flow of fluids.

2.7. Closing Thoughts

This chapter is an overview of petroleum reservoirs and pore-scale issues and does not delve into the details of mass and energy transport, relevant thermodynamics, mathematical models and solution methods. Fortunately, mathematical models and numerical solution methods for reservoir engineering applications have advanced to a state of maturity and computational efficiency [19]. There is, however, an emergence of new models based on discrete particles dynamics, including the lattice Boltzmann model (LBM), where computation is made at the mesoscopic scale (Chapter 9 of this book). The LBM should fill the gap between molecular dynamics and fluid mechanics in pore-scale modeling where conventional formulations encounter problems adhering to the underlying physics and thermodynamics. In addition to numerical modeling, micro and meso-scale experiments provide new insights on pore-scale flow and transport issues (e.g., Chapter 8).

2.8. Nomenclature

Base symbols and units

F	Force $[Newton]$
L	Length $[m]$
M	Mass $[kg]$
P	Pressure $[Pa]$
T	Time $[s]$

Symbols, units, and dimensions

b	Rate decline exponent, dimensionless
b_l	Linear flow equation intercept, $Pa/(m^3s^{-1})$ $[FL^{-2}/(L^3T^{-1})]$
b_{bl}	Bilinear flow equation intercept, $Pa/(m^3s^{-1})$ $[FL^{-2}/(L^3T^{-1})]$
B	Formation volume factor, $res\ m^3/surf\ m^3$ $[L^3/L^3]$
B_g	Gas volume factor, $res\ m^3/surf\ m^3$ $[L^3/L^3]$
B_o	Oil volume factor, $res\ m^3/surf\ m^3$ $[L^3/L^3]$
B_w	Water volume factor, $res\ m^3/surf\ m^3$ $[L^3/L^3]$
c_g	Gas compressibility, Pa^{-1} $[L^2F^{-1}]$
c_ϕ	Rock compressibility, Pa^{-1} $[L^2F^{-1}]$
c_t	Total reservoir fluid compressibility, Pa^{-1} $[L^2F^{-1}]$
$c_{t,f}$	Total fracture compressibility, Pa^{-1} $[L^2F^{-1}]$
$c_{t,m}$	Total matrix compressibility, Pa^{-1} $[L^2F^{-1}]$
c_w	Water compressibility, Pa^{-1} $[L^2F^{-1}]$
d	Diameter, m $[L]$
D_i	Initial decline rate, s^{-1} $[T^{-1}]$
h	Formation thickness, m $[L]$
k	Permeability, m^2 $[L^2]$
k_f	Fracture permeability, m^2 $[L^2]$
$k_{f,\text{eff}}$	Effective reservoir fracture permeability, m^2 $[L^2]$
k_{hf}	Hydraulic fracture permeability, m^2 $[L^2]$
k_m	Matrix permeability, m^2 $[L^2]$
L_x	Average fracture spacing in the x direction, m $[L]$
L_y	Average fracture spacing in the x direction, m $[L]$
m_l	Slope of the linear flow, $Pa.s^{-1/2}/(m^3s^{-1})$ $[FL^{-2}.T^{-1/2}/(L^3T^{-1})]$
m_{bl}	Slope of bilinear flow, $Pa.s^{-1/2}/(m^3s^{-1})$ $[FL^{-2}.T^{-1/2}/(L^3T^{-1})]$
n_{hf}	Total number of hydraulic fractures
p	Pore pressure, Pa $[FL^{-2}]$
p_i	Initial reservoir pressure, Pa $[FL^{-2}]$

p_{wf}	Flowing bottom-hole pressure, Pa $[FL^{-2}]$
q	Production rate, m^3s^{-1} $[L^3T^{-1}]$
q_g	Gas production rate, m^3s^{-1} $[L^3T^{-1}]$
q_i	Initial oil production rate, m^3s^{-1} $[L^3T^{-1}]$
q_o	Oil production rate, m^3s^{-1} $[L^3T^{-1}]$
q_w	Water production rate, m^3s^{-1} $[L^3T^{-1}]$
s_{hf}^{face}	Skin factor on the face of hydraulic fracture, dimensionless
s_{hf}^{well}	Skin factor at the intersection of well and hydraulic fracture, dimensionless
S_g	Gas saturation, fraction
S_o	Oil saturation, fraction
S_w	Water saturation, fraction
t	Time, s $[T]$
w_f	Fracture aperture, m $[L]$
w_{hf}	Hydraulic fracture width, m $[L]$
y_{hf}	Fracture half-length, m $[L]$

Greek

λ_g	Gas mobility, $(Pa.s)^{-1}$ $[(FL^2T)^{-1}]$
λ_o	Oil mobility, $(Pa.s)^{-1}$ $[(FL^2T)^{-1}]$
λ_t	Total mobility, $(Pa.s)^{-1}$ $[(FL^2T)^{-1}]$
λ_w	Water mobility, $(Pa.s)^{-1}$ $[(FL^2T)^{-1}]$
Δp_{sf}	Skin pressure drop in hydraulic fracture stimulated well, Pa $[FL^{-2}]$
Δp	Well flowing pressure change, Pa $[FL^{-2}]$
ϕ_f	Fracture porosity, fraction
ϕ_m	Matrix porosity, fraction
μ	Viscosity, $Pa.s$ $[FL^{-2}T]$
ω	Storativity ratio, fraction
τ_m	Matrix tortuosity, equal or greater that one

Acknowledgement

I am grateful for support from Colorado School of Mines (CSM), Marathon Center of Excellence for Reservoir Studies (MCERS), Unconventional Natural Gas and Oil Institute (UNGI), and Saudi Aramco. I also thank my current and past graduate students: Najeeb Alharthy, Ilkay Eker, Basak Kurtoglu, and Mehmet Torcuk for their contributions.

References

[1] S. Succi, *The Lattice Boltzmann Equation for Fluid Dynamics and Beyond*. Oxford Science Publications (2001).

[2] M. C. Sukop and D. T. J. Thorne, *Lattice Boltzmann Modeling*. Springer (2010).

[3] M. Kiani. *Field Pilot Testing of EOR Surfactants in Dual-porosity Naturally Fractured Reservoirs: Single-well vs. Multiple-well Approach*. PhD thesis, Colorado School of Mines, Golden, CO (2011).

[4] Q. R. Passey, K. M. Bohacs, W. L. Esch, R. Klimentidis, and S. Sinha. From oil-prone source rock to gas producing shale reservoirs–geologic and petrophysical characterization of unconventional shale gas reservoirs. In *International Oil and Gas Conference and Exhibition in China*, Beijing, China (8-10 June, 2010). SPE-131350-MS.

[5] S. J. Pirson, *Oil Reservoir Engineering*, 2 edn. McGraw-Hill (1958).

[6] P. T. R. C. (PTRC). Weyburn field. URL `http://www.prtc.ca/weyburn_statistics.php` (2012).

[7] L. Koottungal, 2014 worldwide EOR survey, *Oil & Gas Journal*. pp. 79–91 (April 7, 2014).

[8] S. J. Fathi and T. Austad. Water-based enhanced oil recovery (EOR) by "smart water" in carbonate reservoirs. In *SPE EOR Conference at Oil and Gas West Asia*, Muscat, Oman (16-18 April, 2012). SPE 154570.

[9] B. Kurtoglu. *Integrated Reservoir Characterization and Modeling in Support of Enhanced Oil Recovery for Bakken*. PhD thesis, Colorado School of Mines, Golden, CO (2013).

[10] B. Tepper, G. Baechle, J. Keller, and R. Walsh. Petrophysical evaluation of shale oil & gas opportunities in emerging plays; some examples and learning from the Americas. In *IPTC*, Beijing, China (26-28 March, 2013).

[11] T. W. Patzek, F. Male, and M. Marder, Gas production in the barnett shale obeys a simple scaling theory, *Proceedings of the National Academy of Sciences* (2013).

[12] J. Lacayo and J. Lee. Pressure normalization of production rates improves forecasting results. In *SPE URC*, the Woodlands, TX (1-3 April, 2014). SPE-168974-MS.

[13] M. A. Torcuk, N. Kurtoglu, B. Alharthty, and H. Kazemi, Analytical and numerical solutions for multiple-matrix in fractured reservoirs: Application to conventional and unconventional reservoirs, *SPE Journal*. **18**, 969–981 (2013). SPE-164528-PA.

[14] K. M. Bohacs, Q. R. Passey, M. Rudnicki, W. L. Esch, and Q. R. Lazar. The spectrum of fine-grained reservoirs from shale gas to shale oil/tight liquids: Essential attributes, key controls, practical characterization. In *IPTC*, Beijing, China (26-28 March, 2013).

[15] C. H. Whitson and S. Sunjerga. Pvt in liquid-rich shale reservoirs. In *SPE ATCE*, San Antonio, TX (8-10 October, 2012). SPE 155499.

[16] J. Douglas, Jim, D. W. Peaceman, and J. Rachford, H. H., A method for calculating multi-dimensional immiscible displacement, *SPEJ*. pp. 297–308

(1959). SPE-1327-G.

[17] Y. Keehm, T. Mukerji, and A. Nur, Computational rock physics at the pore scale: Transport properties and diagenesis in realistic pore geometries, *The Leading Edge*. **20**(2), 180–183 (2001). doi: 10.1190/1.1438904.

[18] T. Ramstad, P.-E. ren, and S. Bakke, Simulation of two-phase flow in reservoir rocks using a lattice boltzmann method, *SPE Journal*. **15**, 917–927 (2010). doi: 10.2118/124617-PA. SPE-124617-PA.

[19] H. Kazemi, M. S. Al-Kobaisi, B. Kurtoglu, A. Heris, S. Charoenwongsa, P. Fakcharoenphol, and J. Akinboyewa, Mathematical modelling of petroleum reservoirs, *Exploration and Production: The Oil and Gas Review*. **10**(2), 45–50 (2012).

Chapter 3

Quantitative Nano and Microscale Rock Characterization

Manika Prasad, Saeed Zargari and Milad Saidian

Department of Petroleum Engineering,
Colorado School of Mines,
Golden, CO, USA
mprasad@mines.edu

Rapid development of the unconventional resources has necessitated application of newer techniques to evaluate the hydrocarbon resource contained in such rocks. Given that the use of terminologies and concepts is relatively new in unconventionals, it often lacks in definition and scope of these techniques as is prevalent in literature. In this work, we are revisiting some of these concepts and definitions and have discussed some new methods that are becoming standard characterization techniques for these resources. We also present the current quantitative techniques in characterizing micro to nanoscale porosity, pore size distribution, and matrix mechanical properties. These techniques include Scanning Acoustic Microscopy (SAM), Nanoindentation, Nano Dynamic Mechanical Analysis (NanoDMA®), Atomic Force Acoustic Microscopy (AFAM), Mercury Intrusion Porosimetry (MIP), Nuclear Magnetic Resonance (NMR) and nitrogen gas adsorption. In addition to the quantitative imaging, we show how such techniques can be used to quantify acoustic impedance and moduli as well as porosity and pore-size distribution. We are primarily focused on quantitative measurements of rock physics properties. Auxillary textural characterizations, such as Scanning Electron Microscopy (SEM) and CT imaging are mentioned only briefly due to the subjectivity in the image interpretation.

3.1. Introduction

Unconventional shale plays have become a major source of oil and natural gas production in the United States. In spite of the large quantities of proven and potential reserves, they pose numerous development challenges.

The success so far in unconventional hydrocarbon exploitation is largely attributed to innovative engineering and advances in hydraulic fracturing. However, new reservoir characterization techniques have yet to assume a major role in the development of such resources.

The low-permeability unconventional reservoirs are often called shale reservoirs. They may not necessarily contain any clay minerals or have similar lithology. Some shale reservoirs are carbonate muds while others are siliciclastic. Specifically, shale is defined as fine-grained, fissile, clay- or organic-rich, siliciclastic sedimentary rock with a grain diameter less than 64 microns. In shale reservoirs, organic matter is a critical component of the rock fabric. Although alternate terms, such as unconventional reservoirs, self-resourcing rocks, organic-rich rocks and mudstones, have been used for shale reservoirs, the term shale has endured.

The ***unconventional resources*** and ***self-sourcing systems*** terminology lacks adequate lithological definition. In conventional resources, the source rock and the reservoir are separate; the hydrocarbon migrates from source rock to the reservoir through flow channels while secondary migration is controlled by buoyancy. Unconventional resources have different accumulation processes. In these accumulations, the reservoir, which is often the source rock, is charged by forces of expulsion. The source rock maturation level controls fluid phase distribution where oil is found in shallower parts of the basin and gas in the deeper parts. Self-resourcing systems (SRS) are referred to as a group of adjacent tight formations within an unconventional accumulation that are charged by maturation products from a contained source rock.

Exploration and development of shale reservoirs is an exciting venture that has rejuvenated the energy industry. It has promoted development of micro and nano-scale characterization of shale rock even for basic properties such as porosity and grain stiffness. New methods have been developed as the technology and our understanding has progressed to demonstrate existence of small, nanometer-sized pores in organic matter. Such small pores have even been known to affect seismic-scale phenomena and its modeling through rock physics approximations. However, several areas, such as, scale effects, formation stress, fluid pressure, fluid sensitivity, heterogeneity, anisotropy, and seismic attenuation, need urgent attention before geophysical exploration techniques can be applied to SRS development. Reserve estimation, sweet spot determination and well performance prediction require petrophysical data that are often derived from geophysical techniques and laboratory measurements. Such reserve estimates are bound to be flawed because conventional definitions and determinations of petrophysical properties (such as, effective porosity and fluid flow) are inapplicable to unconventional petroleum systems.

Porosity development and mechanisms of fluid flow in organic matter and clay minerals are poorly understood. Despite several classifications of pore system attributed to wide range of pore-hosting grains (for example, Ref. [1]), there are limited studies on evolution of the pore system throughout burial history and maturation processes. Combinations of several factors also render conventional porosity and permeability measurement techniques inconsistent and time consuming. Such factors include, but are not limited to, fine-grain texture, complex and nano-scale pore network, low permeability, abundance of clay minerals and abundance of organic matter.

The variation in grains surface properties, pore sizes and connectivity results in different contributions to flow and storage capacity. These variations are mainly due to adsorption and absorption of hydrocarbon and intraparticle porosity in the kerogen and clay bound hydrocarbon. As a result the contribution of pore systems in rock's storage and flow capacity could be different. Therefore, a better understanding of the nature of the pores with a better definition of effective porosity is needed for flow simulation and reserve estimation purposes.

This paper introduces implementations of advanced techniques to address the main rock characterization issues and the associated definition for adequate representation of the results. The first part addresses the ***matrix***: measurements that help us study rock components as well as the composite rock and associated changes with in-situ conditions, mineralogy, and maturity. The second part addresses the ***pores***: methods for quantitative analysis of pore sizes and their preferred locations within the matrix, and the changes in pore properties and their mineral associations.

The techniques described here yield quantitative measurements of rock physics properties. Auxiliary characterizations obtained with Scanning Electron Microscopy (SEM) and CT imaging is often used to assess textural anisotropy [2]. Image analysis methods for pore quantification are not included here because they introduce inaccuracies due to subjective interpretation. Such inaccuracies are minimal in quantification of large pores. Given the large surface areas in nanoporous rocks, detection of grain boundaries can introduce bias in pore quantification.

3.2. Matrix Characterization

Understanding elastic behavior of SRS formations and their interrelations with maturity can provide crucial information about the amount of generated hydrocarbon and geomechanical characteristics of the formation from seismic measurements. Since production from SRS is economically feasible only with hydraulic fracturing, knowledge about stiffness and brittleness is vital for a successful stimulation program. When rocks are brittle, they are favorable for hydraulic fracture stimulation. The brittleness factor, a measure of "fracability" [3], is defined by a combination of Young's modulus and Poison's ratio derived from well logs.

The brittleness ratio is often used to determine zones that are better suited for hydraulic fracturing. This brittleness ratio is calculated from dynamic elastic properties, Young's modulus and Poisson's Ratio, determined from sonic well log data [3]. However, stiffness, modulus and velocity are elastic properties, brittleness (or its cousin, the newly minted term "Fracability"), hardness, and toughness are fracture or static deformation properties. Brittleness is the ratio between ***hardness*** and ***toughness***, where hardness is resistance to deformation and toughness is resistance to fracture [4]. In relating the elastic properties to static deformation properties, we need site-specific dynamic to static conversions. Young's modulus and Poisson's ratio are derived from compressional (V_p) and shear (V_s) velocities measured in well logs by assuming an isotropic and homogeneous system. If the reservoir rock is anisotropic, Poisson's ratio calculations will be in error. Also, Young's modulus - Poisson's ratio plots change with saturation state.

Rock models correlate the bulk mechanical properties of the rock to petrophysical properties using textural characteristics and mechanical properties of individual or aggregates of grains. Generating representative rock models to describe behavior of organic rich rocks is challenging due to uncertainties in mechanical properties of some key components and complex textures. Despite previous studies on the texture of organic rich rocks in relation to their elastic properties, anisotropy [5] and mechanical characteristics of the grains [6–10], there is little firm knowledge about the characteristics of some essential components such as clays and organic matter. Moreover, changes in their properties during the course of maturation are poorly understood. In the next section we introduce micro-to nanoscale matrix characterization methods.

3.2.1. *Matrix Characterization Techniques*

3.2.1.1. *Acoustic imaging*

In opaque minerals, such as kerogen and pyrite, optical and SEM methods give only surficial information. In this section, we show how we use acoustic waves to probe subsurface information and provide the 3-D **impedance microstructural information**. The non-destructive technique of scanning acoustic microscopy (SAM) uses high-frequency acoustic waves. A complete description of the application of scanning acoustic microscopy in petrophysics can be found in Ref. [6]. Acoustic transducers in a SAM instrument generate high frequency (10–200 MHz) acoustic waves that are focused on a sample. Given the impedance contrast between the couplant and the sample, these acoustic waves are partly mode converted, partly transmitted into the sample, and partly reflected back to the transducer. In geophysics, SAM is a microscopic analog of side-scan sonar and reflection seismic mapping, while it is an analog of ultrasound sonography in medicine.

The images of surface and sub-surface microstructures are prepared on the basis of reflected acoustic waves, a manifestation of impedance changes in the sample. Reflection coefficient variations control signal intensity and can be calculated from elastic constants of the material. Changes in acoustic impedance influence wave reflection characteristics and can be studied by mapping the reflected waves. Figure 3.1 shows results of acoustic characterization at 1 GHz in a sandstone sample with varying cement stiffness and damage due to stress concentrations. In Fig. 3.1(a), the quartz sandstone has pore-filling cement and intact quartz grains. The grains have inclusions and fractures. Figure 3.1(b) shows compaction damage in quartz grains with deformation bands due to point loading and stress concentrations. Note the abundant microscale intergranular pore space mapped by the acoustic waves in the quartz grains. Figure 3.1(c) shows an example of grain contact that can have detrimental effects on permeability.

3.2.1.2. *Mechanical probing*

Nanoscale mechanical properties can be obtained by nanoindentation, nanoDMA and AFAM. In the following, we provide examples of each technique.

Nanoindentation technique is a quasi-static modulus measurement which involves a cyclic penetration of a sharp tip into the specimen [11]. Displacement is recorded versus loading and unloading force. Stiffness is

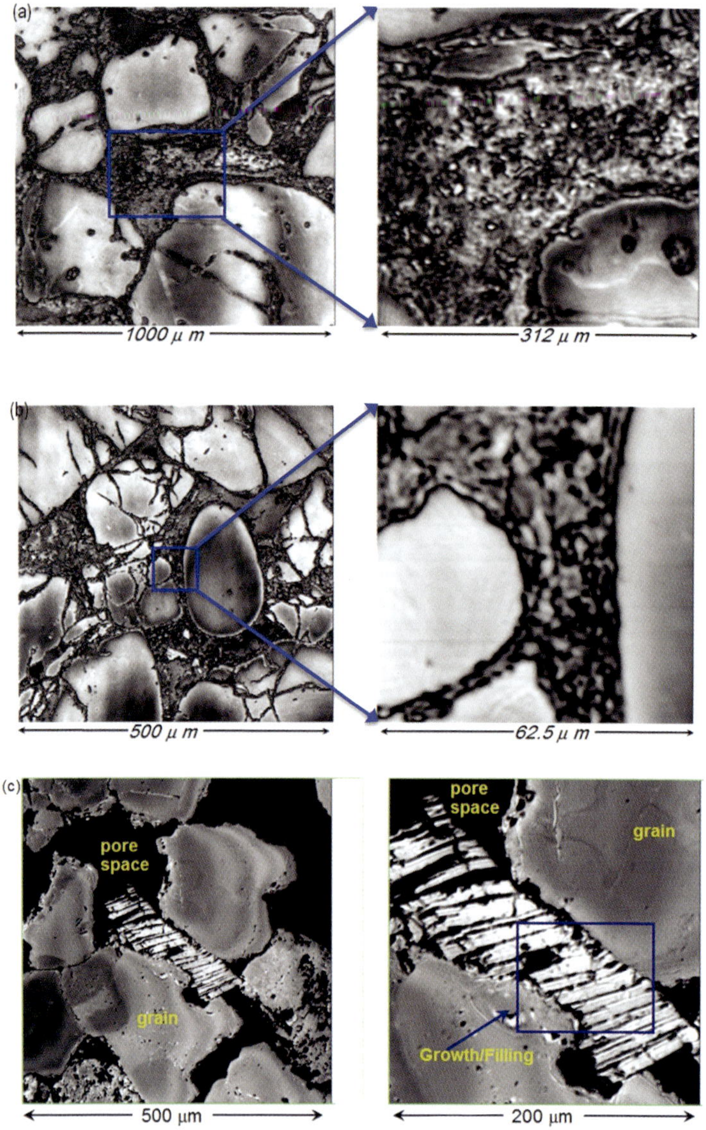

Fig. 3.1. (a) Acoustic images of a quartz sandstone with pore-filling cement (dark area) and intact quartz grains (light area) (b) shows compaction damage in quartz grains in a sandstone with deformation bands due to point loading and stress concentrations (c) shows permeability damage due to an anisotropic cement between the grains.

calculated from the slope of unloading curve after accounting for indentation tip geometry and modulus. Quantitative interpretation of nanoindentation data is usually very inaccurate due to the effect of indentation size, underlying material sensing and comingled particles [12]. When applied to organic rich shales, larger indentation size in comparison with grain sizes increases the chance of probing multiple grains with different physical properties. Thus, the measured values would be an average of sensed particles [13]. This effect is shown in Fig. 3.2(a), an SEM image of an area where an indentation array (with 8000 microNewton load) was performed. The imprints of indentation are visible in topographic SEM image mode (pointed in Fig. 3.2(b)). Comparing the size of indentation with grain sizes reveals that some of the grains in shale are smaller than the imprint of indentation; therefore the results are controlled by mechanical properties of comingled particles. Visual inspection of indentation points is a useful method to screen the data points for extracting mechanical properties of the individual grains.

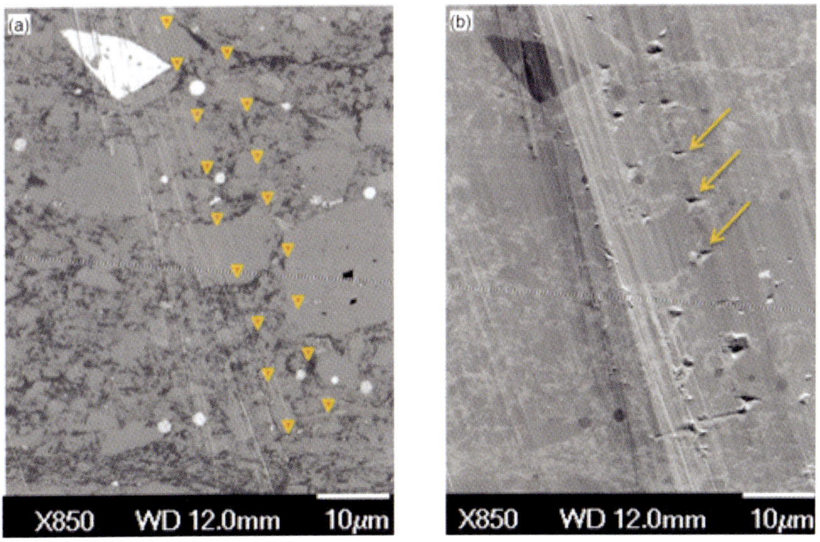

Fig. 3.2. SEM images of a shale sample. The left image (a) shows the indented surface and the probed locations and the right image (b) shows the topography of the same image. By locating the indented area in (b) we are able to determine the mineral groups that were measured under the probe. It is evident that the size of indentation is greater than the shale grain sizes in most indents, causing a comingling of properties.

The nano dynamic mechanical analysis (NanoDMA®) technique is useful to overcome the errors associated with the indentation size and the depth of investigation. NanoDMA® is a probing technique where a special modulus map is produced by analyzing material response to static and overlaid dynamic load. Since the mechanical properties are calculated from displacement amplitude corresponding to the static and dynamic load amplitudes, topography causes error by imposing additional positive or negative displacement amplitude. In composite materials, with components of high mechanical contrast, the response to dynamic load varies depending on stiffness of the probed grains. In order to measure reliable properties for all the components, the dynamic load amplitude must be selected such that the displacement amplitude, while probing all the particles, remains reasonably above noise level. Therefore, another consideration in application of NanoDMA® is the environmental noise level. Figure 3.3 shows an SEM image and a dynamic modulus map of the corresponding ion-milled Bakken shale sample. The SEM image (Fig. 3.3 (a)) shows the morphology and distribution of the main components present in the sample which are calcite, clay and kerogen. The panels in Fig. 3.3 (b)–(d) contains three visualization maps including (from left to right): topography, loss modulus and storage modulus. Note that the modulus map from the region with stiffer grains shows low resolution because the load amplitude was reduced to optimize strain measurements in kerogen particles. To avoid this problem, one must generate a composite map from multiple scans with different load amplitudes, each suitable to measure strains in grains with different stiffnesses. Stiff grains need high loads to deform. Higher loads, however, may cause plastic deformation in more compliant parts (i.e. kerogen and clays).

Atomic Force Microscopy (AFM) is another technique that allows us to image the topography of a sample surface based on deflections of the cantilever tip from its equilibrium position. Atomic Force Acoustic Microscopy (AFAM) is an enhancement of the AFM technique used to measure dynamic elastic moduli. Ultrasonic transducers generate acoustic waves in the sample or the AFM cantilever. Dynamic Young's modulus is determined from contact-resonance frequencies between the AFM cantilever and the sample surface [14]. These resonance frequencies depend, amongst other parameters, on the stiffness of the tip-sample contact, which in turn is a function of the Young's modulus of the sample and the tip, their Poisson's ratios, the cantilever tip radius and shape, the load exerted on the tip, and the geometry of the sample surface. An AFM image and an acoustic image

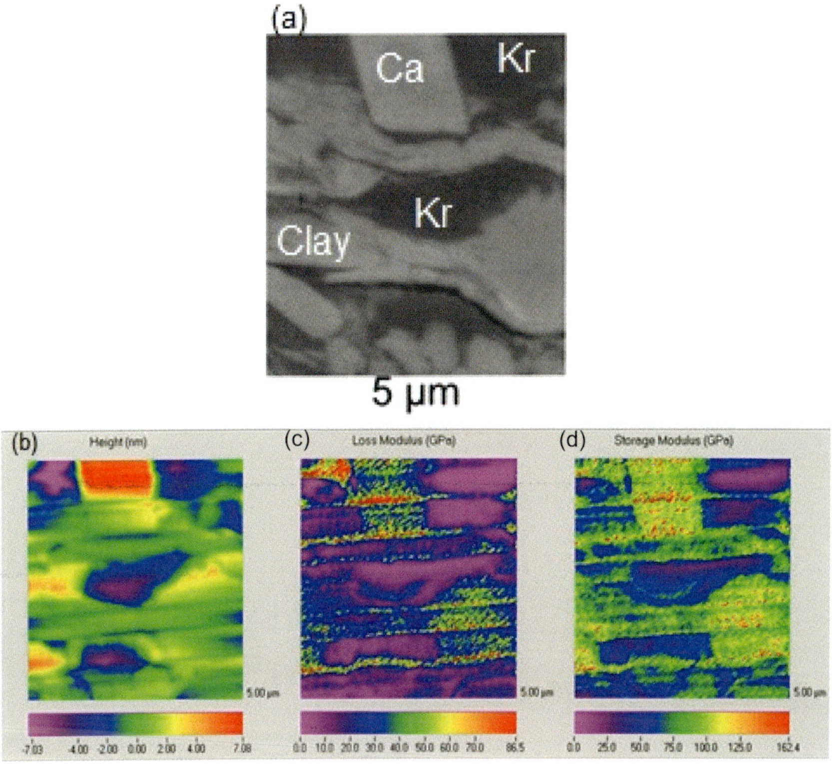

Fig. 3.3. 55 m^2 area of co-located SEM image (a) and dynamic modulus visualization map (b) of a perfectly polished Bakken Shale sample. The SEM image (a) shows the morphology and distribution of the main components present in the sample that are calcite (Ca), clay and kerogen (Kr). Visualization maps include topography (b), loss modulus (c) and storage modulus (d). Note that the modulus map from the region with stiffer grains shows low resolution because the load amplitude was reduced to optimize strain measurements in kerogen particles. To avoid this problem, one must generate a composite map from multiple scans with different load amplitudes, each suitable to measure strains in grains with different stiffnesses.

of the sandstone from Fig. 3.1(c) are shown in Fig. 3.2.1.2 and Fig. ?? respectively. The AFM image (Fig. 3.2.1.2) is made on a 4×4 μm area from the acoustic image marked by the blue box in Fig. ??. Note that the topographic difference between the grains and the cement zone is 1–3 μm.

AFM images of a clay mineral, phlogopite and biotite are shown in Fig. 3.5(a) and Fig. 3.5(b), respectively. In these images, the topographic variations are around 10 nm and 25 nm, respectively. Such low topographic

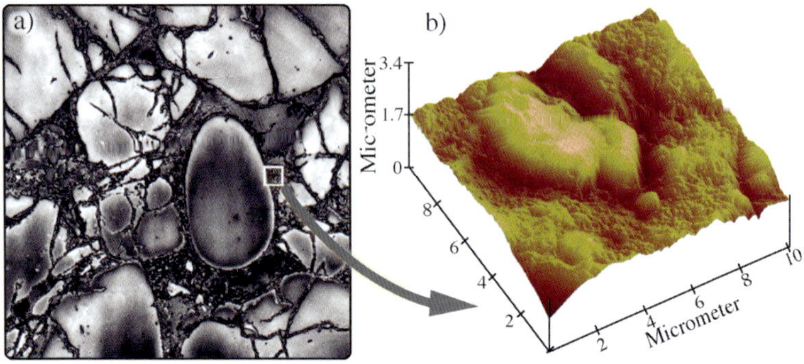

Fig. 3.4. AFM images of the contact zone between quartz grains in a sandstone. (a) Acoustic image. (b) AFM image showing the large topography between grains and cement zone. The topographic scale is 1.7 μm per division.

variations are ideal for making nanoscale AFAM measurements. Typical atomic force acoustic spectra for first and second contact-resonance for phlogopite are compared with those in silicon in Fig. 3.6(a) and Fig. 3.6(b), respectively. As shown in these figures, both first and second contact-resonance spectra for the clay mineral (phlogopite) are at a lower frequency than fused silica. There is also a significant broadening of resonance frequencies in the lossy clay material as compared with the high-Q fused silica.

3.3. Pore Characterization

A major challenge in estimating transport and storage capacity of SRS is the inadequate understanding of pore evolution, distribution, associated minerals, and grain size. Quantification of pore space is a mixed bag of visual (SEM, CT-scans) and non-visual (nitrogen gas adsorption, mercury intrusion, water immersion porosimetry, elastic modulus and acoustic impedance mapping) techniques. Better understanding of the nature of pore morphology, pore throat and network, and total porosity versus effective porosity, is crucial. Pore hosting grains control different aspects of the pore network; Pore size, shape, aspect ratio, and surface properties are mainly controlled by the composition of the rock. These aspects influence the petrophysical properties of the rock as well as our measurement of porosity and pore size distribution (PSD). Applying traditional methods of measuring porosity and PSD to fine grained SRS usually results in inconsistent values [15–17]. Accessibility of displacement fluid to the pore system, wettability and surface properties of the grains are the main controls on acquiring reliable and

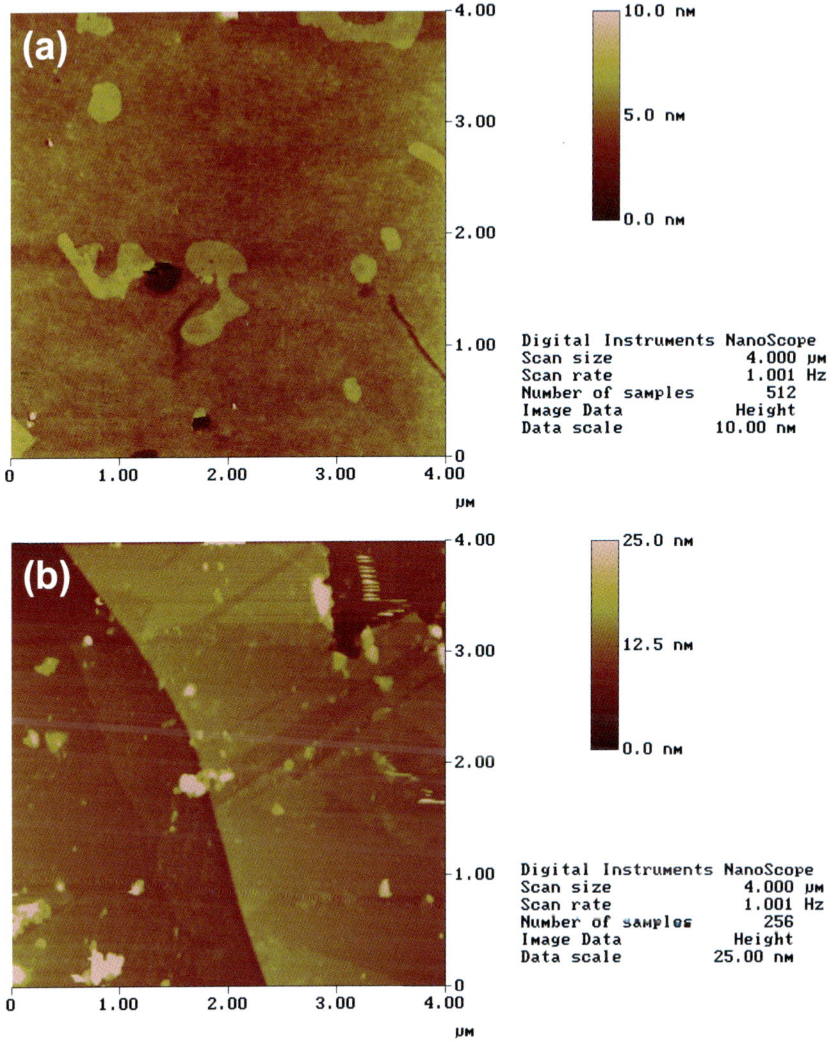

Fig. 3.5. AFM images of a clay mineral, (a) phlogopite and (b) biotite. In these two 4×4 μm images, the topographic variations are around 10 nm and 25 nm, respectively. Such low topographic variations are ideal for making nanoscale AFAM measurements.

consistent laboratory measurements.

Accessibility of fluids to the pore body and mineral surfaces and their interactions are the main issues for reliable and repeatable laboratory measurements. The main issues are small pores, low permeability, complex pore

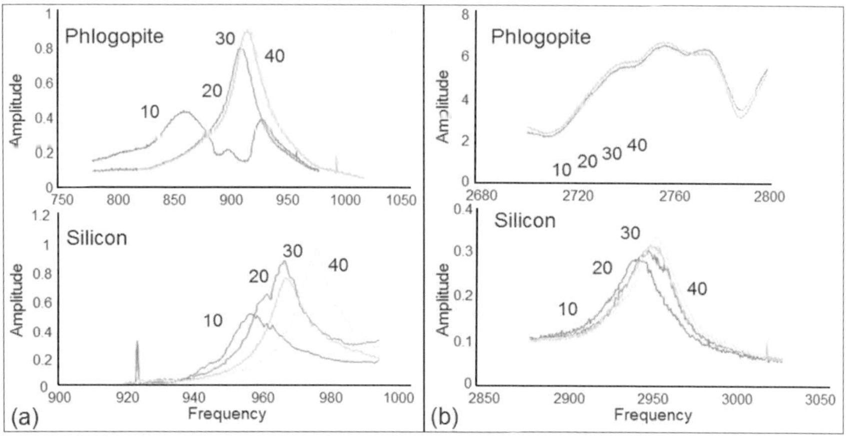

Fig. 3.6. Measured atomic force acoustic resonance frequencies in clay mineral phlogopite and silicon. (a) First contact-resonance frequency (b) second contact-resonance frequency. Note the significant broadening of first resonance frequencies in the lossy clay materialas compared with silicon. The resonance frequencies for phlogopite are lower and broader than those of the high-Q silicon.

structure and heterogeneous mineralogy. For each technique, repeatability can be compromised due to different pretreatment such as grinding and sieving, or laboratory conditions such as relative humidity and temperature. Regardless of using the same measurement technique or comparing different methods, it is important to have uniform pretreatment and ambient conditions when estimating bulk, grain and pore volume.

In the following, we briefly introduce different porosity and pore size distribution measurement techniques. We also discuss the limitations of each technique and suggest some integrated interpretation of the results from different measurements.

3.3.1. *Pore Characterization Techniques*

3.3.1.1. *Porosity*

Porosity measurement techniques such as helium expansion, fluid displacement and immersion, Mercury Intrusion Porosimetry (MICP) and Nuclear Magnetic Resonance (NMR) are well established for conventional applications. Since each measurement technique records different aspects of porestructure, in the following, we briefly discuss advantages and limitations for each:

- Helium expansion, using either crushed or intact samples, measures the pores that are connected and are accessible by the helium gas. In intact nano-darcy permeability rocks, temperature fluctuations can compromise pressure equilibrium and decrease accuracy. However, by crushing the samples, the gas may access "isolated" pores, which are inaccessible otherwise.
- Immersion porosity measurements with water or oil as the saturating fluid can measure pores that can be occupied by the saturating fluid. In this method, the fluid type, pore surface wettability and saturation method affects the porosity values. Reference [18] showed that uncontrolled pretreatment methods and laboratory conditions significantly affect the measured porosities.
- MICP measures porosity and pore throat size distribution. Due to technical limitations, this technique theoretically measures the pore space associated with connected pores with throat sizes bigger than 3.6 nm (at 60000 psi). A significant portion of the pore space in fine grained rocks may be inaccessible by the MICP technique. A reasonable estimate of pore body to throat size ratio may be used to calculate the pore size ranges from MICP throat size distributions. This ratio is not a reservoir rock or sample specific number. It needs to be specified for each pore system and mainly depends on the pore type, mineralogy and cementation. An estimate of the overlooked porosity may be possible by using a complimentary method such as nitrogen adsorption, which is capable of measuring pores smaller than minimum measured pores by MICP.
- Nitrogen adsorption is primarily used for specific surface area measurement using sub-critical nitrogen gas. This technique is also capable of accurately measuring pore sizes in the range of 1.7–200 nm.
- NMR measures the hydrogen nuclei concentration in the rock which directly translates to pore volume and consequently porosity with proper calibrations. Partial saturation of the sample causes underestimation of the porosity. Reliable NMR porosity and PSD measurements depend on surface relaxivity of the pores and tool resolution.

Different methods measure different aspects and size ranges of the pore network. Combining these techniques helps us understand the complex pore structure in these rocks.

3.3.1.2. *Pore-size Distributions*

Some of the methods described above, such as mercury intrusion, nitrogen adsorption and NMR can be used to assess the pore body and throat size distributions. Since each method uses a specific displacement fluid and mechanism to assess the pore size distribution, a combination of all methods may provide a complete spectrum of pore-sizes, as well as provide reasonable estimates of the pore body to pore throat ratio. Due to the wide range of pore sizes in fine-grained rocks, this approach has important implications in studying pore systems in SRS.

As mentioned earlier, mercury intrusion and nitrogen adsorption only cover a limited range of pore throat and pore body sizes. Mercury intrusion is limited to throat sizes bigger than 3.6 nm. Therefore, this method overlooks any pores associated with throats smaller than 3.6 nm, regardless of the pore size. Nitrogen adsorption, on the other hand, measures the pore volume and pore size distribution in the 1.7–200 nm range. A combination of pore-throat-size distribution from MICP data with pore-body-size distribution from nitrogen adsorption can reveal the full spectrum of PSD.

NMR measures the hydrogen nuclei concentration and transverse relaxation time (T_2) distribution. Depending on the tool's resolution and grain surface properties the minimum measurable pore size varies (an NMR tool with echo spacing of 60 μsec can detect the hydrogen nuclei in a spherical pore with a surface relaxivity of 10 μm/sec and a radius larger than 1.8 nm). Due to dominant surface relaxation, smaller pores have faster relaxation compared to the bigger pores. In a typical T_2 distribution, the faster relaxation times on the lower values of the spectra correlate with smaller pores.

Surface relaxivity of the pore is the key to converting NMR time distribution to pore size distribution. The common methodology to calculate surface relaxivity is correlating the mercury intrusion pore-throat size distribution with the NMR T_2 distribution. In this method, we assume consistent proportionality between pore-throat size and pore-body size in the rock. This ratio is reflected in the effective surface relaxivity. This methodology fails in SRS for two reasons. First, mercury intrusion is unable to reach a significant portion of the pore space in nanometer-sized pores. Second, effective surface relaxivity in the mineral-hosted pores will be different from the organic matter-hosted pores. A more accurate way to calculate surface relaxivity consists of two steps. In Step 1, use measured PSD from nitrogen adsorption to calibrate surface relaxivity for the fast relaxation parts of NMR T_2 distribution. Since both methods measure the pore size, the calibration is not biased by any simplistic assumption of pore shape. In Step 2, use measured PSD from mercury intrusion to calibrate surface relaxivity for the long relaxation parts of NMR T_2 distribution.

Figure 3.7 shows the pore size distributions assessed for two samples using all three methods. NMR results for the brine-saturated condition (Fig. 3.7(a)) show a clear difference between the two samples. Sample 1 shows a bimodal distribution representing two sets of pore sizes in this rock, whereas sample 2 shows a uni-modal distribution. A surface relaxivity of 1 μm/sec [19] has been used to convert the time domain to size domain for the NMR distributions. Figure 3.7(b) shows the mercury intrusion for both samples. In this case, mercury intrusion does not capture the bimodal distribution of Sample 1 and the NMR peak corresponding to the mercury intrusion peak is unknown. Figure 3.7(c) shows the pore size distribution from nitrogen adsorption is similar to NMR PSD and both methods capture the amplitude difference in small pores.

3.4. Conclusions

Shale characterization continues to demand high-resolution techniques for analyzing porosity (NMR and MIP), pore size distribution (NMR, MIP, gas adsorption) and matrix mechanical properties (SAM, AFAM, NanoDMA®). We presented a combination of micro and nano-scale characterization techniques to better describe fine-grained tight shale sediments. We also concluded that the burial history and maturation processes affect mechanical and petrophysical properties of formations at different stages of burial.

Acknowledgements

We acknowledge members of OCLASSH consortium at Colorado School of Mines for financially supporting this research. We also thank Drs. M. Kopycinska and Walter Arnold, formerly of the Fraunhofer Institute for Nondestructive test methods, Germany and Drs. M. Kopycinska-Mueller and D.C. Hurley formerly of NIST for the AFAM and acoustic imaging help, Corinne Packard and Taylor Wilkinson for their help with acquiring NanoDMA® scans, and Saul Rivera for providing us with pore size distribution data.

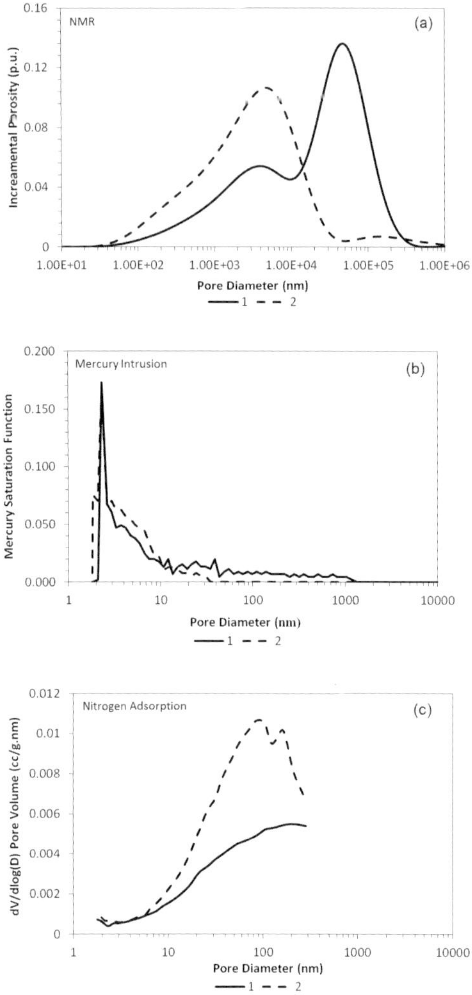

Fig. 3.7. Pore size distribution of two Monetary shale samples measured by (a) NMR with brine saturation [19], (b) mercury intrusion, and (c) nitrogen adsorption [20]. NMR data shows bi-modal and uni-modal distributions for Samples 1 and 2, respectively. Surface relaxivity of 1 μm/sec has been used to convert NMR time distribution to pore size distribution. Mercury intrusion data for both samples is similar and does not capture the differences in pore size distribution. Assuming that fast relaxing components in the NMR spectra represent smaller pores, we can compare the fast relaxing peaks with nitrogen adsorption data. The nitrogen adsorption captures differences between the two samples.

References

[1] R. G. Loucks, R. M. Reed, S. C. Ruppel, and U. Hammes, Spectrum of pore types and networks in mudrocks and a descriptive classification for matrix-related mudrock pores, *AAPG Bull.* **96**, 1071–1098 (2012).
[2] T. Mukerji and M. Prasad. Image processing of acoustic microscopy data to estimate textural scales and anistropy in shales. In ed. M. P. André, *Acoustical Imaging*, vol. 28, pp. 21–29. Springer Netherlands (2007).
[3] R. Rickman, M. Mullen, E. Petre, B. Greiser, and D. Kundert. A practical use of shale petrophysical for stimulation design optimization All shale plays are not clones of the barnett shale. In *SPE Annual Technical Conference and Exhibition*, vol. SPE 115258, Denver, Colorado (2008).
[4] B. R. Lawn and D. B. Marshall, Hardness, toughness, and brittleness an indentation analysis, *J. Amer. Ceram. Soc.* **62**, 7–8 (1979).
[5] L. Vernik and A. Nur, Ultrasonic velocity and anisotropy of hydrocarbon source rocks, *Geophysics.* **57**, 727–735 (1992).
[6] M. Prasad, Mapping impedance microstructures in rocks with acoustic microscopy, *The Leading Edge.* **20** (2001).
[7] J. C. Zeszotarski, R. R. Chromik, R. P. Vinci, M. C. Messmer, R. Michels, and J. Larsen, Imaging and mechanical property measurements of kerogen via nanoindentation, *Geochimica et Cosmochimica Acta.* **68**(20), 4113–4119 (2004).
[8] R. Ahmadov, T. Vanorio, and G. Mavko, Confocal laser scanning and atomic-force microscopy in estimation of elastic properties of the organic-rich bazhenov formation, *The Leading Edge.* **28**(1), 18–23 (2009).
[9] A. Pal, Bathija, H. Liang, N. Lu, M. Prasad, and M. L. Batzle, Stressed swelling clay, *Geophysics.* **74**(4), A47–A52 (2009).
[10] K. Mba, M. Prasad, and M. Batzle. The maturity of organic-rich shales using microimpedance analysis. In *SPE Annual Technical Conference and Exhibition*, vol. SPE 135569, Florence, Italy (2010).
[11] A. C. Fischer Cripps, Critical review of analysis and interpretation of nanoindentation test data, *Surface & Coatings Technology.* **200**, 4153–4165 (2006).
[12] S. Zargari, M. Prasad, K. C. Mba, and E. D. Mattson, Organic maturity, elastic properties, and textural characteristics of self resourcing reservoirs, *Geophysics.* **78**(4), D223–D235 (2013).
[13] G. Constantinides, K. Ravi Chandran, F.-J. Ulm, and K. Van Vliet, Grid indentation analysis of composite microstructure and mechanics Principles and validation, *Mater. Sci. Eng. A.* **430**, 189–202 (2006).
[14] U. Rabe, K. Janser, and W. Arnold, Vibrations of free and surfacecoupled atomic force microscope cantilevers: Theory and experiment, *Rev. Sci. Instrum.* **67**(9) (1996).
[15] T. J. Katsube and N. Scromeda, Effective porosity measuring procedure for low porosity rocks, *Geological Survey of Canada.* **91-1E**, 291–297 (1991).
[16] J. J. Howard, Porosimetry measurement of shale fabric and its relationship to illite / smectite diagenesis, *Clays, Clay Min.* **39**, 355–361 (1991).

[17] J. Dorsch and T. J. Katsube, Effective porosity and pore-throat sizes of mudrock saprolite from the nolichucky shale within bear creek valley on the oak ridge reservation Implications for contaminant transport and retardation through matrix diffusion, *Technical Report Oak Ridge National Lab., Environmental Sciences Division.* **ORNL/GWPO 025** (1996).

[18] K. Utpalendu. *Measurement and Interpretation of Porosity and Pore-Size Distribution in Mudrocks The Hole Story of Shales.* PhD thesis, Colorado School of Mines, CO, USA (May, 2013).

[19] S. Rivera. Petrophysical properties of the Monterey formation, San Joaquin Valley (2013).

[20] J. Godinez, Lemuel. Control factors for fluid saturation and pore size distributions based on gas adsorption, CEC and 2D dielecric microscopy: A case study of the quartz phase porcelanites in the miocene monterey formation (April, 2014).

Chapter 4

Impact of Surface Forces on Coupled Flow, Acoustic and Geomechanical Characteristics in Shale Reservoirs

Azra N. Tutuncu

Department of Petroleum Engineering,
Unconventional Natural Gas and Oil Institute (UNGI)
& Colorado School of Mines, Golden, CO, USA
atutuncu@mines.edu

Earth sediments can be envisioned as mineral grains in contact with neighboring grains to form a solid frame. Yet, in order to model the sediment deformation, acoustic and failure characteristics, we must incorporate of the associated fluid-grain interactions at various stress states. The separation distance between two contacting asperities is controlled by the surface forces and the external forces applied. The fluid-grain interactions have substantial effect in tight reservoirs (particularly in shale and tight gas sands) where microstructure is dominated by smaller grain sizes and larger surface areas impacting geomechanical, petrophysical, transport and failure characteristics. We use a non-Hertzian grain contact model to investigate the role of surface forces on deformation and acoustic properties of shale formations, providing a quantitative analysis of how surface forces influence the deformation and how the pore fluid interaction impacts the frame moduli, deformation and acoustic behavior.

4.1. Introduction

The shale formation mineral composition has an important impact on the shale pore structure, petrophysical characteristics, grain-to-grain and fluid-grain interactions, rock deformation, and failure characteristics. Shale reservoir quality is a function of clay content in the matrix composition as shown in Fig. 4.1. In general, sealing-shale formations contain high surface area clays causing ductile deformation and failure. On the other hand, reservoir shales typically consist of relatively smaller amounts of inert clays along with quartz, feldspar and carbonates. The latter causes a more brittle

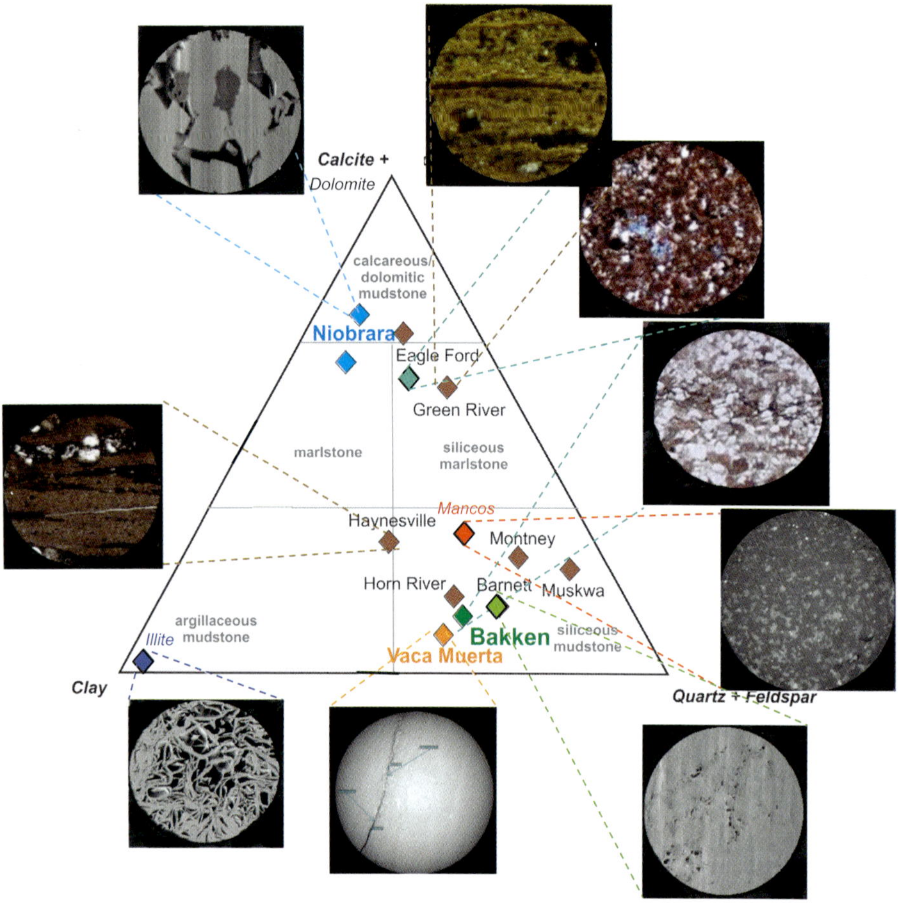

Fig. 4.1. Pore structure change with varying mineralogy in shale formations. CT Scans from several known shale basins are included in the ternary diagram to present interrelation between the mineralogy and pore structures.

behavior with complex induced and natural fracture networks [1]. The large variations in mineralogy observed in CT scans from several shale basins are overlain in the shale mineralogy ternary diagram in Fig. 4.1 to show the relation between the pore structure and mineralogy.

The mineral composition affects geomechanical, acoustic, failure characteristics and the transport properties of shale formations. The predictive capability of the coupled geomechanics, transport, acoustic, and petrophysical models relies on integration of the fluid-grain interactions in pores with the core and reservoir scale data from laboratory, seismic, microseismic, logging, and well testing for economic and environmentally sound development of shale reservoirs. An integrated predictive model should also include fluid type and dissolved salt ionic concentrations to account for their interactions with the matrix, natural and induced fracture network in the reservoir. In the following section, the significance of these interactions on deformation, acoustic and failure characteristics in shale reservoirs will be briefly introduced via the non-Hertzian deformation model.

4.2. Non-Hertzian Grain Contact Adhesion Model

The Hertz contact deformation theory does not take into account the effect of surface forces while the non-Hertzian model introduced in this study (an extension of the Hertz models) accounts for the surface forces coupling their effect into geomechanics calculations through the calculations of deformation, grain contact radius and separation distance between the grains. Hence, when the Hertz contact model is used for the deformation and contact area as a function of applied load between two spheres in contact, it will only include the effect of the mechanical forces as expressed in Eq. (4.1) by Timeshenko and Goodier (1951),

$$a^3 = \frac{3\pi}{4}\left(\frac{1-\vartheta_1^2}{\pi E_1} + \frac{1-\vartheta_2^2}{\pi E_2}\right) RF_0 \qquad \text{with} \qquad \alpha = \frac{a^2}{R} \tag{4.1}$$

where E and ϑ are the static Young's modulus and static Poisson's ratio for the elastic asperities and R is the average grain radius calculated as $(R_1 R_2/(R_1 + R_2))$ for two grains in contact with radius R_1 and R_2 , α is the deformation at the grain contact area and a is the contact radius. The Young's modulus and Poisson's ratio of the shale matrix in the calculations can be determined using Eq. (4.2) and Eq. (4.3), respectively. The input for these equations include XRD analysis of the shale reservoirs of interest and assigning pure mineral moduli with proper volume fractions (f_{clay}, f_{quartz}, f_{calcite}) in the matrix as illustrated in the ternary diagram Fig. 4.1. Young's modulus, Poisson's ratio, acoustic properties, and permeability typically have anisotropic behavior in shale formations which critically impacts experimental and modeling characterization and forecasting studies.

In this chapter, in addition to shale-fluid interactions, we have included shale anisotropic behavior data from experiments on preserved cores, prepared in vertical, horizontal and various inclination angles to the bedding

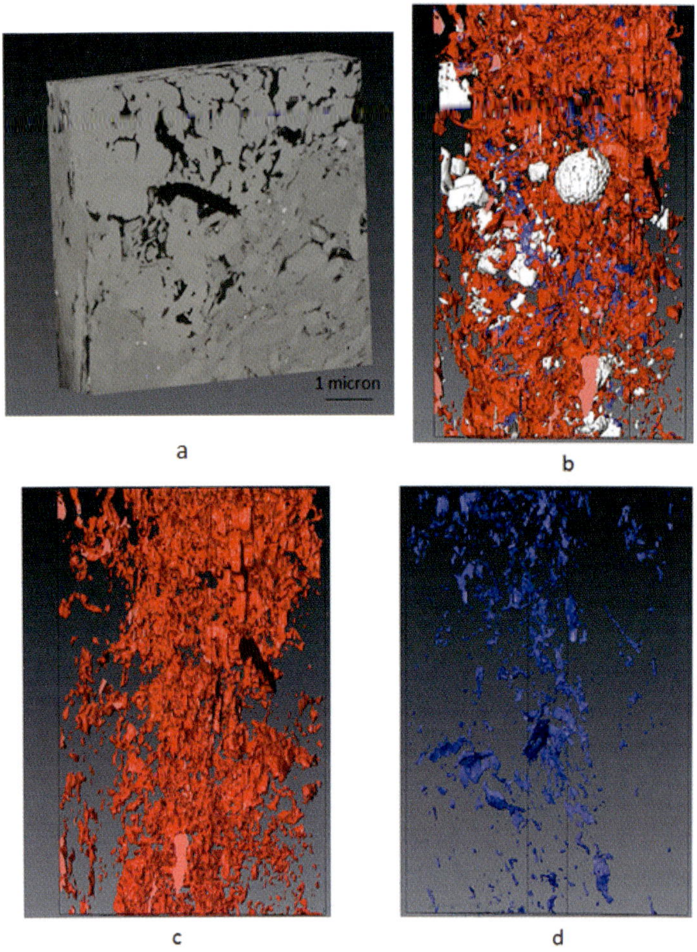

Fig. 4.2. FIB-SEM images of an Eagle Ford preserved shale sample: (a) 10 micron x 10 micron x 10 micron cube analyzed, (b) FIB-SEM image for the full sample, (c) kerogen distribution in the sample, and (d) pore distribution in the sample [2].

planes, in model simulations. Eq. (4.2) and Eq. (4.3) include anisotropy for vertical and horizontal Young's moduli and Poisson's ratio of clay (E_{vclay}, E_{hclay}, ϑ_{vclay}, ϑ_{hclay}) assuming that the anisotropy is the result of clay distribution variability in different directions. If calcite and quartz also contribute to the level of anisotropy in the formation of interest, then the anisotropy contribution from quartz and calcite can be incorporated using

Eq. (4.2) and Eq. (4.3) with the addition of (E_v, E_h) and Poisson ratios (ϑ_v, ϑ_h) for the quartz and the calcite components. Kerogen (that is, TOC) is also present within the matrix as well as in the pore volume based on the SEM and Focused Ion Beam (FIB) SEM measurements. The TOC content is included clays. An example of TOC distribution in an Eagle Ford preserved shale sample is shown by the FIB SEM images in Fig. 4.2 it is evident that nano-pores in the Eagle Ford shale are elongated and parallel to the organic grain boundaries; thus, causing direction dependences.

$$\begin{aligned} E_{\mathrm{v}} &= f_{\mathrm{clay}} E_{\mathrm{vclay}} + f_{\mathrm{quartz}} E_{\mathrm{quartz}} + f_{\mathrm{calcite}} E_{\mathrm{calcite}} \\ E_{\mathrm{h}} &= f_{\mathrm{clay}} E_{\mathrm{hclay}} + f_{\mathrm{quartz}} E_{\mathrm{quartz}} + f_{\mathrm{calcite}} E_{\mathrm{calcite}} \end{aligned} \tag{4.2}$$

$$\begin{aligned} \vartheta_{\mathrm{v}} &= f_{\mathrm{vclay}} \vartheta_{\mathrm{clay}} + f_{\mathrm{quartz}} \vartheta_{\mathrm{quartz}} + f_{\mathrm{calcite}} \vartheta_{\mathrm{calcite}} \\ \vartheta_{\mathrm{h}} &= f_{\mathrm{hclay}} \vartheta_{\mathrm{clay}} + f_{\mathrm{quartz}} \vartheta_{\mathrm{quartz}} + f_{\mathrm{calcite}} \vartheta_{\mathrm{calcite}} \end{aligned} \tag{4.3}$$

Nanoscale laboratory experiments by Ref. [3] show finite contact radius even in the absence of the externally applied load. However, when there is no external load applied, Eq. (4.1) predicts contact radius and deformation to be zero [4, 5]. The surface forces between the grains and asperities are associated with the size of the asperities, the extent of deformation and the surface chemical parameters of the asperities in the intervening fluid. The surface forces have significant influence on coupled deformation, failure and flow properties as the pore throat sizes vary with stress dependent frame moduli and fluid properties. The modified Hertz contact theory is used to

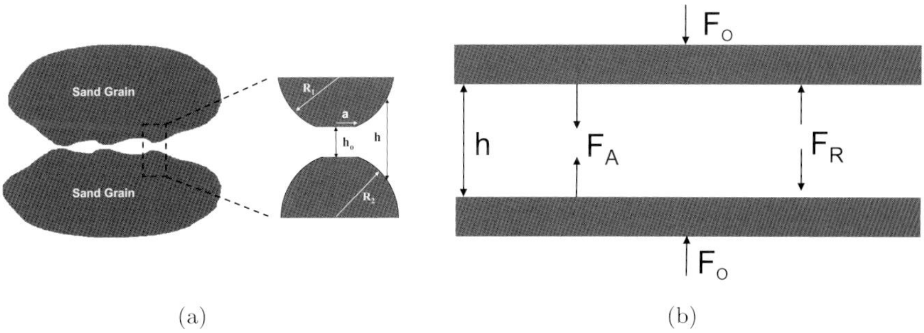

Fig. 4.3. Schematic of grain contact between two sand grains (a), and two clay platelets (b).

calculate deformation pore fluid film thickness at the grain contacts and contact area of deformation for two spherical asperities in contact and subjected to an external load for grains in contact as illustrated in Fig. 4.3(a). In this paper the non-Hertzian grain contact model has been implemented to investigate the frame moduli in shale reservoirs (Fig. 4.3(b)). Individual contributions due to electrostatic repulsion, Born repulsion, structural and van der Waals attractive forces have been added together to determine the total energy of interaction within the contact and non-contact regions for the matrix and associated fluid pressure alterations and deformation in several shale formations. Microcracks are also included in the calculations as they are typically thought to be high aspect ratio grain-to-grain contacts. Each such contact may in reality consist of several asperities in contact and the aspect ratio for individual asperities can be calculated using the non-Hertzian grain contact model. Wettability of shale-oil-brine systems can be described using similar principles as in non-Hertzian contact theory because contact angle, curvature of the pore walls, and capillary pressure are the factors influencing wettability and equilibrium conditions. For such systems, the interface is defined by the augmented Young-Laplace equation, given by Eq. (4.4):

$$P^{\alpha} - P^{\gamma} = \Pi + 2H^{\alpha\gamma}\sigma^{\alpha\gamma} \tag{4.4}$$

where $H^{\alpha\gamma}$ is mean curvature, $\sigma^{\alpha\gamma}$ interfacial tension and $P^{\alpha} - P^{\gamma}$ Laplace or capillary pressure Ref. [6]. Hirasaki states that, in the meniscus region, where the separation of the interfaces is large, $\Pi = 0$ and Eq. (4.4) reduces to the well-known Young-Laplace equation. He also states that in cases where one of the bulk phases is a flat solid, as with a clay platelet or the contact area of the two sand grains, the mean curvature of the fluid-fluid interface is zero because the interfaces are parallel. Thus, capillary pressure becomes equal to the disjoining pressure given by Eq. (4.5).

$$\begin{aligned}\Pi = P_{\text{stotal}} = -\tfrac{A}{6\pi h_o^3} + \left(\tfrac{64nkT}{\kappa}\right)\beta^2\kappa\exp(-\kappa h_o) +\\ K_i\exp\left(-\tfrac{h_o}{l}\right) + \tfrac{1}{45}\tfrac{A\sigma\varphi^6}{h_o^9}\end{aligned} \tag{4.5}$$

where A is the Hamaker coefficient, n is the ionic concentration of the fluid phase, K_1 and l are the structural constants, φ is the collision diameter and κ is inverse Debye-Heckel length parameter.

As external stress is increased on each asperity, they approach each other with a variation in separation distance. Therefore, the following mechanical equilibrium condition has to be satisfied.

$$\frac{\partial F_s}{\partial h} < K_S \tag{4.6}$$

where K_S is the effective spring constant for contacting asperities. In shale with high clay content, the change in separation distance is greater than two sand grains in contact creating larger irreversible deformation, large surface force hysteresis and as a result ductile behavior. Stiffer shale with higher calcite and/or quartz-feldspar content will deform less and hysteresis observed will be much smaller compared to ductile shale unless introduced fluids result larger fluid-grain interaction altering surface forces significantly. The larger the applied external force (F_o), the higher the repulsive surface force is having larger deformation with the smaller separation distance between the platelets and/or grains. Hence, even in the cases where no external force is applied, there will still be a finite deformation due to the surface forces acting on the grains and/or platelets. When the external force is changed, the resulting surface forces will also change for the new equilibrium state. For example, in shale platelets, if the external forces are large, surface forces are mostly repulsive, allowing water to be absorbed in between the clay platelets. If external forces are increased, the resulting equilibrium separation distance between clay platelets becomes smaller. At these small separation distances, the attractive forces are dominant. Thus, fluid, instead of becoming absorbed into shale, will be extracted due to the attractive forces, resulting in shale dehydration. The separation distance between the surfaces is a strong function of the fluid-shale interactions and related ionic concentration and other chemical parameters. One such key parameter is the Hamaker coefficient, a chemical constant controlling the attractive van der Waals forces between the bodies that will complement the externally applied compressive forces. Hamaker coefficient is calculated using the Lifshitz theory incorporating the dielectric constants and refractive indices of the matrix material involved. In this paper, two identical material (medium 1) interacting across fluid medium 3 is used in calculating the sedimentary granular material Hamaker coefficients.

$$A_{131} = A = \frac{3}{4}kT\left(\frac{\varepsilon_1 - \varepsilon_3}{\varepsilon_1 + \varepsilon_3}\right)^2 + \frac{3h\vartheta_e}{16\sqrt{2}}\frac{(n_1^2 - n_3^2)^2}{(n_1^2 + n_3^2)^{\frac{3}{2}}} \tag{4.7}$$

where ε_1 is the dielectric constant of the solid spheres, ε_3 is the dielectric constant of the fluid between the spheres, h is Planck's constant, ν_e is the

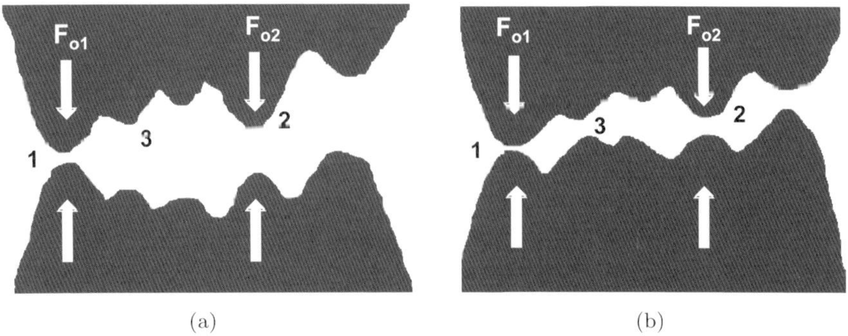

Fig. 4.4. Schematic representation of the effect of stress on deformation and flow at grain contacts. At low applied stress (F_o), the separation distance is smallest in asperity 1 ($F > F_o + F_{\min}$) making asperity 1 a flow barrier (reducing permeability) and experiencing the largest deformation and highest energy loss as a result of surface force influence on asperity 1 (a). Flow restriction and deformation contribution from asperities 2 and 3 is not significant at lower stress and contribution get larger at higher applied stress (b).

absorption frequency of the fluid, n_1 is the refraction index for the solid spheres, and n_3 is the refraction index for the fluid between the spheres.

External stresses alter the contact geometry that, in turn, changes the measured wave velocities. At low stresses, as the grain contacts and/or fracture surfaces are rough, not all the asperities in the given contact area are in actual contact. Therefore, the stiffness is smaller when compared to the corresponding stiffness at higher stresses as shown in Fig. 4.4. When the formation is saturated, the flow of fluid into and out of grain contacts will cause an additional stiffness defined by Ref. [7] as k_{gap}. On the other side, the number of contacting asperities and matrix mineralogy will determine the matrix stiffness ($k_{\text{nmatrix}}; k_{\text{tmatrix}}$). Therefore, the following relations should be satisfied.

$$\begin{gathered} k_{\text{neffective}}(\sigma, \omega, S) = k_{\text{nmatrix}}(\sigma, S) + \gamma k_{\text{gap}}(\sigma, \omega, S), \\ \text{and} \\ k_{\text{teffective}}(\sigma, S) = k_{\text{tmatrix}}(\sigma, S) \end{gathered} \tag{4.8}$$

Above, σ is effective stress, S is the fluid saturation, and γ is the number density of the dissipative contacts. k_{nmatrix} and k_{tmatrix} may be obtained in microscale via the non-Hertzian contact model, Eq. (4.9), and in reservoir scale from the measured dry compressional and shear velocities V_{p} and V_{s}. If shale formation presents strong anisotropy, normal and tangential stiffness are calculated using Eq. (4.9) in vertical and horizontal direction replacing E by E_{h}, E_{V}, and replacing V_{s} with the fast and slow shear wave velocities.

$$k_{\text{nmatrix}}(\sigma, S) = \frac{Ea}{1-\vartheta^2}; \qquad k_{\text{tmatrix}}(\sigma, S) = \frac{2Ea}{(1+\vartheta)(2-\vartheta)}$$

$$k_{\text{nmatrix}}(\sigma, S) = \frac{12\pi\rho_{\text{matrix}} R}{C_n}\left[V_p^2(\sigma, S) - \tfrac{4}{3}V_s^2(\sigma, S)\right] \tag{4.9}$$

$$k_{\text{tmatrix}}(\sigma, S) = \frac{24\pi\rho_{\text{matrix}} R}{C_n}\left[V_s^2(\sigma, S) - \tfrac{1}{3}V_p^2(\sigma, S)\right]$$

In this paper, single phase fluid flow and full saturation of the matrix and pore volume has been assumed. Force exerted on each asperity was calculated using [8] random packing of spheres in Eq. (4.10).

$$F_{o,\text{asperity}} = \frac{4\pi R^2 \sigma_{\text{effective}}}{C_n(1-\phi)} \tag{4.10}$$

where $\sigma_{\text{effective}}$ is the effective stress at which the laboratory and/or field data for velocity is obtained.

4.3. Results and Discussions

Stress and fluid type impact the deformation acoustic characteristics in sedimentary formations. Even for the clean sandstones, the effect of fluid-grain interactions can be incorporated in the deformation characteristics playing a role in both static and dynamic moduli obtained from the data in the field and in the laboratory. The stress and fluid salinity effect is evident for quartz grains in contact in varying salinity water as shown in Fig. 4.5. Associated compressional and shear velocity variations as a function of effective stress and measurement frequency are also determined by coupling the non-Hertzian grain contact and the wave velocity models as presented in Fig. 4.6 and Fig. 4.7 in a shale reservoir for 3% KCl solution.

As discussed in detail by Ref. [9] and shown in Fig. 4.8, the location of the organically rich material has influence on the deformation and failure characteristics as also predicted from the model presented here. In order to test the model reliability with the mineralogical composition variations, laboratory ultrasonic velocity measurements of a low TOC and a high TOC shale reservoir from the same basin have been used. In addition to the quartz, calcite and clay content of the shale matrix, TOC was also included in determining the effective matrix properties for the asperities to be implemented in the model with a representative mixture of the shale material.

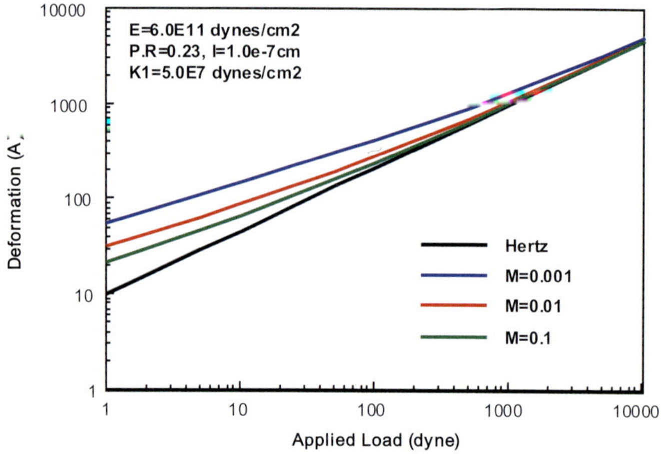

Fig. 4.5. The role of applied load and fluid Molarity on the deformation of two quartz grains of 200-micron diameter immersed in varying salinity brine (after Ref. [5]).

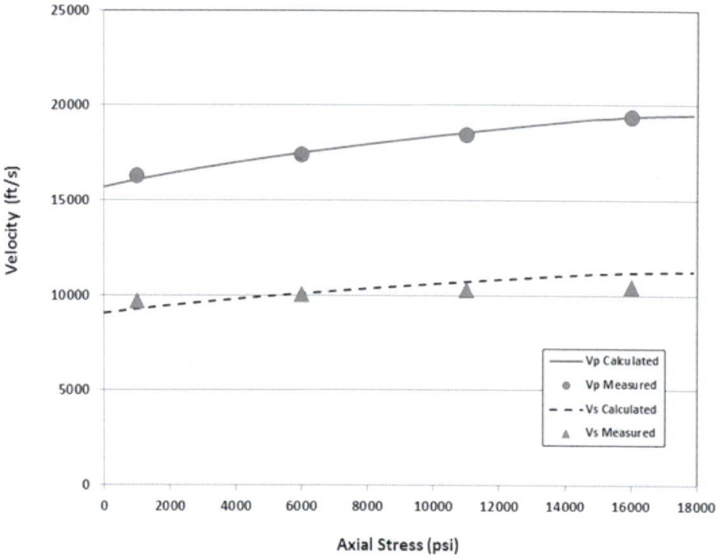

Fig. 4.6. Dependence of wave velocities on effective stress using coupled Modified non-Hertzian grain contact and velocity model in a shale reservoir.

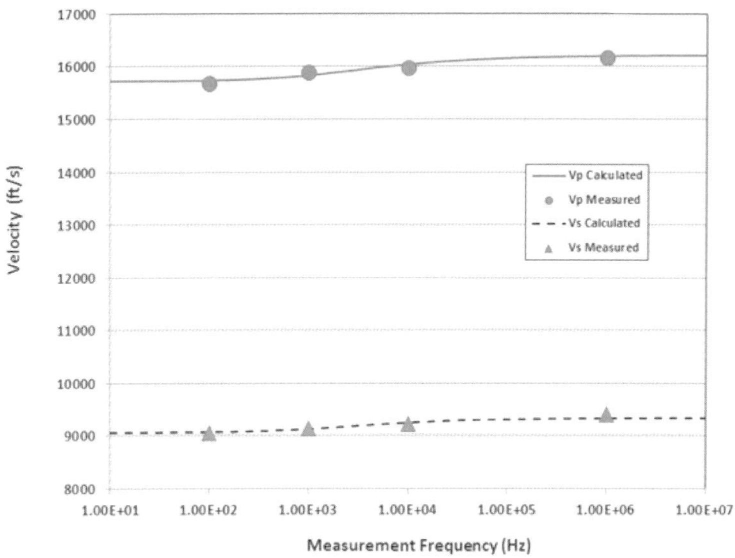

Fig. 4.7. Dependence of wave velocities on measurement frequency using the coupled model. The shale reservoir core measurements used in Fig. 4.6 also collected at log, ultrasonic and low frequencies in the laboratory providing a good agreement with the model results.

The agreement observed between the predicted results of the model and the laboratory measurement agreement is promising as indicated in Fig. 4.9.

4.4. Conclusions

The non-Hertzian grain contact model, developed in this chapter, provides a quantitative analysis of surface forces influence on shale deformation at the grain contact and the shale frame moduli. The model was verified using laboratory ultrasonic compressional and shear wave velocities measured as received, and after it was saturated with brines at various ionic concentrations. In addition to viscosity and density of the saturating fluid, the role of Hamaker coefficients and electrochemical interactions of fluid with the grains and fracture surfaces were determined. In addition to our own laboratory measurements, shale formation data from the literature were used to illustrate the dependence of stress and saturating fluid on wave velocities in the nano and reservoir scale. Using the modified Hertz theory and the grain contact adhesion concept, the velocity changes with varying Hamaker

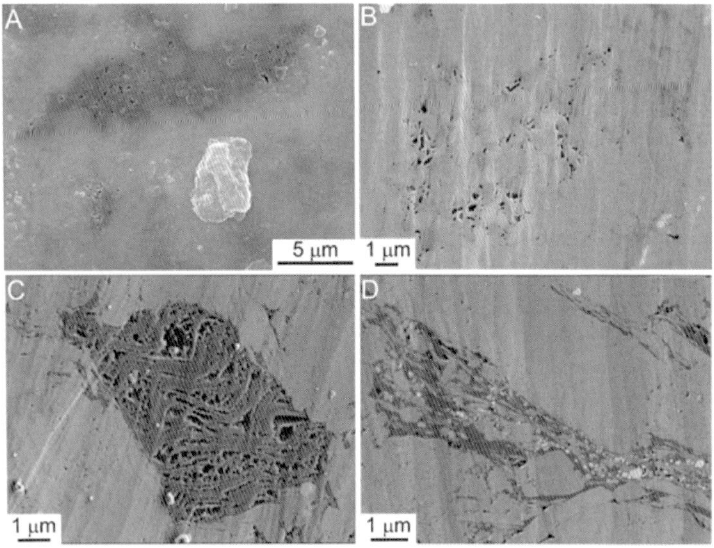

Fig. 4.8. Nano-pores in the organic material in Barnett Shale (after Ref. [9]).

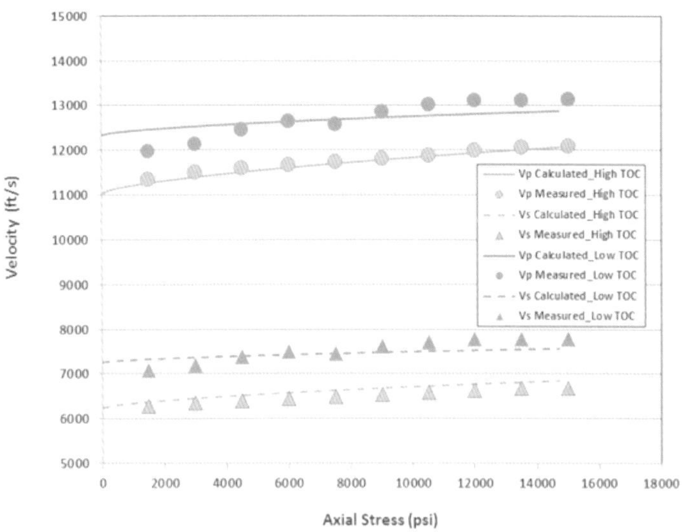

Fig. 4.9. Dependence of wave velocities on stress in a low TOC and high TOC shale reservoir. The stress dependence was predicted using Modified non-Hertzian grain contact model.

coefficient was investigated. Our shale experimental results and model calculations are in good agreement and indicate dependence of compressional and shear wave velocities on stress, frequency, salinity and surface forces.

Acknowledgements

The author would like to express gratitude for the financial support from the sponsors of the Unconventional Natural Gas and Oil Institute (UNGI) CIMMM Consortium (Coupled Integrated Multiscale Measurements and modeling) for shale research, access to shale cores, and field data.

References

[1] A. N. Tutuncu. Anisotropy, compaction and dispersion characteristics of reservoir and seal shales. In *Procedings of the 44th US Rock Mechanics Symposium and 5th US-Canada Rock Mechanics Symposium, Salt Lake City, UT* (2010).
[2] A. N. Tutuncu. Unconventional natural gas and oil institute couple integrated multiscale measurements and modeling (cimmm) annual meeting report (2014).
[3] J. N. Israelachvili, *Intermolecular and Surface Forces*, third edn. Academic Press Inc. (2011).
[4] A. N. Tutuncu. *Velocity Dispersion and Attenuation of Acoustic Waves in Sedimentary Granular Media*. Phd dissertation, The University of Texas at Austin (1992).
[5] A. N. Tutuncu and M. M. Sharma, The influence of fluids on grain contact stiffness and frame moduli in sedimentary rocks, *Geophysics*. **57**, 1571–1582 (1992).
[6] G. J. Hirasaki, Wettability: Fundamentals and surface forces, *SPE Formation Evaluation*. pp. 217–226 (1991).
[7] W. F. Murphy, K. W. Winkler, and R. L. Kleinberg, Acoustic relaxation in sedimentary rocks: Dependence on grain contacts and fluid saturation, *Geophysics*. **51**, 757–766 (1986).
[8] H. Brandt, A study of the speed of sound in porous granular media, *J. Applied Mech.* **22**, 479–486 (1955).
[9] G. Louck, R. Reed, S. C. Ruppel, and D. M. Jarvie, Morphology, genesis and distribution of nanometer-scale pores in silicious mudstones of the mississippian barnett shale, *J. Sediment. Res.* **79**, 848–861 (2009).

Chapter 5

Microporosity Characterization of Mud-Rich Carbonate Rocks

T. D. Jobe* and J. F. Sarg

Department of Geology & Geological Engineering, Colorado School of Mines, Golden, CO, USA

jsarg@mines.edu

Matrix related microporosity is an important control on the transmissivity and storage capacity of hydrocarbons in mudrocks. The matrix pore network feeds the vugs and fractures through diffusive transport. There is a significant portion of micro- to nanoscale porosity that is not captured by traditional optical microscopy. Microporosity in carbonate mudrocks was evaluated using five distinct low permeability reservoir facies from a Lower Cretaceous-aged reservoir. Mineralogy was determined and porosity was estimated for each facies using petrographic image analysis and a quantitative SEM tool (QEMSCAN®). Estimated porosities were compared to measured porosities from a CMS-300® automated permeameter. Porosity and pore throat distributions were determined by mercury porosimetry and gas adsorption experiments in order to capture both micro- and nano-pore distributions. Results showed distinct differences in porosity, permeability, surface area and tortuosity among the five facies, despite their similar mudrock textures. Pore size distributions indicate bimodal pore distributions that are in the micro to nanoporosity range. In general the porosities reported from mercury porosimetry, gas adsorption and CMS 300® experiments agree with those determined by QEMSCAN® analysis, and all are significantly higher than those reported by petrographic image analysis. Pore size distributions as well as tortuosity vary across the different lithofacies, probably related to subtle differences in physical rock properties. We believe these differences strongly affect fluid flow in these rocks.

*Now at Aramco Services

5.1. Introduction

Carbonate reservoir heterogeneity is complex, due to ternary porosity distribution composed of matrix, vugs, and fractures. With the advancement of completion technologies for low-permeability reservoirs, quantifying the matrix-related micro-porosity, understanding pore size and pore throat distributions as well as tortuosity is critical. Traditional methodologies for porosity characterization developed for conventional reservoirs are inadequate for low permeability, microporous reservoirs. Within carbonate reservoirs, the matrix porosity contribution is often overshadowed by the relative contribution from vugs and fractures, yet it is the matrix pore network that effectively "feeds" the vugs and fractures through diffusive transport.

This study focused on microporosity characterization in carbonate mudstones and wackestones (mud supported rock with greater than 10% macrograins). Low permeability core plugs from a Cretaceous aged reservoir were evaluated and five distinct reservoir facies were identified based on detailed core and thin section description. Porosity was estimated for each facies by petrographic image analysis as well as a new technique for porosity determination from QEMSCAN® (quantitative evaluation of minerals and porosity by scanning electron microscopy) analysis [1]. Estimated porosities were compared to measured porosity from a CMS-300® (core measurement system) automated permeameter. Porosity and pore throat distributions were determined by mercury porosimetry and gas adsorption experiments in order to capture both micro and nanopore distributions. Results of the study show distinct differences in porosity, permeability, surface area and tortuosity among the facies, despite their seemingly similar mudstone to wackestone textures.

5.2. Background Information

Carbonate reservoirs are known for their heterogeneity, large pore volumes and highly permeable vug and fracture networks. The focus of this study was the often ignored matrix characteristics of porosity, permeability, pore size and pore shape distributions, surface area and tortuosity. It is these components that limit fluid flow in reservoirs, effectively controlling transmissivity and more importantly storage capacity of hydrocarbons. Understanding how these parameters relate to lithofacies is an important step in understanding and characterizing fluid flow in reservoirs. Microporosity is recognized as a significant contributor to reservoir volumes in the Middle East, particularly in the Jurassic and Cretaceous carbonate reservoirs of the Arabian Platform [2–9]. These reservoirs are volumetrically small compared to many of the highly porous and permeable mega-reservoirs that exist, however, as the worlds oil and gas supply depletes, understanding the controls on these lower permeability reservoirs is becoming critical.

Mudrock microporosity is recognized as a significant contributor to reservoir storage and much work has been done in identifying, characterizing and classifying the spectrum pore types that exist [10–14]. There is still the question of how to adequately quantify microporosity and to determine its contribution to effective storage capacity of hydrocarbons. This study presents a best practices approach for evaluating microporosity in carbonate as well as clastic mudrock systems. A variety of analytical techniques are presented each of which has specific analytical resolution limitations. Both qualitative and quantitative data were collected by traditional techniques including petrographic image analysis [15–21], CMS-300® permeameter measurements [22–25], and Mercury intrusion/extrusion [26, 27]. These data were then compared to non-traditional analytical methods involving QEMSCAN®-BSE mode analysis [1] and Nitrogen gas adsorption experiments [28–37]. In addition, tortuosities were calculated for each sample [38].

Data for this investigation of microporosity in mudstones and wackestones has come from a series of subsurface core samples collected as part of a larger study being conducted on a giant oil field in the offshore United Arab Emirates. The carbonates studied are dominantly mudstones to wackestones with matrix rocks that have very low permeabilities. Table 5.1 outlines the rock textures, biota and diagnostic character of the lithofacies from which the samples for this study were taken.

5.3. Results

The mineral compositions, textures and pore types were estimated qualitatively from thin sections. Samples are predominantly calcite with variable original textures ranging from foraminiferal and bioclastic wackestones to algal boundstones. Samples F5a and F6 contain partially crystalline textures where there has been recrystallization to micro-rhombic, sucrosic dolomite. Dolomitization in these samples is a local phenomenon preferentially concentrated within Thallasinoides burrows. Matrix, in all samples, consists of micritic mudstone and associated mudstone micropros-

Table 5.1. Summary of lithofacies and rock characteristics for the six samples used in this study. The lithofacies are based on rock texture, biota present, diagnostic character and dominant pore type.

Sample / Rock Texture	Biota	Diagnostic Character	Pore Type
F3: Rudist bivalve wacke-to-packstone	Abundant skeletal debris (rudist bivalves, echinoderms, rare corals), benthic forams (*Miliolida*, *Textularia*, *Orbitolina*)	Vuggy macrofossils & burrows	Micro, Macro
F4: *Lithocodium Bacinella* mudstone / wackestone	Common *Lithocodium Bacinella* & sekeltal debris, abundant benthic forams (*Miliolida*, *Textularia*, *Orbitolina*), some pelagic forams	Stylolitic / heterogenous fractures & burrows	Micro, Macro
F5a & **F5b**: *Lithocodium Bacinella* boundstone	Abundant *Lithocodium Bacinella*, common echinoderm, coral, bivalve skeletal debris, benthic forams (*Miliolida*, *Textularia*, *Orbitolina*)	Heterogenous bioclastic boundstone	Micro, Macro
F6: *Lithocodium Bacinella* Wackestone	Abundant oncoidal *Lithocodium Bacinella*, common benthic forams (*Miliolida*, *Textularia*), no *Orbitolina*	Dolomitic burrows	Micro, Macro, Fracture
F7: Rudist Bivalve Wackestone	Common rudist bivalves, peloids, skeletal, coral, echinoderm debris, rare benthic forams	Intense fracturing	Micro, Fracture

ity. Dominant pore types include intercrystalline mudstone microporosity, moldic micropores, interparticle and intercrystalline meso/macro pores and minor micro-fracture porosity. Table 5.2 summarizes the porosity values that were determined from each analytical technique.

Representative photomicrographs and QEMSCAN®-BSE mode porosity maps are shown in Fig. 5.1. Quantitative porosity estimates from petrographic image analysis and QEMSCAN®-BSE mode porosity analysis are also shown. Petrographic image analysis in general showed very little visible porosity in the samples; with 2 samples (F6 & F7) reporting no visible porosity (NVP). In contrast, the QEMSCAN®-BSE mode porosity maps indicate a much larger amount of porosity, particularly in the samples that showed no visible porosity.

In addition to porosity (%ϕ, Table 5.2) the QEMSCAN®-BSE mode porosity maps report a Maximum Porosity (%ϕ Max) value. This values is inclusive of the porosity value (%ϕ) as well as a "pore mineral transi-

Table 5.2. Comparison of results from this study. Shown are the porosity percentages from petrographic image analysis, Porosity and Maximum Porosity values from QEMSCAN®-BSE mode porosity maps, as well as porosity determined from CMS 300®, Mercury porosity and Nitrogen gas adsorption experiments.

Sample/ Rock Fabric	Dominant Pore Type	PIA %ϕ	QEMSCAN-BSE mode analysis %ϕ	QEMSCAN-BSE mode analysis %ϕ max	CMS 300 %ϕ	Mercury porosimetry %ϕ	Gas Adsorption %ϕ	Mercury + Gas %ϕ	Tortuosity
F3: Rudist Wackestone	Inter- & intra-particle macro- to micropores	4.4	10.8	18.3	19.1	17.5	0.5	18.0	2.7
F4: *Lithocodium Bacinella* Wackestone	Inter- & intra-particle macro- to micropores	1.3	9.6	21.6	21.8	18.0	1.4	19.4	2.5
F5a: *Lithocodium Bacinella* Boundstone	Inter- & intra-particle macro- to micropores	2.1	6.7	13.2	17.6	15.6	1.0	16.6	2.8
F5b: *Lithocodium Bacinella* Boundstone	Inter- & intra-particle macro- to micropores	0.9	5.8	16.6	17.0	12.5	1.2	13.7	2.9
F6: Bioclastic Wackestone	Mudstone micropores	NVP	13.3	32.6	23.7	27.1	1.3	28.4	2.4
F7: Rudist Wackestone	Fractures, mudstone micropores	NVP	12.0	28.2	20.8	22.7	1.6	24.3	2.6

tion value" PMT value (% PMT), measured during the QEMSCAN®-BSE mode analysis. The PMT value represents a transitional zone between what is confidently called porosity and what is confidently mineral [1]. While there is a certain degree of uncertainty associated with the PMT values, their addition provides an estimate of the upper limit of porosity or maximum porosity. Maximum porosity values are considerably higher than the porosities estimated from petrographic image analysis.

Pore size and shape distributions were also calculated from QEMSCAN®-BSE mode porosity maps for samples F3, F4, F5a, F5b and F6. Meaningful results could not be determined for F7 because of computational limitations. Results from the pore size and shape analysis are shown in Fig. 5.2.

Distributions are calculated directly from the QEMSCAN®-BSE mode porosity maps by digitally extracting individual pores, measuring the long and short axes and calculating a dimensionless aspect ratio. Pore sizes in the samples were on the whole less than 70 micro-meters, with the exception of samples F3 and F5a that reported greater than 20 volume percent of pores that were greater than 70 micro-meters. The majority of the pore sizes reported are below 10 micro-meters, particularly in samples F4, F5b and F6. The majority of porosity in these samples falls within the micropore range [13, 14].

All samples exhibited non-Gaussian pore shape distributions and no two samples had the same pore shape distribution (Fig. 5.2.). With the exception of F5b, samples reported approximately 30 volume percent of pores with aspect ratios of less than 10 (near circular). Sample F5b reports an even greater 52 volume percent of its pores having aspect ratios less than 10. In contrast, sample F5a reports approximately 50 volume percent of pores that have aspect ratios greater than 100 (long, irregularly shaped pores). The variability in pore shape distributions and the non-Gaussian nature of the distributions indicate heterogeneity both within individual samples as well as between samples.

Mercury porosimetry intrusion-extrusion curves, calculated pore size distributions and porosities are shown in Fig. 5.3. Intrusion-extrusion curves show six distinct pore volumes and hysteresis expressions, indicating a clear distinction between samples. Pore size distributions are distinguishable between samples, and they all share a roughly bimodal distribution. The major peaks for each sample range from 0.2 to 0.9 micro-meters, while the minor peaks range from 0.005 to 0.01 micro-meter (5–10 nm). Pore sizes for these samples fall within the nano to micropore range as defined

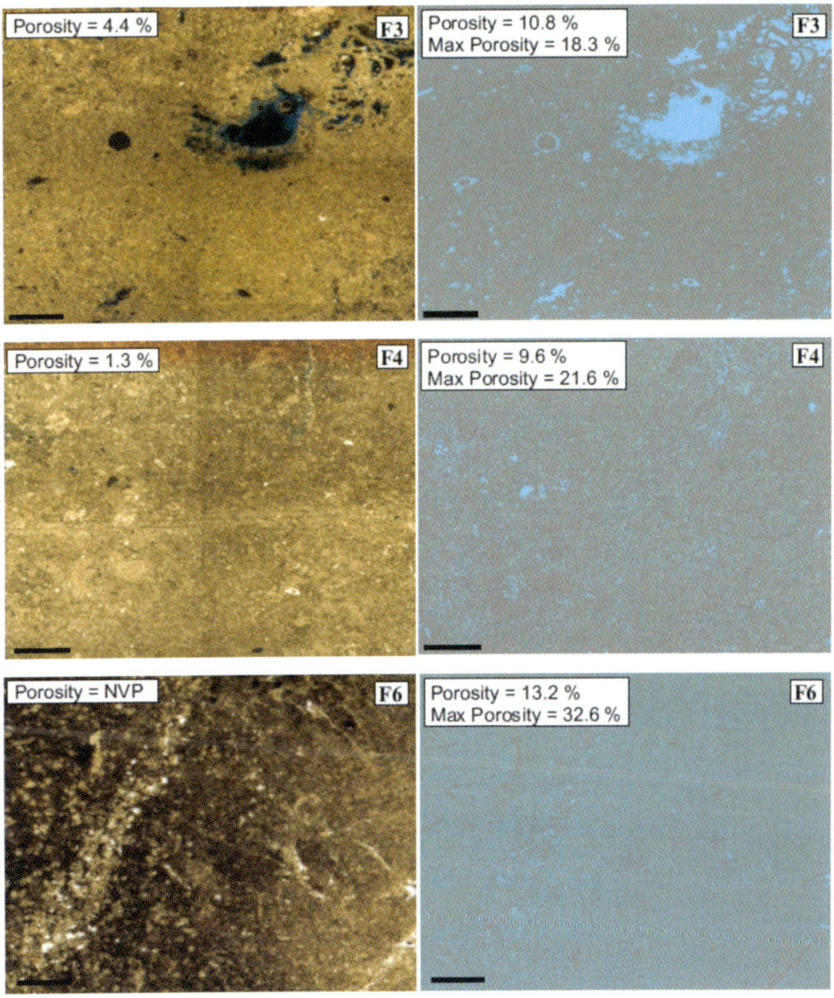

Fig. 5.1. Comparison of photomicrographs and QEMSCAN®-BSE mode porosity maps. Pixel counting was used to calculate QEMSCAN®-BSE modal abundances, where grey = mineral, blue = porosity and green = pore-mineral transition. The Maximum Porosity value represents addition of the blue and green pixels. Scale bars = 1 mm. Shown are (F3) Rudist wackestone with inter- and intraparticle macro- to micropores. Note the large vug with abundant blue epoxy. (F4) Lithocodium-Bacinella wackestone with inter- and intraparticle macro- to micropores, some blue epoxy present within skeletal fragments and microfractures. (F6) Rudist wackestone with dolomitic burrow (left side) and mudstone micropores with no visible blue epoxy present.

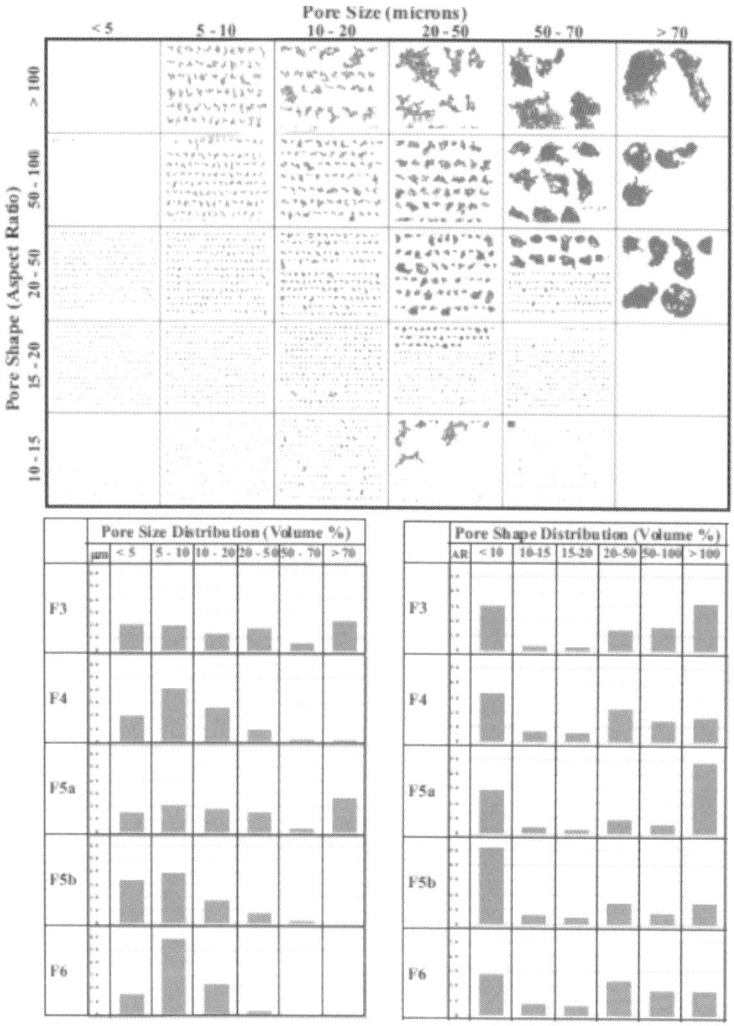

Fig. 5.2. Example of a digital pore extraction from sample F4 (top) as well as calculated pore size (lower left) and shape distributions (lower right). Pore sizes in microns range from < 5, 5–10, 10–20, 20–50, 50–70, and > 70, while volume % scale ranges from 0–70. Dimensionless aspect ratios range from < 10, 10–15, 15–20, 20–50, 50–100 and > 100, while volume % ranges from 0–60. Note the linear relationship between pore size and shape. Large pore spaces have large aspect ratios due to irregular shapes, while small pore spaces have small aspect ratios as they approach roundness. Distributions are shown graphically for comparison between samples. Note that no data is available for sample F7 due to computational limitations.

by Loucks and others [14].

Nitrogen gas adsorption experiments were conducted in order to capture information on the nanopore range. Fig. 5.4 shows the Nitrogen gas adsorption-desorption isotherms, calculated BET surface areas, porosities as well as the BJH-Halsey pore size distributions. The resulting adsorption-desorption isotherms show a unique pattern and hysteresis for each sample. Porosities determined by Nitrogen gas adsorption are less than 2 percent for all samples, which may seem small, but when compared to the bulk porosities, these account for 1/10 or greater of the total porosity. Calculated surface areas for these samples are very low, only samples F6 and F7 reach surface areas of greater than $1\mathsf{m}^2$, indicating that grain size fractions for these samples are indeed smaller than the others. Pore size distributions show a roughly bimodal distribution. The major peaks range from 0.06 to 0.15 micro-meters, while the minor peaks all show the same pore size fraction at 0.004 micro-meters (4 nm).

Measured porosities from CMS-300® experiments as well as calculated tortuosities are also presented in Table 5.2 for comparison.

5.4. Discussion

In this chapter we have introduced a variety of analytical techniques to help improve reservoir evaluation at the micro- to nano- pore scale. The data collected covers a large range of analytical resolution. The challenge then becomes integrating this information and understanding the significance of the data. The following discussion outlines the meaningful information that can be gained from each analysis, compares the porosity, pore size distribution and tortuosity results, and relates these findings to lithofacies.

5.4.1. *Analytical Considerations*

Of all of the analyses presented in this study, the one most traditionally used by geologists for porosity evaluation is optical microscopy and petrographic image analysis from thin sections. This technique is the lowest resolution for porosity determination. Petrographic image analysis consistently under reported porosity values and in the case of samples F6 and F7, there was no visible porosity detected. This occurs because the resolution of an optical microscope is about ≈ 60 micro-meters [13, 14, 39], and anything less than 60 micro-meters will not be detected, making optical microscopy and petrographic image analysis unreliable, particularly for microporosity

Samples	F3	F4	F5a	F5b	F6	F7
Porosity %	17.5	18.0	15.6	12.5	27.1	22.7

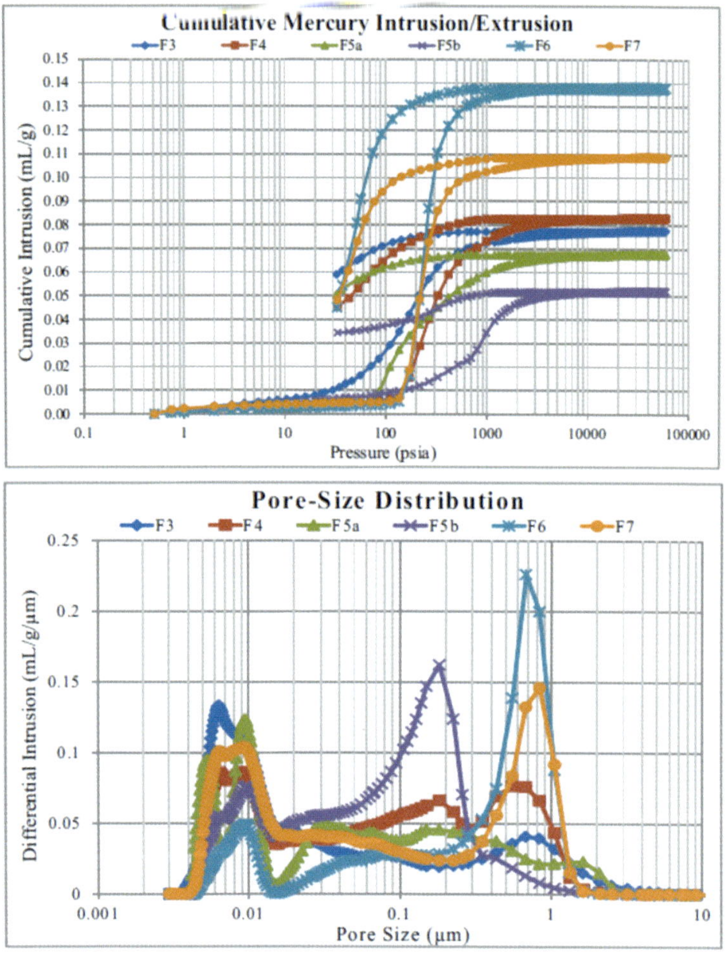

Fig. 5.3. Results of the mercury porosimetry analysis. Of importance is the relative magnitude of the curves, there is a strong variance between samples with respect to the amount of mercury that was intruded, as well as shape of the curves. Curve shapes and hysteresis are slightly different for each sample, which has implications for the grain size distributions. A clear difference is seen in among samples F5b, F6 and F7, which show bimodal distribution, while F3, F4 and F5a are more broadly distributed, indicating there is a range in pore size and increased heterogeneity.

evaluation.

Samples	F3	F4	F5a	F5b	F6	F7
Porosity %	0.48	1.43	1.0	1.21	1.28	1.65
BET Surface Area (m^2/g)	0.337	0.845	0.679	0.949	1.024	1.301

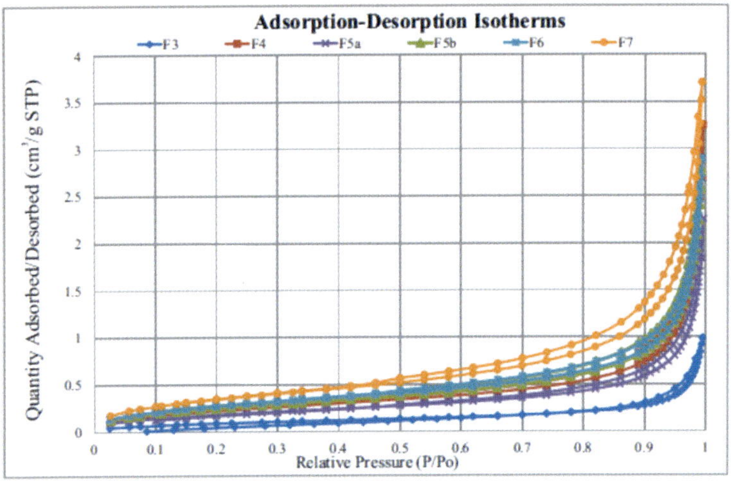

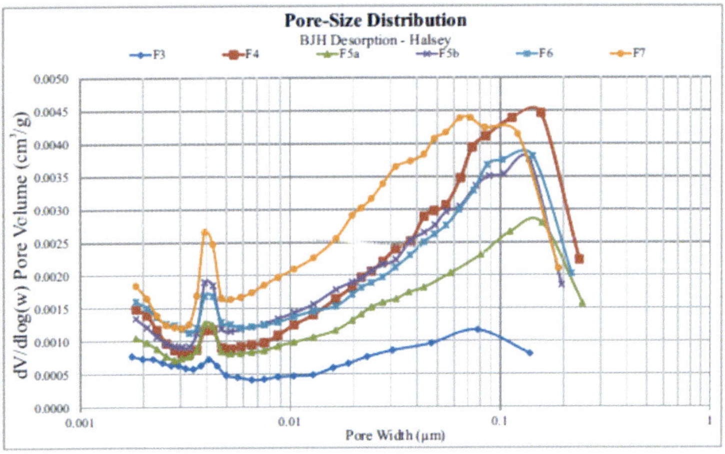

Fig. 5.4. Results of the Nitrogen gas adsorption experiments. As with the mercury experiments, there is a variance in the data between samples and they can be easily distinguished from one another. Pore size distributions show bimodal distributions for all samples, with the major peaks ranging from 0.07 to 0.15 micro-meters and the minor peaks all aligning at 0.004 micro-meters (4 nm). Porosities reported are less than 2%, but it should be noted that this is the accounts for the nano- to microporosity present in the samples. The fractions account for roughly 10% of the total porosity present.

Porosity values reported by the QEMSCAN®-BSE mode porosity maps are much higher than those reported by petrographic image analysis. These porosity values represent a quantifiable amount of pore space that is unquestionably present in the samples. The QEMSCAN®-BSE mode analysis works on a micro-meter scale and has a much higher working resolution than an optical microscope. Because of this, the high resolution scans are capable of detecting pore spaces below the optical resolution (60 micro-meters). The QEMSCAN®-BSE mode analysis is capable of resolutions as low as $^1/_2$ a micro-meter, making it ideal for most microporosity evaluations.

The QEMSCAN®-BSE mode analysis reports two values, a Porosity value and a Maximum Porosity value. The Porosity value is the amount of porosity present in the sample that has been measured with absolute confidence, while the Maximum Porosity value is uncertain because it contains some fraction of porosity that is below the 2 micro-meter scan resolution used in this study. Although uncertain, the Maximum Porosity values in this study more closely match the values generated by the CMS-300® measurements, mercury porosimetry and Nitrogen gas adsorption experiments. This indicates that there is a significant portion of microporosity that is below the two micro-meter scan resolution. Increasing the resolution and reducing the scan rate to 0.5 micro-meters reduces the uncertainty associated with the porosity and Maximum Porosity values. Porosity values from the present study are meaningful, however, when calibrated with other results. In this study, at the present resolution of 2 micro-meters, the Maximum Porosity values are a better estimate of porosity in the samples.

CMS-300® and mercury porosimetry measurements are standard practices for obtaining porosity, permeability and capillary pressure information in the oil and gas industry. Both of these techniques are well suited for traditional microporous to macroporous samples, but care must be taken when considering nanoporous materials. The CMS-300® uses permeability and a modified Darcys Law to calculate porosity, and when low permeability samples are measured, care must be taken when interpreting the results. For example, the measured permeability in samples F6 and F7 was less than 2 millidarcies for each sample. The calculated porosities from CMS-300® were slightly lower than estimated by the other techniques. The low permeability of these samples accounts for the lower porosity values from CMS-300® experiments.

All of the samples in this study contain a significant proportion of micro- to nanoporosity, particularly samples F6 and F7. The small pore size contributes to low permeability of the samples and dramatically increases the associated capillary pressure. Capillary pressures in reservoir rocks govern how immiscible fluids (i.e. water and oil) behave in pores (capillaries). Mercury porosimetry experiments directly measure the capillary pressures and the associated pores sizes. Mercury porosimetry experiments are traditionally used for microporous to macroporous systems and typically the measured pressures are in the low pressure spectrum. Mercury porosimetry results presented here all contain a considerable amount of nanoporosity with pore sizes less than 0.01 micro-meters (10 nm). This pore size fraction corresponds to capillary pressures greater than 18,000 psia. These pressures are well beyond those typically measured by mercury porosimetry. There is therefore a higher degree of confidence in the low pressure results than in the high pressure regime and independent validation of the apparent nanoporosity is necessary.

Nitrogen gas adsorption experiments were carried out in order to characterize any nanoporosity present in the samples. In order to do these experiments successfully, a critical amount of surface area per sample (≈ 1 m^2) is needed (ISO 15901-2:2006). The resulting calculated BET surface areas indicate that samples F3, F4, F5a, and F5b did not contain the critical amount of surface area for the experiment. This is consistent with all other results and indicates that these samples have relatively large grain size and pore sizes; and given the fact that surface area is inversely proportional to grain size, this result is not surprising. Samples F6 and F7, however, each contained more than a square meter of surface area. These two samples contain the highest porosities and the most proportion of nanoporosity according to both QEMSCAN®-BSE mode analysis and mercury porosimetry. These results suggest that Nitrogen gas adsorption experiments are better suited for microporous carbonates with high surfaces areas, including micritic mudstones, chalks and carbonates containing calcareous nanofossils.

5.4.2. *Porosity Comparison*

Comparison of the CMS-300®, Maximum Porosity, and the Mercury + Nitrogen porosity values (Table 5.2) shows subtle differences in porosities reported. The Maximum Porosity values agree with both the CMS-300® and the Mercury + Nitrogen results for samples F3 and F4. These are the samples with the largest pore size fractions. Samples F5a and F5b, however, have the poorest agreement among porosity estimates. These are the
samples with the largest pore size fractions. Samples F5a and F5b, how-

ever, have the poorest agreement among porosity estimates. These samples are from the "Lithocodium-Bacinella Boundstone" lithofacies, which has an abundance of bioclastic fragments of all sizes and shapes and therefore is highly heterogeneous. The discrepancy between porosity estimates is most probably attributed to sampling inadequacies. For example, even though the thin sections come from the same core plugs, there is a chance that the thin section will fail to capture all of the inherent heterogeneity within a sample. Because the CMS-300® uses an entire core plug for measurement, there is a possibility that it could encounter more porosity (i.e. a macropore or vug) that would not be captured in the thin section. This could account for the anomalous porosity results in samples F5a and F5b and should be taken into consideration when analyzing heterogeneous samples.

Samples F6 and F7 have relatively good agreement among porosity estimates, yet the Maximum Porosity values are consistently higher than those estimated by the CMS-300® and the Mercury + Nitrogen values. These are the two samples that reported no visible porosity from petrographic image analysis meaning that all porosity is microporosity. This accounts for the very high Maximum Porosity values. Both the CMS-300® and the Mercury + Nitrogen report significant amounts of porosity but the values are lower than the Maximum Porosity value. This is due to the uncertainty associated with the Maximum Porosity. Scanning at a higher resolution will reduce this uncertainty and should be considered when analyzing microporous samples.

5.4.3. *Pore Size Distributions*

Pore size distributions determined from QEMSCAN®-BSE mode porosity maps are in qualitative agreement with those calculated from mercury porosimetry and Nitrogen gas adsorption experiments; meaning that the distribution peaks do not directly correspond, but rather the distribution shapes agree. Sample F3, for example, shows a wide, even distribution of pore sizes among all three analyses. The broad distribution of pore sizes signifies a great deal of heterogeneity in this sample. This is in contrast to sample F6, which shows a very pronounced peak in all distributions, meaning that its pore sizes are of a uniform size. This is not surprising as F6 comes from the Lithocodium-Bacinella Wackestone lithofacies. This lithofacies is characterized as having abundant burrows filled with sucrosic dolomite. This type of dolomite forms equidistant micro-rhombic dolomite crystals and therefore, the pore spaces between these micro-rhombs are uniform. The homogeneous pore spacing accounts for the large fraction of single mode pore sizes in sample F6.

The difference in magnitudes between the pore size distributions is a function of the analytical resolution at which they were sampled. QEMSCAN®-BSE mode analysis sampled a much lower resolution than the mercury porosimetry and nitrogen gas adsorption experiments and the QEMSCAN®-BSE mode analysis results indicate much larger pore sizes. The large pore sizes are interpreted to be a combination of connected pores which are smaller than the QEMSCAN®-BSE mode sampling resolution. The mercury porosimetry and nitrogen gas adsorption experiments are at much higher resolution, are dependent upon physical measurement, and are capable of detecting the smaller pore sizes. These methods more accurately represent the true pore size distributions in microporous samples.

5.4.4. *Tortuosity*

Calculated tortuosities range from 2.3 to 2. The highest tortuosity values belong to samples F5a and F5b while the lower tortuosity vales belong to samples F6 and F7. Sample F6, the "Lithocodium-Bacinella Wackestone" with dolomitic burrows reports the lowest tortuosity. This is related to the connectivity in the burrows and uniform grain size which creates a less tortuous pathway. This is in contrast to samples F5a and F5b, which are "Lithocodium-Bacinella Boundstone", which are comprised of skeletal and bioclastic fragments with a wide range or pore sizes and shapes. This heterogeneity ultimately results in a more tortuous, less direct pathway.

5.4.5. *Pore Architecture*

Table 5.1 outlines the rock textures, biota and diagnostic character of the lithofacies from which the samples for this study were taken. The following discussion relates the lithofacies and physical rock properties to their experimental responses. Two end member lithofacies, representing a heterogeneous (F3) and a homogenous (F6) sample, are selected for further discussion.

Sample F3, shown in Fig. 5.5, comes from a ***Vuggy Rudist-Bivalve Wackestone to Packstone***. It is heavily oil stained and contains abundant peloids and skeletal debris including rudists, other bivalves, gastropod, brachiopod and echinoderm fragments and rarely coral fragments. Poorly preserved *Thalassinoides* burrows are also present, as well as large intact macrofossils, most commonly rudists. This lithofacies also contains abundant moldic and common vuggy porosity that may or may not be cemented. The QEMSCAN®-BSE mode porosity maps show large easily identified macropores (vugs/burrows) as well as abundant matrix related microporosity. SEM imagery shows a wide variety of grain sizes as well as associated pore sizes. This is consistent with the broad pore size distributions indicated by QEMSCAN®, mercury porosimetry and nitrogen gas adsorption data. The heterogeneity in this lithofacies is also reflected in the calculated tortuosity of 2.7, which is relatively high compared to other lithofacies.

Sample F6, shown in Fig. 5.6, comes from a ***Burrowed Lithocodium-Bacinella Wackestone***, characterized by an abundance of dolomitic *Thalassinoides* burrows. These burrows are pervasive throughout this lithofacies and are filled with sucrosic dolomite. This lithofacies is a massive wackestone containing abundant oncoidal Lithocodium-Bacinella, and is heavily oil stained with micro-fractures and stylolites throughout. The QEMSCAN®-BSE mode porosity maps indicate macroporsity related to the dolomitic burrows as well as matrix related microporosity. Mercury porosimetry and nitrogen gas adsorption data indicate a bimodal pore size distribution with significant contributions from pores that are in the nano to micropore size range. SEM imagery shows a uniform distribution of grain size as well as pore sizes in this sample. The homogeneity of this sample is striking, particularly when compared with SEM image of the F3 sample. The majority of the connected pores is in the 1 micro-meter size range and is related to the dolomitic burrows. This lithofacies also has the lowest tortuosity, related to the "path of least resistance" through the well connected, equi-dimentional, dolomitic pore network. This lithofacies accounts for the majority of the production in this reservoir.

5.5. Conclusions

This chapter has integrated a variety of investigative techniques, with differing analytical resolutions, to evaluate microporosity in a series of carbonate mudstones and wackestones. Specifically this study assimilates several data sets in order to (1) quantify the matrix related microporosity in the system, (2) understand the pore size distributions (3) determine the tortuosity (4) relate these parameters to lithofacies. Conclusions from this study are as follows:

(1) The results of all of the analytical procedures presented show a clear

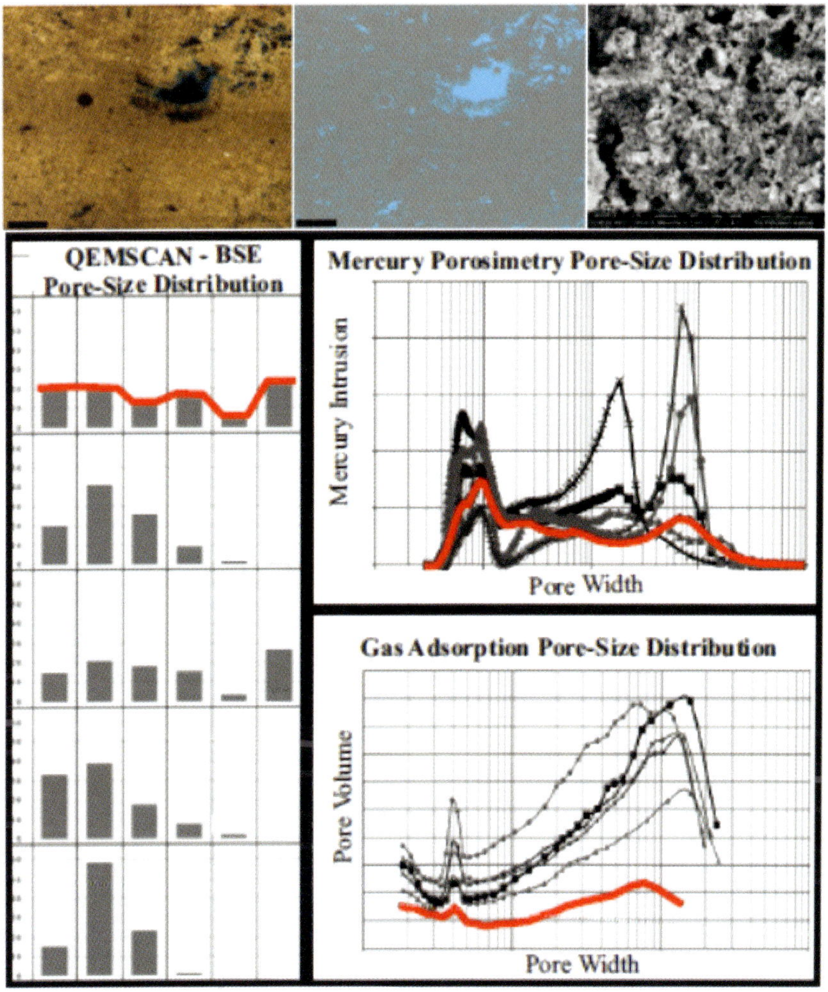

Fig. 5.5. Comparative results of all analysis for sample F3, highlighted in red. Larger versions of all graphs and scales presented in Figs. 5.1–5.4. Sample F3 is a rudist wackestone with inter- and intraparticle macro- to micropores. Note the large vug with abundant blue epoxy. The QEMSCAN®-BSE mode porosity map also shows an abundance of matrix related pores that are not visible in the photomicrograph. SEM imagery indicates a variety of pore types with a large range in size. This wide range of pore size is also reflected in the pore distributions, which all show a low broad distribution of pore sizes, indicating a high degree of heterogeneity within the sample.

distinction between the samples, indicating that each lithofacies can be directly related to distinctly different and defining values of porosity,

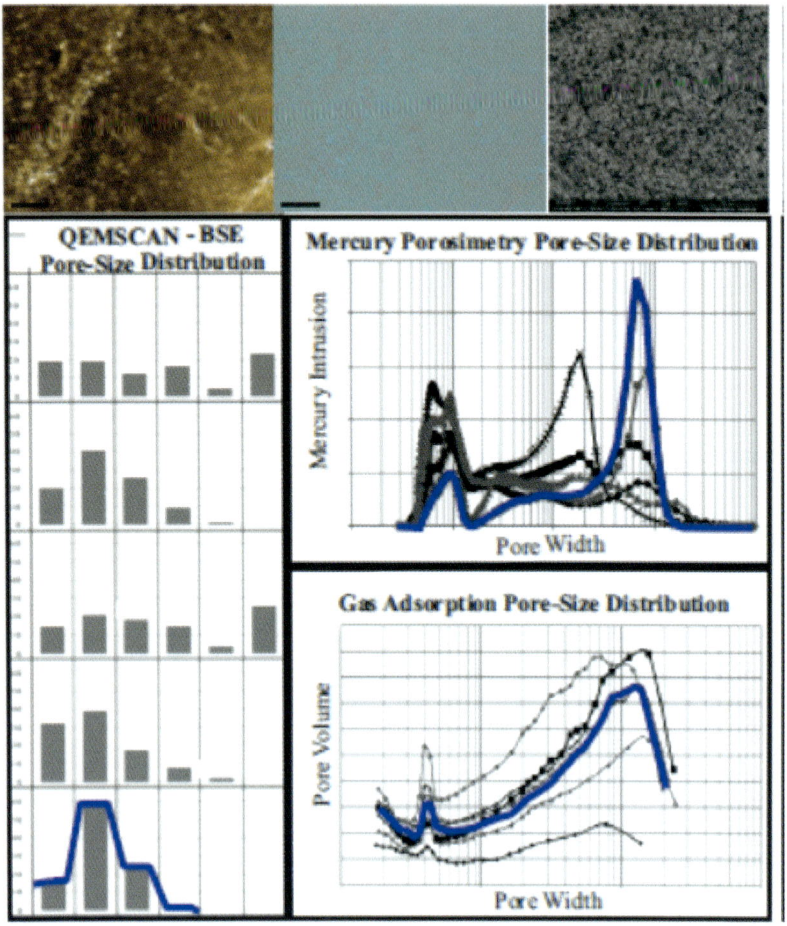

Fig. 5.6. Results of all analysis for sample F6, highlighted in blue. Larger versions of all graphs and scales presented in Figs. 5.1–5.4. Sample F6 is a rudist wackestone with dolomitic burrowing (left side) and abundant mudstone micropores. This sample shows no visible blue epoxy in the photomicrograph while the QEMSCAN®-BSE mode porosity map shows an abundance of matrix related pores. SEM imagery indicates a fairly uniform grain size as well as pore size within the sample. The homogeneity in this sample is also indicated by the unimodal pore size distributions. The pore size distribution from QEMSCAN® analysis indicates an average pore size of 5–10 micrometers which is related to the larger more connected pores within the dolomitic burrows, while the mercury and nitrogen pore size distributions indicate an average pore size of $\sim$ 1 micro-meter which is related to matrix micrite.

permeability, surface area, size/shape distributions, capillary pressures and tortuosity.

(2) Petrographic image analysis and optical microscopy are inadequate for microporosity evaluation because the optical resolution of most microscopes cannot detect pore sizes less than about 60 micro-meters.

(3) The maximum porosity values from QEMSCAN®-BSE mode analysis, although uncertain, are a more robust estimate of porosity in this study.

(4) The Maximum Porosity values are in agreement with the porosity values reported by CMS-300® and mercury + nitrogen. Over estimation of porosity in samples F6 and F7 indicate that there is a significant amount of microporosity which is below the current 2 micro-meter resolution. Scanning at a higher resolution will reduce this uncertainty and should be considered when analyzing microporous samples.

(5) Mercury porosimetry is ideal for measuring microporous to macroporous systems in the low pressure realm and caution should be exercised when interpreting high pressure data. The presence of nanopores in the high pressure realm should be independently validated.

(6) Nitrogen gas adsorption techniques are ideal for nanoporous systems with pore sizes less than 50 nm (0.05 micro-meters). This technique would be particularly useful for carbonate mudstones with high surface areas such as micritic mudstones, chalks and mudstones containing calcareous nanofossils.

(7) Porosity estimates from all techniques are in agreement and indicate that there is a significant amount of microporous and nanoporous storage capacity within these samples.

(8) Pore size and shape distributions agree qualitatively across all analytical techniques and the differences are a function of the analytical resolution of the techniques used.

(9) Samples showed consistent variability in tortuosity; samples with the most nanoporosity always had the lowest tortuosity, while the most heterogeneous samples consistently had the highest tortuosities, despite changing the method of calculation and/or the source of the porosity input data.

Results of this microporosity evaluation indicate that each lithofacies has a unique set of values for porosity, pore size distribution and tortuosity. These parameters are what control fluid flow in the matrix. Different lithofacies (i.e., different rock textures) will have different fluid flow responses in a reservoir. Understanding the controls on low permeability microporous reservoirs, such as the one described here, is critical to field developments as the worlds hydrocarbon supply diminishes. This chapter presents a variety of techniques for microporosity evaluation that can and should be used to evaluate the pore architecture and flow characteristics of other low permeability, microporous reservoirs.

Acknowledgement

This work was completed as part of a joint research agreement between the Colorado School of Mines and the Petroleum Institute of Abu Dhabi. The Abu Dhabi National Oil Company and their affiliated companies supported this work and provided the data for this study. We thank ADNOC for giving permission to publish this paper. We thank Dr. Katharina Pfaff and others at the Colorado School of Mines QEMSCAN facility for their help in data collection, reduction and processing. We also thank Waleed Alameri, Tedesse Teklu and Dr. Hossein Kazemi for help with core preparation, measurement and for providing a better understanding of the methodology and standards of operation.

References

[1] T. D. Jobe. *Sedimentology, Chemostratigraphy and Quantitative Pore Architecture in Microporous Carbonates: Examples from a Giant Oil Field Offshore Abu Dhabi, U.A.E.* Ph.d dissertation, Colorado School of Mines, Golden, CO (2013).

[2] D. A. Budd, Micro-rhombic calcite and microporosity in limestones: A geochemical study of the Lower Cretaceous Thamama group, U.A.E., *Sedimentary Geology.* **63**, 293–311 (1989).

[3] S. O. Moshier, Microporosity in micritic limestones: A review, *Sedimentary Geology.* **63**, 191–213 (1989).

[4] S. O. Moshier, Development of microporosity in a micritic limestone reservoir, Lower Cretaceous, Middle East, *Sedimentary Geology.* **63**, 217–240 (1989).

[5] R. D. Perkins, Origin of micro-rhombic calcite matrix within Cretaceous reservoir rock, West Stuart City Trend, Texas, *Sedimentary Geology.* **63**, 313–321 (1989).

[6] D. L. Cantrell and R. M. Hagerty, Microporosity in Arab Formation Carbonates, Saudi Arabia, *GeoArabia.* **4**(2), 129–154 (1999).

[7] A. A. Al-Awar and J. D. Humphrey, *Diagenisis of the Aptian Shuaiba Formation at Ghaba North Field, Oman*, In eds. A. S. Alsharahan and R. W. Scott, *Middle East models of Jurassic/Cretaceous carbonate systems*, pp. 173–184. SEPM Special Publication 69. SEPM, Tulsa (2000).

[8] L. Lambert, C. Durlet, J. P. Loreau, and G. Marnier, Burial dissolution of micrite in Middle East carbonate reservoirs (Jurassic-Cretaceous): Keys for recognition and timing, *Mar. Pet. Geology.* **23**, 79–92 (2006).

[9] E. A. Clerke, H. W. Mueller III, E. C. Phillips, R. Y. Eyvazzadeh, D. H. Jones, R. Ramamoorthy, and A. Srivastava, Application of Thomeer hyperbolas to decode the pore systems, facies and reservoir properties of the Upper Jurassic Arab D limestone, Ghawar field, Saudi Arabia: A "Rosetta Stone" approach, *GeoArabia.* **13**(4), 113–160 (2008).

[10] F. Javadpour, Nanopores and apparent permeability of gas slow in mudrocks (shales and siltstones), *J. Can. Pet. Technol.* **48**(8), 16–21 (2009).

[11] G. R. Chalmers, R. M. Bustin, and I. M. Power, Characterization of gas shale pore systems by porosimetry, pycnometry, surface area, and field emission scanning electron microscopy/transmission electron microscopy image analyses: Examples from the Barnett, Woodford, Haynesville, Marcellus, and Doig units, *AAPG Bulletin.* **96**(6), 1099–1119 (2012).

[12] M. E. Curtis, C. H. Sondergeld, R. J. Ambrose, and C. S. Rai, Microstructural investigation of gas shales in two and three dimensions using nanometer-scale resolution imaging, *AAPG Bulletin.* **96**(4), 665–677 (2012).

[13] R. G. Loucks, R. M. Reed, S. C. Ruppel, and D. M. Jarvie, Morphology, genesis, and distribution of nanometer-scale pores in siliceous mudstones of the Mississippian Barnett Shale, *J. Sediment. Res.* **79**, 848–861 (2009).

[14] R. G. Loucks, R. M. Reed, S. C. Ruppel, and U. Hammes, Spectrum of pore types and networks in mudrocks and a descriptive classification for matrix-related mudrock pores, *AAPG Bulletin.* **96**(6), 1071–1098 (2012).

[15] F. S. Anselmetti, S. M. Luthi, and G. P. Eberli, Quantitative characterization of carbonate pore systems by digital image analysis, *AAPG Bulletin.* **82**(10), 1815–1836 (1998).

[16] R. Ehrlich, S. K. Kennedy, S. Crabtree, and R. L. Cannon, Petrographic image analysis, I. Analysis of reservoir pore complexes, *J. Sediment. Petrol.* **54**, 1365–1378 (1984).

[17] R. Ehrlich, S. J. Crabtree, K. O. Horkowitz, and H. J. P., Petrography and reservoir physics I: Objective classification of reservoir porosity, *AAPG Bulletin.* **75**(10), 1547–1562 (1991).

[18] J. B. Ferm, R. Ehrlich, and G. A. Crawford, Petrographic image analysis and petrophysics: Analysis of crystalline carbonates from the permian basin, west texas, *Carbonates and Evaporites.* **8**, 90–108 (1993).

[19] A. Lønøy, Making sense of carbonate pore systems, *AAPG Bulletin.* **90**(9), 1381–1405 (2006).

[20] C. A. McCreesh, R. Erlich, and S. J. Crabtree, Petrography and reservoir physics II: Relating thin section porosity to capillary pressure, the association between pore types and throat size, *AAPG Bulletin.* **75**, 1563–1578 (1991).

[21] T. T. Mowers and D. B. Budd, Quantification of porosity and permeability reduction due to calcite cementation using computer-assisted petrographic image analysis, *AAPG Bulletin.* **80**(3), 309–322 (1996).

[22] D. K. Keelan. Automated core measurement system for enhanced core data at overburden conditions. SPE 15185, Society of Petroleum Engineers (1986). Presented at the Rocky Mountain Regional Meeting of the Society of Petroleum Engineers, Billings, Montana, USA, May 19-21, 1986.

[23] V. V. Petunin, A. N. Tutuncu, M. Prasad, H. Kazemi, and X. Yin. An experimental study for investigating the stress dependence of permeability in sandstones and carbonates. ARMA 11-279, American Rock Mechanics Association (2011). Presented at the 45th US Rock Mechanics/Geomechanics Symposium, San Francisco, CA, June 26-29, 2011.

[24] V. V. Petunin, X. Yin, and A. N. Tutuncu. Porosity and permeability changes in sandstones and carbonates under stress and their correlation to rock texture. Canadian Society for Unconventional Gas/Society of Petroleum Engineers 147401, Society of Petroleum Engineers (2011). Presented at the Canadian Unconventional Resources Conference, Calgary, Alberta, Canada, November 15-17, 2011.

[25] Y. Cho, O. G. Apaydin, and E. Ozkan. Pressure-dependent natural-fracture permeability in shale and its effect on shale-gas well production. SPE 159801, Society of Petroleum Engineers (2012). Presented at the Society of Petroleum Engineers Annual Technical Conference and Exhibition, San Antonio, Texas, USA, October 8-10, 2012.

[26] E. W. Washburn, The dynamics of capillary flow, *Phys. Revi.* **17**(3), 273–283 (1921).

[27] H. L. Ritter and D. L. Drake, Pore-size distribution in porous materials, *Ind. Eng. Chem.* **17**, 782–791 (1945).

[28] I. Langmuir, The adsorption of gases on plane surfaces of glass, mica and platinum, *J. Amer. Chem. Soc.* **40**, 1361–1402 (1918).

[29] S. Brunauer, P. H. Emmett, and E. Teller, Adsorption of gasses in multimolecular layers, *J. Amer. Chem. Soc.* **60**, 309–319 (1938).

[30] P. H. Emmett, *Surface area measurements: A new tool for studying contact catalysts*, In ed. W. G. Frankenburg, *Advances in Catalysis*. Academic Press, New York, NY, USA (1948).

[31] S. J. Gregg and K. S. W. Sing, *Adsorption, Surface Area, and Porosity*, 2nd edn. London, Academic Press (1982).

[32] K. S. Sing, Reporting physisorption data for gas/solid systems with special reference to the determination of surface area and porosity, *Pure and Applied Chemistry*. **57**, 603–619 (1985).

[33] G. Leofanti, G. Padovan, M. Tozzola, and B. Venturelli, Surface area and pore texture of catalysts, *Catalysis Today*. **41**, 207–219 (1998).

[34] F. Rouquerol, J. Rouquerol, and K. S. W. Sing, *Adsorption by Powder and Porous Solids: Principles, Methodology and Applications*. Academic Press, London (1999).

[35] R. Richards, ed., *Surface and Nanomolecular Catalysis*. CRS Press, Taylor and Francis Group, Boca Raton, FL, USA (2006).

[36] E. P. Barrett, L. G. Joyner, and P. P. Halenda, The determination of pore volume and area distributions in porous substances — (i) computations from nitrogen isotherms, *J. Amer. Chem. Soc.* **73**, 373–380 (1951).

[37] P. A. Webb and C. Orr, *Analytical methods in fine particle technology*. Micromeretics Instrument Corporation, Norcross, GA, USA (1997).

[38] W. O. Winsauer, H. M. Shearin Jr., P. H. Masson, and M. Williams, Resistivity of brine saturated sands in relations to pore geometry, *AAPG Bulletin*. **36**(2), 253–277 (1952).

[39] P. W. Choquette and L. C. Pray, Geologic nomenclature and classification of porosity in sedimentary carbonates, *AAPG Bulletin*. **54**, 207–244 (1970).

Chapter 6

New Frontiers: Hydrates in Porous Media

Luis E. Zerpa

Department of Petroleum Engineering,
Colorado School of Mines, Golden, CO, USA
lzerpa@mines.edu

Carolyn A. Koh

Department of Chemical & Biological Engineering,
Colorado School of Mines, Golden, CO, USA
ckoh@mines.edu

Gas hydrates enclathrate gas molecules (e.g. methane) in a cubic crystal lattice comprised of a hydrogen-bonded network of water cages. The energy trapped within naturally occurring methane hydrates found in sediments in oceanic and arctic regions can be an order of magnitude greater than that contained in known fossil fuel reserves. Advanced understanding of the pore-scale interactions of methane hydrate and sediment grains during hydrate formation and dissociation as a function of thermodynamic driving force, hydrate saturation, and sediment grain type will be key to improving reservoir models that can be used to reliably evaluate the technical feasibility of producing gas from methane hydrate deposits. This new knowledge will also contribute to assessments of the environmental impacts of methane gas release during production of hydrates or during natural occurrences. This chapter presents a summary of the current state of knowledge on natural hydrate systems and identifies opportunities for continuing research and development.

6.1. Introduction

Gas hydrates are crystalline inclusion compounds that are formed when water and gas molecules (e.g. methane, propane, carbon dioxide) come into contact at elevated pressures and low temperatures. The ability of hydrates to concentrate gas by a factor of around 164, leads to a number of industrial

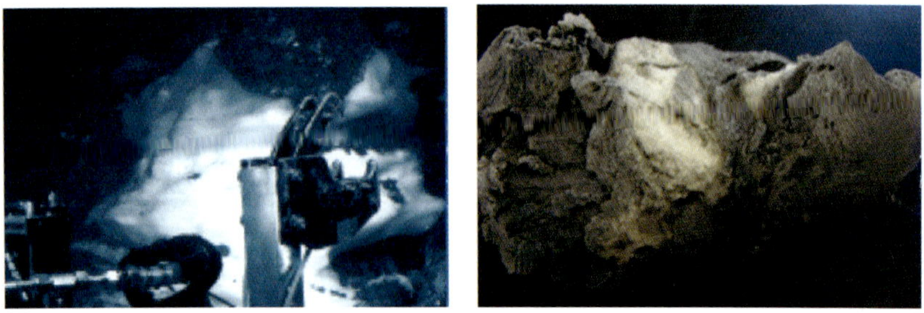

(a) (b)

Fig. 6.1. Naturally occurring hydrates: (a) in-situ photograph of a seafloor deposit found in Barkley Canyon, Off Vancouver Island (courtesy P. Walz, MBARI; cover art of Ref. [2]); (b) recovered hydrate in sediment core from under the ocean floor in the Godavari Basin, India (courtesy M. Walsh).

and technological energy applications for these compounds, as well as significant growing interest in the energy potential (an order of magnitude greater than fossil fuel reserves) of gas hydrates occurring naturally in porous media (Fig. 6.1). Gas hydrates are formed by water molecules arranging themselves around the guest molecules to form a hydrogen-bonded three-dimensional lattice (host) comprising polyhedral water cages that trap the guest molecules. There is typically a maximum of one guest molecule per water cage. These compounds are non-stoichiometric, which means that not all the water cages need to be occupied by a guest molecule. Gas hydrates can be approximated as solid solutions, since there are only weak van der Waals interactions between each guest molecule and a water cage, with the assumption that a guest molecule in one cage will not interact with another guest molecule in a neighboring cage [1, 2].

The main criterion for whether a hydrate will form is that the temperature and pressure conditions are within the thermodynamic hydrate stability region. Figure 6.2(a) illustrates the pressure and temperature dependence of natural gas hydrates. Furthermore, a molecule can be incorporated into the water cages of a hydrate if the size of the guest molecule can fit into the specific cage, e.g. methane has a diameter of around 0.436 nm and so can fit in both the large and small cages of sI (Fig. 6.2(b)). The large and small cage radius of sI is 0.395 nm and 0.433 nm, respectively. Conversely, propane has a diameter of around 0.628 nm and therefore is too large to be accommodated in the sI cages, and instead will fit into the

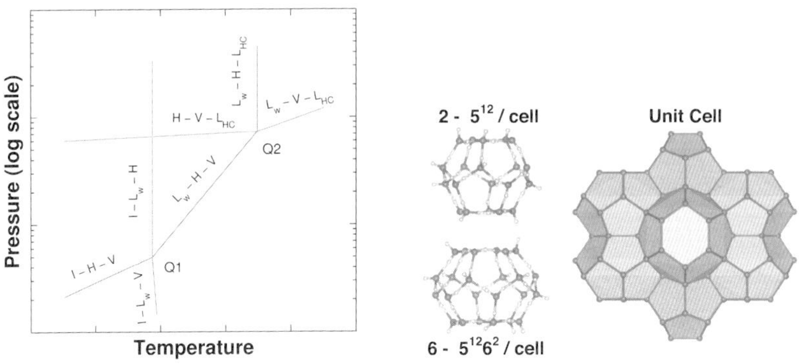

(a) (b)

Fig. 6.2. Hydrocarbon + water phase diagram showing the hydrate stability dependence of pressure and temperature, and specifically the phase boundaries for Ice-Hydrate-Vapor (I-H-V), Ice-Liquid water-Hydrate (I-LW-H), Ice-Liquid water-Vapor (I-LW-V), Liquid water-Hydrate-Vapor (LW-H-V), Hydrate-Vapor-Liquid hydrocarbon (H-V-LHC), Liquid water-Hydrate-Liquid hydrocarbon (LW-H-LHC), Liquid water-vapor-Liquid hydrocarbon (LW-V-LHC). Q1 and Q2 are the quadruple points (Q1 for methane is at 272.9 K and 2.563 MPa; no Q2 for methane) [2]. (b) Structure I (sI) hydrate comprising six large $5^{12}6^2$ cages (i.e. twelve pentagons + two hexagons in the faces of each large cage) + two small 5^{12} cages (i.e. twelve pentagons in the faces of each small cage) per unit cell. The unit cell of the sI crystal contains 46 water molecules in the hydrogen-bonded framework (modified from Refs. [3] and [2]).

large cage (radius of 0.473 nm) of sII. sI and sII are the most common gas hydrates structures, however, only sI hydrate will be considered in this review, since naturally occurring gas hydrates formed in porous media, the focus of this article, are comprised primarily of methane hydrate (sI; derived from biogenic, bacterially generated methane), with sII being less common (gas derived from thermogenic sources). Therefore, when referring to the naturally occurring hydrates in porous media, the term methane hydrate will typically be used.

It should be noted that the fundamental thermodynamic properties of gas hydrates need to be understood before kinetic, time-dependent properties (e.g. rate of formation and dissociation) are considered. Though the former are well established for bulk hydrates, these are often less well defined for hydrates in porous media, and the latter kinetic properties are even less understood in both hydrate types. Furthermore, while the structure and physical properties of bulk gas hydrates are well-established and have been extensively studied since the first discovery of gas hydrates by Sir Davy in 1910, there are still significant uncertainties on the physico-chemical properties of gas hydrates formed in porous media. These differences in the extent of work performed in these two distinct areas may be explained by the timescales on which their applications were realized. Specifically, the formation of bulk gas hydrates in pipelines was first discovered in 1934 by Hammerschmidt [4] and studied widely and continuously over the last eight decades to mitigate pipeline blockages due to gas hydrates, with a hydrate plugging model developed at CSM over the last decade [2, 5, 6]. Conversely, although naturally occurring gas hydrates (formed in porous media) were first discovered in Messoyakha, Russia in 1965 [7], studies of methane hydrates in porous media (also sometimes referred to as hydrate bearing sediment) have been sporadic, with two notable global resource estimates from the 1970s, seven in the 1980s, five in the 1990s [2], followed by at least three global and six regional/national (USA, Canada, India, Japan, GoM, ANS) estimates in the last decade [2, 3, 8, 9]. Significant widespread studies have been underway in the last decade, with the emergence of national hydrate programs notably in Japan, India, China, Korea, New Zealand, Canada, and the US. An illustration of the growing interest in natural gas hydrates is the growth of research papers presented on this topic at the international conference on gas hydrates (ICGH) first initiated by E.D. Sloan in 1993 compared to 2011, i.e. at ICGH1: 30 papers and 50% of total papers (1993) versus at ICGH7: 400 papers and 70% of total papers (2011) were related to hydrates in nature.

The realization of vast amounts of energy trapped in naturally occurring gas hydrates in sediments (porous media) in oceanic and arctic locations along the continental margins worldwide, and the advancing technology, through proof-of-concept production tests, have been the major driving forces for the substantial international interest in energy recovery from these deposits. Most notable pilot-scale production tests of gas recovery from methane hydrates are the Malik Well, MacKenzie Delta (in 2010) and Ignik Sikumi Well, Alaskan North Slope (in 2012) production tests, which were for hydrates below the permafrost, and the Nankai Trough, Japan offshore production test in 2013 [10]. Accompanying the interest for energy recovery from gas hydrates is the concern for potential environmental impacts due to these deposits and their exploitation [2, 3]. This review article will examine the state-of-the-art of gas hydrates in porous media in the context of their formation in natural environments, and the outstanding issues for the technologically feasible and safe recovery of energy from these deposits.

Table 6.1. Typical Multi-dimensional tools used in modern methane hydrate expeditions (adapted from Ref. [12]).

Seismic	2D , 3D, and high resolution 3D seismic data Electromagnetic surveys
Downhole tools	Formation pressure and temperature and thermal conductivity measurements
Drilling	Coring methane hydrate samples (including pressure cores) Logging while drilling (LWD) and/or wireline Physical property measurements on cores ex-situ and in-situ High resolution fluid chemistry

6.2. Methane Hydrate Accumulations in Nature

6.2.1. *Locations & Resource Estimates of Methane Hydrate Accumulations*

Figure 6.3 illustrates the different locations of naturally occurring gas hydrate accumulations that have been identified in sediments (porous media) under the ocean floor along the continental margins, and under the permafrost in arctic regions [3]. The accumulations have been identified either from various drilling expeditions, including international (IODP, IDP) and national (US, Canada, India, Japan) ocean drilling programs from 1965 to 2010, or inferred from seismic studies from 1970 to 2012 [11]. Identifying gas hydrate locations from mainly seismic studies, in particular Bottom Simulating Reflectors (BSRs) can lead to false positives and negatives, i.e. methane hydrate accumulations though predicted from BSRs, resulted in subsequent drilling finding no evidence of gas hydrates, and visa versa. Therefore, multiple methods need to be implemented to provide reliable identification of possible methane hydrate accumulations. The multi-dimensional modern approach is illustrated in Table 6.1, which lists examples of the range of seismic, imaging, and coring methods that are needed to provide more accurate resource identifications.

Having identified the location of methane hydrates, assessment of the resource amount for the particular accumulation is far more challenging, and in addition to the multi-dimensional and multi-tool approach, needs to be combined with reservoir and pore-scale modeling to provide confidence in the accuracy of the resource assessment estimate. Table 6.2 shows the variations in resource estimates over the last several decades, where

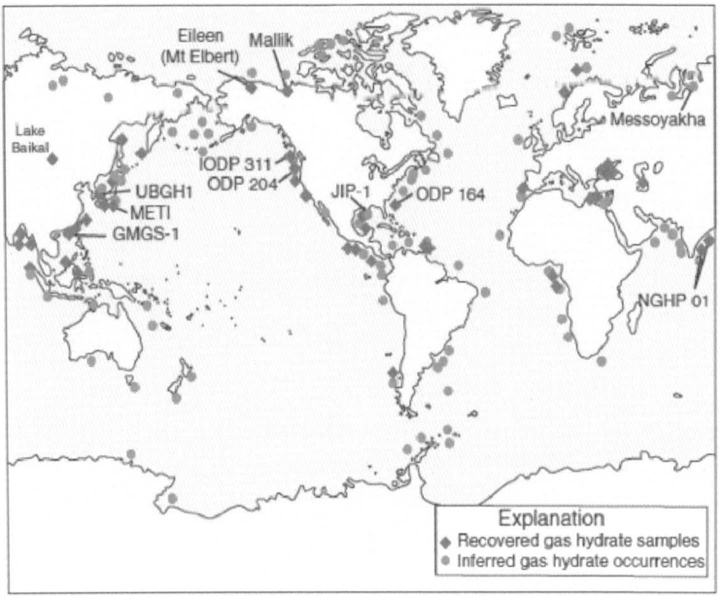

Fig. 6.3. Global locations of naturally occurring gas hydrates in porous media identified by drilling expeditions in which gas hydrate core samples are recovered and tested for the presence of gas hydrates, or inferred through seismic signatures. GMGS: Guangzhou Marine Geological Survey; IODP: Integrated Ocean Drilling Program; JIP: Joint Industry Project supported by the US DOE Methane Hydrate Program; METI: Ministry of International Trade and Industry of Japan; NGHP: National Gas Hydrate Program of India; ODP: Ocean Drilling Program; UBGH: Ulleung Basin Gas Hydrate. Different project or drilling expeditions are indicated by the numbers. (Reference [3], modified from Kvenvolden, 1988).

these estimates have been obtained from different modeling approaches. As Table 6.2 illustrates, the discrepancies between the different resource estimates are significant. Therefore, there is a need to improve the reliability of the resource estimates using an integrated multi-dimensional measurement and modeling approach. Within the major national and international gas hydrate programs, the resource location, accumulation assessment, and ultimately the production of energy (see Sec. 6.3) from methane hydrate accumulations are major long-term goals. A critical limitation in the accuracy

Table 6.2. Examples of global methane hydrate resource estimates [2, 3].

Proposed by:	Permafrost methane hydrate (10^{14} m^3)	Oceanic methane hydrate (10^{16} m^3)
Trofimuk et al., 1977	0.57	0.5 - 2.5
McIver, 1981	0.31	0.31
Dobrynin et al., 1981	340	760
Makogon, 1988	1.0	1.0
MacDonald, 1990	7.4	2.1
Milkov, 2004	15.0 (for both permafrost & oceanic regions)	
Klauda & Sandler, 2005	120.0 (for both permafrost & oceanic regions)	

and reliability of methane hydrate resource and production assessments is the lack of knowledge on the critical inputs required for the reservoir and pore-scale models. Specifically, the key parameters on the physical properties (e.g. porosity, permeability, tensile strength) of the methane hydrates in porous media are uncertain in many cases, with the response of these physical properties to in-place variables (such as hydrate saturation, sediment grain size/type) severely lacking [12]. Such parameters currently are largely arbitrary with limited (validated) physical basis. As discussed in the next section, in-situ field experiments and monitoring and controlled and systematic laboratory studies are required to address these issues.

Although prior discussion has been given to the resource location and assessment of methane hydrate and the multi-dimensional approaches required for these evaluations, the thermodynamic stability of methane hydrate in permafrost or oceanic locations should be the first fundamental assessment to be addressed before considering more complex issues of the characteristics of methane hydrate formation in porous media. That is, will the hydrate be stable/exist at the sediment/ocean depths and temperatures and pore water chemistry encountered for a particular permafrost or oceanic location? Figure 6.4 gives example profiles of the thermodynamic phase equilibria boundaries of methane hydrate in the different natural environments. The methane hydrate stable regions are enveloped by the methane hydrate phase boundary; at higher temperatures or lower depths the hydrate will be unstable. To produce gas from methane hydrate in porous media deposits, one can simply remove the system from the ther-

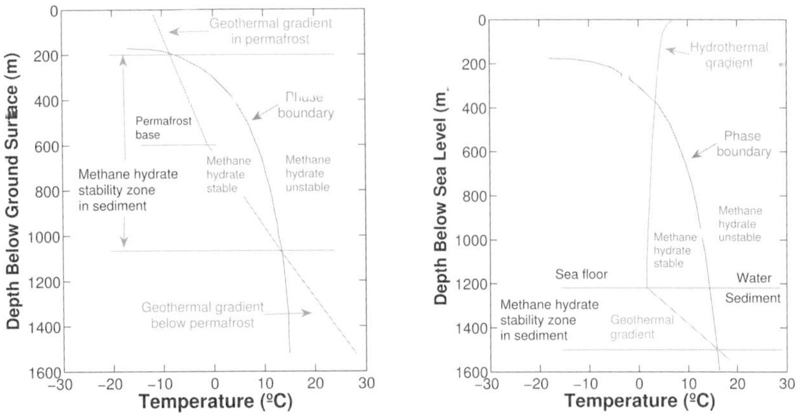

(a) (b)

Fig. 6.4. Thermodynamic phase equilibria profiles of methane hydrates in porous media: a) in permafrost, b) in oceanic locations (redrawn from Refs. [3] and [13]).

modynamic stability region by increasing the temperature or depressurizing to release gas from the dissociated deposit, now comprising water and gas, as illustrated below. The dissociation of methane hydrate is an endothermic process, which is a consideration that is particularly relevant for permafrost hydrates, since this endothermic dissociation process can result in ice formation, which may limit the continued gas recovery from methane hydrate [2].

$$CH_4\,(H_2O)_6 \rightarrow CH_4 + 6H_2O;\ \Delta h_{diss} = +54.2\ \text{kJ/mol (gas)}$$

Studies of the economic viability of producing natural gas from promising gas hydrate reservoirs were performed by Ref. [14]. The results suggested that economic gas production from gas hydrates were not significantly higher than market gas prices in N. America (in 2009). These economic evaluations would, however, depend significantly on well performance, sediment type, initial gas in-place, reservoir thermodynamic conditions, and existing infrastructure access. The study also concluded that it is reasonable to consider gas hydrates as contributing to the global energy portfolio.

6.2.2. *Environmental Aspects*

The environmental aspects of uncontrolled release of methane from methane hydrates and the subsequent slope instability resulting in the dissociation of methane hydrates needs to be considered, particularly during the production of gas from methane hydrate accumulations. This is of particular concern when one considers that methane is around 20 times more potent a greenhouse gas than carbon dioxide. The environmental aspects associated with methane hydrate are uncertain, largely due to the lack of data available for the assessment of these potential impacts. As recommended by the national academies committees assessing the methane hydrate programs for the US [3] and Canada [15], environmental monitoring needs to be performed during (long term) production field tests. The potential release of methane hydrate during production is illustrated in Fig. 6.5. This type of scenario can also be a potential hazard during conventional petroleum production in a hydrate stability zone where oil/gas is being produced from a well in close proximity to a natural hydrate accumulation. Such potential hazards have been severe enough to cause operating companies to move/alter the production location on realizing operation and production will be within a hydrate stability region/hydrate deposit, even at the mature stages towards bringing the well online. In these cases, the risk of hydrate dissociation and subsequent slope instability results in hydrate locations being avoided for conventional petroleum production. To accurately assess the environmental impacts of methane release from gas hydrate deposits during gas production from methane hydrates a combined field and laboratory measurement and modeling approach is required, involving considerations of physical properties and geomechanical stability of the hydrate in porous media deposits.

6.3. Experimental Approaches to Characterize Gas Hydrate Bearing Sediments

6.3.1. *Models for Distribution of Methane Hydrates in the Pore Space*

Accurate pore-scale models describing the distribution of methane hydrate in porous media are critical to reliable assessment of the location, resource amount, and gas production of these deposits. The pore-scale models based on work by Refs. [17] and [18] are shown in Fig. 6.6, including hydrate growth away from grain contacts and surfaces and into the pore space, also

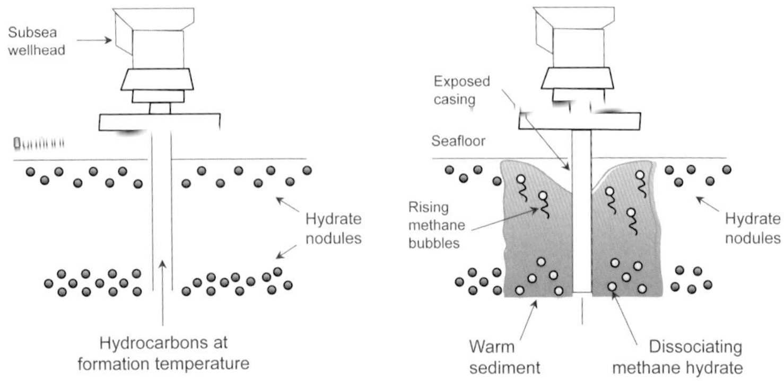

Fig. 6.5. Potential release of methane hydrate at the wellhead due to gas production from methane hydrate deposits, or during conventional petroleum production within a hydrate stability zone/in the presence of a natural hydrate accumulation [3, 16].

called pore filling (Fig. 6.6(a)); and hydrate formation at the gas-water interface and at grain contacts, also called grain cementing (Fig. 6.6(b)). The type of model representing a particular methane hydrate deposit has a major effect on the physical properties and hence the evaluation of the resource and gas production. Specifically, the geomechanical stability of the hydrate deposit will be significantly weakened on gas production from grain cementing hydrates compared to pore filling hydrates. Therefore, establishing the type of pore scale model which applies to a particular accumulation is important, and has been a major focus of a number of experimental research programs, as detailed in the next section.

6.3.2. *Laboratory-scale Synthetic Hydrates*

Advancements and validation of pore-scale models of methane hydrate in porous media and the effect that parameters such as hydrate saturation, sediment type, P, T have on the hydrate-porous media interactions affecting the porosity and permeability of the sample are needed for reservoir model development and implementation. Specifically, this is critical for reliable predictions/assessments of hydrate resource locations and amount, and production feasibility and environmental impact assessment. Approaching the needs for advancement of pore-scale models via laboratory measurements is an attractive and cost-effective strategy, since the opportunities to perform field measurements are severely limited and laboratory synthetic samples enable one to perform well-controlled and systematic measurements.

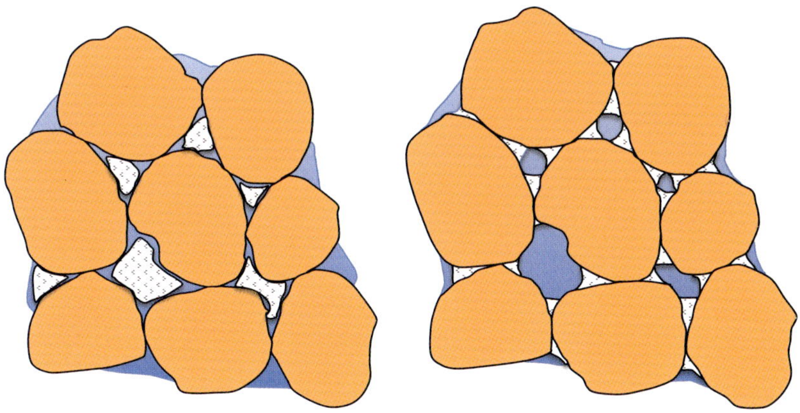

(a) Pore filling (b) Grain cementing

Fig. 6.6. Pore-scale models. (Adapted from Refs. [3] and [17]).

A major challenge for laboratory-scale methane hydrate experiments is to produce reproducible and representative hydrate samples. The issue of reproducibility should be more easily addressed than attempting to produce representative methane hydrate samples that attempt to mimic the natural hydrate samples that have been formed over geologic timescale. As illustrated by an inter-laboratory study conducted by key hydrate laboratories in N. America (CSM, GaTech, LBNL, USGS, U. Calgary, CoP), Europe (U. Southampton) and Asia (KIGAM) [19], attempts to synthesize a methane hydrate sample and perform fundamental characterization measurements on the synthesized sample led to significant variability between the datasets. The experimental synthesis protocol was specified to all laboratories to help establish similar synthesis conditions in all the participating laboratories, and the same porous material was used (Ottawa sand F110 was distributed from one batch to all laboratories). However, the measurement data on the synthetic samples showed variability likely due to heterogeneous pore-filling distributions and differences in the measurement techniques. This inter-laboratory study was the first of its kind and has been most useful to highlight the need for detailed sample preparation procedures to be provided when reporting hydrate in porous media laboratory data. The study also emphasized the need to combine imaging of the hydrate in porous media samples with other physical measurements. By imaging the samples (e.g, using X-ray computed tomography; [20]), it is possible to determine the extent of heterogeneity in the sample and also in some cases obtain further information on the hydrate-pore-scale interactions.

The ability to produce hydrate in porous media samples synthetically in the laboratory that are representative of naturally occurring hydrate samples is far more challenging and may not be feasible in many cases. However, one can consider tailoring the hydrate in porous media synthesis such that different pore-scale models can be tested and investigated in a systematic manner to provide much needed data to aid in calibrations and interpretations of complex field data. Synthesis methods have been developed for methane hydrate formation for a range of hydrate bearing sediment systems [17]. These synthesis methods include hydrate formation from dissolved gas which leads to pore filling hydrates; from partial water saturation which results in hydrate formation at grain contacts [12, 17]. Hence, using these different synthesis methods, one can attempt to control/bias the pore-scale habit/hydrate formation mechanism. The synthesized hydrated sediment samples can then be analyzed using physical property characterization tools including: CT-X-ray imaging, permeability measurements, hydrates distribution, and ultrasonic velocity. These measurements can be performed before, during, after gas production from methane hydrate as a function of temperature and pressure conditions, sediment type, hydrate saturation, etc. The following chapter by Schindler and Batzle presents a detailed example of hydrate formation in synthetic porous media samples, which includes CT-X-ray imaging and ultrasonic velocity measurements.

There is a growing effort to also perform measurements using pressurized recovered (natural) core samples to avoid the issues of trying to synthesize representative hydrate samples and also the disturbance of the natural hydrate from its original environment. Clearly, it is optimum if the laboratory synthesized sample studies can be performed in parallel with these pressurized core studies for cross comparison, since the availability of the pressurized (natural) cores are limited both in terms of amount and type. Significant progress has been made in development of pressure core sampling devices (e.g. PCS, HYACINTH, Fugro, PCCT), which enable a natural hydrate core to be recovered under pressure, thereby minimizing the dissociation of the hydrate core as it is retrieved to the surface [12]. An even further advancement is the development of these pressurized core sampling devices to comprise of multiple characterization chambers which allow in-situ analysis of the core under pressure (and under effective stress) to measure "hydrological, thermal, chemical, biological, and mechanical properties " [11]. These advancements in pressure coring will be important to better understand the natural hydrate in porous media structure and behavior, reducing the large uncertainty of the changes in the integrity of the sample that would occur on retrieval without pressure coring.

6.4. Modeling Fluid Flow in Gas Hydrate Bearing Sediments

Modeling gas production from gas hydrate bearing sediments involves the combination of a series of physico-chemical mechanisms, such as, hydrate formation and dissociation reactions, mass transfer between different phases, multiphase fluid flow in porous media, and conductive and convective heat flow. These mechanisms require the implementation of a compositional numerical scheme for the solution of the governing conservation equations of mass and energy. The basic components considered for modeling gas hydrate systems are water and methane; other additional components that can be considered are salt, chemical inhibitors, and other hydrate-forming gases (e.g., ethane, propane, CO_2, nitrogen). These components can be distributed among different phases, such as, a liquid aqueous phase, a gaseous phase, an immobile solid hydrate phase, and an immobile solid ice phase. In order to establish the initial distribution of gas hydrate within a sedimentary layer, the equilibrium conditions of gas hydrate in terms of pressure and temperature should be determined using phase equilibria calculations (Sec. 6.4.1). The local geothermal gradient controls the vertical extension of the hydrate stability zone, as shown in Fig. 6.4. The amount of hydrate formed can be obtained from equilibrium reaction relations (assuming stoichiometry) or from intrinsic kinetics, discussed in Sec. 6.4.2. The modeling approaches implemented in state-of-the-art numerical models for fluid flow in hydrate bearing sediments are discussed in Sec. 6.4.3, presenting the fundamental governing equations and its constitutive submodels, along with its assumptions and limitations.

6.4.1. *Phase Equilibria Calculations of Gas Hydrate*

The phase envelope of a bulk gas hydrate phase is defined by equilibrium conditions in terms of pressure and temperature. These can be determined by thermodynamic flash calculations based on the van der Waals and Platteeuw (1959) model with excellent engineering accuracy [21]. Several commercial thermodynamic programs (e.g., Multiflash® from Infochem Computer Services, Ltd., and PVTSim® from Calsep) implement a modified version of this model for the prediction of gas hydrate properties [2]. Another program developed at the Colorado School of Mines, based on the Gibbs energy minimization method (CSMGem), can also be used to estimate the gas hydrate equilibrium conditions [22]. The composition of the gaseous and aqueous phases is used as input to these programs to obtain the hydrate equilibrium conditions.

Experimental measurements of gas hydrate dissociation temperature in small pore radii indicate a shift of the hydrate equilibrium temperature toward lower values. In sufficiently large pore sizes, greater than 60 nm, the gas hydrate equilibria shift is negligible; i.e., hydrate equilibrium conditions equals that of bulk hydrate [23]. Given a hydrate stability zone, which vertical extension is controlled by the geothermal gradient, gas hydrate will form preferentially in larger pore spaces. The possibility of hydrates being present in small pores will depend on the sediment temperature being lower than the shifted hydrate equilibrium temperature. This correlates with observations from cores recovered during ocean drilling expeditions, where gas hydrates were found preferentially within coarse grain pores or filling fractures in the sediments. These observations associate the hydrate saturation to the reservoir quality of the sediments, and imply that the hydrate saturation and distribution in the porous media within the hydrate stability zone could be estimated if the pore size distribution is known.

6.4.2. *Equilibrium and Kinetic Models of Hydrate Formation and Dissociation*

The reactions of gas hydrate formation and dissociation govern the mass transfer between the immobile hydrate phase and the aqueous and gaseous mobile phases. During hydrate formation, water and hydrate-forming gas molecules will be consumed from the aqueous and/or gaseous phases to form hydrate. Conversely, during hydrate dissociation, water and hydrate-forming gas molecules will be released from the hydrate phase and transferred to the aqueous and/or gas phases. Two types of models for hydrate formation and dissociation are typically used to model hydrate bearing sediments: equilibrium reaction relations and intrinsic kinetics [24].

Equilibrium reaction relations assume stoichiometry, where a hydrate-forming gas molecule will react with N_H water molecules to form hydrates under appropriate thermodynamic equilibrium conditions according to,

$$g\,(G) + N_H\mathrm{H_2O}\,(W) \leftrightarrow g \cdot N_H\mathrm{H_2O}\,(H) \qquad (6.1)$$

where N_H is the hydration number (water to gas ratio in the hydrate phase), g is the hydrate-forming gas molecule (e.g., CH_4), and G, W and H refer to gaseous, aqueous and hydrate phases, respectively. For methane hydrate the hydration number is approximately 6. The double arrow in Eq. (6.1) indicates that this equilibrium reaction is also used for hydrate dissociation. The phase transitions are determined by the availability of the components and the current thermodynamic conditions (P and T) relative to the equilibrium conditions.

The intrinsic kinetics model represents the rate of methane gas consumption during hydrate formation or gas release during hydrate dissociation (r_m) as a function of a kinetic rate constant (K_0); surface area for the reaction (A); system temperature (T); and the reaction driving force given by the difference in the values of fugacity in the gas phase at system P and T (f) and at equilibrium on the hydrate surface (f_e), and is given by the following equation,

$$r_m = K_0 \exp\left(\frac{-E}{RT}\right) A\left(f_e - f\right) \tag{6.2}$$

where E is the hydration activation energy, and R is the universal gas constant. The intrinsic kinetics model assumes immediate hydrate nucleation for conditions inside the hydrate formation region, and immediate decomposition outside the hydrate formation region, similar to the equilibrium reaction model. The rate of hydrate formation or dissociation is a fundamental component for estimating the mass transfer between different phases in the mass conservation equation, and for estimating heat generated by the exothermic hydrate formation reaction or heat consumed by the endothermic hydrate dissociation reaction in the energy conservation equation.

Usually, the equilibrium reaction model is preferred over the intrinsic kinetics model because of its simplicity and low computational requirements. Also, Ref. [24] observed indistinguishable results from the equilibrium reaction and the intrinsic kinetics models using numerical simulation in long-term large-scale production cases, while there exist important differences between the results obtained with these models for short-term core-scale simulation cases. These two models represent the best approximations available to represent hydrate formation and dissociation in gas hydrate bearing sediments. However, the dissociation of gas hydrates in sediments is a phenomenon that still is not completely understood.

6.4.3. *Governing Equations for Fluid Flow in Gas Hydrate Bearing Sediments*

The fundamental governing equations for fluid flow in gas hydrate bearing sediments consist of the mass balance equation for each component and the energy equation. These governing equations are constrained by distribution of geological strata; initial and boundary conditions; fluid and rock properties; and mass and heat transfer between the phases. The model solves the system of equations to obtain phase pressure, phase saturation and composition, and temperature.

Here we present the mass balance equations for the two basic components, water and methane, encountered in gas hydrate systems to illustrate the current approach to modeling fluid flow in hydrate bearing sediments. The mass balance equation for the water component could be written in terms of the transient mass accumulation in an infinitesimal control volume; advection of aqueous and gaseous phases described by Darcy's Law; molecular diffusion of vapor water in the gaseous phase; external source and sink term; and hydrate reaction related mass transfer rate.

$$\begin{aligned}\frac{\partial}{\partial t}\left(\sum_{\beta=A,G,I,H}\phi S_\beta \rho_\beta X_\beta^w\right) &= \nabla\cdot\left[\frac{kk_{rA}}{\mu_A}\rho_A X_A^w\left(\nabla P_A-\gamma_A\nabla D\right)\right]\\ &\quad+\nabla\cdot\left[\frac{kk_{rG}}{\mu_G}\rho_G X_G^w\left(\nabla P_G-\gamma_G\nabla D\right)\right]\\ &\quad+\nabla\cdot\left[\phi S_G \tau_G D_G^w \rho_G \nabla X_G^w\right]\\ &\quad+\sum_{\beta=A,G} X_\beta^w+\psi^w \end{aligned}\tag{6.3}$$

The hydrate reaction related mass transfer rate of the water component is considered only when the intrinsic kinetics model is used to represent the hydrate formation/dissociation reaction, and is given by

$$\psi^w = N_H r_m \tag{6.4}$$

A similar mass balance equation can be written for the methane component,

$$\frac{\partial}{\partial t}\left(\sum_{\beta=A,G,I,H}\phi S_\beta\rho_\beta X_\beta^m\right) = \nabla\cdot\left[\frac{kk_{rA}}{\mu_A}\rho_A X_A^m\left(\nabla P_A-\gamma_A\nabla D\right)\right]$$
$$+\nabla\cdot\left[\frac{kk_{rG}}{\mu_G}\rho_G X_G^m\left(\nabla P_G-\gamma_G\nabla D\right)\right]$$
$$+\nabla\cdot\left[\phi S_A\tau_A D_A^m\rho_A\nabla X_A^m\right]$$
$$+\sum_{\beta=A,G}X_\beta^m+\psi^m \tag{6.5}$$

The hydrate reaction related mass transfer rate of the methane component is given by

$$\psi^m = r_m \tag{6.6}$$

The energy conservation equation considers the transient heat accumulation in an infinitesimal control volume of all the phases and the rock matrix; the heat conduction in the rock and all the phases; convective heat transfer in the aqueous and gaseous phases; heat removal or addition associated with fluid injection or production; and an energy source/sink related to the hydrate reaction.

$$\frac{\partial}{\partial t}\left(\sum_{\beta=A,G,H,I}\phi S_\beta\rho_\beta u_\beta\right)+\frac{\partial}{\partial t}\left[(1-\phi)\left(\rho_R C_R T\right)\right]$$
$$=\nabla\cdot\left\{\left[(1-\phi)K_R+\phi\left(S_AK_A+S_GK_G+S_HK_H+S_IK_I\right)\right]\nabla T\right\}$$
$$+\nabla\cdot\left[\left(\frac{kk_{rA}}{\mu_A}\rho_A h_A\right)\left(\nabla P_A-\gamma_A\nabla D\right)\right]$$
$$+\nabla\cdot\left[\left(\frac{kk_{rG}}{\mu_G}\rho_G h_G\right)\left(\nabla P_G-\gamma_G\nabla D\right)\right]$$
$$+\sum_{\beta=A,G}h_\beta q_\beta+S_{E,\mathrm{hyd}} \tag{6.7}$$

The energy source/sink term related to the hydrate formation/dissociation reaction is given by,

$$S_{E,\mathrm{hyd}} = r_m\Delta h_H \tag{6.8}$$

where Δh_H is the heat of hydrate formation or dissociation.

The governing equations are constrained by following constitutive relations: the sum of the mass fraction of the components in each phase equals one,

$$\sum_i X_\beta^i = 1; \text{ for each phase } \beta, \text{ and } i = w, m \tag{6.9}$$

the sum of phase saturations equals one,

$$\sum_i S_\beta = 1; \text{ for each phase } \beta \tag{6.10}$$

the pressures of the flowing phases are related to the capillary pressure, which is a function of saturation,

$$P_c = P_G - P_A \tag{6.11}$$

The differential governing conservation equations of mass and energy can be discretized spatially using the integral finite difference method and temporally using a first order finite difference, resulting in a set of coupled nonlinear algebraic equations. These nonlinear equations are solved simultaneously at each control volume using a fully implicit numerical scheme. This solution approach has been implemented in the available state-of-the-art codes specially develop for hydrate bearing sediments [25, 26]. Simulation results from available hydrate reservoir simulators have been compared in an inter-code comparison study, concluding that all the codes consistently capture the basic mechanisms of mass and heat transfer and hydrate dissociation.

A good understanding of the reservoir flow properties in the presence of gas hydrates is necessary for reliable modeling of methane production from gas hydrate accumulations. The presence of an immobile hydrate phase reduces the effective porosity for fluid flow, and affects the intrinsic permeability and capillary pressure of the porous media. Current gas hydrate reservoir modeling approaches assume that hydrate, as a generic pore-filling solid, will reduce overall porosity and absolute permeability of a porous medium [27]. Depending on the micro-structural arrangement of gas hydrate within the sediment, the impact of gas hydrate saturation on permeability and relative permeability can be significantly [28]. From the comparison of permeability measurements with model predictions, pore-filling hydrate models seems to provide the best estimates of permeability [29], but this observations were made using limited data available for single phase hydrate-bearing systems. The multiphase flow nature of hydrate bearing sediments, containing water, gas and hydrate, requires the development of additional relationships governing the relative permeability of each flowing phase as function of saturation. Ref. [28] performed laboratory studies of multiphase flow through hydrate bearing sediments, and used X-ray CT images to develop permeability models based on the pore space presence of hydrate and its effect on gas and water flow. Their experiment study only involved one experimental realization of hydrate formation in an unsaturated sand sample. A key need to calibrate field data and assess gas production rates from methane is to determine the permeability as a function of gas hydrate saturation and grain size distribution/clay content during gas production from hydrate bearing sediments. Such permeability measurements will be critical to the interpretation of production field data and development and validation of prediction models.

6.5. Concluding Remarks

The high energy density stored in natural gas hydrate accumulations, in the form of a clean-burning fuel (i.e., methane), has motivated the consideration of this unconventional energy resource as a future energy supply. The feasibility of producing methane gas from natural hydrate bearing sediments relies on the understanding of key phenomena at the pore scale; phenomena that are still under research. These phenomena revolve around the mechanisms of gas hydrate formation and dissociation, which control the distribution of gas hydrates in porous media and the interactions of gas hydrates with other phases and the sediment grains, affecting porous media parameters such as porosity, permeability and geomechanical integrity of the sedimentary layers. Understanding of these phenomena is critical for reliable assessments and predictions of gas hydrate resource locations and amount, production feasibility and environmental impact. In this chapter we have presented a summary of the current state of knowledge on natural hydrate systems and identified opportunities for continuing research and development to advance the technologies and fundamental knowledge required for hydrate production and environmental assessments.

Nomenclature

A Surface area, m^2

C_R	Heat capacity of dry rock, $J/kg/K$
D	Depth, m
D_β^w	Molecular diffusion coefficient of water in phase β in the absence of a porous medium, m^2/s
E	Hydration activation energy, J/mol
f	Fugacity of gas phase at system P and T, Pa
f_e	Fugacity of gas phase at equilibrium on the hydrate surface, Pa
k	Permeability, m^2
K_0	Kinetic rate constant, $kg/m^2/Pa/s$
K_β	Thermal conductivity coefficient of phase β, $J/m/K/s$
$k_{r\beta}$	Relative permeability of phase β, fraction
h_β	Enthalpy per unit mass of phase β, J/kg
N_H	Hydration number (water to gas ratio in hydrate phase), dimensionless
P	Pressure, Pa
P_β	Pressure of phase β, Pa
P_c	Capillary pressure, Pa
q_β	Volumetric flow rate of phase β, m^3/s
R	Universal gas constant, $J/mol/K$
r_m	Rate of methane gas mass consumption per unit volume during hydrate formation, $kg/m^3/s$
S_β	Saturation of phase β, fraction
$S_{E,\text{hyd}}$	Energy source term related to exothermic hydrate reaction, $J/kg/s$
T	Temperature, K
t	Time, s
u	Specific internal energy of phase β, J/kg
X_β^m	Mass fraction of methane in phase β, fraction
X_β^w	Mass fraction of water in phase β, fraction

Greek

Δh_H	Enthalpy change of hydrate formation, J/kg
γ_β	Hydrostatic pressure gradient of phase β, Pa/m
ρ_β	Density of phase β, kg/m^3
ϕ	Porosity, fraction
ψ^κ	Mass transfer rate of component κ, $kg/(m^3 s)$

μ_β Viscosity of phase β, $Pa.s$
τ_β Tortuosity of phase β, equal or greater that one

References

[1] C. A. Koh, Towards a fundamental understanding of natural gas hydrates, *Chem. Soc. Rev.* **31**(3), 157–167 (2002). doi: 10.1039/B008672J.
[2] E. Sloan and C. Koh, *Clathrate Hydrates of Natural Gases*, 3rd edn. CRC Press, Boca Raton, FL (2007). ISBN 9780849390784.
[3] C. Paull, W. Reeburgh, S. Dallimore, G. Enciso, C. Koh, K. Kvenvolden, C. Mankin, and M. Riedel, Realizing the energy potential of methane hydrate for the united states, *National Research Council of the National Academies: Washington, DC* (2010).
[4] E. G. Hammerschmidt, Formation of gas hydrates in natural gas transmission lines, *Ind. & Eng. Chem.* **26**(8), 851–855 (1934). doi: 10.1021/ie50296a010.
[5] E. D. Sloan, C. A. Koh, and A. K. Sum, *Natural Gas Hydrates in Flow Assurance.* Gulf Professional Publishing, Burlington, MA, USA (2011). doi: 10.1016/B978-1-85617-945-4.00010-8.
[6] L. E. Zerpa. *A practical model to predict gas hydrate formation, dissociation and transportability in oil and gas flowlines.* PhD thesis, Colorado School of Mines, Golden, CO (2013).
[7] I. Makogon, *Hydrates of Hydrocarbons.* PennWell Books (1997). ISBN 9780878147182.
[8] T. S. Collett, M. W. Lee, M. V. Zyrianova, S. A. Mrozewski, G. Guerin, A. E. Cook, and D. S. Goldberg, Gulf of mexico gas hydrate joint industry project leg ii logging-while-drilling data acquisition and analysis, *Marine and Petroleum Geology.* **34**(1), 41–61 (2012). doi: 10.1016/j.marpetgeo.2011.08.003.
[9] M. Frye, W. Shedd, and R. Boswell, Gas hydrate resource potential in the terrebonne basin, northern gulf of mexico, *Marine and Petroleum Geology.* **34**(1), 150–168 doi: 10.1016/j.marpetgeo.2011.08.001.
[10] R. Boswell and T. S. Collett, Current perspectives on gas hydrate resources, *Energy Environ. Sci.* **4**(4), 1206–1215 (2011). doi: 1754-569210.1039/C0EE00203H.
[11] T. Collett, M. Frye, D. Goldberg, J. Hasubo, C. A. Koh, M. Malone, C. Shipp, and M. Torres. Historical methane hydrate project review. Report, Consortium for Ocean Leadership (2013). URL `http://oceanleadership.org/scientific-programs/methane-hydrate-field-program/`.
[12] T. S. Collett, J. J. Bahk, M. Frye, D. Goldberg, J. Husebo, C. A. Koh, M. Malone, C. Shipp, M. Torres, D. Myers, G.and Divins, and M. Morell. Marine methane hydrate field research plan. Topical report, Consortium for Ocean Leadership (2013). URL `http://oceanleadership.org/wp-content/uploads/2013/01/MH_Science_Plan_Final.pdf`.

[13] K. A. Kvenvolden, Methane hydrates and global climate, *Global Biogeochem. Cycles.* **2**, 221–229 (1988). doi: 10.1029/GB002i003p00221.

[14] M. R. Walsh, S. H. Hancock, S. J. Wilson, S. L. Patil, G. J. Moridis, R. Boswell, T. S. Collett, C. A. Koh, and E. D. Sloan, Preliminary report on the commercial viability of gas production from natural gas hydrates, *Energy Economics.* **31**(5), 815 – 823 (2009). doi: 10.1016/j.eneco.2009.03.006.

[15] J. Grace, T. Collett, F. Colwell, P. Englezos, E. Jones, R. Mansell, J. Meekison, R. Ommer, M. Pooladi-Darvish, M. Riedel, J. Ripmeester, C. Shipp, and E. Willoughby. Energy from gas hydrates—Assessing the opportunities and challenges for Canada (2008).

[16] W. S. Borowski and C. K. Paul. The gas hydrate detection problem: Recognition of shallow-subbottom gas hazards in deep-water areas. In *Offshore Technology Conference* (1997). doi: 10.4043/8297-MS. OTC-8297-MS.

[17] W. F. Waite, J. C. Santamarina, D. D. Cortes, B. Dugan, D. N. Espinoza, J. Germaine, J. Jang, J. W. Jung, T. J. Kneafsey, H. Shin, K. Soga, W. J. Winters, and T. S. Yun, Physical properties of hydrate-bearing sediments, *Reviews of Geophysics.* **47**(4), RG4003 (2009). doi: 10.1029/2008RG000279.

[18] M. B. Rydzy. *The effect of hydrate formation on the elastic properties of unconsolidated sediment.* PhD thesis, Colorado School of Mines, Golden, CO (2014).

[19] W. Waite, J. Santamarina, M. Rydzy, S. Chong, J. Grozic, K. Hester, J. Howard, T. Kneafsey, J. Lee, and S. Nakagawa. Inter-laboratory comparison of wave velocity measurements in a sand under hydrate-bearing and other set conditions. In *7th International Conference on Gas Hydrates* (2011).

[20] T. J. Kneafsey, Y. Seol, A. Gupta, and L. Tomutsa, Permeability of laboratory-formed methane-hydrate-bearing sand: Measurements and observations using x-ray computed tomography, *SPE Journal.* **16**(1), pp. 78–94 (2011). doi: 10.2118/139525-pa.

[21] L. Ballard and E. D. Sloan, The next generation of hydrate prediction iv: A comparison of available hydrate prediction programs, *Fluid Phase Equilibria.* **216**(2), 257–270 (2004). doi: 10.1016/j.fluid.2003.11.004.

[22] A. L. Ballard and E. D. Sloan, The next generation of hydrate prediction: Part III. gibbs energy minimization formalism, *Fluid Phase Equilibria.* **218** (1), 15–31 (2004). doi: 10.1016/j.fluid.2003.08.005.

[23] D. J. Turner, R. S. Cherry, and E. D. Sloan, Sensitivity of methane hydrate phase equilibria to sediment pore size, *PPEPPD 2004 Proceedings.* **228-229** (0), 505–510 (2005). doi: 10.1016/j.fluid.2004.09.025.

[24] M. B. Kowalsky and G. J. Moridis, Comparison of kinetic and equilibrium reaction models in simulating gas hydrate behavior in porous media, *Energy Convers. Manage.* **48**(6), 1850–1863 (2007). doi: 10.1016/j.enconman.2007.01.017.

[25] H. Hong and M. Pooladi-Darvish, Simulation of depressurization for gas production from gas hydrate reservoirs, *J. Can. Pet. Tech.* **44**(11) (2005). doi: 10.2118/05-11-03.

[26] G. Moridis, Numerical studies of gas production from methane hydrates,

SPE Journal. **8**(4), 359–370 (2003). doi: 10.2118/87330-pa.

[27] G. J. Moridis, M. B. Kowalsky, and K. Pruess. Tough-fx/hydrate v.1.0 user's manual: A code for the simulation of system in hydrate-bearing geologic media. Report, Lawrence Berkeley National Lab. (2005).

[28] Y. Seol and T. J. Kneafsey, Methane hydrate induced permeability modification for multiphase flow in unsaturated porous media, *Journal of Geophysical Research: Solid Earth.* **116**(B8), B08102 (2011). doi: 10.1029/2010JB008040.

[29] R. L. Kleinberg, C. Flaum, D. D. Griffin, P. G. Brewer, G. E. Malby, E. T. Peltzer, and J. P. Yesinowski, Deep sea nmr: Methane hydrate growth habit in porous media and its relationship to hydraulic permeability, deposit accumulation, and submarine slope stability, *Journal of Geophysical Research: Solid Earth.* **108**(B10), 2508 (2003). doi: 10.1029/2003JB002389.

Chapter 7

Pore-Scale Investigations of Hydrate Bearing Sediments

Mandy Schindler and Michael L. Batzle

Department of Geophysics,
Colorado School of Mines,
Golden, CO, USA

manschin@mines.edu, mbatzle@mines.edu

Naturally occurring gas hydrates contain significant amounts of natural gas which might be produced in the future. Thus, it is necessary to understand the pore-space characteristics of hydrate reservoirs, especially the pore-scale distribution of hydrates and their interaction with sediments. The goal of this study is to determine how the hydrate formation process influences the distribution of gas hydrates in the pore space, how the pore-scale distribution of hydrates affects seismic velocities and sonic logs and whether we can improve our estimate of hydrate saturation from seismic velocities and sonic logs.

Laboratory measurements were conducted to obtain information about the distribution of gas hydrate in the pore space by using a synthetic sediment (glass beads). Tetrahydrofuran (THF) was used as a guest molecule since THF hydrate is a proxy for naturally occurring hydrate. We performed micro X-ray computed tomography (MXCT) and ultrasonic velocity measurements on laboratory formed glass-bead samples. Both, MXCT images and ultrasonic velocity measurements, indicate that THF hydrate forms in the pore space with only a small portion contacting the grain surfaces. At higher hydrate saturations, the material begins to bridge and support the sediment grains. These hydrate-bearing sediments appear to follow a load-bearing model. This information is helpful for decisions regarding production of methane from gas hydrates.

7.1. Introduction

Gas hydrates occur naturally in shallow sediments in Arctic permafrost regions and beneath the seafloor along continental slopes. These environments exhibit hydrate stability conditions of low temperature and high pressure. The estimated amount of natural gas, mainly methane, stored in hydrate reservoirs exceeds the amount of natural gas stored in conventional resources by at least one order of magnitude [1–3]. The vast quantity of methane contained in gas hydrates and their widespread geographic occurrence attracts attention from scientists, governmental institutions and major energy companies. References [4] and [5] demonstrated that gas-hydrate production can be developed with existing oil and gas production technology. Physical properties of gas-hydrate bearing sediments need to be known to assist in identifying and producing methane from hydrate reservoirs. The most common geophysical methods used to characterize and quantify gas hydrates in nature are seismic surveying and well logging. In order to calibrate and interpret these field measurements, laboratory studies are necessary to determine the bulk physical properties of hydrate-bearing sediment. Currently, it is possible to predict the existence of gas hydrates from geophysical measurements. However, the techniques used to estimate hydrate saturation based on either seismic data or well logs require further development [6]. Thus, the amount of hydrate stored in a reservoir remains uncertain in some cases. Information about the distribution of gas hydrates in the rock is necessary to determine hydrate saturation.

Gas hydrates occur as massive units and lenses in fractures of fine-grained material (shale) as well as distributed in the pore space of coarse-grained porous media. Different distributions of gas hydrate in the pore space affect the physical properties of the hydrate-bearing sediment. In Chap. 6, Zerpa and Koh point out that the physical properties of bulk gas hydrates are well understood compared to the physical properties of hydrate-bearing sedimentsHydrate-bearing sediment. Thus, it is imperative to ascertain hydrate-sediment interactions rather than physical properties of pure hydrate [7]. Hydrate distribution in the pore space depends mainly on the formation method. In nature, gas hydrates are formed in two ways: (1) from methane dissolved in water or (2) from free methane gas [3]. Hydrate formed from methane gas tends to form at the grain surface and grain contacts [8, 9] while hydrate formed from methane dissolved in water tends to form in the pore body with little or no contact to the sediment grains [10, 11].

In this study, we focus on the pore-scale distribution of gas hydrates in synthetic coarse-grained porous media. Coarse grained sand reservoirs, as they exist in permafrost regions and marine environments, are relatively easy to access and thus more likely to permit methane production from gas hydrates than hydrate reservoirs in shale [12].

Our goal is to determine the pore-scale distribution of hydrate formed from gas dissolved in water. It is often assumed that gas hydrate in coarse grained, marine sediments is more likely to form from methane dissolved in water, whereas gas hydrate in permafrost regions is associated with formation from free methane gas [3]. Micro X-ray computed tomography (MXCT) images were utilized to image hydrate distribution in the pore space. In addition, we investigated the influence of hydrate saturation and distribution on ultrasonic velocities. The relation between pore-scale distribution, hydrate formation method and ultrasonic velocities is essential to determine the amount of hydrate in a reservoir from seismic data and acoustic logs.

7.2. Background

7.2.1. *Pore Scale Distribution Models*

Theoretical models infer different pore-scale hydrate distributions and their impact on ultrasonic velocities. One such model is the effective medium theory applied to hydrate-bearing sediments by Refs. [10] and [13].

Figure 7.1 shows four different pore-scale distrisbutions according to this

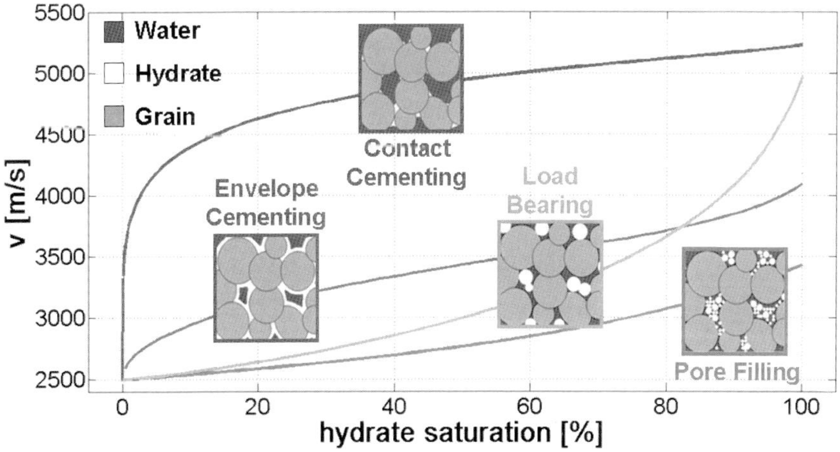

Fig. 7.1. Effective medium theory after Refs. [10] and [13]: different distributions of hydrate in the pore space and their influence on ultrasonic velocities, conceptual figures from Ref. [14].

effective medium model for gas hydrates in sediment: contact cementing, envelope cementing, pore filling and load bearing. The envelope cementing model shows a drastic increase in ultrasonic velocities even for small amounts of hydrate in the pore space whereas hydrate formed according to the pore filling model shows a much smaller influence on the ultrasonic velocities. Hydrate formation from free gas appears to create grain-cementing hydrate, which resembles the contact or envelope-cementing models [8, 9]. Hydrate formed from gas dissolved in water seems to have little or no contact with the sediment grains and thus conforms to the load-bearing or pore-filling model [14]. More laboratory measurements are necessary to determine the correlation of hydrate distribution in the pore space with ultrasonic velocities and to corroborate previous results.

7.2.2. *Proxy for Natural Gas Hydrate*

The most common hydrate in nature is methane hydrate. The experimental setup inside our MXCT scanner does not yet provide a control of pressure and temperature, thus scanning of stable methane hydrate samples is not possible. Methane hydrate either needs to be cooled to −78.7°C to be stable under atmospheric pressure [15] or requires a pressure of 4 MPa to be stable at 4°C [16]. Therefore, to simplify our experiments, we used Tetrahydrofuran (THF) which forms hydrates at atmospheric pressure and temperatures of 4°C [15].

The controlled synthesis of methane hydrate from the aqueous phase is additionally difficult, mainly because of the low solubility of methane in water (1.5×10^{-3} mol methane per 1 mol water at 5 MPa and 25°C) [17]. This low solubility necessitates extensive pore fluid circulation and long experimental time for hydrate formation from the aqueous phase [18]. THF is a heterocyclic ether with the molecular formula C_4H_8O. THF is completely miscible in water [15] and thus widely used to resemble hydrate formation from methane dissolved in water [7, 19, 20]. Reference [21] demonstrated that THF and methane hydrates exhibit similar macroscale mechanical, electrical and thermal characteristics. THF has the advantage of providing close control on the hydrate saturation by varying the stoichiometric THF-H_2O mixture without having the long formation history of methane hydrate. However, it should be noted that THF hydrate forms structure-II hydrates whereas pure methane forms structure-I hydrates. Hydrates in nature, especially when formed from thermogenic gas, often contain longer chain hydrocarbons (e.g. ethane, propane) in addition to methane and thus can form structure-II hydrates [15]. Therefore, THF hydrate is used as a proxy for naturally occurring gas hydrate.

7.3. Experimental Methodology

7.3.1. *Micro X-Ray Computed Tomography*

The MicroXCT-400 apparatus from XRadia was used for the MXCT measurement. A sample holder consisting of two dry-ice containers at the top and bottom and a plastic tube for the sample in between has been used to perform the measurements (Fig. 7.2). This setup provides temporary cooling of the sample to keep the hydrate stable long enough to complete the required scans. Data acquisition for one tomographic image takes about 20 minutes, allowing the examination of hydrate dissociation.

THF hydrate was formed outside of the CT scanner prior to the image acquisition. The samples were formed in cylindrical plastic containers with a volume of 1.1 ml. The samples consisted of unconsolidated borosilicate

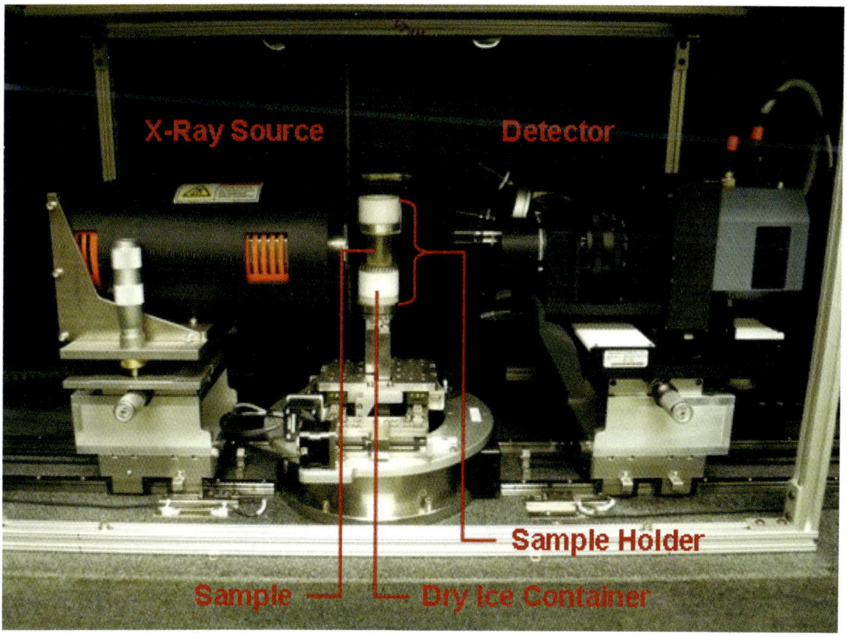

Fig. 7.2. Experimental setup inside the CT scanner.

glass beads (diameter: 1 mm), de-ionized water, THF and barium chloride. The specimen exhibited a porosity of approximately 0.35. The porosity was determined from CT images. H_2O-THF ratios resulting in 20%, 40%, 60%, 80% and 100% hydrate saturation were injected into glass bead samples. After combining all sample compounds, the sample was sealed to prevent the THF from volatizing and water from evaporating. The samples were then stored in a freezer, where they were slowly cooled to −25°C over a time period of 48 hours. Subsequently, the samples were cooled to −80°C for one hour. Temperatures well below the THF hydrate equilibrium temperature of 4.3°C were necessary to keep the hydrate stable long enough to obtain several scans and to avoid immediate dissociation.

MXCT measurements allow the distinction between materials with different bulk densities. They provide insight into the distribution of different sample components in the pore space. In contrast to sediment grains, glass beads have a uniform X-ray absorbance value, shape and density (2.23 g/cm^3). These properties make it easier to interpret the MXCT images.

5 wt% barium chloride were added to the fluid mixture to enhance the density contrast between residual water and hydrate. For example, the density contrast for a sample with 60% hydrate saturation is increased five times when barium chloride is added. As the hydrates form, barium chloride is excluded from the hydrate structure and remains within the water phase. Note that due to the low temperatures used for the hydrate formation, our samples initially contain ice and hydrate which cannot be distinguished based on their density (Table 7.1). Further, differential scanning calorimetry has shown that the addition of 5 wt% barium chloride decrease the dissociation temperature of ice to −7.9°C and the hydrate equilibrium temperature to 3.8°C. Thus, we expect the ice to melt before the THF hydrate dissociates.

The dissociation process was monitored by taking six consecutive tomographic scan. For each scan, 200 images were taken with an angular increment of 1°. The low number of images and the large angular increment were chosen to minimize the scanning time. An X-ray source voltage of 90 keV and a power of 7 Watt were used for the tomographies. The scanning time for one series of six tomographies was approximately 2 hours.

The output data of the MXCT scanner are gray scale images. Different gray values represent materials with different bulk densities. Dark gray areas indicate low bulk density, light gray to white areas correspond to high bulk density.

7.3.2. *Ultrasonic Velocity Measurements*

The basic principle behind pulse transmission ultrasonic velocity measurements is to send an acoustic signal through a sample and measure the signals travel time. An example setup of a sample with an ultrasonic transducer at each end is illustrated in Fig. 7.3.

The transducer contains piezoelectric lead zirconate titanate (PZT) crystals to generate compressional wave (P-wave) and two orthogonally polarized shear waves (S1 and S2-wave). The signal is excited by a pulse generator which sends an electric impulse to the PZT crystal. The electric impulse generates a mechanical strain wave with a frequency of about 1 MHz that propagates through the sample and is recorded by a transducer

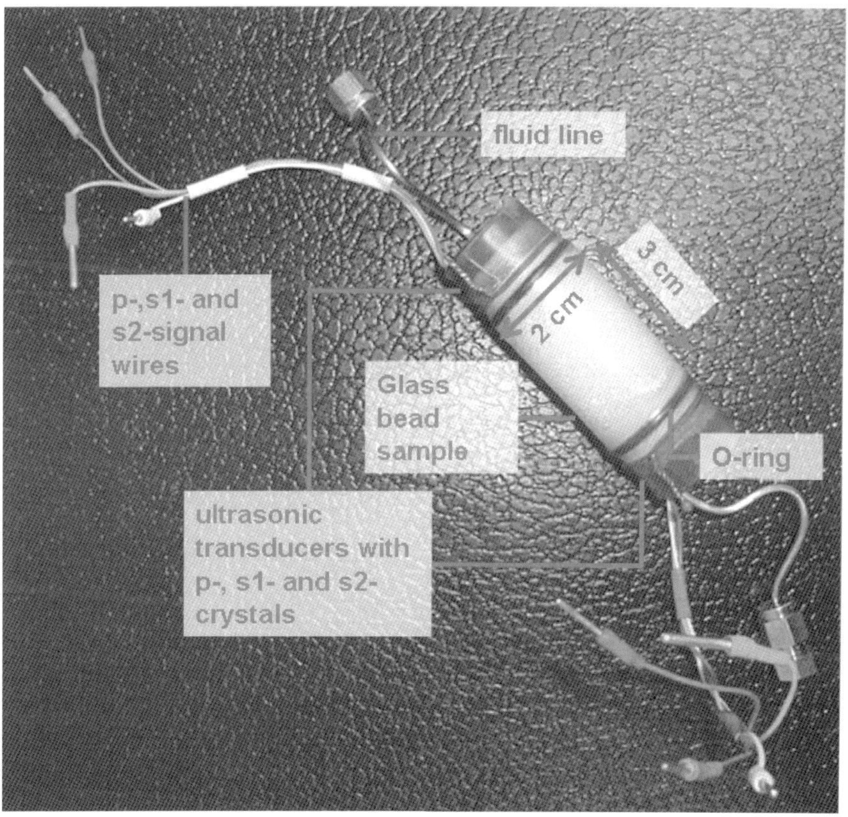

Fig. 7.3 Sample assembly for ultrasonic velocity measurements.

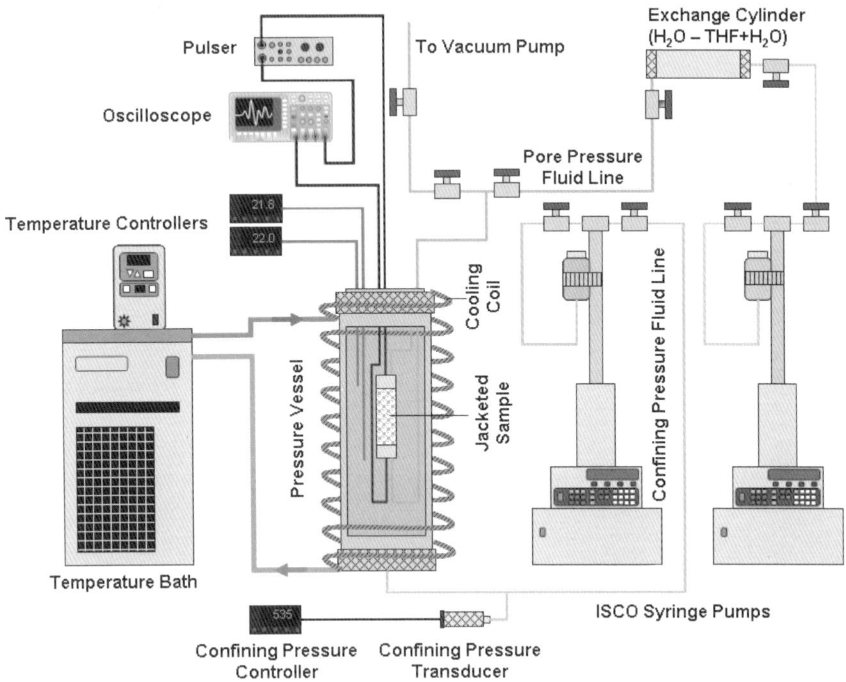

Fig. 7.4. Experimental setup for ultrasonic velocity measurements.

at the other end of the sample and then visualized on a digital oscilloscope before transfer to a computer. Each sample has a total volume of 9.4 ml and composition is similar to the samples used for MXCT measurements.

The ultrasonic experiments were performed in a thermally insulated pressure vessel (Fig. 7.4). Two thermocouples recorded the temperature inside the pressure vessel but outside of the sample at its top and bottom. The pressure vessel was encased by a cooling coil connected to a temperature bath for regulating the temperature of the sample. THF-water-barium-chloride mixture was pumped into the sample with a pore pressure of 0.7 MPa. The confining pressure was increased to 3.7 MPa. The resulting differential pressure was 3 MPa. Measurements were conducted on samples with THF-hydrate saturations of 40%, 60%, 80% and 100%.

After the hydrate-forming mixture was pumped into the sample, the temperature was gradually decreased below hydrate stability temperature (4.3°C for 100% THF hydrate saturation). The cooling process was faster in the beginning (4.5°C per hour) but the cooling rate was decreased to

0.7°C per hour when the hydrate equilibrium temperature was reached. Waveforms for all three wave types (P, S1, S2) were recorded during the cooling process for each 1°C interval. At temperatures close to the hydrate stability zone, waveforms were recorded more frequently. The cooling process was continued until hydrate formed in the sample which was indicated by a significant increase in all three velocities. When the hydrate formation started, the temperature was kept constant until the velocities stabilized. After hydrate formation, the sample was cooled further until the remaining water in the pore space froze. The ice formation was indicated by an additional increase in velocities.

First arrival times were picked from waveforms. The sample length was determined from MXCT images. Ultrasonic velocities can be determined from arrival time and sample length:

$$v = \frac{l}{t - t_0} \tag{7.1}$$

where l is the length of the sample, t is the first arrival time and t_0 is the dead time (the travel time of a pulse from one transducer assembly to the other without a sample in between).

7.4. Results and Discussion

7.4.1. *MXCT Imaging*

The reconstructed CT images (Fig. 7.5) contain information about the distribution of different components of the sample but for a quantitative information, further steps are required. Different gray values in the CT images represent different densities in the sample. Because of its low density (Table 7.1), THF hydrate and ice appear as low gray-value areas in the CT image whereas barium chloride brine and precipitated barium chloride show higher gray values caused by their higher density (Table 7.1).

Figure 7.5 shows a series of CT images for a sample with 80% hydrate saturation. The images capture the dissociation of THF hydrate. The first two tomographic scans indicate that THF hydrate and ice is arranged in the pore body and has few contact points with the grain surfaces. Precipitated barium chloride covers most of the grain surfaces and forms bridges between neighboring grains. During the imaging process, the ice melts. Consequentially, water is present in the pore space and mixes with barium chloride (scan Tomo3).

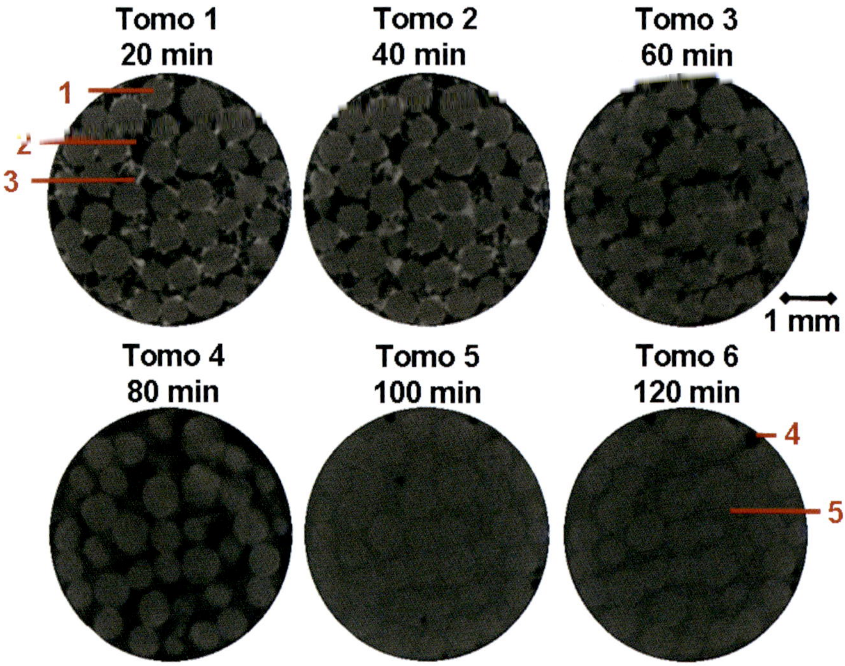

Fig. 7.5. Series of CT images of a glass bead sample with initial hydrate saturation of 80%; 1: glass bead, 2: ice/THF hydrate, 3: $BaCl_2$, 4: air bubble, 5: mixture of $BaCl_2$, H_2O and THF.

Later in the data acquisition, THF hydrate dissociates. Dissociation and melting processes in the pore space cause the observed blurring in scan Tomo4. After the dissociation (Tomo5 and Tomo6), the pore space is filled with a liquid mixture of barium chloride, water and THF. This mixture has an intermediate density and exhibits medium gray values (Tomo5 and Tomo6). The dark gray circular spots in Tomo5 and Tomo6 are air bubbles.

Differences between the density of the components before and after dissociation are used to interpret the images. For image analysis, *ImageJ* [a] and *Matlab* were used. Image processing was applied to determine the gray value range in the CT images representing different components of the sample. Pore space which was initially filled with THF hydrate, ice (low density) and barium chloride precipitate (high density) becomes filled with mixture of water, barium chloride and THF by the end of the scanning process (in-

[a]ImageJ is a free image processing package mainly used for medical imaging. It is available at http://imagej.nih.gov/ij/

Table 7.1. Density of sample components for hydrate saturation $S_h = 60\%$ and 5 wt% $BaCl_2$.

Component	Density [g/cm³]
ice	0.92
water	1.00
$BaCl_2$	3.84
$BaCl_2$ brine (5 wt%)	1.49
THF hydrate	0.97
THF-water-$BaCl_2$ mixture	1.11

termediate density). Two tomographic images were subtracted from each other to obtain a gray value difference and hence a density difference. Table 1 shows the expected density values for the sample components. Density differences are negative between THF hydrate and mixture and between ice and mixture. Density differences are positive between barium chloride and mixture and between barium chloride brine and mixture. Thus, images acquired before and after hydrate dissociation were subtracted to better identify the location of ice, barium chloride and THF hydrate in the images (Figs. 7.6 and 7.7).

An *ImageJ* plugin called *Mines Bilateral Filter* was used to reduce noise. This filter smoothes gray scale differences below a chosen limit yet preserves larger differences, such as the contact between pore space and glass beads (Figure 7.6). A threshold of gray scale values representing the glass beads is produced with *ImageJ*. This threshold is used to remove the glass beads from the image by setting their gray value to zero (Figure 7.6). The image subtraction was done with *Matlab* by reading two images as 8-bit matrices. They are then converted to double matrices to prevent data loss during subtraction by allowing negative density differences. These two double matrices are subtracted. A *Matlab* figure showing negative (blue) and positive (red) gray value differences is created (Fig. 7.6). Then, the differential matrix is converted back to a matrix of positive integers which is exported as an 8-bit image for further analysis in *ImageJ*.

Figure 7.7 shows histograms and differential images referenced by subtracting Tomo6 for a whole tomography series. The histogram median is shifting from low gray values to medium gray values during the scanning process. The number of pixels representing THF hydrate and ice (blue) and pure barium chloride (red) decreases. Pixels with intermediate gray values (125 ± 20) appearing white in Fig. 7.7 are considered either barium chloride brine, mixture of water, barium chloride and THF or indistinct

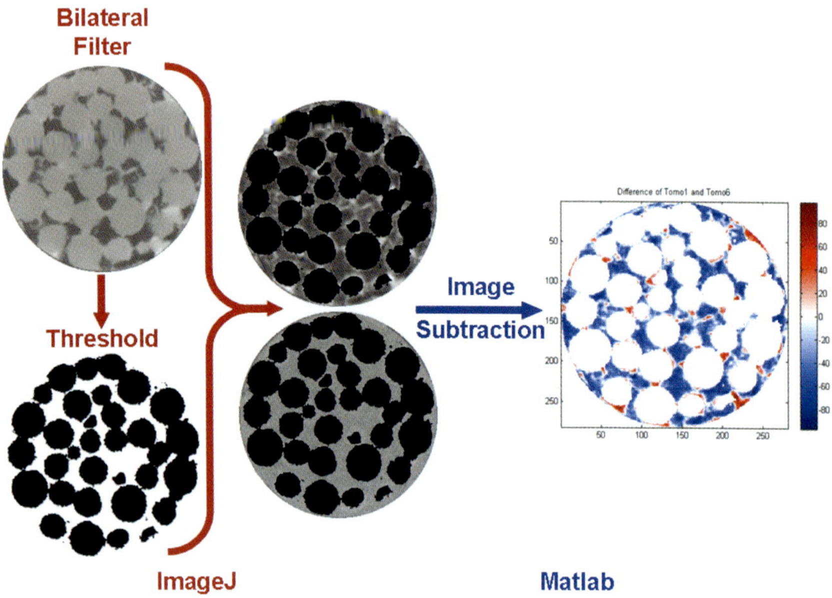

Fig. 7.6. Image processing workflow for MXCT images, red: positive density difference, i.e. barium chloride, blue: negative density difference, i.e. ice or THF hydrate, white: density difference of zero, i.e. glass beads.

due to misinterpretation of pixels at the interface of high and low density areas. Figure 7.7 further shows that the number of pixels with intermediate gray values increases and the number of pixels with high (> 150) and low (< 100) gray values decreases during the scanning time. Areas of high gray value (red) and low gray value (blue) disappear with time due to dissociation of hydrate and melting of ice. The pore space appears mainly white for Tomo5 meaning all hydrate is dissociated. The remaining blue areas in Tomo5 are air bubbles.

Figure 7.7 shows that hydrate and ice are found in the bulk pore volume while the grain surfaces and grain contacts are mainly covered with barium chloride precipitate and later in the dissociation process barium chloride brine. It further appears that the dissociation process propagates from the grain surfaces into the bulk pore volume. The hydrate in the center of big pores dissociates last.

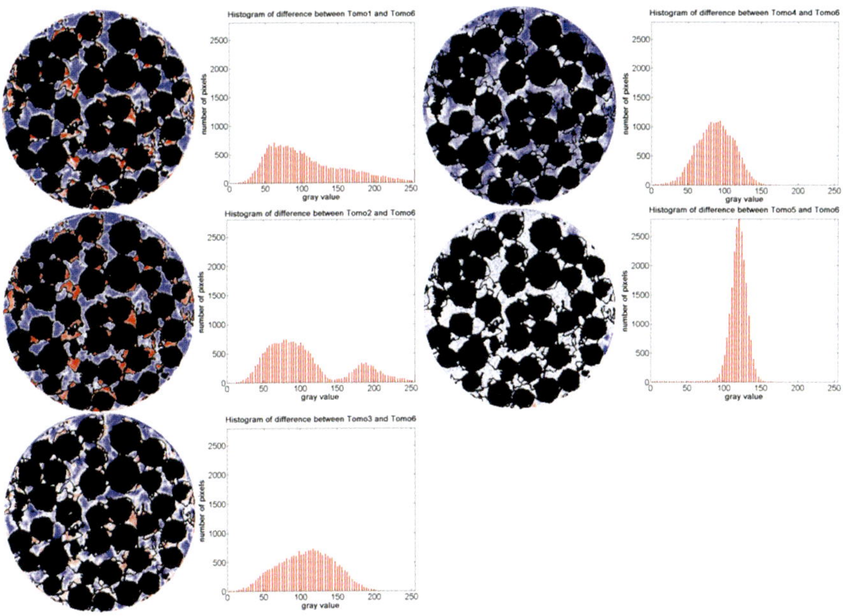

Fig. 7.7. Histograms of gray values and differential CT images for a sample with 80% initial hydrate saturation; black: glass beads, blue: ice and THF hydrate, red: $BaCl_2$ precipitate, white: barium chloride brine or mixture of H_2O, $BaCl_2$ and THF.

7.4.2. *Ultrasonic Velocity Measurements*

Figure 7.8 shows the P and S-wave velocities recorded during the cooling of a sample from 23 °C to -8 °C with 40% hydrate saturation. The errors for P and S-wave velocities are 10% to 11% and 16% to 19%, respectively. Errors in velocity were calculated by propagating the uncertainties in sample length and picking of arrival times:

$$\sigma_v = v \cdot \sqrt{\sigma_l^2 \left(\frac{\partial v}{\partial l}\right)^2 + \sigma_t^2 \left(\frac{\partial v}{\partial t}\right)^2 + \sigma_{t_0}^2 \left(\frac{\partial v}{\partial t_0}\right)^2} \tag{7.2}$$

where v is the velocity, l is the length of the sample, t is the arrival time, t_0 is the dead time, σ_l is the uncertainty in sample length, σ_t is the uncertainty in arrival time, σ_{t_0} is the uncertainty in dead time, and σ_v is the absolute error in velocity.

The arrival time for the compressional wave is picked manually at the first significant divergence from zero in amplitude. Since the manual pick is subjective, an error of $\sigma_t = 10^{-6}s$ is assumed. The shear wave is picked at an amplitude maximum since the first divergence from zero for the shear wave is not visible in the waveform. This introduces an error of $\sigma_t = 5 \times 10^{-6} s$. However, the arrival times were picked at the same location for each of the waveforms. Thus, the relative error is much smaller than the absolute error displayed in Figs. 7.8 and 7.9.

Hydrate formation started at temperatures lower than equilibrium temperature for THF hydrates of 4.3°C due to the addition of 5 wt% barium chloride and the subcooling required to trigger hydrate formation. Furthermore, 4.3°C is the equilibrium temperature for 100% hydrate saturation. For lower hydrate saturations, the equilibrium temperature decreases [15]. Although the hydrate formation started at temperatures below 0°C, the simultaneous formation of ice in the sample can be ruled out since the formation temperature of ice is decreased to −7.9°C due to the addition of 5 wt% barium chloride.

The velocities remain constant until hydrate formation starts. Hydrate formation results in a significant increase of velocities for both, P and S-wave. The second increase in velocities is caused by the formation of ice in the sample upon further cooling. Hydrate stability and ice formation temperatures are both decreased due to the addition of barium chloride (Figure 7.8). The presence of barium chloride causes an increase of velocities of about 300 m/s for the compressional wave and 200 m/s for the shear wave. This has been determined by comparing velocities of two samples with similar hydrate saturation, one with and one without barium chloride.

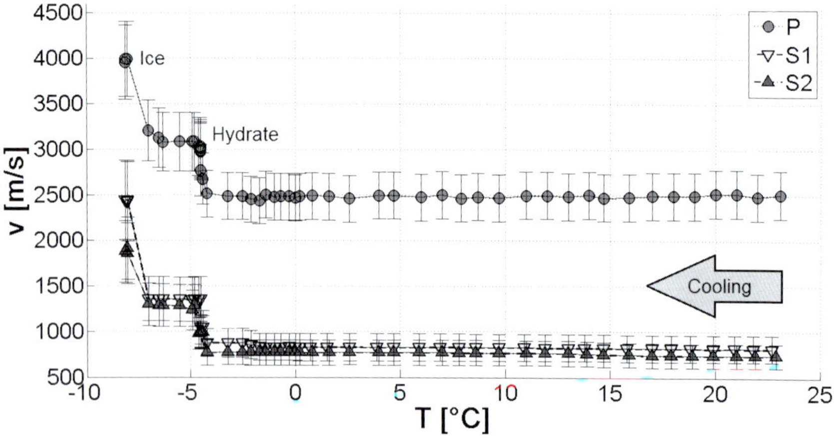

Fig. 7.8. Velocities of P-, S1- and S2-wave for a sample with 40% hydrate saturation during the cooling process.

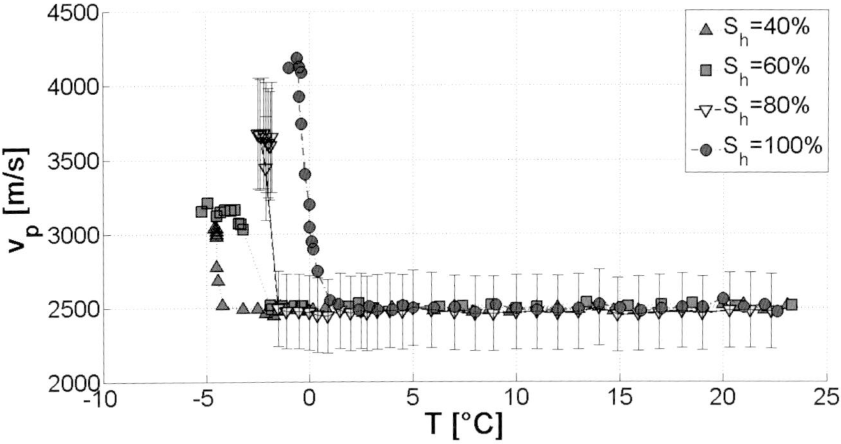

Fig. 7.9. P-wave velocities for hydrate saturations of 40%, 60%, 80% and 100% during the cooling process. Error bars are shown exemplary for one data row with 80% hydrate saturation.

Figure 7.9 displays velocities recorded during the cooling process for four different hydrate saturations (40%, 60%, 80% and 100%). The velocity increase is directly proportional to hydrate saturation: the more hydrate in the pore space, the higher the velocity increase (Table 7.2). Hydrate formation starts at lower temperatures for lower hydrate saturation (Figure 7.9). This observation confirms that hydrate equilibrium temperature decreases with decreasing hydrate saturation [22]. Note that the sample with 60% hydrate saturation shows overall lower velocity than the other samples. The difference is within the margin of error but causes the velocity after hydrate formation to be lower than the velocity of the sample with 40% hydrate saturation.

Table 7.2. Ultrasonic velocities after hydrate formation.

S_h [%]	v_p [m/s]	v_s [m/s]
0	2500 ± 250	820 ± 123
40 ± 5	3034 ± 315	1356 ± 242
60 ± 5	3162 ± 327	1310 ± 219
80 ± 5	3677 ± 376	1555 ± 281
100 ± 5	4185 ± 482	1801 ± 351

Figures 7.10 and 7.11 demonstrate how the measured ultrasonic veloci-

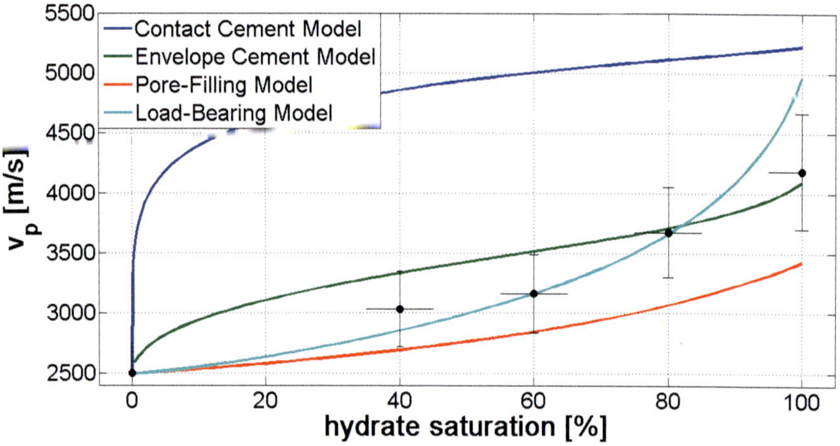

Fig. 7.10. Effective medium model after Refs. [10] and [13] with compressional wave velocities obtained by ultrasonic measurements.

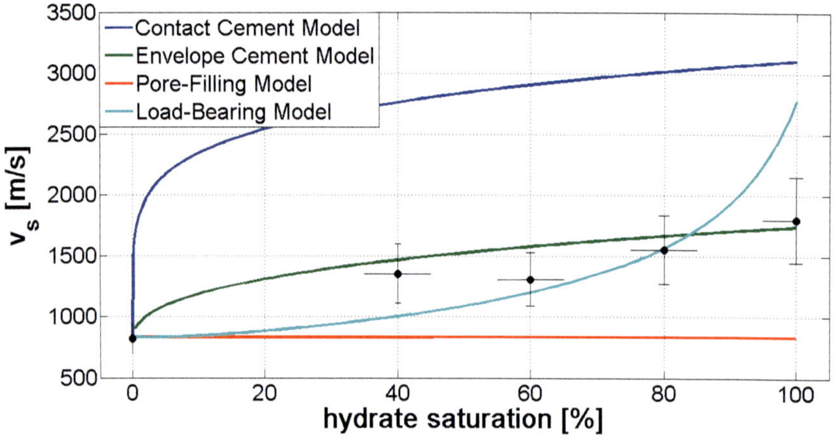

Fig. 7.11. Effective medium model after Refs. [10] and [13] with shear wave velocities obtained by ultrasonic measurements.

ties coincide with the effective medium model by Refs. [10] and [13] . The uncertainties in hydrate saturation are assumed to be 5%. P and S-wave velocities are close to the load bearing and the envelope-cementing model. The MXCT images, however, showed that the hydrate does not form according to the envelope-cementing model.

The margin of error for all measured hydrate saturations covers both, envelope cementing and load bearing models. An explanation for the increased velocities could be a combination of two different formation mechanisms. The CT images showed that the hydrate has some contact points with the grains. Thus, it is possible that part of the hydrate formed on the grain surfaces perhaps due to impurities of the glass bead surfaces. This mixed formation mechanism could explain why the velocities are located between the load-bearing and the envelope-cementing model.

7.5. Conclusions

Micro X-ray computed tomography observations indicated that hydrate from water-THF-solution forms in the pore body, mainly away from the grain surfaces with small amounts of hydrates contacting the grain surfaces. From these observations we conclude that THF hydrate can contribute to the velocity as explained by a combination of both, the load-bearing model and the envelope-cementing model of the effective medium theory. To ascertain this conclusion, ultrasonic velocities were measured and compared to the effective medium models.

Given that THF hydrate is a proxy for one possible texture of natural occuring hydrate, we conclude that natural gas hydrate in coarse grained sediments is primarily non-cementing when formed from natural gas dissolved in water. This conclusion is essential for the assessment of gas hydrate saturation from seismic and well log data. Any workflow for calculating hydrate saturation from seismic and well log data requires reliable rock physics transforms - describing the relationship between measured or inverted properties, such as sonic/seismic velocities or acoustic impedance, and hydrate saturation [23].

Editors' Note

Professor Mike Batzle, Baker Hughes Distinguished Chair of Petrophysics and Borehole Geophysics, died January 2015. Mike enjoyed an outstanding career in industry and academia with a worldwide reputation for teaching and research. He was a great colleague and we miss him.

The Editors

References

[1] R. F. Meyer, Speculations on oil and gas resources in small fields and unconventional deposits, *Long-term Energy Resources.* 1, 49–72 (1981).

[2] V. M. Dobrynin, V. P. Korotajev, and D. V. Plyuschev, *Gas Hydrates — A Possible Eenergy Resource.* Pitman Publishers, Boston (1981).

[3] T. S. Collett, A. H. Johnson, C. C. Knapp, and R. Boswell, Natural gas hydrates: A review, *Natural gas hydrates — Energy resource potential and associated geologic hazards: AAPG Memoir.* **89**, 146–219 (2009).

[4] B. J. Anderson, J. W. Wilder, M. Kurihara, M. D. White, G. J. Moridis, S. J. Wilson, M. Pooladi-Darvish, Y. Masuda, T. S. Collett, R. B. Hunter, et al., Analysis of modular dynamic formation test results from the mount elbert 01 stratigraphic test well, milne point unit, north slope, Alaska, *British Columbia, Canada* (2008).

[5] S. R. Dallimore, J. F. Wright, F. M. Nixon, M. Kurihara, K. Yamamoto, K. Fuji, M. Numasawa, M. Yasuda, and Y. Imasato. Geologic and porous media factors affecting the 2007 production response characteristics of the jogmec/nrcan/aurora mallik gas hydrate production research well. (2008).

[6] T. S. Collett and M. W. Lee, Well Log Characterization of Natural Gas Hydrates 1. *Petrophysics* **53**(5), 348–367 (2012).

[7] T. S. Yun, F. M. Francisca, J. C. Santamarina, and C. Ruppel, Compressional and shear wave velocities in uncemented sediment containing gas hydrate, *Geophy. Res. Lett.* **32**(10) (2005).

[8] W. F. Waite, W. J. Winters, and D. H. Mason, Methane hydrate formation in partially water-saturated ottawa sand, *Am. Mineral.* **89**(8-9), 1202–1207 (2004).

[9] J. A. Priest, A. I. Best, and C. R. I. Clayton, A laboratory investigation into the seismic velocities of methane gas hydrate-bearing sand, *J. Geophy. Res.: Solid Earth.* **110**(B4) (2005).

[10] C. Ecker, J. Dvorkin, and A. Nur, Sediments with gas hydrates: Internal structure from seismic AVO, *Geophysics.* **63**(5), 1659–1669 (Sept., 1998).

[11] D. C. Kunerth, D. M. Weinberg, J. W. Rector, C. L. Scott, and J. T. Johnson, Acoustic laboratory measurements during the formation of a thf-hydrate in unconsolidated porous media, *J. Seism. Explor.* **9**(4), 337–354 (2001).

[12] R. Boswell and T. S. Collett. The gas hydrates resource pyramid Fire in the Ice: (2006).

[13] M. B. Helgerud, J. Dvorkin, A. Nur, A. Sakai, and T. S. Collett, Elastic-wave velocity in marine sediments with gas hydrates: Effective medium modeling, *Geophy. Res. Lett.* **26**(13), 2021–2024 (jul, 1999).

[14] M. B. Rydzy and M. L. Batzle. Rock physics characterization of THF hydrate-bearing sediment. In *Proceedings of the 7th International Conference on Gas Hydrates* (2011).

[15] E. D. Sloan and C. A. Koh, *Clathrate hydrates of natural gases.* CRC press (2008).

[16] J. Carroll, *Natural gas hydrates: a guide for engineers.* Gulf Professional Publishing (2009).

[17] D. R. Lide and H. P. R. Frederikse, *CRC handbook of chemistry and physics* (1995).

[18] E. Spangenberg, J. Kulenkampff, R. Naumann, and J. Erzinger, Pore space hydrate formation in a glass bead sample from methane dissolved in water, *Geophy. Res. Lett.* **32**(24) (2005).

[19] C. Pearson, J. Murphy, and R. Hermes, Acoustic and resistivity measurements on rock samples containing tetrahydrofuran hydrates: laboratory analogues to natural gas hydrate deposits, *Journal of Geophysical Research: Solid Earth.* **91**(B14), 14132–14138 (1986).

[20] T. S. Collett and J. Ladd, Detection of gas hydrate with downhole logs and assessment of gas hydrate concentrations (saturations) and gas volumes on the blake ridge with electrical resistivity log data. Procs. of the Ocean Drilling Program **164**, 179–191 (2000).

[21] J. Y. Lee, T. S. Yun, J. C. Santamarina, and C. Ruppel, Observations related to tetrahydrofuran and methane hydrates for laboratory studies of hydrate-bearing sediments, *Geochemistry, Geophysics, Geosystems.* **8**(6) (2007).

[22] T. Makino, T. Sugahara, and K. Ohgaki, Stability boundaries of tetrahydrofuran+ water system, *J. Chem. and Engin. Data.* **50**(6), 2058–2060 (2005).

[23] Z. Zhang, D.-h. Han, and D. R. McConnell, Characterization of elastic properties of near-surface and subsurface deepwater hydrate-bearing sediments, *Geophysics.* **78**(3), D169–D179 (2013).

Chapter 8

Microfluidic and Nanofluidic Porous Media Analogs for Oil and Gas Applications

Jeong Tae Ok, Keith B. Neeves, Wei Xu and Xiaolong Yin

Department of Chemical & Biological Engineering,
Department of Petroleum Engineering,
Colorado School of Mines,
Golden, Colorado 80401, USA
xyin@mines.edu, kneeves@mines.edu

Microfluidic / nanofluidic porous media analogs (μPMA / nPMA) are important tools to visualize flow, transport, and phase behavior in porous media. In this article, we present the fabrication of μPMA / nPMA and their use in studying enhanced oil recovery and capillary retention of water in sub-micron pores in a gas-water drainage process. Using intermediate-wet polydimethylsiloxane (PDMS) based μPMA, we studied the effect of pore geometry on water-oil drainage efficiency. The porosity and permeability of μPMA are 0.19 and about $2 \times 10^{-13}\text{m}^2$, respectively. We found that both pore size distribution and vugs decrease the drainage efficiency. While application of surfactant increased the drainage efficiency in all geometries, the gain was less for heterogeneous geometries with vugs. Using reactive ion etching, nPMA was fabricated on silicon substrates and then anodically bonded to pyrex covers. Multiphase flow experiments were conducted in nPMA to characterize the critical gas pressure needed to remove imbibed liquids from nanochannels. The measured high pressure indicates that it is very difficult to recover injected hydraulic fracturing fluid once they are imbibed into the low-permeability formation.

8.1. Introduction

Transparent network models, collectively known as micromodels in the porous media literature, have been particularly useful in visualizing multiphase flows [1]. The first glass micromodel was developed in 1961 to visualize water flood of an oil-saturated network [2]. In the late 1970s and 1980s, micromodels were significantly improved by employing photolithography to define pore sizes down to approximately 20 μm scale. During this time, micromodels supported an explosion of experimental studies on multiphase flows [3, 4]. Micromodels were used to study critical multiphase flow phenomena such as invasion percolation [5, 6] and capillary and viscous fingering [7]. Since then, micromodels have been applied to study a diverse set of applications including enhanced oil recovery [8], solute and colloid transport [9–11], microorganism transport [12], and fuel cells [13]. In the past decade, micromodels have undergone another transformation with the introduction of fabrication techniques such as LIGA [14], deep reactive ion etching [15], and soft lithography [11]. A related approach is to pack beads into microfluidic channels to form the solid phase of a porous media [16]. Taken together, these techniques allow for almost any periodic, random, or fractal geometry down to a pore size of approximately 1 μm [17] as well as recreations of rock geometry from imaging data [18].

Development of fuel cell, membranes, and recently emerged unconventional oil / gas resources brings in additional fundamental needs to visualize and measure fluid flow and transport on the nanometer level. In these nanoporous materials, the large surface-to-volume ratio amplifies the roles of surface energy and surface forces. In addition, the small size of nanopores makes the continuum assumption of the pore fluids questionable, especially for macromolecules. Transparent nanofluidic devices fabricated using wet / dry etching, lithography, or imprinting [19] have become very useful research tools.

In this paper, we present the fabrication of microfluidic porous media analogs (μPMA) and nanofluidic porous media analogs (nPMA) and their use in studying chemically enhanced water flooding and capillary retention of liquids in sub-micron pores.

8.2. Fabrication Methods

8.2.1. *Fabrication of μPMA*

We designed three μPMAs with random pore geometries based on the Voronoi diagrams [20], and eight μPMAs with regular arrangement of triangles, squares, diamonds or hexagons as shown in Fig. 8.1. Of the three random geometries, one has a Gaussian channel size distribution in the range of 4–8 μm, with the peak of the Gaussian at 6 μm. To compare with random geometries and to study the effect of the coordination number, periodic pore geometries with hexagon, square, diamond, and triangle grains were also designed. Both random and regular pore geometries include heterogeneous networks, where randomly selected grains are removed to create vugs. All other pore geometries have uniform channel width. The dimension of the μPMA is 30 $\times$ 3 mm (length $\times$ width).

Microfluidic networks as shown in Fig. 8.1 were defined in 3-inch diameter Si wafer by optical photolithography and transferred into polydimethylsiloxane (PDMS) by soft lithography. A negative tone photoresist (KMPR 1010, MicroChem, Newton, MA) was used for PDMS mold with a depth of approximately 15 μm. Following photolithography, PDMS with a 10:1 ratio of silicon rubber base to curing agent was molded on KMPR-Si master using standard soft lithography techniques [21]. Prior to PDMS molding, KMPR-silicon master is treated by gas phase trichlorosilane (Tridecafluoro-1, 1, 2, 2-Tetrahydro-octyl-Trichlorosilane, Gelest, Inc. PA) to keep the surface hydrophobic for the facile demolding procedure. Inlet and outlet holes were first punched in PDMS devices, then sonicated in 1 M HCl for 8 minutes, acetone for 5 minutes, and ethanol for 5 minutes before being dried in a convection oven at 70°C for 1 hour. The devices and the glass were exposed to oxygen plasma (30 W for 45 s) and bonded together. Special care is required due to the shallow depth of the device (15 μm). PDMS surface is lightly pushed on the glass substrate by a roller. The plasma bonded PDMS-glass chips were put in vacuum chamber with

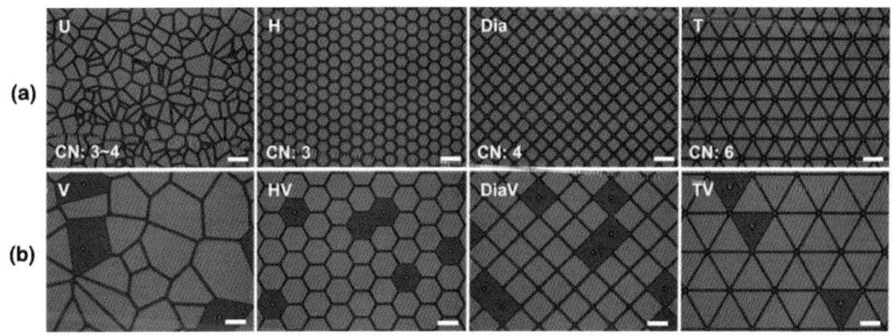

Fig. 8.1. Eight of the eleven pore geometries studied for water flooding and surfactant flooding. U: Randomly connected channels of *uniform* width (6 μm); H: Channels (6 μm) arranged in *hexagonal* patterns; Dia: Channels (6 μm) arranged in *diamond* patterns; T: Channels (6 μm) arranged in *triangular* patterns; V: Randomly connected channels (8 μm) with *vugs*; HV, DiaV, TV: Channels (6 μm) arranged in hexagonal, diamond, and triangular patterns with vugs. The three patterns not included in this figure are: Square (S), Square with vugs (SV), and randomly connected channels with a Gaussian channel size distribution (4–8 μm) (D). CN stands for the coordination number of the geometries.

gas phase trichlorosilane to obtain hydrophobic surface on both PDMS and glass for 4 hours.

8.2.2. *Fabrication of nPMA*

We fabricated nanofluidic devices with different geometric configurations. Here, the general method is presented, together with selected Scanning Electron Microscope (SEM) images of the devices. Microchannels and nanochannels of nanofluidic chips were formed in double side polished $\langle 100 \rangle$ silicon wafers (thickness = 250 mum) with low-stress silicon nitride on both sides using methods described in the literature [22]. First, the nanoscale features are defined by reactive ion etching using photoresist and silicon nitride as the etching mask on the front side of the wafer. For the device shown in the left four graphs of Fig. 8.2, an array of 100 nanochannels spaced 10 μm edge-to-edge were defined. Each channel had cross-sectional dimensions of 100 nm (depth) by 5 μm (width), and was 200 μm long. For the device shown in the right graph of Fig. 8.2, a network of nanochannels is defined on an area of 600 μm $\times$ 400 μm. The nanochannels are 3 μm wide and 300 nm deep; the porosity of the network is 19%. Typical root-mean-square horizontal surface roughness by reactive ion etching is known to be lower than 1 nm on a 1 μm^2 zone. Next, two microchannels were defined on the sides of the nanochannel array. The microchannels are used to control

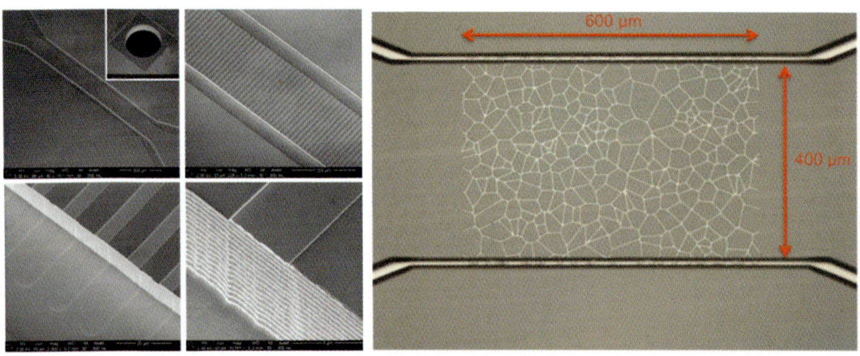

Fig. 8.2. The four graphs on the left show the SEM images of a nanofluidic device at different magnifications. Clockwise from upper left: Image of the whole device with inlet / outlet hole shown in the insert; Nanochannels arranged in parallel between the microchannels; two graphs showing the junction between a microchannel and the nanochannels. The graph on the right shows a network of nanochannels etched between the microchannels, observed through a microscope.

the pressure drop across the nanochannels [23]. Finally, inlet and outlet holes were defined at the four ends of the two microchannels by a back side deep reactive ion etching through the wafer. The fabricated device was removed from the wafer by cleavage along crystal planes. The front side of each device (40 mm × 20 mm × 0.25 mm) was anodically bonded (330°C, 1 kV, 1 hour) to a think Pyrex coverslip (Pyrex 7740, 40 mm × 20 mm × 0.3 mm, Newport Industrial Glass, Inc. Stanton, CA). Prior to the Si-Pyrex bonding, silicon nitride film was completely removed by a hot phosphoric acid etching (85 wt. % H_3PO_4, 2 hours at 150 °C) procedure. In order to maintain clean surface conditions, the whole nanofluidic chip was rinsed with reagent grade water, methanol and dried by nitrogen gas prior to use.

8.3. Experimental Procedures

8.3.1. *Water Flooding and Surfactant Flooding in μPMA*

Prior to the two-phase experiments, the hydraulic permeability of the networks Fig. 8.1 was measured by infusing deionized water at a constant pressure and measuring the flow rate through the device. The experimental apparatus included compressed nitrogen gas to drive the flow, a relief valve, a regulator to modulate the inlet pressure, a piezoelectric pressure sensor, a pressure gauge, and a glass capillary tube that served as a fluid reservoir. The glass capillary tube was connected to the device with small bore (0.01 inch) Tygon tubing (Fisher). The Tygon tubing was attached to the PDMS μPMAs. The flow rate through the device at a given pressure was calculated by measuring the time it took for the air-fluid meniscus to move 2 mm in the capillary tube. The measurement was repeated three times for each pressure between 4 and 20 psig.

In the two-phase experiments, the displacing fluid was either an NaCl solution (1.5% wt.) or an NaCl solution (1.5% wt.) with surfactant (0.5% wt. NEODOL 91–8 ethoxylated alcohol (EA)). Light mineral oil was used as the oleic wetting phase. Blue food dye (FD&C #1, Spectrum Chemical) was added to the displacing fluids (1.0% wt.) to visualize the drainage processes. Viscosity measurements were performed using cone-plate geometry on TA Instruments stress and strain controlled DHR-3 (Discovery Hybrid Rheometer-3, New Castle, DE, USA) at 25 ± 0.1°C, with temperature control provided by a Peltier element. Interfacial tension (IFT) between brine and mineral oil phases was measured by a spinning drop tensiometer

(SITE100, Krüss). Density of fluids was measured using a precision pipette and a precision top-loading balance (Denver Instruments, PI-6001). In this set of experiments, the μPMAs are treated with trichlorosilane, leading to a uniform hydrophobic surface with a contact angle of about 111°.

The experimental setup for water flooding and surfactant flooding is identical to that for the permeability measurement. In the experiments, the displacing fluid was first injected at 15 psi to fill the porous media. Next, mineral oil was injected at 15 psi for 20 minutes to establish the irreducible brine saturation. The third step is to displace the mineral oil with brine or surfactant solution. The displacing fluid was injected at 1.5 psi. Images were taken during injection and at the time of water breakthrough (BT).

As it is difficult to measure the volume of the displaced liquid due to the small size of μPMAs, an optical method was developed to measure the saturation of the injected brine or that of the surfactant solution. For each μPMA, seven consecutive pictures with 20X magnification were taken by a motorized inverted microscope (Olympus IX81) to cover the entire μPMA. First, images were taken of the μPMA filled with water. These images were used to calculate the total area available for flow and the porosity. Analysis of images during fluid displacements yielded an area filled with the dyed injected fluid as a function of time. The saturation of injected fluid was calculated by dividing the area taken by the injected fluid by the total area available for flow.

8.3.2. *Multiphase Flow Experiments in nPMA*

Similar to the experimental setup for μPMA, the compressed nitrogen gas is used to drive the multiphase flow in nPMA. These devices can operate at pressures of up to 1,500 psi) when outfitted with nanoport assemblies (Upchurch Scientific, 6-32 Coned Nanoport Assemblies) and capillary PEEK tubing (Upchurch Scientific, outer diameter of 360 m). Multiphase flow experiments in nPMA are visualized on an upright microscope (Olympus BX60F), with 10X objective, and CCD camera (Lumenera Infinity2-1C). Standard reflection illumination with a halogen lamp gives a good contrast between gas and liquid phase for depths down to 30 nm. Gas-liquid flow in hydrophilic nanochannels was achieved by controlling the pressure difference between the two microchannels [23]. Initially the desired liquids filled both microchannels using a pressure difference (5 psi) between the inlet and outlet, and no pressure gradient across the nanochannels. The bridging nanochannel network across microchannels was spontaneously filled by capillary action. In order to have nitrogen gas displace the liquids in nPMA, gas pressure difference in one microchannel was varied in the range of 0–150 psi while maintaining constant pressure difference (5 psi) in the other microchannel (liquid-filled).

8.4. Results

8.4.1. *μPMA Experiments*

Table 8.1 shows the porosities and permeabilities of the μPMAs. Table 8.2 shows the properties of the fluids used in the experiments. The density of the light mineral oil is 0.85 g/cc, less than that of the displacing fluid. The viscosity, however, is much higher, leading to high mobility ratios of 45 and 38, respectively, for water and surfactant flooding experiments. Application of 0.5 wt. % of EA significantly reduces the IFT of light mineral oil to the displacing fluid. The IFT of oil to brine is 28.37 mN/m; the IFT of oil to the surfactant solution is 3.57 mN/m.

Table 8.1. μPMA properties. Numbers in the parentheses indicate standard deviations from three repeats.

Chip Name	No. of Grains	Porosity	Permeability (μm^2)
U	25626	0.190	0.187 (0.011)
D	25629	0.191	0.167 (0.007)
V	3413	0.185	0.153 (0.006)
H	25000	0.190	0.182 (0.007)
HV	7344	0.185	0.133 (0.005)
S	29088	0.192	0.215 (0.009)
SV	6250	0.189	0.161 (0.006)
Dia	25170	0.189	0.268 (0.011)
DiaV	6568	0.187	0.195 (0.009)
T	19074	0.190	0.225 (0.011)
TV	4913	0.187	0.148 (0.003)

Table 8.2. Fluid properties [24].

Fluid	Density (g/cc)	Viscosity (cP)	IFT with Oil (mN/m)
Mineral Oil	0.848	42.46	–
DI Water	0.998	0.89	–
Brine	1.006	0.94	28.37
Surfactant Solution	1.007	1.13	3.57

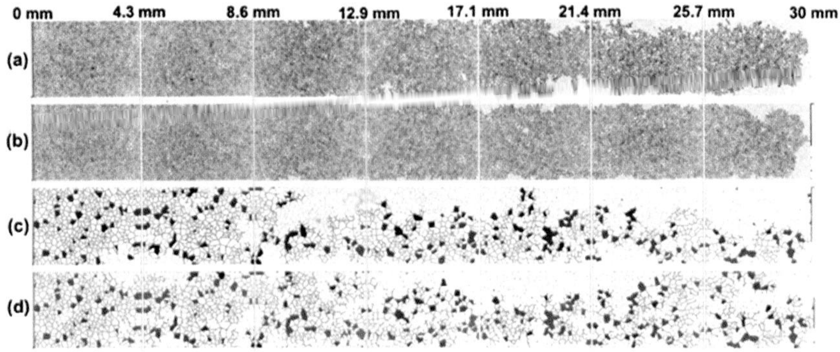

Fig. 8.3. Typical drainage patterns of water flooding and surfactant flooding in μPMAs: (a) Water flooding in U; (b) Surfactant flooding in U; (c) Water flooding in V; (d) Surfactant flooding in V.

A measure of the displacement efficiency is the saturation of the non-wetting phase at the point of breakthrough (BT). The BT times and the non-wetting phase (brine or surfactant solution) saturations at BT measured from the experiments are listed in Table 8.3 and Table 8.4. Not surprisingly, surfactant increased the recovery efficiency for all μPMAs (see Fig. 8.3 for two examples). Reduced IFT also leads to reduced BT time. Close examination of the data reveals the differences made by the geometries. Physical heterogeneities such as pore size distribution or vugs always reduce the displacement efficiency. The efficacy of the surfactant, when applied to heterogeneous μPMAs, is generally not as good as when it is applied to homogeneous μPMAs.

Drainage efficiency as studied in these experiments is influenced by the capillary instability which is generally measured by the capillary number $Ca = \mu U/\sigma$ where μ is the viscosity of the displacing fluid, U is the velocity of injection, and σ is the IFT. In our experiment, the injection is driven by a constant pressure (1.5 psi) and the rate of injection decreases with increasing time. The capillary number is therefore defined using the BT time which reflects the average rate of injection. These capillary numbers are also included in Table 8.3 and Table 8.4.

8.4.2. *nPMA Experiments*

We measured the pressure required to displace liquids by gas in nanochannels. Fig. 8.4 shows a typical bright field image sequence of the phase dis-

Table 8.3. Water flooding: breakthrough (BT) time, water saturation at BT, and capillary number Ca based on the BT time [24].

Chip Name	BT Time (10^3 s)	BT Water Saturation	Ca (10^{-7})
U	2.84 ± 0.22	0.859 ± 0.017	3.51 ± 0.27
D	3.17 ± 0.60	0.754 ± 0.029	3.20 ± 0.58
V	2.66 ± 0.17	0.655 ± 0.035	3.75 ± 0.25
H	4.28 ± 0.41	0.810 ± 0.004	2.33 ± 0.23
HV	3.03 ± 0.17	0.713 ± 0.018	3.28 ± 0.19
S	2.74 ± 0.15	0.679 ± 0.025	3.63 ± 0.21
SV	2.98 ± 0.42	0.578 ± 0.032	3.38 ± 0.46
Dia	2.71 ± 0.21	0.654 ± 0.072	3.68 ± 0.28
DiaV	2.84 ± 0.10	0.628 ± 0.075	3.50 ± 0.12
T	2.50 ± 0.36	0.679 ± 0.025	4.02 ± 0.54
TV	2.41 ± 0.60	0.622 ± 0.042	4.27 ± 0.93

Table 8.4. Surfactant flooding: breakthrough (BT) time, water saturation at BT, and capillary number Ca based on the BT time [24].

Chip Name	BT Time (10^3 s)	BT Water Saturation	Ca (10^{-7})
U	1.80 ± 0.20	0.908 ± 0.023	53.3 ± 6.0
D	1.72 ± 0.13	0.849 ± 0.014	55.5 ± 4.3
V	1.26 ± 0.15	0.734 ± 0.010	76.4 ± 9.4
H	2.05 ± 0.23	0.927 ± 0.012	46.8 ± 5.4
HV	1.74 ± 0.09	0.752 ± 0.029	54.6 ± 2.8
S	1.72 ± 0.13	0.900 ± 0.028	55.3 ± 4.3
SV	1.48 ± 0.08	0.707 ± 0.049	64.1 ± 3.6
Dia	2.62 ± 0.09	0.860 ± 0.012	36.2 ± 1.2
DiaV	2.52 ± 0.20	0.803 ± 0.034	37.9 ± 3.1
T	2.34 ± 0.01	0.768 ± 0.030	40.5 ± 2.5
TV	2.47 ± 0.61	0.752 ± 0.011	39.8 ± 8.7

tribution in micro- and nanochannels, when the pressure difference through the nanochannel array was 130 psi for NaCl solution (1.5% wt.) with surfactant (0.5% wt. NEODOL 91-8 ethoxylated alcohol (EA)). The light gray in both micro- and nanochannels represent gas phase, whereas the dark gray indicates the liquid phase. Once gas breakthroughs the nanochannels, bubbles form at the interface between micro- and nanochannels immediately flow toward the outlet. As pressure difference between two and microchannels (ΔP) increases, the number of nanochannels filled by nitrogen gas proportionally increase. The size and frequency of gas bubbles also increase with (ΔP). For IPA (isopropyl alcohol 99%), gas-breakthrough in nanochannels was observed at 75 psi of ΔP. For water, there was no displacement up to 145 psi of ΔP.

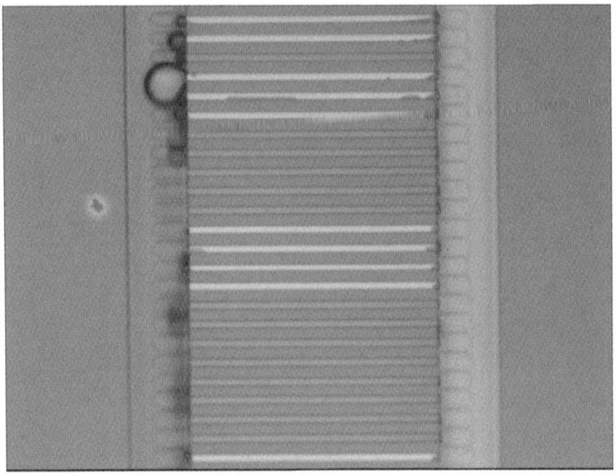

Fig. 8.4. A nitrogen-water drainage experiment conducted in a nanofluidic device. A capillary pressure of about 145 psia is needed for the nitrogen to breakthrough 100 nm nanochannels.

8.5. Summary

Fundamental transport phenomena of the two-phase flow in micro- and nanoscale porous media were studied by using PDMS microfluidic and Si-Pyrex nanofluidic devices. Fabrication methods for both micro- and nanofluidic chips were well established by using conventional fabrication techniques such as optical/soft lithography, reactive ion etching, oxygen plasma glass-PDMS bodning, and anodic pyrex-silicon bonding. In microfluidic experiments, there were significant pore-geometrical effects on both water and surfactant flooding. In these experiments the water phase is the non-wetting phase. Surfactant reduces the interfacial tension, leading to more rapid breakthrough and higher displacement efficiency. The nanofluidic experiment shows that, during hydraulic fracturing operations, it is extremely difficult to recover imbibed hydraulic fracturing fluid into the nanopores.

Acknowledgement

XY and KBN thank the support from Research Partnership to Secure Energy for America 09122-29 and from the American Chemical Society Petroleum Research Fund. Support in the experimental facility from the Oak Ridge National Laboratory (S. T. Retterer) and discussions with B.

Bai and Y. Ma from Missouri University of Science and Technology are gratefully acknowledged.

References

[1] N. K. Karadimitriou and S. M. Hassanizadeh, A review of micromodels and their use in two-phase flow studies, *Vadose Zone J.* **11**(3) (2012).
[2] C. C. Mattax and J. R. Kyte, Ever see a water flood?, *Oil Gas J.* **59**, 115–128 (1961).
[3] W. E. Soll, M. A. Celia, and J. L. Wilson, Micromodel studies of 3-fluid porous-media systems – pore-scale processes relating to capillary-pressure saturation relationships, *Water Resour. Res.* **29**(9), 2963–2974 (1993).
[4] R. Lenormand, Liquids in porous-media, *J. Phys. Condens. Mat.* **2**(SA), 79–88 (1990).
[5] R. Lenormand and C. Zarcone, Invasion percolation in an etched network – measurement of a fractal dimension, *Phys. Rev. Lett.* **54**(20), 2226–2229 (1985).
[6] R. Lenormand and C. Zarcone, Capillary fingering – percolation and fractal dimension, *Transport Porous Med.* **4**(6), 599–612 (1989).
[7] M. Ferer, C. Ji, G. S. Bromhal, J. Cook, G. Ahmadi, and D. H. Smith, Crossover from capillary fingering to viscous fingering for immiscible unstable flow: Experiments and modeling, *Phys. Rev. E.* **70**(1), 016303 (2004).
[8] M. I. J. Van Dijke, K. S. Sorbie, A. Sohrabi, and A. Danesh, Simulation of wag floods in an oil-wet micromodel using a 2-d pore-scale network model, *J. Pet. Sci. Eng.* **52**(1–4), 71–85 (2006).
[9] Y. Corapcioglu, S. Chowdhury, and S. E. Roosevelt, Micromodel visualization and quantification of solute transport in porous media, *Water Resour. Res.* **33**(11), 2547–2558 (1997).
[10] J. Wan and J. L. Wilson, Colloid transport in unsaturated porous-media, *Water Resour. Res.* **30**(4), 11–23 (1994).
[11] M. Auset and C. J. Werth, Pore-scale processes that control dispersion of colloids in saturated porous media, *Water Resour. Res.* **40**(3), W03503 (2004).
[12] J. Wan, J. L. Wilson, and T. L. Kieft, Influence of the gas-water interface on transport of microorganisms through unsaturated porous-media, *Appl. Environ. Microbiol.* **60**(2), 509–516 (1994).
[13] A. Bazylak, V. Berejnov, B. Markicevic, D. Sinton, and N. Djilali, Numerical and microfluidic pore networks: Towards designs for directed water transport in gdls, *Electrochim. Acta.* **53**(26), 7630–7637 (2008).
[14] C. Tsakiroglou and D. Avraam, Fabrication of a new class of porous media models for visualization studies of multiphase flow processes, *J. Mater. Sci.* **37**(2), 353–363 (2002).
[15] C. Chomsurin and C. J. Werth, Analysis of pore-scale nonaqueous phase liquid dissolution in etched silicon pore networks, *Water Resour. Res.* **39** (9), 1265 (2003).
[16] M. Grumann, M. Dobmeier, P. Schippers, T. Brenner, C. Kuhn, M. Fritsche,

R. Zengerle, and J. Ducree, Aggregation of bead-monolayers in flat microfluidic chambers simulation by the model of porous media, *Lab Chip.* **4**(3), 209–213 (2004).

[17] F. Javadpour and D. Fisher, Nanotechnology based micromodels and new image analysis to study transport in porous media, *J. Can. Pet. Technol.* **47**(2), 30–37 (2008).

[18] V. J. Niasar, S. M. Hassanizadeh, I. J. Pyrak-Nolte, and C. Berentsen, Simulation drainage and imbibition experiments in a high-porosity micromodel using an unstructured pore network model, *Water Resour. Res.* **45**, W02430 (2009).

[19] R. B. Schoch, J. Han, and P. Renaud, Transport phenomena in nanofluidics, *Rev. Mod. Phys.* **80**(3), 839–883 (2008).

[20] M. Wu, F. Xiao, R. M. Johnson-Paben, S. T. Retterer, X. Yin, and K. B. Neeves, Single- and two-phase flow in microfluidic porous media analogs based on Voronoi tessellation, *Lab Chip.* **12**(2), 253–261 (2012).

[21] J. C. McDonald, D. C. Duffy, J. R. Anderson, D. T. Chiu, H. Wu, O. J. Schueller, and G. M. Whitesides, Fabrication of microfluidic systems in poly(dimethylsiloxane), *Electrophoresis.* **21**(1), 27–40 (2000).

[22] P. Mao and J. Han, Fabrication and characterization of 20 nm planar nanofluidic channels by glass-glass and glass-silicon bonding, *Lab Chip.* **5** (8), 837–844 (2005).

[23] Q. Wu, J. T. Ok, Y. Sun, S. T. Retterer, K. B. Neeves, X. Yin, B. Bai, and Y. Ma, Optic imaging of single and two-phase pressure-driven flows in nano-scale channels, *Lab Chip.* **13**(6), 1165–1171 (2013).

[24] W. Xu. Effect of pore geometry on water and surfactant flooding by pdms microfluidic micromodels. Master's thesis, Colorado School of Mines, Golden, Colorado, USA (2012).

Chapter 9

Direct Numerical Simulation of Pore-Scale Fluid Flow and Transport

Michael S. Newman, Feng Xiao and Xiaolong Yin

Petroleum Engineering Department,
Colorado School of Mines,
Golden, Colorado 80401, USA
xyin@mines.edu

Predictive modeling of oil and gas production requires fundamental understanding of fluid flow and transport through porous media. Pore-scale direct numerical simulation is an important tool to visualize flow and transport, to identify and isolate the mechanisms, and to upscale the pore-scale processes to the REV (representative element volume) level. In this article, we present pore-scale direct numerical simulation methods that are being developed and used to study pore-scale flow and transport. Following an introduction to the lattice Boltzmann method and its variants to simulate slip flow, multiphase flow, and transport, we will present applications in studying Forchheimer flows through porous media and modeling of porosity and permeability reduction during stress compaction.

9.1. Introduction

In this paper, we describe pore-scale simulation methods and their applications in the fundamental studies of Forchheimer flows and stress-induced porosity and permeability change. Both applications are especially relevant to the development of unconventional oil and natural gas resources. In 2010, shale and tight gas sand combined offered 49% of the total US natural gas production [1]. These resources are termed "unconventional" due to their very low permeability ($10^{-16}m^2$ to $10^{-20}m^2$). Production is enabled by horizontal wells and hydraulic fracturing. Near the well bore, produced gas flows through propped fractures at high speeds, generating strong inertial effects that lead to nonlinear, non-Darcy relation between the pressure drop and the flow rate and a significant increase in the flow resistance. The permeability of fractures is stress-sensitive. Pore pressure reduction during production can lead to closure of the fractures and loss of productivity. It is important to model these phenomena on the pore scale to provide physical insights and useful correlations.

We first present pore-scale simulations of Forchheimer flows in two-dimensional porous media. The fluid flow was solved using a lattice-Boltzmann method (LBM). We focus on the effect of pore geometry on the point-of-transition from the Darcy / Stokes flow regime to the Forchheimer / inertial flow regime and on the inertial resistance factor β that describes the inertial correction to the Darcys law

$$-\nabla p = \frac{\mu}{k} u + \beta \rho u^2 . \tag{9.1}$$

Although β is generally regarded as a property of porous medium, the effect of pore structure on this inertial correction factor has not been characterized.

Pore geometry also affects the stress-sensitivity of porosity and permeability. It is well known that both porosity ϕ and permeability k change with stress [2–5]. The relation between the change in porosity and the change in permeability due to stress can be described by a power-law relation

$$\left(\frac{\phi}{\phi_0}\right)^A = \frac{k}{k_0} , \tag{9.2}$$

where ϕ_0 and k_0 represent the porosity and permeability measured at a reference stress state and exponent A is a parameter that varies with rock type and sometimes the range and history of the applied stress. The values of A in the literature vary significantly [6, 7]. Differentiation of the Carman-Kozeny equation, a common and well known porosity-permeability relation developed from homogeneous packed beds, A should be close to three [7, 8]. Real porous media are not homogeneous packed bed of spheres. High values of A observed in the experiments have been attributed to a division of the net connected porosity into "effective" and "non-effective" portions. When the "effective" porosity that controls the permeability (pore throats) is much more stress-sensitive than the "non-effective" porosity (pore bodies), A is expected to be higher than three [8]. Using a discrete-element method (DEM), we will mechanistically simulate the uniaxial compression of a heterogeneous three-dimensional porous medium. The change in the structure of porous medium will be passed to the LBM flow solver to quantify the exponent A.

9.2. Methodology

9.2.1. *Geometry Models for Pore-scale Simulation*

A porous medium is a two-phase disordered material in which one phase is solid and another phase is connected void; distribution of the solid and the void is critical for flow and transport. To the first order, the pore structure (phase distribution) is characterized by the void fraction, namely the porosity. However, as many studies have shown, porosity alone is not sufficient. Early works [9–11] introduced specific surface area, tortuosity and pore size or grain size as additional parameters. They are still widely used today because they can all be measured from experiments. Many additional statistical measures have been developed, such as correlation functions, local porosity distribution, and Minkowski functionals [12–15]. How to relate these parameters to the flow and transport properties of porous media, however, is a question that has not been addressed. There is a need to identify the essential geometric parameters of geological porous medium that have the most influence on the various flow and transport processes. As geometries made up by spheres, tubes, or polyhedrons are easy to create and modify, we are motivated to use stochastic combinations of these simple shapes to approach this complicated problem.

In this work, Voronoi tessellation [16] is used to create stochastic models of porous media using polyhedrons or tubes. Starting from a set of random points (Poisson points), Voronoi tessellation subdivides space into convex polygons (2D) or polyhedrons (3D). 2D porous media models, as shown in Fig. 1, do not reflect the 3D connectivity of real porous media; however, they can be used as computationally affordable alternatives to 3D geometry models to study the qualitative effect of pore geometry [17]. In addition, 2D textures can be directly molded into Polydimethylsiloxane (PDMS) and bounded to glass to build high-precision microfluidic / nanofluidic porous media analogs [18]. 3D porous media geometry models are constructed similar to the 2D models. As exemplified in Fig. 9.1, 3D models follow two basic conformations: granular and tubular. The pore space in the granular models is made up by interconnected fractures; the pore space in the tubular models, on the other hand, is made up by a random network of tubes. Due to the analytical nature of the geometry models, anisotropy, pore size distribution, and heterogeneities in the forms of fractures and vugs can be easily implemented. With this method, the geometric features of porous media can be precisely characterized and controlled.

Fig. 9.1. Upper row: 2D Voronoi diagrams generated from Poisson points and the formation of the pore network. Lower row: 3D porous media models. From left to right: a homogeneous fracture network, a porous medium with vugs, a network of capillaries, a dual-porosity system with a large fracture and a matrix.

9.2.2. *Lattice-Boltzmann Method*

We use a lattice-Boltzmann method to simulate single-phase flow in the pore space to determine the permeability k in the Darcy / Stokes regime and the apparent permeability k_{app} in the Forchheimer / inertial regime. Different from traditional computational methods such as finite difference, finite element, or finite volume, the lattice Boltzmann method does not directly solve the nonlinear Navier-Stokes equation. Instead, it solves a simplified kinetic model for the fluid molecular velocity distribution, the moments of which recover the Navier-Stokes equations. It handles no-slip boundaries using a simple bounce-back scheme, which is especially useful for porous media. The method has been successfully applied to a large number of fluid-dynamics problems, including single-phase flow, multiphase / multicomponent flow, reactive flow, particulate suspension, compressible flow, turbulent flow, thermal flow, viscoelastic flow and micro flow [19, 20].

The lattice-Boltzmann method consists of two processes: advection and collision. In the advection step, the fluid molecular velocity distributions, defined on a square (2D) or cubic (3D) lattice, are moved to the nearest neighbors. In the collision step, the fluid molecular velocity distributions at each lattice point are relaxed toward local equilibrium to simulate the viscous relaxation of stresses. In this study, we use the propagation models of D2Q9 and D3Q19. The collision model is the multiple-relaxation-time (MRT) scheme developed in [21, 22]. When propagation moves a fluid distribution across a fluid-solid interface, a simple bounce-back scheme is applied to recover the no-slip boundary condition [23]. The bounce-back condition can be modified to simulate slip flows on gas-liquid interfaces and in microchannels / nanopores [24]. When multiple components and their interactions are implemented, the method can be used to simulate phase separation and multiphase flows in porous media [25, 26]. Parallel computing is critical for simulating 3D porous media flows of meaningful sizes. A parallel code has been developed using MPI (Message Passing Interface) for distributed-memory clusters. As shown in Fig. 9.2, the parallelized simulator shows very good speedup and efficiency up to 128 nodes (1024 cores) which is the maximum number of nodes tested.

9.2.3. *Discrete-element Modeling of Porous Media Compaction*

In a recent experimental study [7], we conducted porosity-permeability measurements under isostatic (hydrostatic) stress condition and characterized exponent A in sandstones, carbonates, and naturally and artificially fractured cores. Most of the carbonate data and all of the fractured rock data showed $A \gg 3$. In order to verify the experimental data and to achieve a

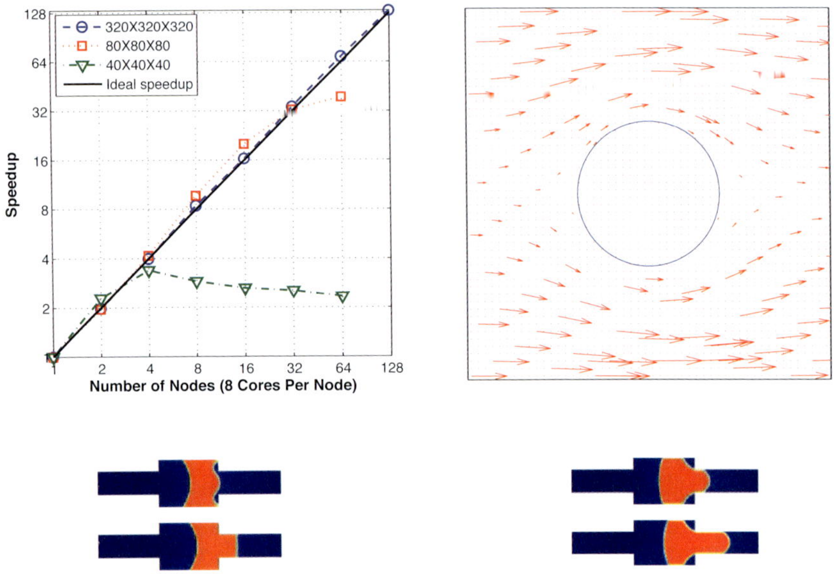

Fig. 9.2. Parallel performance and applications of the lattice Boltzmann method. First row from left to right: (1) Parallel speedup of the lattice-Boltzmann simulator; (2) Simulation of slip flow through a periodic array of cylinders. Second row from left to right: (3) Wetting fluid in a pore body entering a pore throat; (4) Non-wetting fluid in a pore body entering a pore throat. In (1), the speedup is reduced as number of nodes (1 node = 8 cores) becomes too large. This is due to the increasing load of inter-core communication.

fundamental understanding of the effect of pore structure on the pressure sensitivity of permeability, we conduct a mechanistic pore-scale simulation of a uniaxial compaction using DEM. DEM with bonded particle models has been used to simulate the mechanical deformation of granular solids [27, 28]. In this method, the material is represented by a set of interacting particles connected by bonds. An elasticperfectly brittle model [29] is used, for which the bonded particles interact linearly elastically characterized by the normal and tangent stiffness constants given in Table 9.1. The tangential force is also constrained by the Coulomb friction.

A heterogeneous porous media model was constructed by filling spherical particles into a cubic domain of 9 cm × 9 cm × 9 cm using a dense packing algorithm [30]. The particle size obeys a Gaussian distribution: the smallest particle has a diameter of 0.44 mm; the largest particle has a diameter of 2.88 mm; the number-averaged particle size is 1.64 mm and the

Table 9.1. Particle-particle interaction parameters for DEM an porous medium bulk properties

Particle normal stiffness (MN/m)	60.37	Young's modulus (GPa)	18.7
Particle tangential stiffness (MN/m)	11.47	Poisson's ratio	0.21
Particle friction coefficient	0.5	Compressive strength (GPa)	1.27
Bond normal strength (kN)	2.81	Tensile strength (GPa)	0.12
Bond tangential strength (kN)	1.12		

Sauter mean diameter of the particles is 1.77 mm. The bulk mechanical properties of the generated media are also listed in Table 1. The particle-filled cubic domain has a net porosity of 0.246. To simulate a vuggy heterogeneity, a cylindrical pore (diameter = 3 cm) that passes through the center of the xy plane was generated, see Fig. 9.3. This increases the net porosity to 0.322; the porosity in the cylinder is therefore 0.076. For flows in the x and y directions, the cylindrical cavity acts as an isolated vug.

9.3. Simulation Results

In this section, the simulation results on the Forchheimer flows and on the permeability compaction are presented. The Forchheimer flow results are summarized into a formula for the inertial resistance factor. The simulations of permeability compactions showed a power-law type of correlation between porosity and permeability during compaction. The power-law exponent is clearly structure-dependent.

9.3.1. *Forchheimer Flows*

To investigate the impact of pore structure on inertial Forchheimer flows, eleven sets of geometries were created. Each set contains five stochastic geometries generated with the same input parameters but slightly different porosities due to statistical fluctuations. These geometries covered porosities from 8.8% to 30.3%, and contain both homogeneous structures and heterogeneous, vuggy structures. For each structure, approximately twenty simulations were conducted with different driving pressure gradients to achieve different flow rates and reach different flow regimes.

The transition from Darcy to Forchheimer flow is analyzed using dimensionless plots $k^* = k_{app}/k$ versus the Reynolds number $\mathrm{Re}_k = u\sqrt{k}/\nu$ (Fig. 9.4). We observe that transitions from homogeneous pore geometries that are completely channelized generally take place when Re_k is between

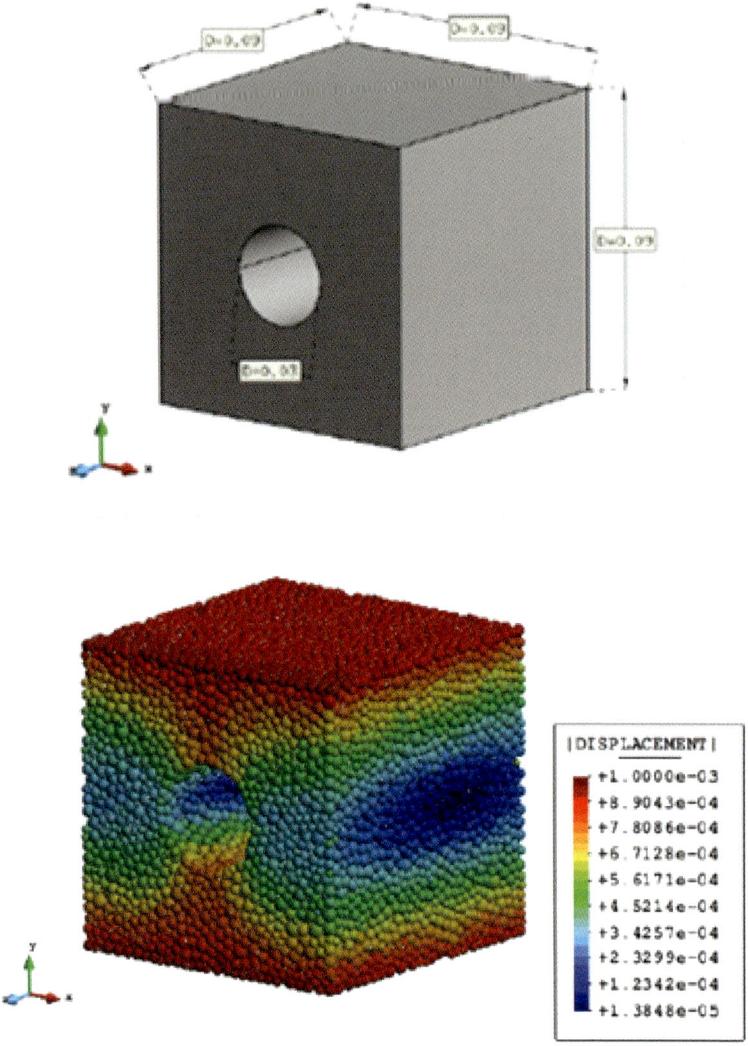

Fig. 9.3. Construction of a heterogeneous porous medium in DEM and displacement under a uniaxial compression.

0.3 and 0.4. For heterogeneous vuggy geometries, transitions occur when Re_k is between 0.1 and 0.2. Our explanation for the difference in the point of transition between the two types of geometries is as follows: The con-

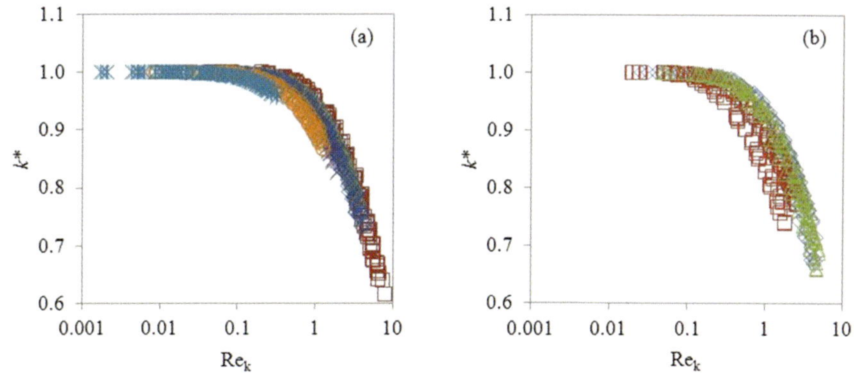

Fig. 9.4. k^* plotted against Re_k showing the transition from the Darcy / Stokes to Forchheimer / inertial regime. Left: Homogeneous geometries; Right: Heterogeneous geometries with vugs.

trast between pore body and pore throat triggers a rapid change in the fluid velocity that amplifies the inertial effects. As fluid velocity increases, this contrast causes an early transition from viscous flows in the Darcy / Stokes regime to inertia-dominated flows in the Forchheimer regime.

Figure 9.5 shows the dimensionless inertial resistance factor $\beta\sqrt{k}$ as a function of porosity. It can be observed that $\beta\sqrt{k}$ decreases with increasing porosity which agrees with many experimental results and correlations. $\beta\sqrt{k}$ obtained from homogeneous geometries (blue diamonds) have a very low scatter, indicating that they can be well correlated as a function of porosity. $\beta\sqrt{k}$ obtained from heterogeneous geometries (red squares and green triangles), however, are significantly higher. These data show that a single porosity is not adequate to correlate the β factor. In order to design a "universal" correlation that can be applied to both homogeneous and heterogeneous geometries, one must incorporate multiple length scales of porous medium into the correlation. After combining the permeability k and the specific surface area s, we obtained a general correlation

$$\beta\sqrt{k} = 0.00993\frac{\phi}{s\sqrt{k}\,(1-\phi)}, \tag{9.3}$$

This equation fits the numerical data with an average deviation of 16% and the maximum deviation is 47%.

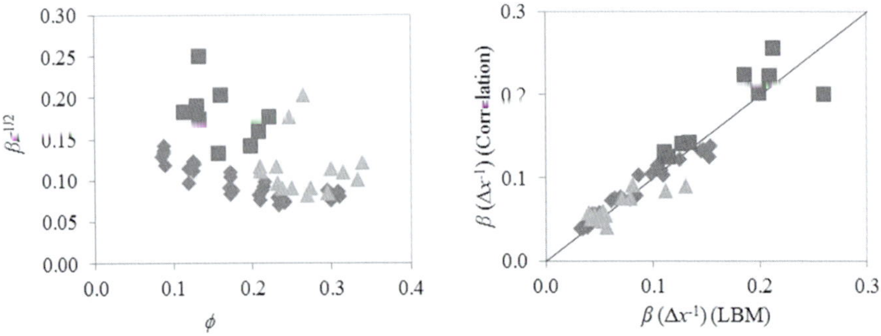

Fig. 9.5. Dimensionless inertial resistance factor $\beta\sqrt{k}$ plotted as a function of porosity (left) and parity plot showing the quality of the developed correlation for $\beta\sqrt{k}$.

Porous media flows in the Darcy / Stokes regime are microscopically very different from those in Forchheimer / inertial flow regimes. As shown in Fig. 9.6, flows in the Stokes / Darcy regime are fully reversible. The streamlines spread out evenly in the entire pore space. With increasing Re_k, inertia of the fluid becomes important and then becomes dominant, leading to circulations in the pore and high speed streams that are often referred to as the "inertial cores". The viscous dissipation in the Forchheimer flow regime becomes much more localized [17].

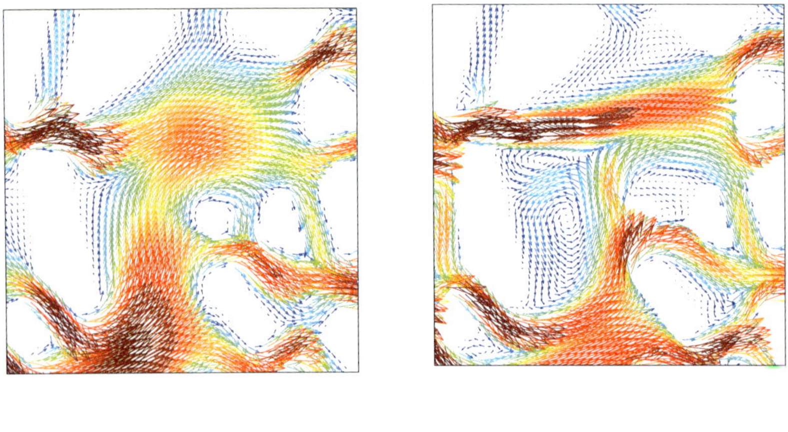

Fig. 9.6. Fluid streamwise velocity distributions in a heterogeneous porous medium. Left: Darcy / Stokes flow regime; Right: Forchheimer / inertial flow regime.

9.3.2. *Porosity / Permeability Compaction*

The heterogeneous porous media model made up by spherical particles, shown in Fig. 9.3, was compressed in the y direction till the strain reached 2%. The change in permeability was characterized using the parallelized 3D lattice-Boltzmann flow simulator presented earlier. The cubic domain was discretized into 1000^3 voxels; the size of an individual voxel is 90 μm. The voxels were marked as fluid or solid using sphere radii and center locations from DEM. The digitalized porous media geometry has a porosity of 0.329.

Because stress-induced deformations are generally very small (approximately 0.1% strain), it is not trivial to capture the permeability change using the LBM. When the porous media is digitalized by 1000^3 voxels, a strain of 0.1% (common for elastic deformations) only changes the voxel map by 1 voxel, which means that the change in the geometry is not adequately resolved. Even though our LB program is parallelized, it is not possible to further increase the domain size by an order of magnitude. As such, we chose the DEM parameters such that the compression yielded a strain of 2%. With this setting, the dimension of the computational domain in the y direction was reduced by 20 voxels, leading to a much improved resolution of the change in the porous media geometry. Although this setting makes our porous medium model more compliant than real rocks, it allows a qualitative assessment of the effect of pore geometry on the geomechanical response. Simulations were conducted using 64 parallel cores, with an average run time of about 72 hours.

The porosity before and after compaction is 0.3290 and 0.3173, respectively; the permeability in the x direction before and after compaction is $3.77 \times 10^{-9} m^2$ and $2.88 \times 10^{-9} m^2$, respectively; the permeability in the y direction before and after compaction is $3.77 \times 10^{-9} m^2$ and $2.98 \times 10^{-9} m^2$, respectively. When k_x/k_{x0} is used in (9.2), the power-law exponent $A = 7.4$; when k_y/k_{y0} is used, $A = 6.5$. These numbers show that the isolated vug indeed increases the power-law exponent, which indicates higher stress-sensitivity of the permeability and is in good agreement with the experimental data published in Ref. [7]. It should be noted, however, that the current numerical model is not a *quantitative comparison* to the experiments. The combination of DEM and lattice Boltzmann, however, is a promising approach. Future studies that employ more realistic stress conditions and geometries should provide more quantitative connections between pore structure and the geomechanical response of porosity and permeability.

9.4. Summary

By conducting Forchheimer flow simulations in stochastically constructed 2D geometries, we show that the pore structure has important effects on the flow resistance. The inertial resistance of a heterogeneous geometry is up to four times higher than that of a homogeneous geometry of the same porosity. Using data obtained from a large number of simulations, a correlation that honors the presence of multiple length scales in heterogeneous porous media was developed. It offers a unified description of the inertial resistance factor for both homogeneous and heterogeneous geometries.

Using a combination of discrete-element modeling and lattice-Boltzmann simulations, we modeled the stress-sensitivity of permeability of a three-dimensional heterogeneous porous medium. The heterogeneous structure of the medium increased the stress-sensitivity which is in good agreement with experimental data.

Acknowledgements

This research is supported by Research Partnership to Secure Energy for America 09122-29. XY acknowledges Carlos Labra and Jin Sun from University of Edinburgh for providing DEM-generated particle assemblies. We thank the Golden Energy Computing Organization at the Colorado School of Mines for the computational resources.

References

[1] U. S. E. I. Administration. Review of emerging resources: US shale gas and shale oil plays. Technical report, US Department of Energy (July, 2011).
[2] M. A. Biot, General theory of three-dimensional consolidation, *J. Appl. Mech.* **24**, 594–601 (1957).
[3] J. Geertsma, The effect of fluid pressure decline on volumetric changes of porous rocks, *Trans. AIME.* **210**(2), 331–340 (1957).
[4] R. F. Sigal, The pressure dependence of permeability, *J. Petrophys.* **43**(2), 92–102 (2002).
[5] The effect of stress-dependent permeability on gas production and well testing, *SPE Formation Eval.* **1**(3), 227–235 (1986).
[6] Y. Bernabé, Pore geometry and pressure-dependence of the transport-properties in sandstones, *Geophysics.* **56**(4), 436–446 (1991).
[7] V. V. Petunin, X. Yin, and A. N. Tutuncu. Porosity and permeability changes in sandstones and carbonates under stress and their correlation to rock texture. In *The Canadian Unconventional Resources Conference*, pp. CSUG/SPE–147401, Calgary, Alberta, Canada (Nov, 2011).

[8] Y. Bernabé, U. Mok, and J. B. Evans, Permeability-porosity relationships in rocks subjected to various evolution processes, *Pure Appl. Geophys.* **160** (5–6), 937–960 (2003).
[9] J. Bear, *Dynamics of Fluids in Porous Media.* Elsevier (1972).
[10] P. M. Adler, *Porous Media: Geometry and Transports.* Butterworth-Heinemann (1992).
[11] F. Dullien, *Porous Media — Fluid Transport and Pore Structure.* Academic Press (1992).
[12] D. Stoyan, W. S. Kendall, and J. Mecke, *Stochastic Geometry and Its Applications*, 2nd edn. John Wiley & Sons (1995).
[13] B. Biswal, C. Manwart, and R. Hilfer, Three-dimensional local porosity analysis of porous media, *Physica A.* **255**(3–4), 221–241 (1998).
[14] C. H. Arns, M. A. Knackstedt, and K. Mecke, *Characterising the Morphology of Disordered Materials*, In eds. K. Mecke and D. Stoyan, *Morphology of Condensed Matter: Physics and Geometry of Spatially Complex Systems*, vol. 600, *Lecture Notes in Physics*, pp. 37–74. Springer-Verlag Berlin (2002).
[15] S. Torquato, Statistical description of microstructures, *Annu. Rev. Mat. Res.* **32**, 77–111 (2002).
[16] G. F. Voronoi, Nouvelles applications des paramètres continus à la théorie des formes quadratiques, *J. reine angew. Math.* **133**, 97–178 (1907).
[17] M. S. Newman and X. Yin, Lattice Boltzmann simulation of non-Darcy flow in stochastically generated 2D porous media geometries, *SPE J.* **18**(1), 12–26 (2013).
[18] M. Wu, F. Xiao, R. M. Johnson-Paben, S. T. Retterer, X. Yin, and K. B. Neeves, Single- and two-phase flow in microfluidic porous media analogs based on Voronoi tessellation, *Lab Chip.* **12**(2), 253–261 (2012).
[19] S. Chen and G. D. Doolen, Lattice Boltzmann method for fluid flows, *Annu. Rev. Fluid Mech.* **30**, 329–364 (1998).
[20] C. K. Aidun and J. R. Clausen, Lattice-Boltzmann method for complex flows, *Annu. Rev. Fluid Mech.* **42**, 439–472 (2010).
[21] P. Lallemand and L. S. Luo, Theory of the lattice Boltzmann method: Dispersion, dissipation, isotropy, Galilean invariance, and stability, *Phys. Rev. E.* **61**(6), 6546–6562 (2000).
[22] D. d'Humières, I. Ginzburg, M. Krafczyk, P. Lallemand, and L.-S. Luo, Multiple-relaxation-time lattice Boltzmann models in three dimensions, *Phil. Trans. R. Soc. Lond. A.* **360**(1792), 437–451 (2002).
[23] U. Frisch, D. d'Humi'eres, B. Hasslacher, P. Lallemand, and P. Yves, Lattice gas hydrodynamics in two and three dimensions, *Complex Sys.* **1**, 649–707 (1987).
[24] X. Yin, D. L. Koch, and R. Verberg, Lattice-Boltzmann method for simulating spherical bubbles with no tangential stress boundary conditions, *Phys. Rev. E.* **73**(2), 026301 (2006).
[25] X. Shan and H. Chen, Lattice Boltzmann model for simulating flows with multiple phases and components, *Phys. Rev. E.* **47**(3), 1815–1819 (1993).
[26] P. Yuan and L. Schaefer, Equations of state in a lattice Boltzmann model, *Phys. Fluids.* **18**(4), 042301 (2006)

[27] D. O. Potyondy and P. A. Cundall, A bonded-particle model for rock, *Int. J. Rock Mech. Min.* **41**(8), 1329–1364 (2004).
[28] C. Labra. *Advances in the Development of the Discrete Element Method for Excavation Processes*, PhD thesis, Universitat Politecnica de Catalunya, Barcelona, Spain (Jun, 2012).
[29] J. Rojek, C. Labra, O. Su, and E. Oñate, Comparative study of different discrete element models and evaluation of equivalent micromechanical parameters, *Int. J. Solids Struct.* **49**(13), 1497–1517 (2012).
[30] C. Labra and E. Oñate, High-density sphere packing for discrete element method simulations, *Comm. Numer. Meth. En.* **25**(7), 837–849 (2009).

PART 2

Hydrogeology

Chapter 10

From Pore to the Field: Upscaling Challenges and Opportunities in Hydrogeological and Land–Atmospheric Systems

Tissa H. Illangasekare and Kathleen M. Smits

CEE, Colorado School of Mines, Golden, CO, USA

tissa@mines.edu, ksmits@mines.edu

Radek Fučík

FNSPE, Czech Technical University in Prague, Prague, Czech Republic

Hossein Davarzani

BRGM, Direction Eau, Environnement et Ecotechnologies, Orléans, France

Numerical simulation of hydrogeologic systems requires parameters characterizing fundamental physical, chemical and biological processes as model inputs. These processes at a very fundamental level occur within the inter-granular pore spaces of the porous medium. In the traditional applications in hydrogeology, the smallest scale where continuum formulations are developed is the scale of representative elemental volume (REV). The fundamental pore–scale processes that occur in a discontinuous space consisting of grains and pores do not get captured accurately at the imposed REV scale. Some of the emerging problems related to climate change and energy development require us to understand and characterize basic processes and their interactions taking into account physical and chemical heterogeneities and dynamics of mass transfer at fluid interfaces within pores. Up–scaling research involves investigation, development and validation of methods to relate the parameters that characterize these processes at the pore and macro-scale to the grid–scale in field scale simulators. Emerging problems involving multiphase flow where up–scaling issues are of central importance is presented. An example of how mass transfer that occurs at two fluid interfaces at the pore-scale is characterized and up-scaled to field systems and how multi–scale physical modeling approaches can be used for validation is presented. Results demonstrate how pore–scale physics combined with geologic parameters of field systems can be used to obtain effective field scale.

10.1. Introduction

One of the major challenges in hydrogeological sciences and reservoir engineering is upscaling when field scale behavior, for example, needs to be simulated using parameters obtained from laboratory scale cores or a limited number of in-situ field measurements. These parameters are often included in numerical models with significantly larger grid sizes. However, the nonlinearity of the hydrological processes and/or the variability across measurement scales in the soil properties themselves lead to model prediction error when parameters determined at much smaller scales are used in larger models grid blocks. Knowledge gaps exist in the basic theoretical approaches needed for upscaling as well as selecting the best approach suited for the type of application.

Hydraulic, geo–bio–chemical and thermal parameters of soils are usually measured at the small scale and it has been long recognized (e.g., [1]) that the natural variability of those parameters at the field scale is vast. The question remains how information obtained from small scale measurements can be used to predict large scale flow and transport behavior. The issue of scale–transfer or moving across scales in the subsurface is very relevant to many conventional hydrogeologic and geo–environmental problems as well as emerging problems related to climate change and unconventional energy development. These include the behavior of subsurface chemical plumes, evapotranspiration, subsurface storage of CO_2 and potential leakage, methane gas emissions from subsurface sources and loading to the atmosphere, to name a few. All these problems, parameters characterizing liquid and gas flow at the pore–scale must be transferred through upscaling for field–scale modeling to be used in design, prediction and decision–making. Even though many advances have been made in both the hydrological sciences and energy/reservoir engineering, many challenges still remain in developing and implementation of such upscaling methods. Specifically, knowledge gaps exist in upscaling of multiphase fluid flow and processes that couple the subsurface to the atmosphere.

A primary challenge of upscaling comes from factoring in the complexities of naturally present heterogeneity that is manifested at all length scales from the pore to the field. This chapter discusses conceptual issues involved in upscaling associated with the above mentioned problems and presents an example of how mass transfer that occurs at water and non–aqueous phase liquid interfaces at the pore–scale is characterized and upscaled to field systems. Results demonstrate how pore–scale physics combined with geologic parameters of field systems can be used to obtain effective field scale parameterizations.

A second example on soil water evaporation and gas flow is used to highlight and discuss some challenges in upscaling processes controlled by land/atmospheric interaction.

This section includes four other companion chapters focusing on a number of aspects of hydrogeological processes that are of central importance in geosciences and engineering. A summary of these chapters is provided here.

Chapter 11. *Pore-Scale Chemical Reactions in Diffusion-Limited Environments at the Pore Scale (D. Benson).* This chapter addresses an issue related to mixing at the pore scale and its contribution to chemical reactions. This is of importance in many problems in the subsurface that involve fluid flow and chemical reactions. Two examples that are of current interest are groundwater contamination and mineralization of dissolved CO_2 in geologic formations. The classical differential equations of chemical reactions assume perfect mixing among reactants. This chapter presents a method to rectify the problem associated with the well mixed assumption in both theoretical and applied settings.

Chapter 12. *Porosity in reactive geochemical systems (A. Navarre-Sitchler, G. Rother, & J. Kaszuba).* This chapter addresses the fundamental issue of reactive fluids in pore spaces; reactive fluids can create a dynamic system that affects the network of pores and hence the flow through the system. The author makes the argument that many fundamental geologic processes, the physical characteristics of pore networks in rocks in geologic materials are not well understood due to their dynamic nature and features at different length scales. Rock weathering and saprolite formation and geochemical reactions within CO_2 geologic sequestration are used as two examples to demonstrate how geochemical reactions affect porosity. Both of these examples have important implications for Earth Science.

Chapter 13. *Quantifying the heterogeneity of hydrologic properties of rocks in core floods (R. Pini & S. Benson).* This chapter discusses the effect of sub-core scale heterogeneities on multiphase flow by considering both experimental and modeling studies. Novel experimental methods that allow non-destructive measurements of core- and sub–core–scale hydrogeological properties during conventional core–flooding tests are presented. These methods allow for the measurement of porosity, permeability and capillary pressure vs saturation relationships needed in multiphase flow modeling. This approach represents a significant improvement from conventional core analysis to investigate complex hydrogeological flows in the subsurface.

Chapter 14. *Monte Carlo Simulation of Distribution of Multiphase Capillarity in a Porous Medium (B. Zeidman, N. Lu, & D. Wu).* This chapter presents an approach to determine the spatial distribution of various fluid components such as air, water, oil, or gas hydrate in multiphase porous media systems. This method based on Coarse–Grained Monte Carlo simulation is well–suited to handle complicated pore space geometries, exploiting the advances in imaging technologies such as X-ray micro CT for obtaining high–resolution maps of pore space.

Chapter 15. *Coupled Thermo–hydrogeologic Processes in Enhanced Geothermal Systems (Y.-S. Wu, Y. Xiong, & H. Kazemi).* Heat extraction from fractured geothermal systems is subjected to complex interactions of high temperature, multiphase flow, rock deformation and chemical reactions. This chapter presents the formulation of a conceptual model of these thermal–hydrological–mechanical–chemical interaction. A numerical model based on this conceptualization is used to conduct an example simulation of a prototype enhanced geothermal reservoir.

10.2. Issues of Upscaling in Emerging Problems

Climate change and unconventional energy development have highlighted the importance of a number of porous media science and technology issues with upscaling implications. In this section, climate and unconventional energy development related problems are briefly introduced to identify the fundamental issues related to pore–scale processes and upscaling for field applications.

10.2.1. *Carbon Capture and Storage*

To address the technical challenge of reducing greenhouse gas loading to the atmosphere from fossil fuel combustion, carbon capture and storage have a large potential. The storage process involves the injection of captured CO_2 into deep subsurface formations such as depleted hydrocarbon reservoirs and deep saline aquifers. The goal is to trap the CO_2 gas in supercritical liquid form ($ScrCO_2$) in the pores of the formation that eventually dissolves into the formation water and mineralizes, resulting in stable and long–term sequestration.

During injection, the $ScrCO_2$ that is lighter than the formation water migrates laterally below low permeability cap rock. Based on our understanding of the behavior of non–aqueous phase liquid (NAPL) in subsurface formations, it can be conceptualized that in heterogeneous systems, the ScrCO_2 will preferentially migrate into higher permeability zones and pool under the interface of the confining low permeability layers due to capillary barrier effects (very high entry pressure needed for the non-wetting fluid to displace the wetting fluid) [2–4]. The formation brine with dissolved CO_2 is heavier and will tend to migrate downwards resulting in unstable finger development contributing to convective mixing [5]. This dissolution process in the long–term is expected to contribute to permanent trapping as a fraction of the dissolved CO_2 mineralizes.

A related problem associated with geologic CO_2 sequestration where gas dissolution is of interest is when the CO_2 leaks from the deep formation either through caprock fractures, faults or defective casings in abandoned wells. If the liquid ScrCO_2 leaks, it will be carried upwards in the overlaying formation until it encounters another confining layer. Again, the possibility exists for the liquid to encounter other leakage points and flow to shallow formations where the ScrCO_2 passes its critical point and becomes a gas or dissolves into the formation water due to its high solubility. It has been shown in laboratory studies that when the water with dissolved CO_2 encounters texture transitions in the overlaying formations, the gas exsolves and accumulates below low permeability layers in the shallow subsurface [6]. Gas bubbles tend to form due to accumulation of gas molecules in the pores of the rock or rough surfaces of the solid grains where nucleation of the bubbles mostly occurs [7]. Nucleation rate is a function of solubility and cavity size distribution that controls the sizes of forming bubbles. Once the gas phase is formed, it migrates upward due to buoyancy and expands due to decreasing pressure until it encounters an impermeable layer. Accumulated gas can then migrate laterally until it finds a preferential pathway and continues to migrate upward reaching the shallow subsurface, thus potentially affecting large areas of the aquifer and eventually the land surface. When fresh water passes through these gas-entrapped zones, mass transfer occurs across the gas/water interfaces contributing to CO_2 gas dissolution. The combined effects of gas trapping and dissolution help to attenuate the leaking CO_2 thus potentially reducing atmospheric loading. All processes that are relevant to trapping, dissolution and gas migration occur at the pore–scale and assessment of trapping efficiency and leakage risks in the field require upscaling.

10.2.2. *Methane Gas Leakage*

In unconventional energy development, methane gas and light oils stored in the pores in shale are extracted through hydraulic fracturing. Gas extraction has the potential to create a secondary problem of leakage, resulting in gas migration through the overlaying geologic media. Leakage can occur through gas production sites, distribution pipelines, or natural fractures in the geologic formation above the shale layer. The primary motivation for understanding fugitive emissions is increased CH_4 gas loading that contributes to global warming with significant climate change implications. The global warming potential of CH_4 is 72 times greater than CO_2 for a 20–year time period and 25 times greater than CO_2 over a 100–year period [8]. Multiple field studies are assessing the local impacts of methane leakage associated with hydraulic fracturing technology used for gas production through air and water quality monitoring. Methane emission from natural gas systems was estimated to be 221.2 million metric tons of CO_2 equivalent in 2009 [9]. Understanding the coupled process of gas migration in the shallow unsaturated zone of the subsurface and the atmospheric boundary layer becomes critical in developing models for the assessment of risk of atmospheric loading and development of effective leak detection technologies. The pathways that carry the gas through the unsaturated zone close to the land surface from the source (both point and distributed) are developed through the air spaces of partially water saturated soil. The spatial distribution of relative permeability of the gas/air phase that depends on the soil–water saturation determines these connected pathways. In natural media, the spatial distribution of soil properties has a direct correlation to the distribution of the soil water saturations. One of the driving forces that controls the gaseous flow rate in the shallow subsurface is the pressure at the land–atmospheric interface. Hence, any pressure fluctuations in the atmosphere can potentially affect the gas movement and correspondingly the leakage signal that is detected near the land surface. The effect of the wind speed and pressure fluctuations at the soil surface and how air is transmitted through the soil pores is not well understood. Many atmospheric and land surface conditions (e.g. micro topography, surface roughness, vegetation etc.) contribute to this process. For example, in the liquid phase, the primary mechanisms of advection, molecular diffusion, and hydrodynamic dispersion contribute by different degrees to mass transport in the gas phase. In some settings, small gradients in gas pressure resulting from fluctuations in the pressure at the land-atmospheric interface can result in advective flux much larger than diffusive flux (e.g. atmospheric pumping).

Poulsen and Møldrup [10] evaluated the effects of wind–induced pressure fluctuations on CO_2 migration and emissions at a landfill site. They analyzed the impact of wind induced gas emission as a function of the standard deviation in pressure variations. They found that the soil–air permeability and the pressure fluctuation amplitude significantly affected gas emission and that 40% of the total gas emissions flux came from wind turbulence–induced gas transport. Maier et al. [11] also found that turbulence induced pressure pumping reached up to 60% of the diffusive flux rates.

The upscaling related challenges of this problem come from first determining what pore–scale processes are important in coupling multi-phase flow in the subsufrace with the free flow in the atmosphere. Second, how these processes are properly upscaled when the soil conditions change both laterally and vertically. The effects of topography of the land surface and vegetation also have to be factored in.

10.2.3. *Vapor Intrusion*

Chlorinated solvents that are in the form of NAPLs when introduced to the subsurface through accidental spills or improper disposal are prevalent at industrial waste sites. These chemicals that are suspected or known carcinogens if ingested through contaminated water or inhaled as vapor results in major health risks. After a spill, NAPL can persist as a separate phase in the vadose zone and in the saturated zone below the water table entrapped within the soil pores. The gas phase NAPL in the vadose zone can readily disperse into air, into the air spaces within soil or underneath a structure, leading to vapor intrusion, or the entry of a volatile chemical to indoor air from underlying contamination in soil and groundwater [12]. It has been reported that the average American spends more than 21 hours per day indoors and roughly 18 hours indoors for every hour spent outdoors [13]. Although it is not conclusively known whether vapor intrusion is a widespread problem with respect to long–term exposure at the very low levels expected in enclosed spaces such as in buildings and basements, several cases have received national attention. In a U.S. Environmental Protection Agency (EPA) review article [14] state, due to difficulty in conclusively identifying the soil–to–indoor pathway via indoor sampling, researchers have suggested moving the focus of vapor intrusion investigations outside the home. The processes that govern the vapor transport in the heterogeneous subsurface outside the home are complex, and the sampling to assess potential pathways is subjected to spatial and temporal variability. Spatial variability is a result of a number of factors that include changing soil and soil moisture conditions and temporal variability controlled by the transient heat, wind, atmospheric pressure and a water flux boundary conditions at the land–atmospheric interface. In addition, a number of physical and geo–bio–chemical processes may attenuate the vapor in the subsurface along the pathways from the sources to the building. The uncertainty resulting from the lack of fundamental scientific understanding of these processes and the inability to fully characterize the pathways through effective sampling impact the prediction of exposure risks and design of effective mitigation strategies. Without an understanding of the partitioning between the NAPL/water and NAPL/air interfaces in the vadose zone under realistic conditions, site managers, for example, cannot generate accurate estimates on remediation efficiency and/or source longevity. Understanding vapor transport and attenuation in the unsaturated zone is paramount to our understanding of the remediation alternatives of the subsurface and to our ability to characterize risk to human health through exposure pathways. The same issues of moving across scales from pore to field are of relevance as the basic processes involved are related to flow of fluids in multiphase systems with the added complexity of effects of heat and mass transfer processes at the land surface that contribute to preferential vapor pathway development in the subsurface.

10.2.4. *Land Mine Detection*

Countermine technology has become the subject of global interest for both military and humanitarian mine–clearing operations. The United Nations (UN) and the US Department of State declared landmines to be one of the most widespread, lethal, and long lasting forms of pollution [15], costing over $33 billion to clear the approximately 100 million landmines strewn throughout 64 countries [16]. Although there is a wide range of sensors available for the detection of buried landmines and many of the sensors perform well, there is general agreement that none of the sensors can reliably detect landmines while also maintaining a low false-alarm rate [17]. One main reason for the high false alarm rate is the variety of landmines that are used as well as the extreme variability of the environment in which the mines are placed. Detecting small mines in large areas is especially difficult when the area is highly heterogeneous with features that can mask the presence of the mine. Because landmine sensors (e.g. ground penetrating radar and thermal imagery) exploit soil and environmental conditions to discern between mines and other objects, all current mine detection technologies require that the spatial and temporal variability of key environmental conditions such as climate, vegetation, soil type, depth of ground water table, and topography be understood. If these factors and the ability to model them in a variety of domains become well defined, then sensor and algorithm simulations can more realistically be tailored to particular operational scenarios and technologies [18]. However, research efforts on mine detection are generally geared toward sensor development and sensor fusion while very little effort has been made to evaluate the environmental conditions that affect sensor performance [19–21]. Although many numerical and experimental investigations have been performed, they focus on the response/effectiveness of the technology or neglect important parameters like soil heterogeneity, and the temporal and spatial variability of the soil moisture and/or thermal properties. Thus, there is a knowledge gap between the signal processing technique and fundamental processes that occur in shallow subsurface zones as affected by near surface boundary conditions. Increasing our knowledge of the effects of geohydrologic/thermal properties and behaviors on the landmine signature is needed to fill the knowledge gap in order to better understand, model/simulate, and predict the environmental conditions that are most dynamic to mine detection performance. The shallow subsurface soil moisture processes as controlled by the mass and heat flux boundary conditions at the land–atmospheric interface are central to the problem of detection of land mines using sensing technologies. Hence, in developing models for signal interpretation in the field, the issues of multiphase flow parameter up–scaling in the shallow subsurface becomes centrally relevant.

10.3. Multiphase Flow

Fundamental to the behavior of water, supercritical CO_2, NAPLs and gases in the systems that are of relevance to the problems introduced in the previous section is the processes that govern multiphase flow in porous media. In this section, the role of heterogeneity, continuum representation and the governing equations of multiphase flow are first presented. This is followed by the introduction of a commonly used constituative model and a brief review of upscaling theories as applied to constitutive models.

10.3.1. *Heterogeneity*

Much of the uncertainty in our understanding and parametrization of liquid and gas flow through soils can be attributed to soil heterogeneity. Both field and laboratory experiments have shown that the soil heterogeneity controls the flow and transport, including preferential flow. As seen in Fig. 10.1, most subsurface formations are heterogeneous with different spatial correlations in all directions.

Figure 10.1(a) shows an example from a diesel contaminated site to demonstrate how texture variability associated with mixtures of sands, silts and clays results in subsurface heterogeneity. Figure 10.1(b) (from a road cut) shows an alluvial formation where sands of different grain sizes are deposited in layers creating heterogeneities and well-defined lithological variations. When core samples are taken and the soil physical parameters determined, the parameter values depend on the size of the sample. Using porosity as a characterization parameter, how this variability is factored in defining the representative elementary volume (REV) for continuum representation of porous media flow is presented in Fig. 10.2. The porosity is defined as

$$\phi(\mathbf{x}_0) = \frac{1}{V_r} \int_{V_r} \chi(\mathbf{x}) \mathrm{d}\mathbf{x}, \tag{10.1}$$

Fig. 10.1. Heterogeneity manifested at field scale defined by lithological variations.

where $\chi(\mathbf{x})$ $[-]$ is the void space indicator function and the volume V_r $[L^3]$ is a sphere of radius r $[L]$ centered around the spatial point $\mathbf{x}_0$ $[L]^3$. The REV is such volume V_r for which exist radii r_{micro} and r_{macro} such that the porosity given by Eq. (10.1) is independent of the radius r within the range as shown in Fig. 10.2. The microscale description ($r < r_{micro}$) focuses on the behavior of a large number of molecules of the present phases (e.g., liquid and gas phases through the soil solid matrix). The equations describing their flow are those of the continuum mechanics within the pores (e.g. direct numerical simulations). The continuum (Fig. 10.2) or REV scale is defined as the scale in which the mean is a constant deterministic quantity and the variance approaches zero [22]. At this scale, individual pores or phase interfaces are no longer noticeable. It is at this scale that is referred to as macroscopic scale the Darcy equation is applicable. At this macroscopic scale, the description of the flow of phases introduces new equations which are the transposition of the mass balance, momentum and energy microscale balances. For example, the equation of Darcy is the momentum balance at the macroscopic scale which can be deduced from Navier–Stokes equations. In these macroscopic equations appear effective properties, as the permeability in Darcy's law, the relative permeability and capillary pressure in the multiphase case, etc. These effective properties can be theoretically obtained from microscopic properties using upscaling techniques.

The field scale (Figs. 10.1 and 10.2) is defined as the spatial dimension where the soil properties become nonstationary [23]. Natural field scale media are in fact generally heterogeneous and contributes to both spatially and temporally non-linear multiphase flow behavior. At this scale, the soil properties are rarely the same at every point of the medium. When solving the governing equations of flow or transport using numerical schemes, where the domain is discretized into computational blocks or grids, it is possible to take into account the effect of these heterogeneities using block sizes smaller than the characteristic size of the heterogeneities. As this may not be feasible due to high computational demands, a second upscaling stage is needed to describe the field scale properties from the macroscale description.

10.3.2. *Processes and Constitutive Models*

Multi–phase porous media systems consist of fluids that exist in separate phases in the inter–granular pore spaces. Figure (10.3) schematically

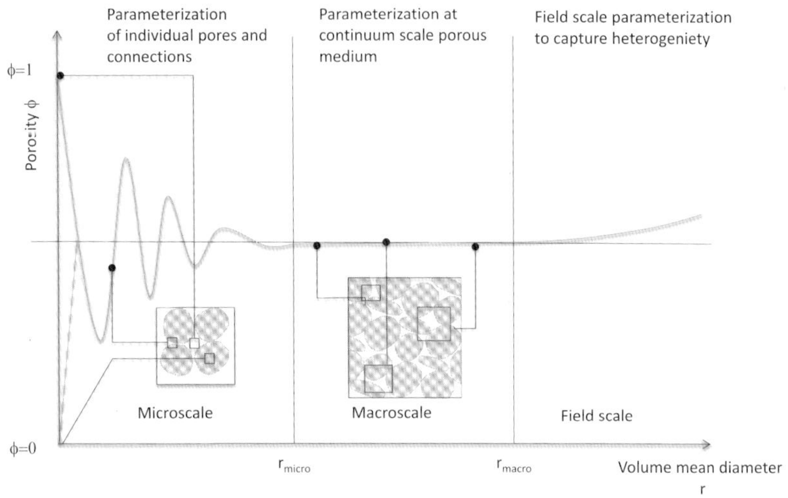

Fig. 10.2. Porosity as a function of the volume mean radius r. Heterogeneity and scale. Here the averaging of porosity over the different length scales from pore, homogeneous continuum scale to heterogeneous field scale is shown.

shows such systems encountered in hydrogeological applications. Two–phase problems of water flow in the unsaturated zone above the water table associated with rainfall infiltration, aquifer recharge and irrigation involve water and air as the wetting and non–wetting fluids, respectively. Below the water table, the soil pores are fully saturated with water, thus describing the flow as a single-phase problem. In industry, related problems involving accidental releases of partially immiscible fluids (NAPLs) such as industrial solvents, wood treating agents and petroleum products, the fluid displaces some of the non–wetting air phase resulting in a three–phase system. When NAPLs that are heavier than water penetrate the water table, the non–wetting phase displaces some of the pore–water resulting in a two–phase system. Interfacial tension at wetting/non–wetting fluid interfaces within pores introduces additional force of capillarity that is not present in saturated single–phase flow systems.

Each fluid in a multiphase system is characterized by its own pressure state. The differences in pressure arise from imbalances of molecular forces at fluid interfaces [24]. In a two–phase system, the capillary pressure p_c $[ML^{-1}T^{-2}]$ that depends on the wetting phase saturation S_w $[-]$ is defined as the pressure difference between wetting and non-wetting phases

$$p_c(S_w) = p_{nw} - p_w, \tag{10.2}$$

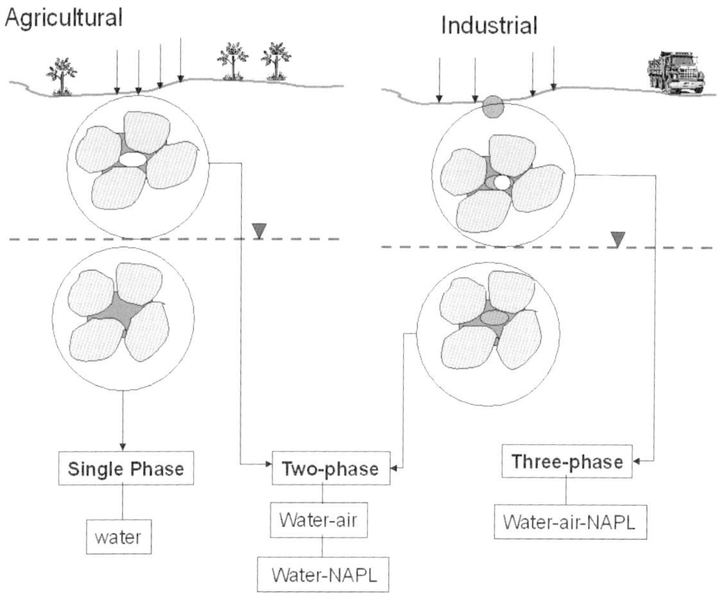

Fig. 10.3. Multi-phase porous media systems in hydrogeology.

where p_{nw} $[ML^{-1}T^{-2}]$ is the non-wetting phase pressure and p_w $[ML^{-1}T^{-2}]$ is the wetting phase pressure. The relationship of capillary pressure and fluid content is referred to as the capillary pressure function (that is referred to as the retention function in unsaturated flow systems). As given in Eq. (10.2), capillary pressure is usually expressed as a function of the saturation of the wetting fluid. This relationship is an intrinsic property of a given porous medium and the two-fluid system. Data on capillary pressure as a function of saturation are obtained experimentally and fitted with mathematical functions to obtain constitutive models for multiphase flow. Two commonly used constitutive models for capillary pressure in soil physics and geohydrologic applications are presented by Brooks and Corey [25] and van Genuchten [26]. The Brooks–Corey model for the retention function is,

$$p_c(S_w) = p_d S_e^{\frac{1}{\lambda}}, \quad \text{for} \ \ p_c \geq p_d, \tag{10.3}$$

where λ $[-]$ is a fitting parameter, and p_d $[ML^{-1}T^{-2}]$ is the pressure at which the non-wetting fluid first enters the pores when the non-wetting place builds up and is called *displacement* or *entry* pressure. The effective saturation S_e is defined as,

$$S_e = \frac{S_w - S_{r,w}}{1 - S_{r,w} - S_{r,nw}}, \tag{10.4}$$

where $S_{r,w}$ $[-]$ and $S_{r,nw}$ $[-]$ are the residual or minimum saturation wetting and non–wetting fluid, respectively.

When multiple fluid phases are present within the pore space, the ability of the medium to conduct a given fluid within a pore will not depend on the geometry of the pore space only, but also on fluid properties, geometry of the fluid–filled part of the pore space and phases volume fraction. Darcy's law that was originally developed and applied for single phase flow is adopted for multiphase flow through the use of the concept of relative permeability [27] as,

$$\mathbf{q}_\alpha = -\frac{k_{r,\alpha}(S_\alpha)}{\mu_\alpha}\mathbf{K}_i\left(\nabla p_\alpha - \rho_\alpha \mathbf{g}\right), \quad \text{for } \alpha = w, nw, \tag{10.5}$$

where $\mathbf{q}_\alpha$ $[LT^{-1}]^3$ is the apparent macroscopic velocity of the phase α, μ_α $[ML^{-1}T^{-1}]$ is the dynamic viscosity of the phase α, $\mathbf{K}_i$ $[L^2]^{3\times 3}$ is the intrinsic permeability tensor, $k_{r,\alpha}$ $[-]$ is the α–phase relative permeability function, ρ_α $[ML^{-3}]$ is the density of the phase α, and $\mathbf{g}$ $[LT^{-2}]^3$ is the gravitational acceleration vector. The relative permeability $k_{r,\alpha}$ of fluid phase α ranges between zero and one and depends on the fluid saturation S_α.

By extension, the hydraulic conductivity tensor $\mathbf{K}$ $[LT^{-1}]^{3\times 3}$ also becomes a function of saturation of the phase α,

$$\mathbf{K} = \mathbf{K}(S_\alpha) = \frac{k_{r,\alpha}(S_\alpha)\rho_\alpha g}{\mu_\alpha}\mathbf{K}_i, \tag{10.6}$$

where g $[LT^{-2}]$ is the scalar gravitational acceleration constant.

The Brooks–Corey constitutive model for the relative permeabilities of the wetting and the non–wetting phases are

$$k_{r,w} = S_e^{-\frac{2+3\lambda}{\lambda}}, \tag{10.7a}$$

$$k_{r,nw} = (1 - S_e)^2\left(1 - S_e^{-\frac{2+\lambda}{\lambda}}\right). \tag{10.7b}$$

Darcy's law for multiphase flow (Eq. (10.5)) and the mass balance for each phase are combined to derive the governing equations for multiphase flow in two-phase system as,

$$\phi\frac{\partial(\rho_\alpha S_\alpha)}{\partial t} - \nabla\cdot\left(\frac{\rho_\alpha k_{r,\alpha}}{\mu_\alpha}\mathbf{K}_i\left(\nabla p_\alpha - \rho_\alpha \mathbf{g}\right)\right) = F_\alpha, \tag{10.8}$$

where F_α $[ML^{-3}T^{-1}]$ is the source/sink term and fluid phase $\alpha = w, nw$. Equation (10.8) written for each phase in the multiphase flow system, in combination with the constitutive models for capillary pressures (Eq. (10.3)) and relative permeabilities (Eq. (10.7)) provide the full formulation of the mathematical model to solve for the phase pressures and saturations for given initial and boundary conditions and source terms. These governing equations are derived for the continuum macro–scale (or Darcy scale). The parameters that appear in these equations have to be upscaled to computational grid scale when used in simulations. The upscaling of constitutive models of capillary pressure and relative permeability are discussed in the next section.

10.3.3. *Upscaling Constitutive Parameters*

It is well supported in literature that the highly non–linear distribution of hydrologic processes often limits the ability to track interactions from scale–to–scale and across space and time. The primary question driving the upscaling problem is: can laboratory scale measurements successfully be applied to properly describe larger scale flow and transport behavior? We know that one can directly apply laboratory measured soil hydraulic properties as inputs for lab scale studies and simulation models. However, in developing upscaling methods, it is necessary to recognize that it is not practical to take measurements at all points in the subsurface system to determine the needed parameters. Hence, the development of theoretical foundations for any upscaling method leads to a practical question: how can soil hydraulic parameters determined for homogeneous samples collected at a limited number of locations or the laboratory scale be used to determine the parameters of the discretized grid blocks of models (grid–block scale) that simulate the field scale behavior? One option is to spatially distribute laboratory scale soil hydraulic properties across larger scales. Alternatively, the lab or discrete in–situ scale properties can be used as initial estimates and improved upon by using inverse modeling or during model calibration. Inverse modeling attempts to minimize differences between observation and simulation using analytical or numerical solutions that include constitutive relationships that contain estimated parameters [28].

The texture variability that contributes to changes in porosity with scale also affects the parameters that depend on the pore sizes and pore size distribution. These include permeability and hydraulic conductivity in saturated groundwater flow, soil–water retention and relative permeability in unsaturated flow and constitutive model parameters of multiphase flow (Eqs. (10.3) and (10.7)). These parameters are obtained using homogeneous test samples and defined at the REV scale to be used in continuum modeling.

What we know as a single phase porous medium hydraulic conductivity K $[LT^{-1}]$ is an effective property which can be induced from upscaling of pore to the macroscale flow. In the simple case of the two–layered heterogeneity, the upscaled hydraulic conductivity is an average determined using the two values of K of the two layers by simple application of Darcy's law. Any analytical solution method applicable for layered systems require the knowledge of the K for each layer and their thicknesses, thus limiting their practical utility in upscaling applications. More rigorous methods have been suggested to determine upscaled hydraulic conductivity. An upscaled parameter of K referred to as the *effective hydraulic conductivity* is estimated by considering the small–scale variation of K as a random space function (RSF). The effective conductivity is estimated independent of the boundary conditions from the spatial correlations and variances of the RSF [29, 30]. These formulations assume statistical homogeneity of the system that is modeled. Considerable knowledge exists on upscaling single phase flow in porous media using effective parameters. Another approach referred to as *equivalent hydraulic conductivity* assumes that the domain to which the variations of K are upscaled to have an equivalent value of K that preserves the mean flux of the heterogeneous formation for a given head gradient [31]. As the equivalent K depends on the boundary conditions that control flow, any uncertainty in the boundary conditions results in non–unique estimates of the upscaled parameter. However, the equivalent K value approaches the effective K value when the aquifer size is larger than the correlation range of K [32].

Figure 10.4 shows the laboratory–simulated migration of a light NAPL in a heterogeneous porous medium. The test tank was packed using five highly characterized soils to represent a spatially correlated random field with known geostatistical parameters (mean of $\log K$, variance of K and correlation lengths in the x and z directions) [35]. The migration of the non–wetting fluid that occurs through the pores by displacing the wetting fluid is controlled by the pore–scale soil parameters as well as the properties of the wetting and the non–wetting fluids. The upscaling problem that needs to be addressed is how the parameters of the multiphase systems affected by the soil and fluid properties are properly represented in an example grid–scale model shown and compared to a numerical model that uses such

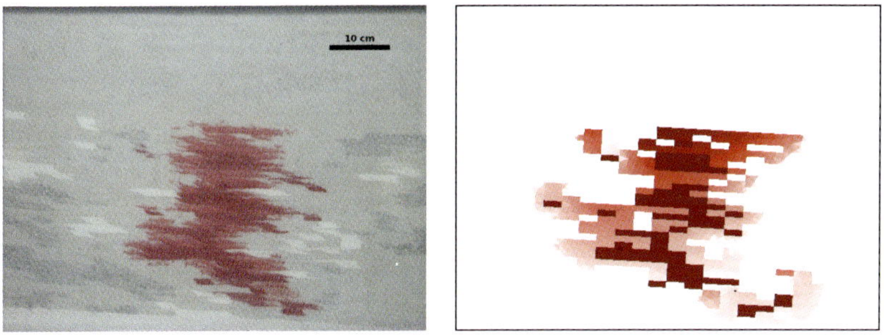

(a) Laboratory experiment (b) Numerical simulation

Fig. 10.4. Laboratory simulation of multiphase flow in a heterogeneous porous medium (a) and numerical simulation using the upscaled parameters (b). The numerical solution was obtained using the mixed hybrid finite element method described in Refs. [33, 34].

estimated upscaled parameters in Fig. 10.4.

At the REV scale, the multiphase system is characterized by the constitutive relationships that include capillary pressure and relative permeability as functions of saturation (Eqs. (10.3) and (10.7)). The issue of upscaling these constitutive models is common to the types of problems in multiphase flow that were presented in Section 10.2. Parameter upscaling has been extensively studied by petroleum engineering as applied to hydrocarbon reservoirs, soil physicists in irrigation water management and geo–hydrologist for applications in subsurface remediation involving solvents and petroleum wastes. Constitutive models are developed from measurements made in the laboratory using small samples whose sizes correspond to size of cores extracted during field investigations. The basic way these parameters are upscaled in multiphase systems is very different from single-phase flow, because in addition to the subsurface heterogeneity, the non–linearities of the basic processes play a critical role [36]. These non–linarites also contribute to different fluid retention behavior at different length scales.

Early work on upscaling in unsaturated flow systems (water as wetting and air as non–wetting fluids) based on small perturbations (e.g. [37]) is not generally applicable when the size of the field domain is small compared to the length scales of the heterogeneity. In the development of upscaling for two–phase flow, some researchers adopted percolation network models used in pore–scale investigations to upscale constitutive relationships at the macro-scale [38–40]. In small samples that are used to get the retention function in the laboratory, the water content that corresponds to a given capillary pressure is fully determined by the capillary forces. The basis for constitutive models such as by [25, 26] for retention and relative permeability functions is capillary equilibrium. Desbarats [41] focused on the upscaling of moisture retention curves in heterogeneous media under conditions of capillary–gravity equilibrium. After assessment of many developments based on percolation approaches and subsequent work on the topic, Desbarats [42] observed that to properly upscale constitutive relationships at the macroscopic scale, in addition to gravity and capillary forces, the viscous forces that define the shear resistance to flow have to be considered. This raises the question on whether the equilibrium based constitutive models are valid at large scales where the water (wetting fluid) distribution is not only controlled by the pore size distribution but also by heterogeneity. To address these limitations, Desbarats [42] used a three–dimensional numerical model to simulate viscous forces in addition to gravity and capillary forces to identify averaging processes that produced the upscaled retention functions from local functions. The author concluded that the model used to determine upscaled constitutive relationships could not be represented by the same parametric model representing the small–scale constitutive relationships. The slope of the relative conductivity curve in the small water content range was interpreted as an upscaled pore–size distribution parameter (inverse is an upscaled capillary length). The main finding of this study where viscous forces were included in the analysis was that the upscaled pore–size distribution parameter can be approximated by a spatial power average of the corresponding small scale values distributed in the flow domain. Using the same assumption of steady flow, Liu et al. [43] used a practical formulation to determine the large–scale (upscaled) retention curve using the small–scale curves assuming spatially uniform capillary pressure exists in the larger upscaled domain. The expression for the upscaled capillary pressure–saturation relationship was given as

$$S(p_c) = \frac{\int\limits_V s(p_c)\mathrm{d}V}{\int\limits_V \phi \mathrm{d}V}, \tag{10.9}$$

where S $[-]$ and s $[-]$ are the water saturations at large and local (measurement) scales, respectively, V $[L^3]$ is the total medium volume, p_c is the capillary pressure, and ϕ — is the spatially variable porosity defined in Eq. (10.1). These authors pointed out that Eq. (10.9) imply that in both small and large scales, the relationships are determined by the pore–size distribution and are independent of correlation length scales of heterogeneity.

The upscaled hydraulic conductivity was given as

$$K(p_c) = \frac{1}{V} \int_V k(p_c)^\omega \mathrm{d}V, \tag{10.10}$$

where, K $[LT^{-1}]$ and k $[LT^{-1}]$ are the upscaled and sample scale hydraulic conductivities, respectively, and ω $[-]$ is a constant scalar parameter. This expression assumes that even though the sample scale relative permeability primarily depends on the pore–size distribution, the upscaled conductivity does not. This approach, even though practical, is limited to steady–state unsaturated flow in porous media with large air entry values. In petroleum engineering, the problem of upscaling constitutive models was posed in the context of managing the computing efficiency by using large grid blocks. The upscaling methods used pseudo–functions [44] that accounts for the heterogeneity within the large grid–block to replace the multiphase effective permeabilities and capillary pressures [45]. The goal was to use large grid dimensions to utilize available computing power manageable levels and with minimum loss of accuracy due to simplified representation of the heterogeneity. Use of this method still requires fine–grid simulations of a representative reservoir section to determine the appropriate pseudo–functions for the selected parameters. Even though some computing efficiencies have been achieved, these methods were considered to be without strong theoretical foundation [42].

Other techniques have been proposed and used for upscaling two–phase flow. In homogenization [46, 47], stochastic representation of conservation laws in porous medium are used to get non–linear effective equations that are considered to govern the flow behavior of the homogenized equivalent of the randomly heterogeneous porous medium. Another method referred to as the large–scale volume averaging (e.g. [48]), the flow equations and the properties at a larger scale are calculated by averaging from a lower scale. This method has been demonstrated to be more efficient than methods based on pseudo–functions that require full simulations using fine-grid model [45]. The volume averaging technique has been extensively used to predict the macro–scale transport properties for many processes including transport in heterogeneous porous media [49], two-phase flow [50], two–phase inertial flow [51], reactive media [45, 52], solute transport with adsorption [53] multi–component mixtures [54], and coupled heat and mass transfer with Soret effect [55, 56].

The question of using a one– or two–equation model is raised when modeling two phase (or region when working at the Darcy's scale) flow in porous media. The one–equation equilibrium model consists of a single transport equation for both phases (or regions). When the two fields in the two regions are close enough, the transport equations that represent the two-equation model can be added to produce this model. In other words, the principle of local–scale equilibrium is valid. If the local equilibrium assumption does not hold, two separate upscaled equations should be solved. However, for many initial boundary–value problems, the two-equation model shows a time–asymptotic behavior that can be modeled with a non–equilibrium one–equation model [57, 58]. The domains of validity of these three different models, which depend mainly on the Péclet number and a characteristic time, have been already explored [50, 59].

Davarzani et al. [55] showed that for moderate property contrast between phases, the local–equilibrium can predict the flow very well, and the model is not very sensitive to boundary conditions or initial conditions. For higher contrasts, the local–equilibrium model fails during the transient period. While, at steady–state, the local–equilibrium model offers again a very good prediction [55].

10.4. Dissolution in Multiphase Systems

10.4.1. *Mass Transfer and Rate Coefficients*

The mass transfer that occurs at fluid-fluid interfaces at the pore level is generally approximated using a linear model based on stagnant film theory as shown in Fig. 10.5. As there is no mass storage within the film, the concentration gradient between the source and the solvent can be assumed to be linear. The mass transfer across the interface between two fluid phases is (based on the linear film theory) generally described through a mass transfer rate coefficient. The rate of mass flux is defined through a linear relationship given by

$$J = k_\ell (C_s - C), \tag{10.11}$$

where J $[ML^{-2}T^{-1}]$ is the mass flux rate from the dissolving phase (can be considered to be the non–wetting phase), k_ℓ $[LT^{-1}]$ is the mass transfer rate coefficient, C_s $[ML^{-3}]$ is the aqueous phase concentration under conditions when the dissolving phase is at solubility limit in water and C $[ML^{-3}]$ is the aqueous phase solute concentration in the bulk solution. The subscript ℓ denotes that the driving force acts along the longitudinal direction of flux.

When upscaling the stagnant film model to the representative elemental volume (REV) scale in porous media, it is necessary to define an over-

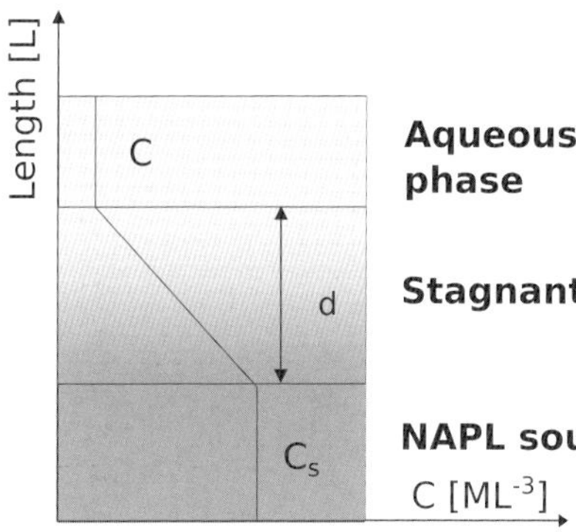

Fig. 10.5. Stagnant layer model with linear concentration profile between the NAPL source concentration C_s and the concentration C the in the bulk phase.

all mass transfer rate coefficient. By extending the single film theory, a linear driving force model similar to Eq. (10.11) can be used to describe mass flux from entrapped non-wetting phase (NAPL, ScrCO_2 or CO_2 gas plume) sources in porous media. This is accomplished by introducing a lumped mass transfer rate coefficient K_c $[T^{-1}]$ [60]. The interface mass rate J' $[ML^{-3}T^{-1}]$ in porous medium takes the form

$$J' = K_c(C_s - C). \tag{10.12}$$

The pore-scale mass transfer coefficient k_ℓ and the lumped mass transfer coefficient K_c are related by

$$K_c = k_\ell \frac{A_{nw}}{V} \tag{10.13}$$

where A_{nw} $[L^2]$ is the total NAPL–water surface area within the REV of volume V $[L^3]$. As A_{nw} cannot be directly measured, in practical applications in porous media, K_c is treated as an empirically determined parameter for a specific multiphase system.

10.4.2. *Gilliland–Sherwood models*

Phenomenological models that are used to predict the mass transfer rate coefficient k_ℓ or the lumped mass transfer coefficient K_c are referred to as Gilliland–Sherwood models and are represented by dimensionless Sherwood number Sh and modified Sherwood number Sh$'$. The Sherwood number Sh is related to the mass transfer rate coefficient k_ℓ as

$$\text{Sh} = k_\ell \frac{d_p}{D_\ell}, \tag{10.14}$$

where d_p $[L]$ is the geometric mean of particle diameter and D_ℓ $[L^2T^{-1}]$ is the diffusion coefficient in the free liquid. The modified Sherwood number Sh' that involves the lumped mass transfer rate K_c and is suitable for use in porous media applications is defined as

$$\mathrm{Sh}' = K_c \frac{d_p^2}{D_\ell}. \tag{10.15}$$

Therefore instead of K_c, the dimensionless Sherwood number Sh' is determined empirically in order to describe the mass transfer process under various physical and chemical conditions in the porous media.

Building on the concept of upscalable Gilliland-Sherwood model for porous media, an empirical model for REV scale mass transfer was proposed by Ref. [61] in a general form

$$\mathrm{Sh}' = \alpha \mathrm{Re}^{\beta} \mathrm{Sc}^{\gamma} \left(\frac{\theta_n d_{50}}{\tau L} \right)^{\delta}, \tag{10.16}$$

that involves four dimensionless fitting parameters α, β, γ, and δ, dimensionless Reynolds and Schmidt numbers Re $[-]$ and Sc $[-]$, respectively, dissolving NAPL content θ_n $[-]$, d_{50} $[L]$ is the particle diameter such that 50% of the porous media are finer by weight (median particle size), τ $[-]$ is the tortuosity factor of the flow path, and L $[L]$ is the dissolution length. The corresponding model that was fitted by Saba and Illangasekare [61] to dissolution data obtained in a two–dimensional flow configuration in a small scale test tank is $\alpha = 11.34$, $\beta = 0.28$, $\gamma = 0.33$, $\delta = 1.037$. Saba and Illangasekare [61] compared other Gilliland–Sherwood models that have appeared in literature and showed that the dimensionality of water flow has to be taken into consideration when upscaling the models based on one–dimensional systems to multi–dimensional flow systems in the field. An overview of other Gilliland-Sherwood models for the modified Sherwood number Sh' applicable for various dissolution configurations within the porous medium is given in [62]. Liu et al. [63] developed a Gilliland–Sherwood model for a synthetic porous medium consisting of random–size spheres and solving the groundwater flow and mass transfer at the pore–scale using a multi–physics simulator COMSOL Multiphysics®. This model is given as

$$\mathrm{Sh} = 3.81 \mathrm{Re}^{0.57} \mathrm{Sc}^{0.33}. \tag{10.17}$$

As the flow is laminar and the kinematic viscosity is constant and the mass transfer is dependent on advection and diffusion, [63] was able to represent

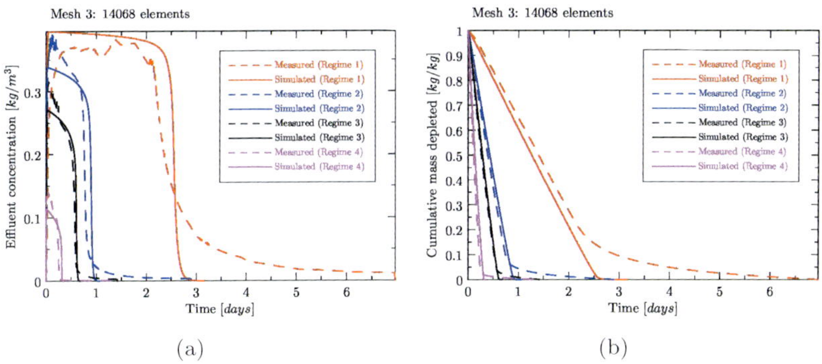

Fig. 10.6. TCE concentration (a) and cumulative TCE content depletion (b) temporal profiles compared to laboratory measured data (dotted lines), [64].

the Sh number as only a function of the Péclet number Pe as

$$\mathrm{Sh} = 0.094\mathrm{Pe}^{0.56}. \tag{10.18}$$

Petri et al. [64] investigated a complete volatilization of a volatile organic compound (VOC) pool in the context of generation of contaminant vapor plumes in heterogeneous porous medium. In the experiment, an immobile NAPL pool of trichlorethylene (TCE) was created in the source zone placed inside a small tank. In one of the scenarios, the pool was exposed to the flowing 100% humidified air under four different velocities studied. A numerical model was developed and used to determine the Gilliland–Sherwood model hypothesized in the general form proposed by [61] in the form

$$\mathrm{Sh}' = 0.0011\mathrm{Pe}^{0.05}\left(\frac{\theta_n}{\theta_n^{ini}}\right)^{0.2} d_0^{1.68}, \tag{10.19}$$

where θ_n^{ini} [−] is the initial NAPL content in the source zone. The small exponent of the Péclet number suggests that the velocity of the flowing air has a negligible effect on the rate of volatilization. In Fig. 10.6, the experimentally measured effluent concentration and the cumulative TCE content depletion are compared to the results of the mathematical model.

10.4.3. *Upscaling of Mass Transfer Rate Coefficients*

The upscaling of dissolution of trapped non–aqueous phase (NAPLs) has been studied for applications in remediation of sites contaminated with solvents and petroleum waste [62, 65].

The goal of upscaling dissolution is to determine what parameters of the field systems have to be included to determine the effective mass transfer rate coefficient. Saenton and Illangasekare [66] hypothesized that as the mass loading to the flowing water occurs at the pores where NAPL is entrapped, the total mass loading at the grid–scale will depend on the saturation distribution of the NAPL. They quantified the distribution (or spread) through a dimensionless second moment $M_{II,z}$ $[-]$. The mass that gets loaded is transported within the grid block by the flowing water and the net mass generation is a result of mixing within the block. This velocity driven mixing was captured through the use of geostatistical parameters of the heterogeneity field and the size of the grid block. Using synthetic data from numerical simulations where various NAPL entrapment architecture were created in correlated random fields, Saenton and Illangasekare [66] obtained an expression for the upscaled mass transfer correlation as

$$\overline{\mathrm{Sh}} = \mathrm{Sh}_0(1+\sigma_Y^2)^{\varphi_1}\left(1+\frac{\Delta z}{\lambda_z}\right)^{\varphi_2}\left(\frac{\hat{M}_{II,z}}{\hat{M}^*_{II,z}}\right)^{\varphi_3} \tag{10.20}$$

where $\overline{\mathrm{Sh}}$ $[-]$ is the upscaled Sherwood number containing the effective mass transfer rate coefficient, σ_Y^2 is the variance of the $\log K$ field, Δz $[L]$ is the vertical dimensions of the simulation grid, λ_z $[L]$ is the vertical correlation length and the last set of terms is the dimensionless second moment of the vertical saturation distribution. This method of upscaling was validated using data from an intermediate scale tank experiment [62] shown in Fig. 10.7.

The example of upscaling of a mass transfer that occurs at the pore–scale to grid scale provides the framework for developing upscaling methods for problems in carbon sequestration. The method was developed for a two-dimensional flow case and validated in a two-dimensional test system. As was discussed earlier, heterogeneity and flow dimensionality will play a critical role in the upscaling process. Hence, further study is needed to evaluate the effects of the third spatial dimensions and use of other parameters of the geologic formations when the assumption of stationarity that is built into the geostatistical parametrization is not valid. Also, the issue of how the dissolved mass diffuses into low permeability formations and how the process gets upscaled needs further study and development [67].

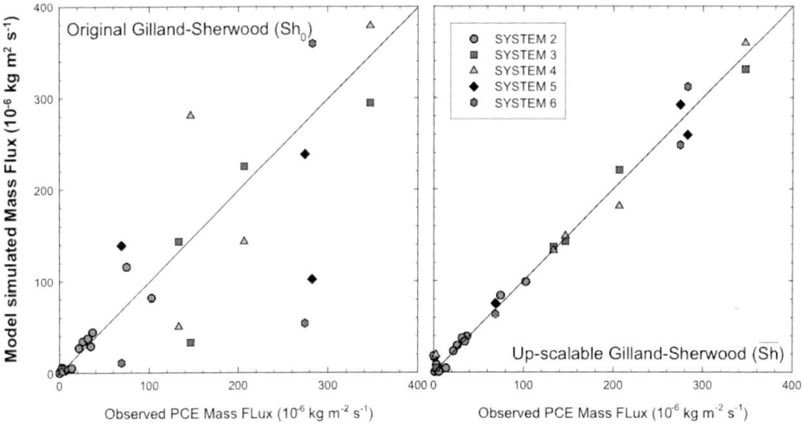

Fig. 10.7. Upscaling of NAPL dissolution: (a) Comparison of observed mass flux and mass flux estimated using the small-scale dissolution model, (b) comparison of observed mass flux and mass flux estimated using the upscaled dissolution model (system 2–5 refer to different grid sizes used in the simulations) [62].

10.5. Land/atmospheric Interactions

A variety of coupled processes and feedbacks between thermal, hydrological, geochemical and biological processes occur at the land-atmospheric interface. These coupled processes and feedbacks significantly influence the energy and mass balances and hence environmental conditions. Understanding heat, mass and momentum fluxes at the land surface at all relevant scales remains a major scientific challenge. Understanding mass and heat fluxes across the land surface at all relevant scales from laboratory to field remains a major scientific challenge. One such exchange process is evaporation, an important process that affects the water and energy balance in the soil and atmosphere, and consequently changes the local and global climatic behavior. In this section a review of the efforts to understand and upscale the processes associate with evaporation are presented.

10.5.1. *Processes and Modeling*

As conceptually shown in Fig. 10.8, the rate of soil evaporation is affected by atmospheric conditions (e.g. humidity, temperature, thermal radiation, wind velocity and turbulent flow regime), and thermal, and hydraulic properties of soil (thermal and hydraulic conductivity, porosity), all of which are strongly coupled. This strong coupling between processes leads to highly dynamic interactions between the atmosphere and soil resulting in dynamic evaporative behaviors [68]. It is recognized that the most important process that determines the coupling between the soil water and heat is the transport of latent heat (the result of phase change) by vapor flux in the unsaturated soil pores and at the interface between the soil and the atmosphere [69]. Models that incorporate these processes have been developed, e.g., [70]; however, as Bittelli et al. [69] note, a detailed experimental verification of vapor movement above the soil surface (i.e., atmospheric boundary layer) has not been conducted. Bittelli et al. [69] suggest that the errors introduced in the vapor flow calculations are due to a number of factors that include lack of proper coupling of the thermal and mass flux processes, deficiencies in the constitutive relationships (e.g. thermal and hydraulic conductivities and soil water content) and difficulty in determining the resistance parameters at the land (soil)–atmospheric interface. For example, a prevalent modeling approach is to derive the aerodynamic and soil surface resistance terms based on semi-empirical or empirical approaches and to adjust the predicted evaporation based on true conditions that depend on ambient conditions such as soil moisture, roughness, and wind speed.

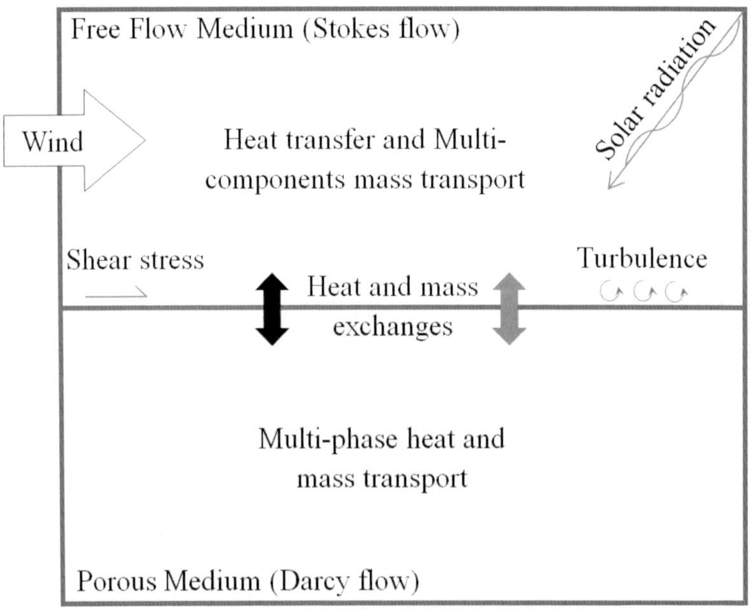

Fig. 10.8. Schematic of the land/atmospheric interaction configuration.

Traditionally, the influences of atmospheric conditions are applied at the soil surface and aerodynamic resistance is applied on the border between the air flow and permeable media (e.g. [69, 71–73]). In these cases, evaporation rate E $[ML^{-2}T^{-1}]$ can be given as

$$E = \frac{1}{r_s + r_v}\left((\rho_v)_{pm} - (\rho_v)_{ff}\right), \qquad (10.21)$$

where r_s $[L^{-1}T]$ is the soil surface resistance for water vapor transport, r_v $[L^{-1}T]$ is the aerodynamic resistance for water vapor, $(\rho_v)_{pm}$ $[ML^{-3}]$ is the vapor density immediately below the soil surface (in porous medium) and $(\rho_v)_{ff}$ $[ML^{-3}]$ is the vapor density immediately above the soil surface (in free medium). The vapor density above the soil surface is calculated from measurement of relative humidity on the boundary of the porous medium domain in the free flow medium.

In Eq. (10.21), the aerodynamic resistance for vapor transport depends on surface roughness properties and wind speed [69, 74]. The soil surface resistance depends on soil surface water content. The relationship between aerodynamic resistance to vapor transport and soil water content is typically expressed in an exponential form; there are many exponential empirical functions used to describe this relationship [71, 73]. Although this approach is widely used, modeling comparison studies have shown significant variation between model parameterizations and evaporative fluxes [75–78]. Recently, with the goal of addressing the issue of coupling the land to the atmosphere, [77] evaluated three different modeling approaches of bare soil evaporation formulated with different land surface boundary conditions and compared modeling results to laboratory generated experimental data. Results demonstrated that no one approach could be deemed most appropriate for every situation, demonstrating that further work focusing on the land/atmospheric interface, properly incorporating the complex interactions between the land and the atmospheric boundary layer is needed to increase the understanding of the processes that control shallow subsurface soil moisture flow that controls bare soil evaporation.

The modeling of non–isothermal single–phase (two–component) transfer in the atmosphere and two-phase (two–component) transfer in porous media have been separately investigated by many authors (e.g. [79–81]). Recently, numerical advances have been made in the coupling of free flow (Navier–Stokes) with porous media flow (Darcy flow) [82–87], however, these models were not adequately validated with experimental data. Mosthaf et al. [85] extended the classical single-phase coupling to two-phase flow in porous media and one phase in the free flow. Their model is based on the continuity of fluxes at the porous medium-free medium interface and use of the Beavers–Joseph boundary condition [85]. Baber et al [83] focused on the numerical concept and its implementation into a local modeling toolbox. The numerical parametric study showed that the proposed model can predict the evaporation phenomenology correctly. They concluded that the variation of permeability influences the duration of the capillary–driven evaporation regime whereas the variation of temperature affects the magnitude of the evaporation rate. They also showed that the choice of the Beavers–Joseph coefficient has a negligible influence on the evaporation rate across the interface [83].

It is well known that a no–slip condition at the free flow and porous domains surface is not a satisfactory assumption requiring the need to consider a slip boundary condition. The slip boundary condition was first obtained experimentally by Ref. [88]. They proposed that the tangential component of the normal stress of the flow at the free flow and porous medium interface is proportional to the jump of the tangential velocity across the interface [88]. The coupling condition was further studied by Saffman [89] who concluded that the filtration velocity in porous media was much smaller than the free–flow velocity and can be neglected. There exist several other formulations for a slip boundary condition to include (a) using a shear stress jump condition by means of the non–local form of the volume averaging technique with an experimentally determined fitting parameter [Ochoa-Tapia and Whitaker, 1997], or (b) using the inertia and boundary effects [90]. Alzami and Vafai [91] compared five different interface conditions between the porous medium and adjacent fluid layer. They concluded that the velocity field is more sensitive to variation in the boundary condition than the temperature field [91]. They showed similar results for all five interface conditions.

10.5.2. *Knowledge of Gaps and Challenges*

Water and energy fluxes in the vadose zone are coupled at the soil surface, which serves as the interface between the land and the atmosphere. Currently, most available upscaling procedures ignore the effects of the land–atmospheric interface [92] resulting in a need to develop upscaling approaches that account for climatic excitations under natural field conditions. Accurate prediction of water distribution and fluxes within the vadose zone is critical for quantifying vapor and energy exchanges between the land and atmosphere during the process of evapotranspiration, assessing groundwater recharge rates, and optimizing water management for agricultural purposes. Despite the importance of these predictions, standard models have limited capabilities to predict water or gas fluxes, flow pathways and water distribution. Even common practices such as understanding evaporation dynamics from homogeneous soils or water distribution after a heavy rainfall has proven to be difficult with standard models (e.g. [77, 93–96]). This can be partially attributed to models not capturing the physical behavior through proper system description/parameterization. As our computational capabilities continue to improve, our ability to describe the added complexity of physical systems should also improve rather than continuing to rely on the standard methods. By understanding the relative contribution of processes at various scales and how the processes can best be implemented at different scales to more accurately predict environmental behaviors remains a challenge. The challenges associated with upscaling mass transport through soil pores close to the land surface comes from the need to parametrize processes that couple Darcian flow in the soil to Stokes flow in the atmosphere. Practical and theoretical limitations of modeling efforts are often magnified at the land–atmosphere interface, where water and energy fluxes are highly dynamic and dramatically influenced by changes in thermal and moisture gradients and direction of flows [97]. However, for most conventional models and practical applications involving vadose zone, the strong coupling between the land and the atmosphere is rarely considered. This is due to the complexity of the problem in field scenarios and the scarcity of field or laboratory data capable of testing and refining energy and mass transfer theories. For most subsurface models, the soil surface serves as the upper boundary to the porous medium domain and is characterized using prescribed flux terms that serve as sources or sinks. Similarly, in most atmospheric models, the vadose zone serves as a lower boundary with prescribed fluxes. Such an approach is a simplification of the interaction processes above and below the soil surface. Although widely used due to its simplicity and ease of use, such an approach has been shown by both atmospheric and hydrogeological scientists to misrepresent flux conditions, resulting in model prediction errors [98]. When considering heterogeneous soils, this is particularly relevant. Heterogeneous soils result in complex flux conditions due to water fluxes from coarse to fine textured soils [94, 99]. This is not captured when prescribing a constant flux over the entire soil surface boundary. In addition, variations in soil surface conditions (e.g. soil type, texture, vegetation) can result in highly dynamic infiltration and evaporation conditions. Surface heterogeneities can affect the air velocity conditions [100], ultimately affecting infiltration rates during rainfall and evaporation, resulting in scale and rate dependence of hydraulic and thermal effective parameters [97].

Remote sensing (i.e. non–contact observational methods) is often used in hydrologic sciences to capture some of the spatial and temporal distributions of hydrologic processes which in turn can be used to model the interactions between the land and the atmosphere. Many past remote sensing studies focus on regression analysis between remotely sensed and observed data and/or comparing aircraft/satellite observations and in-situ observations [101, 102]. However remote sensing alone cannot fully solve the issue of cross-scale interaction as there is a requirement to understand the uncertainties associated with measurements and model predictions from scale to scale [103]. How can we use remotely sensed data at 10–20 km resolutions to make predictions about processes occurring at a local scale? We need to understand the effect of heterogeneity at a large scale as well as a small pore scale to correctly develop methods to synthesize pore scale physics with coarse–scale (e.g remote sensing) measurements. Researchers have observed breaks or transitions in scaling of soil hydraulic properties with spatial scales (e.g. [104, 105]). Nykanen and Foufoula–Georgiou [104] found, for example, transitions between small scale soil moisture samples and aircraft radiometer data, resulting in different relationships between scales. However, uncertainty exists in explaining the reason for these transitions; are they a result of different data sets or are physical processes involved that are not being accounted for at various scales [103]? In addition, how can models properly account for these transitions? Research suggests that if we can properly account for hydrologic variables, like soil moisture in hydrologic models, we can improve our ability to perform hydrologic forecasting (e.g., [106–108]).

10.6. Conclusions

Process up–scaling from pore to field systems still remains a challenge in hydrogeological sciences and reservoirs engineering. In this chapter we identified some of these challenges specifically for emerging problems. The following conclusions are made to help develop future research plans to over come some of these challenges.

Flow of multiple fluids in porous media is basic to the problems that were presented in this chapter. The primary parameters that need up–

scaling in multiphase flow are the relationships between capillary pressure vs saturation and relative permeability vs saturation. The parameter up-scaling methods for these are fundamentally different from that of single phase flow because of the non-linearity of these multiphase parameters. Techniques based on percolation network models, stochastic homogenization, large–scale volume averaging and methods based on pseudo–functions had limited success and the need exist for the development of more generalized approaches that can be applicable to both two–phase and three-phase systems.

The problem of dissolution of trapped non-aqueous liquids (NAPLs) has received recent attention in relation to dissolution trapping of stored supercritical CO_2 in deep geologic formations. Methods have been developed to up–scale the mass transfer rate coefficients to simulate field scale behavior by treating the effective dissolution as a mixing process controlled by the architecture of the NAPL entrapment and geostatistical parameters of the permeability field. This method provides a possible framework to develop dissolution upscaling for supercritical CO_2. However, whether rate limited conditions exist in field settings requires additional evaluation prior to developing such methods. The net mass loading from supercritical CO_2 trapped zones will not only be controlled by dissolution, but also back diffusion from low permeability zones in the heterogeneous formation. Methods are needed to obtain effective parameters that capture both these processes to predict long–term effectiveness of dissolution trapping.

Parameterization of processes that control land–atmospheric interactions is at its early stages of development. The knowledge gaps in the understanding of these processes from pore scale to larger scales, have resulted in models that rarely consider the strong coupling between the land and the atmosphere. Until these knowledge gaps are filled through experimental studies conducted at multiple scales, any validation of such up–scaling theories and methods are not possible.

Field data for validation of upscaling methods are often incomplete and costly to obtain. In field settings, the degree of control that is needed to obtain such data is not adequate. The intermediate laboratory scale offers the ability to study, under controlled conditions, complicated processes in the heterogeneous subsurface in multiple dimensions at different scales.

A conclusion can be made that scaling issues can only be resolved through the integration of theory with experiments, requiring innovative, multidisciplinary research efforts aimed at overcoming our current limited understanding of the influence of small scale processes on larger scale flow behavior.

Acknowledgement

The authors gratefully acknowledge the support of agencies that supported some of the research that was used to develop this chapter. These include, the National Science Foundation (NSF), Strategic Environmental Research for Defense Program (SERDP), Army Research Office, National Security and Science and Engineering Faculty (NSSEFF) Program, Unconventional Natural Gas and Oil Institute (UNGI) at the Colorado School of Mines and Czech Ministry of Education.

List of symbols

Latin symbols

Notation	Units	Description	Page
A_{nw}	$[L^2]$	NAPL–water surface area within REV	183
C	$[ML^{-3}]$	Aqueous phase solute concentration	182
C_s	$[ML^{-3}]$	Aqueous phase solute concentration at solubility limit	182
d_{50}	$[L]$	Median particle diameter	184
d_p	$[L]$	Geometric mean of particle diameter	183
D_ℓ	$[L^2T^{-1}]$	Diffusion coefficient in free liquid	183
E	$[ML^{-2}T^{-1}]$	Evaporation rate	189
F_α	$[ML^{-3}T^{-1}]$	Source/sink term of phase α	176
g	$[LT^{-2}]$	Scalar gravitational acceleration constant	176
$\mathbf{g}$	$[LT^{-2}]^3$	Gravitational acceleration vector	176
J	$[ML^{-2}T^{-1}]$	Mass flux rate	182
J'	$[ML^{-3}T^{-1}]$	Interface mass flux rate	183
K	$[LT^{-1}]$	Single phase porous medium hydraulic conductivity	178
K_c	$[T^{-1}]$	Lumped mass transfer rate coefficient	183
$\mathbf{K}$	$[L^2]^{3\times3}$	Intrinsic permeability tensor	176
k_ℓ	$[LT^{-1}]$	Pore–scale mass transfer rate coefficient	182
$k_{r,\alpha}$	$[-]$	Relative permeability of phase α	176
L	$[L]$	Dissolution length	184
$M_{II,z}$	$[-]$	Second distribution moment	186
p_d	$[ML^{-1}T^{-2}]$	Entry pressure	175
p_c	$[ML^{-1}T^{-2}]$	Capillary pressure	174
p_α	$[ML^{-1}T^{-2}]$	Wetting phase pressure	174
Pe	$[-]$	Péclet number	185
q_α	$[LT^{-1}]^3$	Apparent macroscopic velocity of phase α	176
r_s	$[L^{-1}T]$	Soil surface resistance for water vapor transport	189
r_v	$[L^{-1}T]$	Aerodynamic resistance for water vapor transport	189
Re	$[-]$	Reynolds number	184
S_e	$[-]$	Wetting phase effective saturation	175
$S_{r,\alpha}$	$[-]$	Residual saturation of phase α	176
S_α	$[-]$	Saturation of phase α	176
Sc	$[-]$	Schmidt number	184
Sh	$[-]$	Sherwood number	183
Sh′	$[-]$	Modified Sherwood number	183

Greek symbols

Notation	Units	Description	Page
θ_n	$[-]$	NAPL volumetric content	184
λ	$[-]$	Brooks and Corey fitting parameter	175
λ_z	$[L]$ Vertical correlation length	186	
μ_α	$[ML^{-1}T^{-1}]$	Dynamic viscosity of phase α	176
ρ_α	$[ML^{-3}]$	Density of phase α	176
τ	$[-]$	Toruosity factor	184
Φ	$[-]$	Porosity	172
χ	$[-]$	Void space indicator function	172

References

[1] L. W. Gelhar, Stochastic subsurface hydrology from theory to applications, *Water Resourc. Res.* **22**(9S), 135S–145S (1986).

[2] T. H. Illangasekare, J. L. Ramsey Jr, K. H. Jensen, and M. B. Butts, Experimental study of movement and distribution of dense organic contaminants in heterogeneous aquifers, *J. Contam. Hydrol.* **20**(1), 1–25 (1995).

[3] F. Fagerlund, T. H. Illangasekare, and A. Niemi, Nonaqueous-phase liquid infiltration and immobilization in heterogeneous media: 1. experimental methods and two-layered reference case, *Vadose Zone J.* **6**(3), 471–482 (2007).

[4] F. Fagerlund, T. H. Illangasekare, and A. Niemi, Nonaqueous-phase liquid infiltration and immobilization in heterogeneous media: 2. application to stochastically heterogeneous formations, *Vadose Zone J.* **6**(3), 483–495 (2007).

[5] E. Agartan, L. Trevisan, A. Cihan, J. Birkholzer, Q. Zhou, and T. Illangasekare, Experimental Study on Effects of Geologic Heterogeneity in Enhancing Dissolution Trapping of Supercritical CO2, in review in, *Water Resourc. Res.* (2014).

[6] M. Plampin, T. Illangasekare, T. Sakaki, and R. Pawar, Experimental study of gas evolution in heterogeneous shallow subsurface formations during leakage of stored CO2, *Int. J. Greenh. Gas Cont.* **22**, 47–62 (2014).

[7] I. N. Tsimpanogiannis and Y. C. Yortsos, Model for the gas evolution in a porous medium driven by solute diffusion, *AIChE Journal.* **48**(11), 2690–2710 (2002).

[8] L. Bernstein, P. Bosch, O. Canziani, Z. Chen, R. Christ, O. Davidson, W. Hare, S. Huq, D. Karoly, V. Kattsov, et al., IPCC, 2007: climate change 2007: synthesis report. Contribution of working groups I, II, and III to the Fourth Assessment Report of the Intergovernmental Panel on Climate Change, *Intergovernmental Panel on Climate Change, Geneva* (2007).

[9] US EPA. Inventory of US greenhouse gas emissions and sinks: 1990–2009. Technical report, US Environmental Protection Agency (2011).

[10] T. G. Poulsen and P. Møldrup, Evaluating effects of wind-induced pressure fluctuations on soil-atmosphere gas exchange at a landfill using stochastic modelling, *Waste Manage. Rese.* **24**(5), 473–481 (2006).

[11] M. Maier, H. Schack-Kirchner, M. Aubinet, S. Goffin, B. Longdoz, and F. Parent, Turbulence effect on gas transport in three contrasting forest soils, *Soil Sci. Soc. Am. J.* **76**(5), 1518–1528 (2012).

[12] US EPA. Draft guidance for evaluating the vapor intrusion to indoor air pathway from groundwater and soils (subsurface vapor intrusion guidance). Technical report, US Environmental Protection Agency (2002).

[13] D. A. Olson and R. L. Corsi, Fate and transport of contaminants in indoor air, *Soil Sediment Contam.* **11**(4), 583–601 (2002).

[14] F. D. Tillman and J. W. Weaver, *Review of Recent Research on Vapor Intrusion*. US Environmental Protection Agency, Office of Research and Development Washington, DC 20460 (2005).

[15] GICHD. Guide to mine action. Technical report, Geneva International Centre for Humanitarian Demining (2003).

[16] DOD. Developing science and technologies list: Section 2: Armaments and energetic materials technology. Technical report, US Department of Defense, Defense Threat Reduction Agency, Ft. Belvoir (2002).

[17] C. Bruschini, B. Gros, F. Guerne, P.-Y. Piéce, and O. Carmona, Ground penetrating radar and imaging metal detector for antipersonnel mine detection, *J. Appl. Geophys.* **40**(1), 59–71 (1998).

[18] J. R. Ballard, Jr, G. L. Mason, J. A. Smith, and L. K. Balick. Phenomenological models for landscape signatures: Review and recommendations. Technical report, DTIC Document (2004).

[19] B. S. Das, J. M. Hendrickx, and B. Borchers, Modeling transient water distributions around landmines in bare soils, *Soil Sci.* **166**(3), 163–173 (2001).

[20] J. Hendrickx, S. Hong, T. Miller, B. Borchers, H. Tobin, C. Bristow, and H. Jol. Soil effects on ground penetrating radar detection of buried nonmetallic mines in Ground Penetrating Radar: Applications in Sedimentology, C.S. Bristow and H. M. Jols(eds.), Geological Society, London, (2003).

[21] K. M. Smits, A. Cihan, T. Sakaki, S. E. Howington, J. F. Peters, and T. H. Illangasekare, Experimental and modeling investigation of soil moisture and thermal behavior in the vicinity of buried objects, *Geosci. Remote Sens.* (2013).

[22] J. Bear. *Dynamics of Fluids in Porous Media* (1972).

[23] J. W. Hopmans, D. R. Nielsen, and K. L. Bristow, How useful are small-scale soil hydraulic property measurements for large-scale vadose zone modeling?, *Geophysical Monograph Series.* **129**, 247–258 (2002).

[24] A. T. Corey, *Mechanics of Immiscible Fluids in Porous Media.* Water Resources Publication (1994).

[25] R. H. Brooks and A. T. Corey, Hydraulic properties of porous media, *Hydrology Papers, Colorado State University.* (March) (1964).

[26] M. T. Van Genuchten, A closed-form equation for predicting the hydraulic conductivity of unsaturated soils, *Soil Sci. So. Ame. J.* **44**(5), 892–898 (1980).

[27] A. S. Mayer and S. M. Hassanizadeh, *Soil and Groundwater Contamination: Nonaqueous Phase LiquidsPrinciples and Observations.* vol. 17, American

Geophysical Union (2005).

[28] Y.-K. Zhang and K. Schilling, Temporal scaling of hydraulic head and river base flow and its implication for groundwater recharge, *Water Resourc. Res.* **40**(3) (2004).

[29] G. Dagan et al., *Flow and Transport in Porous Formations.* Springer-Verlag GmbH & Co. KG. (1989).

[30] L. W. Gelhar, *Stochastic Subsurface Hydrology.* Prentice-Hall (1993).

[31] Y. Zhang, C. W. Gable, and M. Person, Equivalent hydraulic conductivity of an experimental stratigraphy: Implications for basin-scale flow simulations, *Water Resourc. Res.* **42**(5) (2006).

[32] P. Renard and G. De Marsily, Calculating equivalent permeability: A review, *Adv. Water Resourc.* **20**(5), 253–278 (1997).

[33] R. Fučík and J. Mikyška, Discontinuous Galerkin and Mixed–Hybrid Finite Element Approach to Two–Phase Flow in Heterogeneous Porous Media with Different Capillary Pressures, *Procedia Computer Science.* **4**, 908–917 (2011).

[34] R. Fučík and J. Mikyška, Mixed-hybrid finite element method for modelling two-phase flow in porous media, *Journal of Math-for-Industry.* **3**, 9–19 (2011).

[35] F. Fagerlund and G. Heinson, Detecting subsurface groundwater flow in fractured rock using self-potential (sp) methods, *Environmental Geology.* **43**(7), 782–794 (2003).

[36] B. Ataie-Ashtiani, S. Hassanizadeh, M. Celia, M. Oostrom, and M. D. White. Effective two-phase flow parameters for heterogeneous porous media. Technical report, Pacific Northwest National Laboratory (PNNL), Richland, WA (US) (2000).

[37] A. Montoglou and L. Gelhar, Effective hydraulic conductivities of transient unsaturated flow in stratified aquifer, *Water Resour. Res.* **23**(1), 57–67 (1987).

[38] I. Chatzis and F. Dullien, Modelling Pore Structure By 2-D And 3-D Networks With Application to Sandstones, *J. Can. Petrol. Technol.* **16**(1) (1977).

[39] B. H. Kueper and D. B. McWhorter, The use of macroscopic percolation theory to construct large-scale capillary pressure curves, *Water Resourc. Res.* **28**(9), 2425–2436 (1992).

[40] H. Rajaram, L. A. Ferrand, and M. A. Celia, Prediction of relative permeabilities for unconsolidated soils using pore-scale network models, *Water Resourc. Res.* **33**(1), 43–52 (1997).

[41] A. Desbarats, Upscaling capillary pressure-saturation curves in heterogeneous porous media, *Water Resourc. Res.* **31**(2), 281–288 (1995).

[42] A. Desbarats, Scaling of constitutive relationships in unsaturated heterogeneous media: A numerical investigation, *Water Resourc. Res.* **34**(6), 1427–1435 (1998).

[43] H. Liu, G. Bodvarsson, and G. Zhang, Scale dependency of the effective matrix diffusion coefficient, *Vadose Zone J.* **3**(1), 312–315 (2004).

[44] T. Lasseter, J. Waggoner, and L. Lake, Reservoir heterogeneities and their

influence on ultimate recovery, *Reservoir Characterization.* **545** (1986).
[45] A. Ahmadi and M. Quintard, Large-scale properties for two-phase flow in random porous media, *J. Hydrol.* **183**(1), 69–99 (1996).
[46] A. Bourgeat, Homogenized behavior of two-phase flows in naturally fractured reservoirs with uniform fractures distribution, *Comput. Methods in Appl. Mech. Eng.* **47**(1), 205–216 (1984).
[47] U. Hornung, *Homogenization and Porous Media.* vol. 6, Springer (1997).
[48] M. Quintard and S. Whitaker, Two-phase flow in heterogeneous porous media: The method of large-scale averaging, *Transport Porous Med.* **3**(4), 357–413 (1988).
[49] M. Quintard and S. Whitaker, Transport in ordered and disordered porous media ii: Generalized volume averaging, *Transport Porous Med.* **14**(2), 179–206 (1994).
[50] M. Quintard, Diffusion in isotropic and anisotropic porous systems: Three-dimensional calculations, *Transport Porous Med.* **11**(2), 187–199 (1993).
[51] D. Lasseux, A. Ahmadi, and A. A. A. Arani, Two-phase inertial flow in homogeneous porous media: A theoretical derivation of a macroscopic model, *Transport Porous Med.* **75**(3), 371–400 (2008).
[52] S. Whitaker, *The Method of Volume Averaging.* vol. 13, Springer (1999).
[53] A. Ahmadi, M. Quintard, and S. Whitaker, Transport in chemically and mechanically heterogeneous porous media: V. two-equation model for solute transport with adsorption, *Advances in Water Resources.* **22**(1), 59–86 (1998).
[54] M. Quintard, L. Bletzacker, D. Chenu, and S. Whitaker, Nonlinear, multicomponent, mass transport in porous media, *Chem. Eng. Sci.* **61**(8), 2643–2669 (2006).
[55] H. Davarzani, M. Marcoux, and M. Quintard, Theoretical predictions of the effective thermodiffusion coefficients in porous media, *Int. J. Heat and Mass Transfer.* **53**(7), 1514–1528 (2010).
[56] H. Davarzani and M. Marcoux, Influence of solid phase thermal conductivity on species separation rate in packed thermogravitational columns: A direct numerical simulation model, *Comptes Rendus Mecanique.* **339**(5), 355–361 (2011).
[57] F. Zanotti and R. Carbonell, Development of transport equations for multiphase systemsii: Application to one-dimensional axi-symmetric flows of two phases, *Chem. Eng. Sci.* **39**(2), 279–297 (1984).
[58] F. Zanotti and R. Carbonell, Development of transport equations for multiphase systemi: General development for two phase system, *Chem. Eng. Sci.* **39**(2), 263–278 (1984).
[59] Y. Davit, M. Quintard, and G. Debenest, Equivalence between volume averaging and moments matching techniques for mass transport models in porous media, *Int. J. Heat and Mass Transfer.* **53**(21), 4985–4993 (2010).
[60] C. T. Miller, M. M. Poirier-McNeil, and A. S. Mayer, Dissolution of trapped nonaqueous phase liquids: Mass transfer characteristics, *Water Resourc. Res.* **26**(11), 2783–2796 (1990).
[61] T. Saba and T. H. Illangasekare, Effect of groundwater flow dimensionality

on mass transfer from entrapped nonaqueous phase liquid contaminants, *Water Resourc. Res.* **36**(4), 971–979 (2000).

[62] T. Illangasekare and C. Frippiat. Miscible and immicible pollutants in subsurface systems. In ed. H. Fernando, *Handbook Of Environmental Fluid Dynamics*. Taylor and Francis Group (2010).

[63] Y. Liu, T. H. Illangasekare, and P. K. Kitanidis, Long–Term Mass Transfer and Mixing–Controlled Reactions of A DNAPL Plume from Persistent Residuals, *J. Contam. Hydrol.* **157**, 11–24 (2014).

[64] B. Petri, R. Fučík, T. H. Illangasekare, K. M. Smits, J. A. Christ, T. Sakaki, and C. C. Sauck, Effect of nonaqueous phase liquid source morphology on mass transfer in the vadose zone: Experimental and modeling study, to appear in *Groundwater J.* (2014).

[65] T. Illangasekare, C. Frippiat, and R. Fučík. Dispersion and mass transfer coefficients in groundwater of near-surface geologic formations. In eds. L. J. Thibodeaux and D. Mackay, *Handbook of Estimation Methods: Environmental Mass Transport Coefficients, Dispersion and Mass Transfer Coefficients*, pp. 413–452. CRC Press (2010).

[66] S. Saenton and T. H. Illangasekare, Upscaling of mass transfer rate coefficient for the numerical simulation of dense nonaqueous phase liquid dissolution in heterogeneous aquifers, *Water Resourc. Res.* **43**(2) (2007).

[67] B. T. Wilking, D. R. Rodriguez, and T. H. Illangasekare, Experimental Study of the Effects of DNAPL Distribution on Mass Rebound, *Groundwater J.* **51**(2), 229–236 (2013).

[68] M. Sakai, S. B. Jones, and M. Tuller, Numerical evaluation of subsurface soil water evaporation derived from sensible heat balance, *Water Resour. Res.* **47**(2) (2011).

[69] M. Bittelli, F. Ventura, G. S. Campbell, R. L. Snyder, F. Gallegati, and P. R. Pisa, Coupling of heat, water vapor, and liquid water fluxes to compute evaporation in bare soils, *J. Hydrol.* **362**(3), 191–205 (2008).

[70] R. S. Jassal, T. A. Black, M. D. Novak, D. Gaumont-Guay, and Z. Nesic, Effect of soil water stress on soil respiration and its temperature sensitivity in an 18-year old temperate douglas-fir stand, *Glob. Change Biol.* **14**(6), 1305–1318 (2008).

[71] P. Camillo and R. Gurney, A resistance parameter for bare-soil evaporation models, *Soil Sci.* **141**(2), 95–105 (1986).

[72] M. D. Novak, Dynamics of the near-surface evaporation zone and corresponding effects on the surface energy balance of a drying bare soil, *Agri. For. Meteorol.* **150**(10), 1358–1365 (2010).

[73] A. A. van de Griend and M. Owe, Bare soil surface resistance to evaporation by vapor diffusion under semiarid conditions, *Water Resour. Res.* **30**(2), 181–188 (1994).

[74] G. S. Campbell and J. M. Norman, *An Introduction to Environmental Biophysics*. Springer (1998).

[75] C. Desborough, A. Pitman, and P. Iranneiad, Analysis of the relationship between bare soil evaporation and soil moisture simulated by 13 land surface schemes for a simple non-vegetated site, *Global Planet. Change.* **13**(1),

47–56 (1996).
[76] H. Schmid, Experimental design for flux measurements: matching scales of observations and fluxes, *Agri. For. Meteorol.* **87**(2), 179–200 (1997).
[77] K. M. Smits, V. V. Ngo, A. Cihan, T. Sakaki, and T. H. Illangasekare, An evaluation of models of bare soil evaporation formulated with different land surface boundary conditions and assumptions, *Water Resour. Res.* **48**(12) (2012).
[78] L. Villagarcia, A. Were, F. Domingo, M. Garcia, and L. Alados-Arboledas, Estimation of soil boundary-layer resistance in sparse semiarid stands for evapotranspiration modelling, *J. Hydrol.* **342**(1), 173–183 (2007).
[79] W. Chao-Yang and C. Beckermann, A two–phase mixture model of liquid–gas flow and heat transfer in capillary porous media: I. Formulation, *Int. J. Heat and Mass Tran.* **36**, 2747–2747 (1993).
[80] J. Niessner and S. M. Hassanizadeh, Mass and heat transfer during two-phase flow in porous media-theory and modeling, *InTechOpen* (2011).
[81] C. Wang and P. Cheng, Multiphase flow and heat transfer in porous media, *Adv. Heat Transfer.* **30**, 93–196 (1997).
[82] H. Davarzani, K. Smits, R. M. Tolene, and T. Illangasekare, Study of the effect of wind speed on evaporation from soil through integrated modeling of the atmospheric boundary layer and shallow subsurface, *Water Resour. Res.* **50**(1), 661–680 (2014).
[83] K. Baber, K. Mosthaf, B. Flemisch, R. Helmig, S. Müthing, and B. Wohlmuth, Numerical scheme for coupling two-phase compositional porous-media flow and one-phase compositional free flow, *IMA J. Appl. Math.* **77**(6), 887–909 (2012).
[84] P. Chidyagwai and B. Rivière, A two-grid method for coupled free flow with porous media flow, *Adv. Water Resour.* **34**(9), 1113–1123 (2011).
[85] K. Mosthaf, K. Baber, B. Flemisch, R. Helmig, A. Leijnse, I. Rybak, and B. Wohlmuth, A coupling concept for two-phase compositional porous-medium and single-phase compositional free flow, *Water Resour. Res.* **47** (10) (2011).
[86] D. Nield, The beavers–joseph boundary condition and related matters: A historical and critical note, *Transport Porous Med.* **78**(3), 537–540 (2009).
[87] U. Shavit, Special issue on transport phenomena at the interface between fluid and porous domains, *Transport Porous Med.* **78**(3), 327–330 (2009).
[88] G. S. Beavers and D. D. Joseph, Boundary conditions at a naturally permeable wall, *J. Fluid Mech.* **30**(1), 197–207 (1967).
[89] P. G. Saffman, On the boundary condition at the surface of a porous medium, *Stud. Appl. Math.* **50**(2), 93–101 (1971).
[90] K. Vafai and S. Kim, Fluid mechanics of the interface region between a porous medium and a fluid layeran exact solution, *Int. J. Heat Fluid Fl.* **11**(3), 254–256 (1990).
[91] B. Alazmi and K. Vafai, Analysis of fluid flow and heat transfer interfacial conditions between a porous medium and a fluid layer, *Int. J. of Heat Mass Trans.* **44**(9), 1735–1749 (2001).
[92] H. Vereecken, R. Kasteel, J. Vanderborght, and T. Harter, Upscaling hy-

draulic properties and soil water flow processes in heterogeneous soils, *Vadose Zone J.* **6**(1), 1–28 (2007).

[93] Z. Wang, A. Tuli, and W. A. Jury, Unstable flow during redistribution in homogeneous soil, *Vadose Zone J.* **2**(1), 52–60 (2003).

[94] P. Lehmann and D. Or, Evaporation and capillary coupling across vertical textural contrasts in porous media, *Phys. Rev. E.* **80**(4), 046318 (2009).

[95] N. Shokri, P. Lehmann, and D. Or, Characteristics of evaporation from partially wettable porous media, *Water Resour. Res.* **45**(2) (2009).

[96] K. M. Smits, A. Cihan, T. Sakaki, and T. H. Illangasekare, Evaporation from soils under thermal boundary conditions: Experimental and modeling investigation to compare equilibrium-and nonequilibrium-based approaches, *Water Resour. Res.* **47**(5) (2011).

[97] P. Lehmann, I. Neuweiler, J. Vanderborght, and H.-J. Vogel, Dynamics of fluid interfaces and flow and transport across material interfaces in porous mediamodeling and observations, *Vadose Zone J.* **11**(3) (2012).

[98] R. Seager, M. Ting, I. Held, Y. Kushnir, J. Lu, G. Vecchi, H.-P. Huang, N. Harnik, A. Leetmaa, and N.-C. Lau, Model projections of an imminent transition to a more arid climate in southwestern North America, *Science.* **316**(5828), 1181–1184 (2007).

[99] U. Nachshon, A. Ireson, G. van der Kamp, and H. Wheater, Sulfate salt dynamics in the glaciated plains of north america, *J. Hydrol.* **499**, 188–199 (2013).

[100] K. Kondo, M. Tsuchiya, and S. Sanada, Evaluation of effect of microtopography on design wind velocity, *J. Wind Eng. Ind. Aerodyn.* **90**(12), 1707–1718 (2002).

[101] J. Li and S. Islam, On the estimation of soil moisture profile and surface fluxes partitioning from sequential assimilation of surface layer soil moisture, *J. Hydrol.* **220**(1), 86–103 (1999).

[102] J. D. Bolten, V. Lakshmi, and E. G. Njoku, Soil moisture retrieval using the passive/active l-and s-band radar/radiometer, *(IEEE) Trans. Geosci. Remote Sensing* **41**(12), 2792–2801 (2003).

[103] W. F. Krajewski, M. C. Anderson, W. E. Eichinger, D. Entekhabi, B. K. Hornbuckle, P. R. Houser, G. G. Katul, W. P. Kustas, J. M. Norman, C. Peters-Lidard, et al., A remote sensing observatory for hydrologic sciences: A genesis for scaling to continental hydrology, *Water Resour. Res.* **42**(7) (2006).

[104] D. K. Nykanen and E. Foufoula-Georgiou, Soil moisture variability and scale-dependency of nonlinear parameterizations in coupled land–atmosphere models, *Adv. Water Resour.* **24**(9), 1143–1157 (2001).

[105] W. T. Crow and E. F. Wood, Multi–scale dynamics of soil moisture variability observed during SGP97, *Geophys. Res. lett.* **26**(23), 3485–3488 (1999).

[106] P. R. Houser, W. J. Shuttleworth, J. S. Famiglietti, H. V. Gupta, K. H. Syed, and D. C. Goodrich, Integration of soil moisture remote sensing and hydrologic modeling using data assimilation, *Water Resour. Res.* **34**(12), 3405–3420 (1998).

[107] W. T. Crow and E. F. Wood, The assimilation of remotely sensed soil

brightness temperature imagery into a land surface model using ensemble Kalman filtering: A case study based on ESTAR measurements during SGP97, *Adv. Water Resour.* **26**(2), 137–149 (2003).

[108] S. Dunne and D. Entekhabi, An ensemble based reanalysis approach to land data assimilation, *Water Resour. Res.* **41**(2), W02013 (2005).

Chapter 11

Chemical Reactions in Diffusion-Limited Environments at the Pore-Scale

David A. Benson

Hydrologic Science and Engineering,
Colorado School of Mines, Golden, CO, USA
dbenson@mines.edu

In many natural systems, the lack of mixing is a primary limiting factor in the rate of chemical reactions. The poor mixing can be imparted by substrate structure, such as heterogeneous pore networks, or self-imposed by reactant segregation. The classical differential equations (DEs) of chemical reaction assume perfect mixing among reactants, and this assumption is transmitted to the most common solution methods. These methods are based on an Eulerian framework, in which space is partitioned into blocks of some size, within which the well-mixed solution is applied. Our recent work aims to rectify the problems associated with the well-mixed assumption in both theoretical and applied settings. First, we examine the effect of sub-grid concentration fluctuations on Eulerian simulations of reaction. Second, we show how a Lagrangian transport and reaction algorithm accounts for the diffusion-limited mixing of reactants. Third, we review the coarse-grained deterministic DEs that include the effects of concentration perturbation growth. This method includes the transport Green function, which can incorporate the specifics of the transport process. Finally, we review the connection between continuum perturbation methods and the numerical (Lagrangian) techniques, providing a map between the new and classical methods to predict chemical reaction rates. Because our methods are derived as scale-independent, virtually any reactive system that suffers from imperfect mixing—whether imposed by underlying structure (i.e., pore networks) or self-imposed by organization of reactants—will benefit from the added realism.

11.1. Introduction

Chemical reactions that take place in systems that are imperfectly mixed share several very subtle but important features. Not only do reactions slow down (in some cases drastically) relative to the well-mixed system, but the functional form of the reaction rate changes as well. As a result, the classical deterministic differential equations of reaction (based on the law of mass action) are not valid in imperfectly mixed systems—even with effective, upscaled parameters. This was first noticed in a theoretical setting over 60 years ago [1, 2], but the late-time asymptotic functional form for simple systems with diffusion-limited mixing was only discovered in the 1980s in research concerning the physics of the early expanding universe [3, 4].

Solutions to the equations of chemical transport and reaction can be approximated by several methods that fall into the broad categories of Eulerian and Lagrangian frameworks. The former divides space and/or time into regular grids and approximates the continuous solution on the fixed grid. The continuous solution is approximated by combinations of lower-order basis functions using, for example, finite-difference approximations of smoothly varying functions, finite-elements that approximate the solution as a sum of (commonly) piecewise-linear basis functions, or finite-volume methods that integrate fluxes into grid elements over finite time intervals. Excellent overviews of Eulerian methods are given by Celia and Gray [5] and Leveque [6]. On the other hand, the Lagrangian method assigns physical properties to "particles" that move with the velocity field. The properties may include mass of a chemical species, and the concentrations are not assigned to cells on a grid. Instead, the particles interact according to appropriate physical laws, and the final solution is obtained by interrogating the positions, densities, and properties of the particles. An overview of the Lagrangian method for non-reactive species is given by LaBolle et al. [7].

Eulerian simulations based on the law of mass action also enforce perfect mixing at the subgrid scale. This is due to the fact that concentration fluctuations cannot be represented within a single cell, which carries only one concentration per species. As grid cells become larger, the smoothing imparted by averaging and loss of information is typically treated by modifying the reaction rate coefficients either by wholesale (empirical) reduction or by some effective kinetic (often called pseudo-kinetic) term that treats poor mixing in an *ad-hoc* way. Eulerian codes and solutions continue to be the workhorse for predictions of radionuclide transport [8–10], CO_2 sequestration [11–13], ore body generation [14, 15], hydrothermal systems [16, 17], and basin-scale diagenesis, reservoir rock and petroleum generation [18–21] to name a few, so these predictions must be considered grid-scale dependent.

Imperfect mixing and the modified dynamics have also been revealed at the nano- to micro-meter scale in space and pico- to micro-second scale in time [22–24]. The smallest scale experiment to date examines the reaction of excited species within a carbon nanotube, revealing the ubiquity of transport-limited reaction, from molecule- to pore- to aquifer- (or atmospheric-) scales [25, 26]. This defines a significant theoretical and practical problem, because the bulk of simulations of reactive transport are based on empirical adjustments to law of mass action, which is built upon the well-mixed assumption.

From a scientific standpoint, the problem of translating a stochastic particle collision/transformation model into the upscaled governing PDEs has been studied for many decades [1–4, 27–31]. Newer particle-reaction simulations that make no assumptions about the scale or degree of mixing are able to match not only the reduced reaction rates, but also the different functional forms that point to completely different governing equations at larger scales. These Lagrangian models (see e.g., [32, 33] and references therein) also reproduce—without additional parameters such as kinetic reaction terms—the nontrivial random time that it takes to bring reactants together. One algorithm recently introduced has some unique advantages, including a direct solution of the chemical Langevin equation [34] with an exact expression for the probabilities of particle interaction [35, 36]. In this chapter, we illustrate the fundamental differences between Eulerian and Lagrangian methods and show the links between the two using analytic stochastic methods.

To highlight the effects of diffusion-limited transport and reactions, we concentrate on the simple bimolecular annihilation reaction $A + B \to \emptyset$ under imperfectly mixed conditions. We also choose a system with as few complicating effects as possible, i.e., low Poclet and high Damkohler numbers. The Peclet number (P_e) measures the ratio of advective to diffusive fluxes and the Damkohler number (D_a) measures the ratio of reaction to diffusion rates. In the simulations that follow, the reactive constituents are emplaced within a 2 mm pore at low velocity with some specified initial fluctuations in the concentration distribution (Fig. 11.1). With molecular diffusion coefficients on the order of 10^{-3} mm^2/s, the assumption of zero velocity is valid for Peclet numbers less than 0.1, which means that the average velocity must be less than approximately 10^{-4} mm/s (≈ 10 mm/d). This condition is often found in creeping flow in aquifer material. The reaction rate is chosen so that D_a is typically greater than 100, another condition that is often found in porous material.

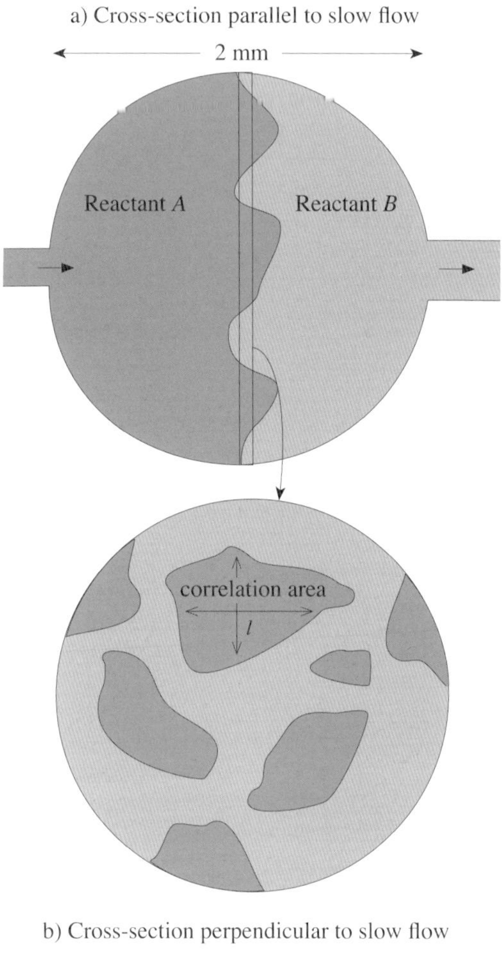

Fig. 11.1. Schematic representation of slow flow in a pore imparting heterogeneity to reacting species. The pore under study here (top) is 2 mm in diameter, and Peclet numbers are small enough to ignore advective flux. The 2-dimensional cross-section at the heterogeneous interface (bottom) holds the relevant heterogeneity that dictates the rates of diffusion-limited mixing and reaction.

11.2. An Illustrative System

Chemical reactions at the pore- to kilometer-scale are most often simulated using an approximation of the law of mass action. For a well-mixed bimolecular annihilation or precipitation reaction $A + B \rightarrow \emptyset$, the reaction rate follows

$$\frac{\partial C_A}{\partial t} = \frac{\partial C_B}{\partial t} = -KC_AC_B, \tag{11.1}$$

where C_i denotes molar concentration of species i. The rate constant K is a thermodynamic coefficient and contains all information about molecular-scale dynamics and energies of combination. The concentrations are not explicitly functions of space in this equation, therefore the concentrations are considered homogeneous, or well mixed. If spatial fluctuations exist, then $C_A = C_A(x, t)$, and the natural laws of transport must be simultaneously considered. If we look at advective and diffusive transport with local mean velocity vector $\mathbf{u}$ and diffusive/dispersive coefficient tensor D, then in the absence of reaction $\partial C_i/\partial t = -\nabla \cdot (\mathbf{u}C_i - D\nabla C_i)$, and commonly the chemical reaction equation is simply added to the transport equation as a mass sink:

$$\frac{\partial C_i}{\partial t} = -\nabla \cdot (\mathbf{u}C_i - D\nabla C_i) - KC_AC_B \tag{11.2}$$

Unfortunately, the concentrations used in equations (11.1) and (11.2) have very different definitions and meanings [37, 38], and this equation (or any other transport/reaction equation) will yield different solutions based on solution technique and assumptions. In short, (11.1) describes the mean concentration without fluctuations, while (11.2) describes the realistic concentrations that vary in space (i.e., within a pore).

11.3. Importance of Concentration Fluctuations

In an analytic sense, we may rectify the inconsistent concentrations in (11.1) and (11.2). Decomposing concentrations as the spatial mean plus some fluctuations: $C_i(x, t) = \overline{C}_i(t) + C'(x, t)$, we may replace the concentrations in (11.2) with these perturbed quantities. For simplicity of notation let the mean velocity be zero. Now taking the mean of (2) with the perturbed concentration we have [38]:

$$\frac{\partial \overline{C}_i}{\partial t} = -K\overline{C}_A\overline{C}_B - K\overline{C'_AC'_B}, \tag{11.3}$$

which shows the importance of the interaction of fluctuations. To explore the behavior of this simple system we shall take the example where the initial mean concentrations of A and B are equal. Under some mild assumptions, the final term is the Green function of transport of the concentration fluctuations [38], which can be quite complex given irregular boundary conditions within a pore. For diffusion in d-dimensions at early time before the boundaries have an influence on global transport, this equation becomes

$$\frac{\partial \overline{C}_i}{\partial t} = -K\overline{C}_A\overline{C}_B - K\frac{\sigma^2 l^d}{(8\pi Dt)^{1/2}}, \tag{11.4}$$

where σ^2 and l^d are the variance and d-dimensional correlation length of concentration perturbations. The correlation length is defined as the d-dimensional integral of the correlation function, so it has units of length in 1-d, area in 2-d, and volume in 3-d. There are several remarkable features of (11.4). The final term is small at early time, so that the well-mixed equation (11.1) is equivalent. At late time, the final term dominates, so that the segregation of species takes over and mixing by diffusion is the driver of reaction [38–40]. There is an analytic solution to (11.4), which transitions from the well-mixed solution to a power-law decline following

$$\overline{C}_i(t) \sim \frac{K\sigma^2 l^d}{2(8\pi D)^{d/4}}\, t^{-d/4}. \tag{11.5}$$

This simple example illustrates an important point. Any transport and reaction code that ignores the influence of subgrid concentration fluctuations should be solving a new equation that will resemble (11.3), which may or may not admit a tractable form with a solvable Green function such as (11.4). Because equations (11.3) and (11.1) are different, it is not proper to simply find an "effective" rate coefficient K (even a time-dependent form) for the original equations (11.1) or (11.2). This is illustrated immediately.

11.4. Eulerian Simulation Methods

This category includes the popular finite-difference (FD), finite-element (FE), and finite-volume (FV) approximations. These methods subdivide the spatial domain into a fixed grid. At scales smaller than a grid cell, all detail is lost, i.e., concentrations and other parameters/variables are considered homogeneous. Therefore, any Eulerian code that solves a traditional transport and reaction equation such as (11.2) will give different solutions using different discretization.

For an example problem, consider a circular dead-end pore (velocity negligible compared to diffusion) in 2-d, initially filled with equal masses of A and B, each with initial mean $\overline{C}_{A0} = \overline{C}_{B0} = 1$ M (Fig. 11.2). The pore is 2 mm in diameter, with $D = 10^{-3}$ mm^2/s, and $K = 100$ (sM)$^{-1}$. Now consider two scenarios, each with concentration perturbations of magnitude $\sigma = 0.1$ M. In the first scenario, the initial correlation of the perturbations is made to be isotropic with 2-d correlation length of 0.1×0.1 mm (Fig. 11.2) so that the "blobs" of initial concentration are roughly 1/20 of the pore diameter. In the second scenario, the perturbations are assumed to be very-fine scale relative to the discretization, so the concentration is modeled as uncorrelated (Fig. 11.3). Transport is by molecular diffusion only, so an explicit finite difference solution to (11.2) was coded in FORTRAN using square elements and constant $\Delta x = \Delta y$ throughout the domain. A series of simulations was run with identical statistical parameters of concentration variance and correlation length of the initial condition while varying the discretization from $\Delta x = 2$ mm (the size of the entire pore) down to $\Delta x = 0.0025$ mm. The analytic solution to the well-mixed problem (11.1) with these conditions is $\overline{C}_A(t) = \overline{C}_B(t) = \overline{C}_{A0}(1 + \overline{C}_{A0}Kt)^{-1}$, which declines like $1/t$ at late time (Fig. 11.4).

As expected, the $\Delta x = 2$ mm solution in either initial correlation-length scenario matches the well-mixed analytic solution because there are no concentration fluctuations inside any single numerical block. As Δx is made smaller, the initial concentration fluctuations eventually engender "islands" composed primarily of one or the other reactant (Fig 11.2, left column). This is due to natural and random (albeit possibly very slight) differences in initial concentrations in certain areas. One or the other reactant is depleted, leaving the islands, and forcing the reactions to take place at the boundaries. Reactants must migrate from the island interiors to the edges for reactions to take place. This leads to a decreased reaction speed dictated by molecular diffusion and mixing. If the discretization is smaller than the assumed initial correlation length of concentration perturbations, then the solution will be accurate for this system, because the islands only grow. Hence the similarity in the $\Delta x = 0.025$ and 0.0025 mm solutions when the initial correlation length is 0.1 mm (Fig. 11.4). Note, however, the error incurred by the $\Delta x = 0.25$ mm solution, which cannot resolve subgrid heterogeneity. At late time, the well-mixed equation may over-predict the reaction by orders-of-magnitude (Fig. 11.4), while the intermediate

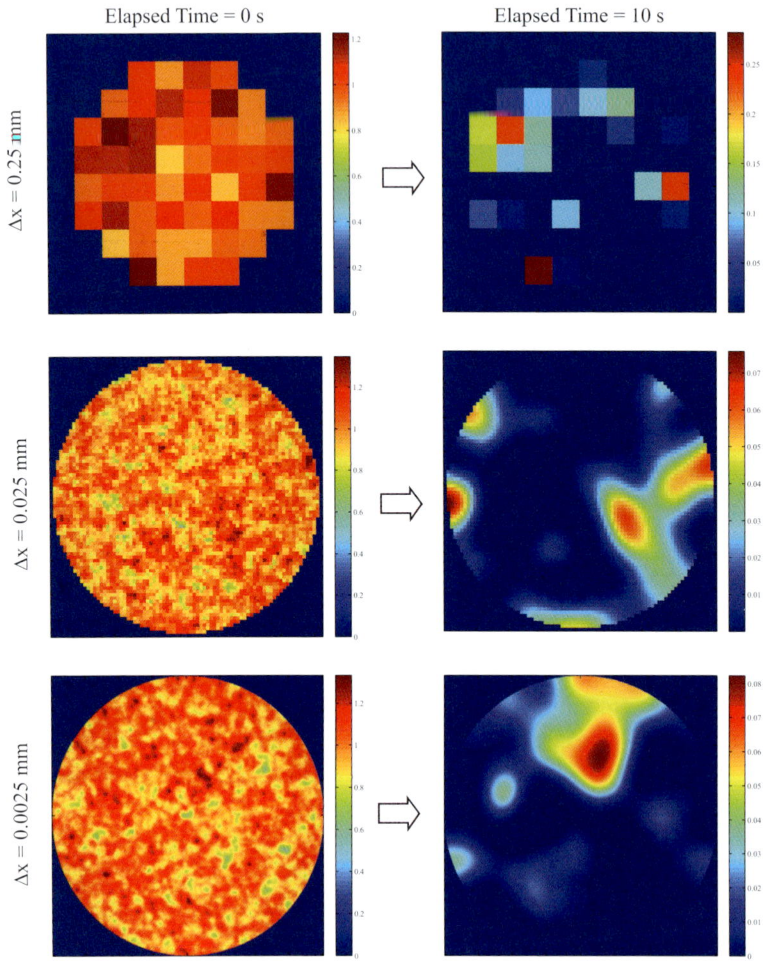

Fig. 11.2. Finite-difference simulation of the diffusion-reaction equation (2) with an initially correlated chemical structure (correlation length = 0.02 mm) using different grid sizes. All other parameters (listed in text) are held constant. All plots show concentrations of reactant A in a 2 mm pore. Left column are at $t = 0$, right column at $t = 10$ s.

discretization ($\Delta x = 0.25$ mm) under-predicts by a similar amount.

It is interesting to note that even "uncorrelated" initial conditions (i.e., Dirac delta function correlation in a numerical sense) self-organize and take on the island structures and give way to diffusion-limited reaction rates (Fig. 11.3, bottom row). If noise is given to the grid cells, then an underly-

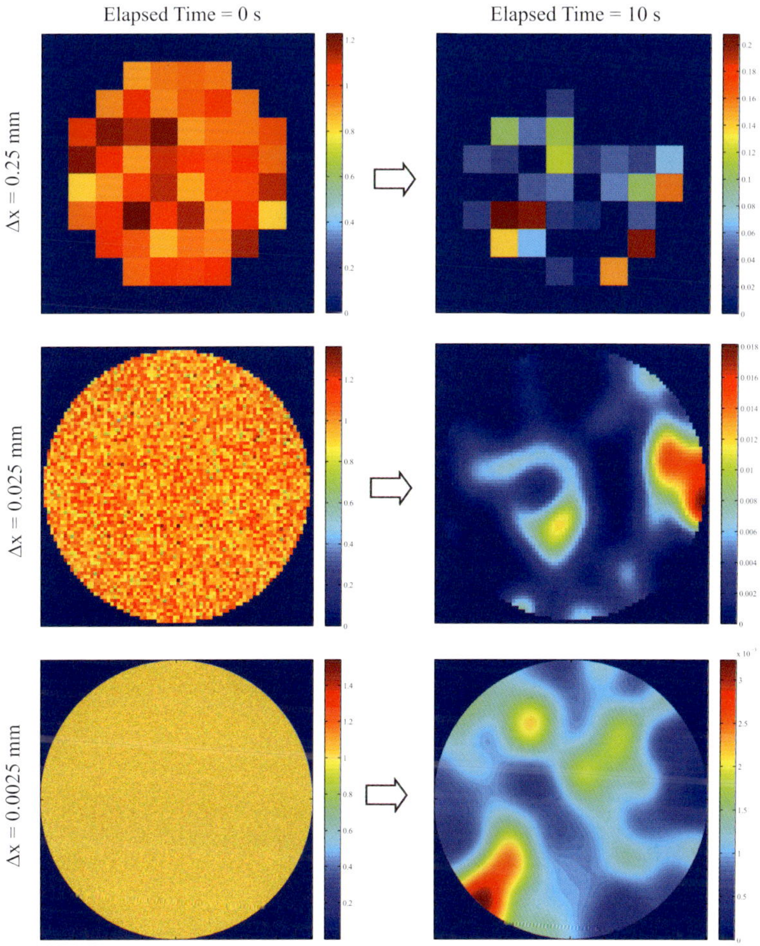

Fig. 11.3. Finite-difference simulation of the diffusion-reaction equation (2) using different grid size and uncorrelated chemical structure. All other parameters (listed in text) are held constant. All plots show concentrations of reactant A in a 2 mm pore. Left column are at $t = 0$, right column at $t = 10$ s.

ing correlation with length Δx is implied because of constant concentrations within each cell. Therefore, every discretization will give different overall reaction rates (Fig. 11.5). When the pore is simulated with more than one element, the solution can change drastically from the well-mixed (single element) solution (Fig. 11.5). There are also more subtle differences, including pore boundary effects that are felt at different times by different

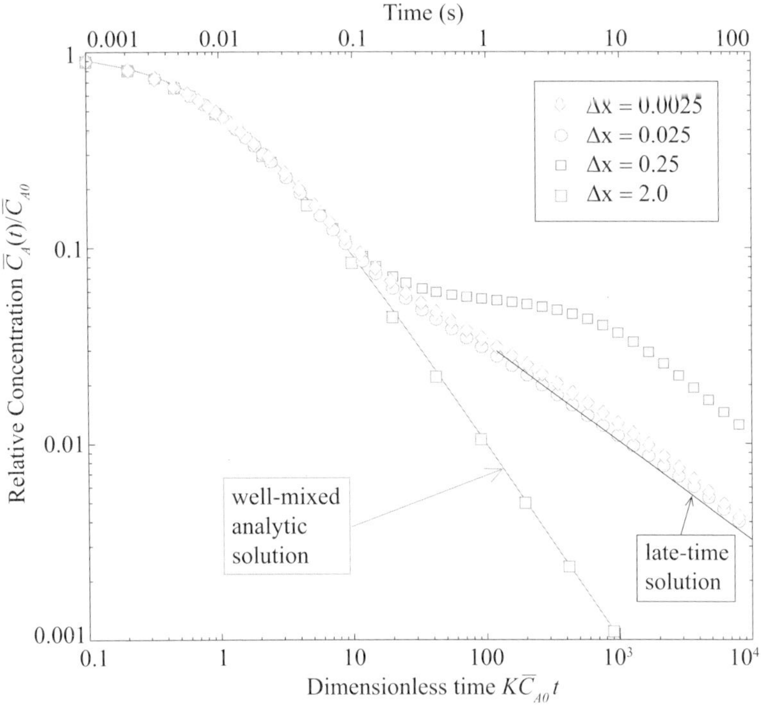

Fig. 11.4. Average concentration of reactant A remaining in pore for different discretizations of finite-difference solutions of the diffusion-reaction equation (11.2) when the initial correlation length = 0.02 mm. Also shown are the well-mixed solution and the late-time asymptotic solution (11.5) based on the concentration perturbation growth Green function.

size meshes. In other words, the shape of the pore will make a difference, but this occurs only at very late time [36].

11.5. Lagrangian Simulation Methods

The primary advantage of Lagrangian (particle-tracking) simulations for solute transport is that the solution is independent of scale and discretization. Particles are moved along advective flowlines and diffused/dispersed by random walks. It is easy to show that the expected number of particles corresponds to the concentration in a transport equation, and also that the particle number variance, hence concentration variance, is proportional to the number of particles used in a simulation. This observation motivated

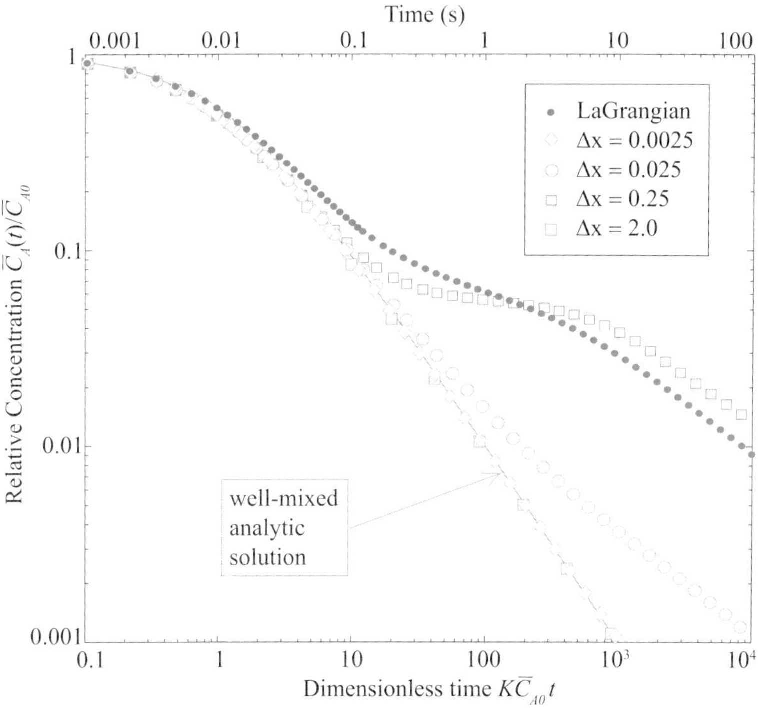

Fig. 11.5. Average concentration of reactant A remaining in pore for different discretizations of finite-difference solutions of the diffusion-reaction equation (11.2) for uncorrelated initial condition. Also included is a Lagrangian simulation with particles approximating the initial condition of the $\Delta x = 0.25$ simulation.

Benson and Meerschaert [39] to derive the probabilistic requirements for particle-based chemical reactions based on the proximity of particles to one another.

The Lagrangian viewpoint has taken several forms, first as particles on a lattice with simple rules of interaction [3, 4]. Because the lattice spacing is directly tied to a diffusion distance, this approach is effectively ruled out for large-scale simulations. Instead, the real-valued position of all particles can be tracked and an exact expression of the particle co-location probability due to local diffusion can be calculated [36, 39, 41]. Denote the positions of two A and B particles by a and b (Fig. 11.6). In some small amount of time the particles will move Δx_A and Δx_B. The probability of moving to any and all positions x is $P(a + \Delta x_A = b + \Delta x_B) = P(\Delta x_A - \Delta x_B = b - a)$. Denote the initial separation $b - a = s$

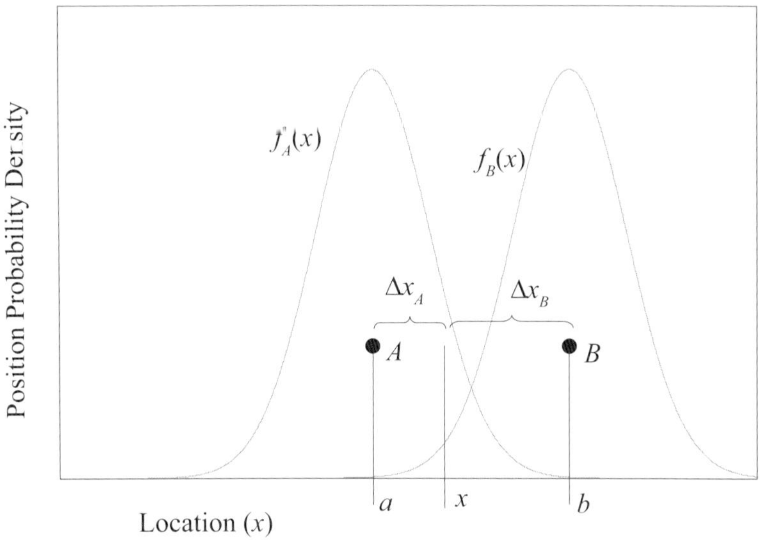

Fig. 11.6. Illustration of the exact calculation of the co-location of two particles. Overlapping location densities imply non-zero probability of co-location. After [39].

and the probability of co-location is probability of the difference between the two random motions. If at least one random motion is symmetric, as in the case of Brownian motion diffusion, then the probability density of collocation is given by the convolution of the motion densities $f_A(x)$ and $f_B(x)$. This convolution is a Gaussian for Brownian motion and is represented by a density function $v(s)$. This probability is combined with the thermodynamic conditional probability that two particles will react given the co-location, i.e., the well-mixed rate. The exact expression for the probability of two particles reacting in the system $A + B \to C$ is [39, 40]:

$$P(\text{forward react}) = K\Delta t \frac{\Omega C_{A0}}{N_0} v(s), \tag{11.6}$$

where Ω is the d-dimensional domain size and N_0 is the initial number of particles. The portion $\Omega C_{A0}/N_0$ is the mass (or moles) of a single particle so that the equation may represent anything from a single molecule to a small parcel of mass that travels through a system. This expression can be viewed as an extension of the Gillespie method and recent improvements [28, 42], which treats the co-location probability density $v(s)$ as equal for all particles, i.e., the well-mixed assumption.

Numerous parametric studies showed that the particle simulations faithfully reproduce the creation of concentration islands and the departure from the well-mixed scenario [35, 39]. Paster et al. [40] recently showed that the only apparently free parameter N_0, is actually directly specified by the initial concentration statistics used to initialize the Eulerian simulations:

$$N_0 = \frac{C_{A0}^2 \Omega}{\sigma^2 l^d}, \tag{11.7}$$

where σ^2 and l^d are the magnitude and d-dimensional correlation length of concentration perturbations in a physical system. The relation (11.7) is based on equating the Dirac-delta correlation of a single particle to the two-point correlation statistics of a smooth concentration field, which may or may not be a valid approximation. As an example, the uncorrelated finite-difference solutions in the previous section enforce a correlation length of Δx because concentrations are constant in any rectangular grid cell. For the $\Delta x = 0.25$ finite-difference simulations, (11.7) dictates that a particle simulation would require $N_0 = (1\ M)^2 \cdot \pi(1\ \text{mm})^2/(0.01\ M^2 \cdot [0.25\ \text{mm}]^2) =$ 5030 particles. But the delta function is a poor approximation for the linear correlation function, so each randomly placed particle was split into 10 that were also randomly placed within a radius of 0.14 (for equal area) from the parent particle. These simulations are reasonably close to the finite-difference solution (Fig. 11.5).

11.5.1. *Advection-dominated Flow*

The Lagrangian reaction method outlined above is extremely flexible, in that there is no requirement of an underlying fine-scale grid. The particles are neither tied to pores nor molecules. To illustrate, Ding et al. [43] took the data from an experiment by Gramling et al. [44], in which one reactant displaced another that completely filled the pore space in a translucent column packed with cryolite sand. All transport parameters previously were determined by analyzing the displacement of an inert tracer. As the displacement of reactive fluids progressed, the reaction created a product that changed the color of the fluid, and could be quantified by a high-resolution CCD camera (Fig. 11.7(a)). Notable in the image is the apparent heterogeneity in the concentrations. Ding et al. [43] analyzed the images, including the inert tracer, and found a correlation length on the order of 1 to 2 cm—spanning several pores—and were able to simulate the reactions without any empirical reaction parameters such as kinetic rate coefficients

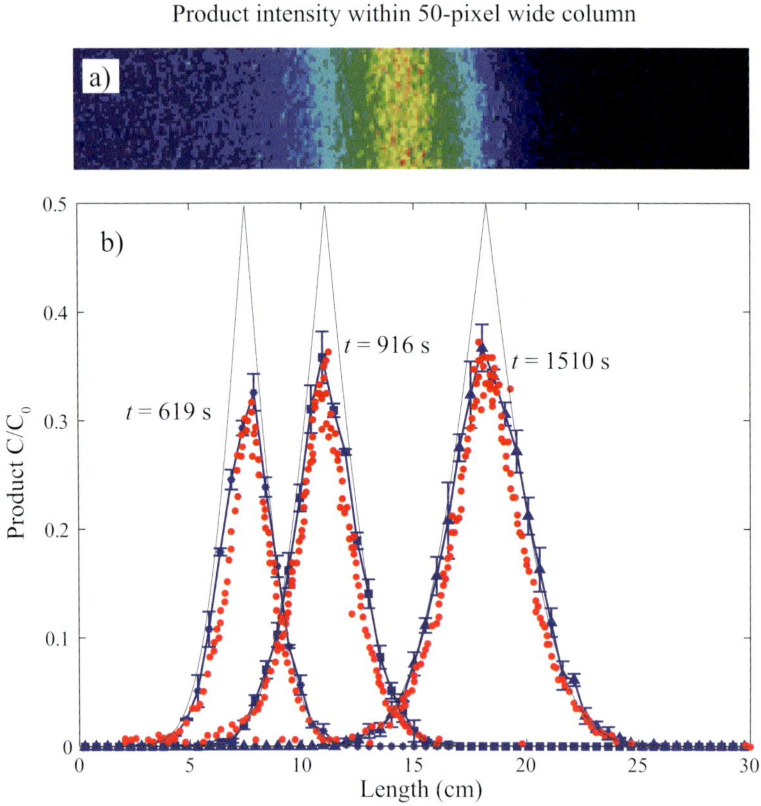

Fig. 11.7. a) False color image of light intensity transmitted through reaction product at a time of approximately 1200 s. b) Red dots are vertically integrated measured product concentrations at three separate times. Blue error bars are predictions from an ensemble of particle transport and reaction simulations. Thin grey curves denote analytic solution of (11.2) (see Ref. [44]). After Ref. [43].

(Fig. 11.7(b)). The required number of particles by (11.7) was 680, which runs extremely quickly on a laptop computer.

The analytic solution for a well-mixed system has similar shape to the measurements and simulations but peaks at a relative concentration of exactly 0.5. The measurements and particle simulations have product concentrations on the order of 30% less than the transport/reaction equation, even for an essentially instantaneous reaction in homogeneous sand.

Another realistic feature that the particle simulations provide relative to the well-mixed solutions is the spatial distribution of the reactants. Each

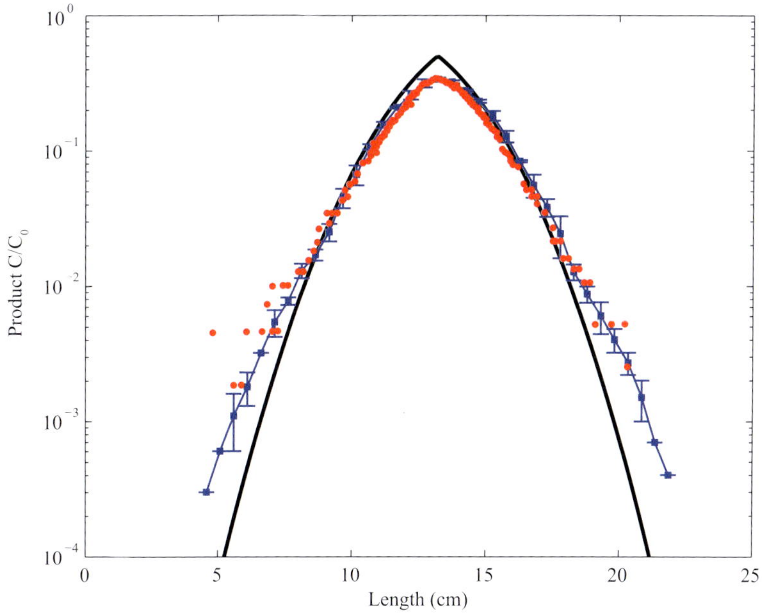

Fig. 11.8. Semi-log plot of measured (red dots), simulated by particles (blue error bars), and predicted by the analytic solution of the classical transport/reaction equation (11.2) (black curve). After [43].

particle has a chance of not reacting until they have ventured some distance into "enemy territory." The tails of the product distribution should be wider than those predicted by (11.2). Indeed, both the measured product distribution and the particle simulations show greater distances traveled by the low concentration tails before reacting out (Fig. 11.8).

11.6. Summary and Future Work

Similar to simulations in other branches of science, a critical component of pore- to field-scale reactive transport is the treatment of sub-grid fluctuations of parameters and dependent variables. Because of the clearly different form of differential equation that arises from upscaling and combining transport and reaction, it is not rendered simple by finding "effective" parameters at a new scale. One is either faced with discretizing at finer and finer scales in an Eulerian approach (down to the chemical correlation length), or solving new and different upscaled equations at different scales, or perhaps using a simulation technique that not tied to any scale or form of an equation. The Lagrangian reaction technique only codes the small scale physics of transport and collision and the thermodynamics of reaction. The fact that the solutions match upscaled equations is purely due to the propagation of probabilities and small-scale physics upward through the scales of observation. Viewed in this way, the particle reaction codes have the potential to reside in simulations of field-scale phenomena—particularly slow reactions and heterogeneous chemistry—that arise from pore-scale processes. This work is underway [45].

Owing to their recent development, the particle reaction simulations created thus far are for very simple reaction chains and simple flow fields and geometries. More complicated reactions can always be re-cast as a series of mono- and bi-molecular intermediate reactions, but the numerical implementation of this process can be slow. We are working on these algorithms [46]. We also intend to incorporate the Lagrangian reaction algorithm into fully 3-dimensional transport codes that can incorporate dynamic (such as turbulent) flow fields. Much work also remains on optimizing the search algorithms implicit in the calculation of particle collision probabilities.

Nomenclature

C_i	Point concentration of species i in d-dimensions (Mol L^{-d})
$\overline{C}_i$	Average concentration of species i in d-dimensions (Mol L^{-d})
C'_i	Concentration fluctuation of species i in d-dimensions (Mol L^{-d})
d	Number of dimensions under study (-)
D	Molecular diffusion coefficient (L^2 T^{-1})
$\Delta x, \Delta y$	Size of Eulerian blocks used in simulations (L)
K	Thermodynamic rate coefficient for n^{th}-order reaction (Mol^{1-n} $\mathrm{L}^{d(n-1)}$ T^{-1})
l^d	d-dimensional correlation volume of the typical fluctuation sizes (L^d)
N_0	Initial number of particles (-)
Ω	Size of domain (L^d)
$P(E)$	Probability of event E (-)
σ^2	Variance of concentration fluctuations (Mol^2 L^{-2d})
$\mathbf{u}$	Mean velocity vector (L T^{-1})

$v(s)$ Probability density of two particles colliding when initially separated by vector s (L^{-d})

References

[1] F. C. Collins and G. E. Kimball, Diffusion-controlled reaction rates, *J. Colloid Sci.* **4**, 425–437 (1949).

[2] T. R. Waite, Theoretical treatment of the kinetics of diffusion-limited reactions, *Phys. Rev.* **107**(2), 463-470 (1957).

[3] D. Toussaint and F. Wilczek, Particle-antiparticle annihilation in diffusive motion, *J. Chem. Phys.* **78**, 2642–2647 (1983).

[4] K. Kang and S. Redner, Fluctuation-dominated kinetics in diffusion-controlled reactions, *Phys. Rev. A.* **32**, 435–447 (1985).

[5] M. Celia and W. Gray, *Numerical Methods for Differential Equations: Fundamental Concepts for Scientific and Engineering Applications.* Prentice Hall (1992).

[6] R. J. LeVeque, *Finite Volume Methods for Hyperbolic Problems.* Cambridge University Press (2002).

[7] E. M. Labolle, G. E. Fogg, and A. F. B. Tompson, Random-walk simulation of transport in heterogeneous porous media: Local mass-conservation problem and implementation methods, *Water Resour. Res.* **32**(3), 583–593 (1996).

[8] R. Ma, C. Zheng, H. Prommer, J. Greskowiak, C. Liu, J. Zachara, and M. Rockhold, A field-scale reactive transport model for u(vi) migration influenced by coupled multirate mass transfer and surface complexation reactions, *Water Resour. Res.* **46**(5), W05509 (2010).

[9] J. Greskowiak, H. Prommer, C. Liu, V. E. A. Post, R. Ma, C. Zheng, and J. M. Zachara, Comparison of parameter sensitivities between a laboratory and field scale model of uranium transport in a dual domain, distributed rate reactive system, *Water Resour. Res.* **46**(9), W09509 (2010).

[10] G. E. Hammond and P. C. Lichtner, Field-scale model for the natural attenuation of uranium at the Hanford 300 area using high-performance computing, *Water Resour. Res.* **46**(9), W09527 (2010).

[11] B. R. Strazisar, C. Zhu, and S. W. Hedges, Preliminary modeling of the long-term fate of CO2 following injection into deep geological formations, *Environ. Geosci.* **13**(1), 1–15 (2006).

[12] P. Audigane, I. Gaus, I. Czernichowski-Lauriol, K. Pruess, and T. Xu, Two-dimensional reactive transport modeling of CO2 injection in a saline aquifer at the Sleipner site, North Sea, *Am. J. Sci.* **307**(7), 974–1008 (2007).

[13] W. S. Han, B. J. McPherson, P. C. Lichtner, and F. P. Wang, Evaluation of trapping mechanisms in geologic CO2 sequestration: Case study of SACROC northern platform, a 35-year CO2 injection site, *Am. J. Sci.* **310**(4), 282–324 (2010).

[14] J. P. Raffensperger, Evidence and modeling of large-scale groundwater con-

vection in precambrain sedimentary basins. In *Basin-Wide Diagenetic Patterns*, vol. 57, pp. 15–26. SEPM (Society for Sedimentary Geology) (1997).

[15] C. Schardt, D. R. Cooke, J. B. Gemmell, and R. R. Large, Geochemical modeling of the zoned footwall alteration pipe, Hellyer volcanic-hosted massive sulfide deposit, western Tasmania, Australia, *Econ. Geol.* **96**(5), 1037–1054 (2001).

[16] P. C. Lichtner, Continuum model for simultaneous chemical reactions and mass transport in hydrothermal systems, *Geochimica et Cosmochimica Acta*. **49**(3), 779–800 (1985).

[17] C. I. Steefel and A. C. Lasaga, A coupled model for transport of multiple chemical species and kinetic precipitation/dissolution reactions with application to reactive flow in single phase hydrothermal systems, *Am. J. Sci.* **294**(5), 529–592 (1994).

[18] M.-K. Lee. Predicting diagenetic effects of groundwater flow in sedimentary basins: a modeling approach with examples. In *Basin-Wide Diagenetic Patterns*, vol. 57, pp. 3–14. SEPM (Society for Sedimentary Geology) (1997).

[19] J. W. Morse, J. S. Hanor, and S. He. The role of mixing and migration of basinal waters in carbonate mineral mass transport. In *Basin-Wide Diagenetic Patterns*, vol. 57, pp. 41–50. SEPM (Society for Sedimentary Geology) (1997).

[20] A. M. Wilson, J. R. Boles, and G. Garven, Calcium mass transport and sandstone diagenesis during compaction-driven flow: Stevens Sandstone, San Joaquin basin, California, *Geol. Soc. Amer. Bull.* **112**(6), 845–856 (June, 2000).

[21] G. D. Jones and Y. Xiao, Geothermal convection in the Tengiz carbonate platform, Kazakhstan: Reactive transport models of diagenesis and reservoir quality, *AAPG Bulletin*. **90**(8), 1251–1272 (2006).

[22] E. Monson and R. Kopelman, Observation of laser speckle effects and nonclassical kinetics in an elementary chemical reaction, *Phys. Rev. Lett.* **85**(3), 666–669 (2000).

[23] E. Monson and R. Kopelman, Nonclassical kinetics of an elementary $A+B \rightarrow C$ reaction-diffusion system showing effects of a speckled initial reactant distribution and eventual self-segregation: Experiments, *Phys. Rev. E*. **69** (2), 021103– (2004).

[24] J. Allam, M. T. Sajjad, R. Sutton, K. Litvinenko, Z. Wang, S. Siddique, Q.-H. Yang, W. H. Loh, and T. Brown, Measurement of a reaction-diffusion crossover in exciton-exciton recombination inside carbon nanotubes using femtosecond optical absorption, *Phys. Rev. Lett.* **111**(19), 197401– (2013).

[25] N. Nakamura. Quantifying inhomogeneous, instantaneous, irreversible transport using passive tracer field as a coordinate. In eds. J. Weiss and A. Provenzale, *Lecture Notes in Physics*, vol. 744, pp. 137–164–. Springer Berlin Heidelberg (2008).

[26] M. Lévy. The modulation of biological production by oceanic mesoscale turbulence. In eds. J. Weiss and A. Provenzale, *Lecture Notes in Physics*, vol. 744, pp. 219–261–Springer Berlin Heidelberg (2008).

[27] M. von Smoluchowski, Versuch einer mathematischen theorie der koagula-

tionskinetik kolloider lösungen, *Z. Phys. Chem.* **92**, 124–168 (1917).
[28] D. T. Gillespie, Exact stochastic simulation of coupled chemical reactions, *J. Phys. Chem.* **81**(25), 2340–2361 (1977).
[29] N. G. van Kampen, *Stochastic Processes in Physics and Chemistry*. Elsevier, Amsterdam (1992).
[30] C. W. Gardiner, *Handbook of Stochastic Methods for Physics, Chemistry and the Natural Sciences, 3rd edition.* Springer, Berlin (2004).
[31] M. Doi, Stochastic theory of diffusion-controlled reaction, *J. Phys. A: Mathematical and General.* **9**, 1479–1495 (1976).
[32] S. A. Isaacson, A convergent reaction-diffusion master equation, *J. Chem. Phys.* **139**(5), – (2013).
[33] D. T. Gillespie, A. Hellander, and L. R. Petzold, Perspective: Stochastic algorithms for chemical kinetics, *J. Chem. Phys.* **138**(17), 170901 (2013).
[34] D. T. Gillespie, The chemical Langevin equation, *J. Chem. Phys.* **113**(1), 297–306 (2000).
[35] D. A. Benson, D. Bolster, and A. Paster, Communication: A full solution of the annihilation reaction $A + B \to \emptyset$ based on time-subordination, *J. Chem. Phys.* **138**(13), 131101 (2013).
[36] A. Paster, D. Bolster, and D. A. Benson, Connecting the dots: Semi-analytical and random walk numerical solutions of the diffusion-reaction equation with stochastic initial conditions, *J. Comput. Phys.* **263**, 91–112 (2014).
[37] D. Raje and V. Kapoor, Experimental study of bimolecular reaction kinetics in porous media, *Environ. Sci. & Tech.* **34**(7), 1234–1239 (APR 1, 2000).
[38] A. M. Tartakovsky, P. de Anna, T. Le Borgne, A. Balter, and D. Bolster, Effect of spatial concentration fluctuations on non-linear reactions in diffusion-reaction systems, *Water Resour. Res.* **48**, W02526 (2012).
[39] D. A. Benson and M. M. Meerschaert, Simulation of chemical reaction via particle tracking: Diffusion-limited versus thermodynamic rate-limited regimes, *Water Resour. Res.* **44**, W12201 (2008).
[40] A. Paster, D. Bolster, and D. A. Benson, Particle tracking and the diffusion-reaction equation, *Water Resour. Res.* **49**, 1–6 (2013).
[41] J. S. van Zon and P. R. ten Wolde, Simulating biochemical networks at the particle level and in time and space: Green's function reaction dynamics, *Phys. Rev. Lett.* **94**(12), 128103– (2005).
[42] Y. Cao, D. Gillespie, and L. Petzold, Efficient step size selection for the tau-leaping simulation method, *J. Chem. Phys.* **124**(2), 044109 (2006).
[43] D. Ding, D. Benson, A. Paster, and D. Bolster, Modeling bimolecular reactions and transport in porous media via particle tracking, *Adv. Water Resour.* **53**, 56–65 (2012).
[44] C. Gramling, C. Harvey, and L. Meigs, Reactive transport in porous media: A comparison of model prediction with laboratory visualization, *Environ. Sci. Technol.* **36**(11), 2508–2514 (JUN 1, 2002).
[45] D. A. Benson, N. Engdahl, T. Aquino, and D. Bolster, An illustration of the fundamental differences between Eulerian and Lagrangian transport and non-linear reaction algorithms, Submitted (2014).

[46] D. Ding and D. A. Benson, Simulating biodegradation under mixing-limited conditions using Michaelis-Menton (Monod) kinetic expressions in a particle tracking model, Submitted (2014).

Chapter 12

Porosity in Reactive Geochemical Systems

Alexis Navarre-Sitchler

Geology and Geological Engineering,
Colorado School of Mines, Golden, CO, USA
asitchle@mines.edu

Gernot Rother

Chemical Sciences Division,
Oak Ridge National Lab., Oak Ridge, TN, USA
rotherg@ornl.gov

John Kaszuba

Department of Geology and Geophysics and School of Energy Resources,
U. of Wyoming, Laramie, WY, USA
john.kaszuba@uwyo.edu

Despite the importance of geochemical reactions in many natural and engineered systems, we have not been able to measure geochemical reaction-rate constants in the laboratory in such a way to be consistent with field observations. Laboratory measured mineral dissolution rates are often two to six orders of magnitude faster than effective reaction rates measured in field systems, partly due to the impact of heterogeneity and fluid flow on geochemical reactions in field systems that are not accurately represented in laboratory experiments. To advance our ability to predict geochemical processes and their products in field systems we need a better understanding of how fluid flow and geochemical reactions are coupled at scales ranging from pores to watersheds. Recent advances in experimental techniques to quantify surface area and porosity at nanometer to micron length scales using neutron scattering are described that provide an opportunity to advance our understanding of coupled reactive transport at the pore scale.

12.1. Introduction

Geochemical reactions play an important role in many natural and engineered systems, from dissolved nutrient and soil production through rock weathering [1–6] to sequestration of CO_2 in mineral form either naturally or in engineered CO_2 sequestration projects [7, 8]. Engineered solutions to many of our energy needs rely on injection of fluids into the subsurface to facilitate energy production. In order to evaluate the various proposed methods of energy production, we need to understand how injected fluids geochemically interact with the minerals that make up rocks in the subsurface, and any changes to the physical, chemical, or mechanical properties of the rocks that result. For example, CO_2 is currently being evaluated as a working fluid in enhanced geothermal energy systems [9], and has been used for enhanced oil recovery for over 30 years. Hydraulic fracturing used to produce shale gas and shale oil from unconventional fields requires the injection of fluids into the subsurface. These fluids can be acidic, leading to coupled geochemical processes in large-scale projects like enhanced oil recovery and carbon capture and storage projects [10]. It is possible that CO_2 can serve as a main component of the fluids used in hydraulic fracturing, combining the two efforts. Thus, the study of geochemical reactions at Earth's surface and in the subsurface plays a vital role in our understanding of many important systems.

Numerical models used to describe and investigate such complex, time-dependent geochemical systems require kinetic rate data to quantify mineral reactions and their products through time. These mineral reaction rates are often based on laboratory mineral dissolution experiments performed in well-mixed batch reactors or homogeneous columns. Our ability to apply experimentally determined mineral reaction kinetics to simulation of natural systems is problematic due to the observed two to six order of magnitude difference in mineral dissolution rates measured in laboratory experiments versus those calculated with field data [8, 11–15]. Well-mixed batch reactors or homogeneously packed columns are typically used to measure reaction rates in laboratory conditions [16]. In rocks, however, mineral reaction rates are influenced by transport of reactive fluids through highly heterogeneous porous material with low permeability zones [17, 18], resulting in a system that is poorly mixed and, not surprisingly, deviates from behavior measured under well-mixed laboratory conditions (Fig. 12.1).

After decades of careful measurement of mineral dissolution kinetics under laboratory conditions [13, 19–29], it became clear that a variation

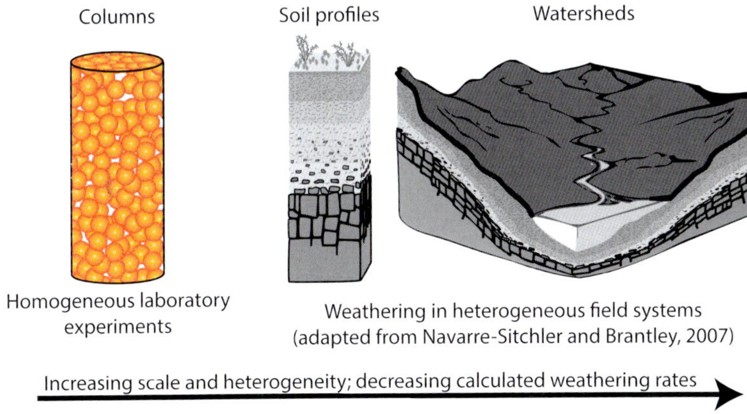

Fig. 12.1. Laboratory experiments used to measure mineral dissolution rates are comprised of homogeneous mineral assemblenges in columns or batch reactors. In natural systems weathering occurs in heterogeneous environments with multi-mineralic assembleges and variations in flow rate that influence calculated weathering rates.

of approximately 2.5 orders of magnitude in measured reaction rates can be attributed to intrinsic factors such as dislocations and other crystal defects, compositional variations, and the stochastic nature of interactions between a fluid and the mineral surface [30]. In addition, extrinsic factors account for some of the difference between laboratory and field measured reaction rates, including development of leach layers or armoring on mineral surfaces [11], availability of mineral surface area [8, 31, 32], reaction affinity [14, 33], composition of reacting fluids [25, 34], physical erosion [35–37], slow precipitation of weathering products [38], and biological activity [39–41]. Observed relationships between solute transport and runoff suggest that fluid transport influences apparent mineral reaction rates in large field systems [42–44]. Pore scale transport and evolution of transport pathways also contribute to differences between laboratory measured mineral dissolution rates and effective reaction rates in field settings [45, 46].

While flow within an individual pore appears to be sufficiently well mixed for most reaction conditions [47], variations in flow at the subcentimeter scale in heterogeneous porous material induce concentration gradients at the pore scale that lead to local transport-control and scale dependence of mineral reaction rates [48–50]. These concentration gradients at the pore scale become increasingly important for reactive transport systems with complex, large-scale heterogeneity [51]. Thus, part of the inability to reconcile field and laboratory rates must be related to the lack of physical and chemical heterogeneity in laboratory systems, that lead to reductions in spatially averaged reaction rates in large field systems. Additionally, our inability to experimentally resolve coupled reactive transport processes in heterogeneous systems at the pore scale prevents the fundamental understanding of how geochemical reactions and fluid flow are coupled in porous materials [52, 53]. This fundamental understanding is needed to develop methods to scale mineral reaction rates from well-mixed laboratory to heterogeneous field systems.

Advances in experimental measurement of mineral dissolution rates in porous materials at scales ranging from pores to watersheds will expand the foundation upon which reaction rates can be scaled from laboratory to field. One key area of research is the quantification of surface area and porosity within porous materials, and the changes to these parameters with geochemical reaction. Traditionally, surface area is measured in Earth materials with nitrogen sorption, Hg porosimetry, or microscopy [8, 11, 54–56]. However, sorption only characterizes accessible, connected pores, while microscopy only images extremely small subsamples [57]. Recent advances in application of neutron scattering to natural samples [58–60] provide a technique capable of producing statistical information about connected and isolated pores, cracks, and other features and structures in rocks at length scales from $\sim$ 1 nm to thousands of nm. Neutrons that pass through a rock sample scatter from particles and interfaces of materials of different scattering contrast (a function of the chemistry and density) at angles that contain information about the length scales of features in the rock. Using established modeling approaches, SANS data provides information about the topology and architecture of pore networks [61–63]. When combined with high-resolution imaging, gas adsorption isotherm analysis, and other techniques, a complete, multi-scale view of the pore network can be obtained [59, 64].

This paper highlights two examples from our ongoing research in field and laboratory experimental studies, where changes in porosity with geochemical reaction were analyzed with neutron scattering: (1) rock weathering and saprolite formation and (2) geochemical reactions within CO_2 sequestration sites.

12.2. Neutron Scattering Methods

The methods used to analyze rocks with neutron scattering techniques have been developed over the last three decades, and are described in detail elsewhere [58, 62, 64, 65]. In particular, the comprehensive review by Radlinski [62] provides a wealth of information about the use of neutron scattering in earth systems science. What follows is a summary of the background of neutron scattering as it applies to analysis of porosity in rocks and the methods used in the studies presented in this paper.

Neutrons passing through a rock sample scatter at interfaces and particles of structures with different coherent scattering length density (SLD) from the adjacent structure (the SLD is a function of the phase chemistry and its mass density). The intensity of scattered neutrons as a function of scattering angle or momentum transfer contains information about the shapes and length scales of features in the rock. Small angle neutron scattering (SANS) data contain statistical information about features and structures in the rock at length scales from $\sim$ 1 nm to 700 nm, and ultra-small angle neutron scattering (USANS) contains information about features and structures at length scales from ca. 100 nm to 15 microns. Using modeling approaches based on the structural properties of the sample, SANS and USANS provide information about the surface area, pore size distribution, topology and architecture of pore networks [58, 61–63, 66]. Pore systems in rocks are commonly highly disordered and scale invariant, and can often be mathematically described by fractal and polydisperse hard-sphere models, which yield the pore size distribution, pore volume, surface roughness, and internal surface area from modeling of scattering data [62, 67].

SANS/USANS analysis was performed along weathering profiles of basalts and granites to quantify nano-scale porosity and internal surface area generation during weathering processes. The experimental details and analysis are described in a previously published paper [64]. Similar techniques were applied for the study of shale samples, collected at potential geologic carbon sequestration sites overlying porous reservoir rock formations. In these studies, caprock porosity and interrelated permeability were measured for samples of unreacted, CO_2-reacted, and brine-reacted Gothic Shale and Marine Tuscaloosa formation. The shale samples were broken into chips < 1 mm and packed into a quartz cuvette with 1-mm path length for SANS analysis. SANS data were collected on the General Purpose SANS (CG-2) and Bio SANS (CG-3) beamlines at the High Flux Isotope Reactor facility at Oak Ridge National Laboratory. Scattering intensity was collected at Q values ranging from 0.0008 to 0.8 Å^{-1} on CG-2 and from 0.0012 to 0.4 Å^{-1} on CG-3. The SANS raw data were reduced and processed using standard data reduction practices, including empty cell and background scattering correction, normalized for transmission, radially averaged, and normalized to absolute intensities [68, 69]. 2D scattering patterns of the granites, basalt, and shales, were always azimuthally symmetric. Scattering from shales when prepared as thick sections is often anisotropic [70], however for the study presented here random packing of shale chips in quartz cuvettes produced the azimuthally symmetric 2D scattering (Fig. 12.2a). The scattering data at high Q values were modeled using equation Eq. (12.1) to determine the intensity of incoherent scattering, which is mainly attributed to hydrogen (in water).

$$I(Q) = bQ^{-m} + c, \tag{12.1}$$

where m is the slope of the scattering curve, and b is a fitting parameter. The incoherent scattering, parameter c in Eq. (12.1), was subtracted from the data to produce corrected I(Q) vs. Q scattering data for further analysis (Fig. 12.2b).

The intensity of coherently scattered neutrons is proportional to difference in scattering length densities ($\Delta\rho_j^*$) between solid phases (minerals and kerogens) and pores. The coherent scattering length density ρ_j^* for a substance j is given by equation Eq. (12.2):

$$\rho_j^* = \frac{\Sigma_{i=1}^{n} b_c^i}{V_m}. \tag{12.2}$$

Here, b_c^i is the bound coherent scattering length of atom i of n atoms of a molecule and V_m is the molecular volume (g mol^{-1}). The values of the scattering length densities (ρ_j^*, Å^{-2}) for the rock phases were calculated from the bulk chemical formulas for each sample. Scattering length densities of the basalts and granites were previously reported [71], and those for the shales range from 4.36×10^{-6} to 4.57×10^{-6} Å^{-2}. Kerogen present in the shale was also taken into consideration. Rarely is the chemical composition of kerogen measured. Rather the type of kerogen is determined from the Hydrogen Index (HI) and Oxygen Index (OI). ρ_j^* for type II-III kerogens ranges from 3.4×10^{-6}Å^{-2} for kerogen composition of H/C = 0.5 and O/C = 0.3 with density of 1.5 g cm^{-3} to 1.4×10^{-6}Å^{-2} for compositions of H/C = 1 and O/C = 0.1 and density of 1 g cm^{-3}. These kerogen ρ_j^* values are similar to calculated ρ_j^* values of the shale samples. Therefore, mineral grains and kerogen phases may have slightly different scattering length den-

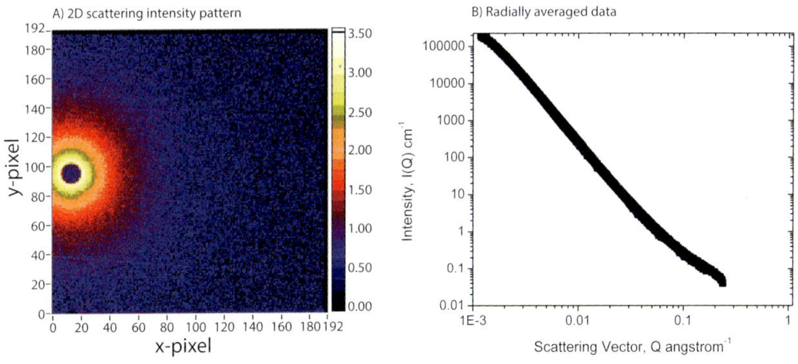

Fig. 12.2. Small-angle neutron scattering data for a Gothic Shale sample. (A) Shows an isotropic, azumuthally symmetric 2D scattering pattern at the detector. (B) Shows the radially averaged data used for analyses

sities that weakly scatter neutrons. However, these differences in scattering length density are negligible compared to the large scattering contrast of mineral-pore or kerogen-pore interfaces, where ρ_j^* of pores is zero. Because pore-mineral and pore-kerogen boundaries scatter with significantly higher intensities than mineral-mineral and mineral-kerogen boundaries, the scattering data can be interpreted using the two-phase approximation (e.g. Refs. [58, 62, 72]).

For all samples, scattering intensity as a function of Q was fit to a polydisperse hard sphere model implemented in the PRINSAS Software package. From this model the size and surface area distributions of pores within the samples and a histogram of pore sizes were calculated using the two-phase approximation and fitting routines in PRINSAS (for details on PRINSAS and the fitting macros see Ref. [67]).

12.3. Nano-scale Porosity Generation During Rock Weathering

Weathering of bedrock contributes to many important earth surface processes. Mineral weathering releases nutrients necessary for vegetation [73], neutralizes acid rain [74], produces erodible material contributing to landscape evolution [75], and plays a role in long-term regulation of atmospheric CO_2 concentrations [76, 77]. However, the rates at which mineral dissolve in

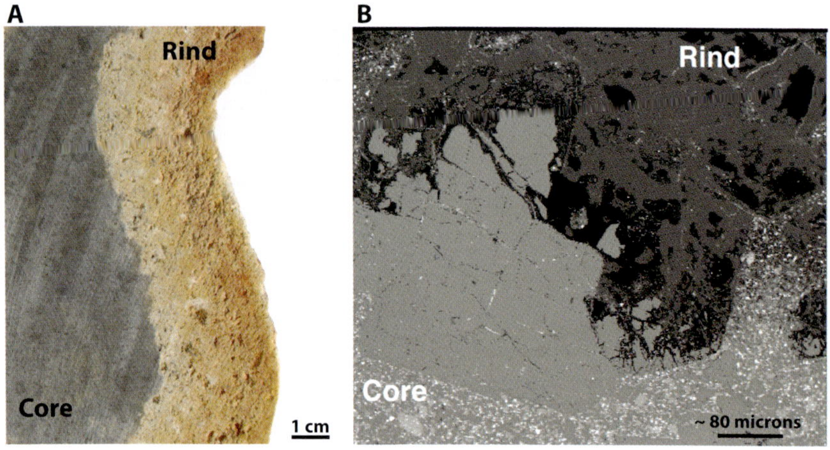

Fig. 12.3. Photograph (A) and scanning electron microscope image (B) of a weathering rind on a basalt clast. Dissolution of plagioclase at the interface (B) opens up porosity that controls the rate of mineral dissolution and advance of the weathering rind into the basalt core.

natural systems are poorly understood, and cannot easily be extrapolated from laboratory mineral dissolution studies [11, 14, 46, 78]. The role that heterogeneity in field systems plays in reducing apparent mineral dissolution rates relative to laboratory derived rates is not well understood [50, 79–81]. The pore network contributes to this natural heterogeneity, that laboratory studies do not often consider (Fig. 12.3(b)).

A series of papers investigating the weathering of basalt and granite and subsequent generation of saprolite presented evidence for generation of porosity as an important control on rates of crystalline rock weathering in field systems [45, 46, 64, 82]. These studies employed detailed chemical and physical characterization of the weathering front with electron microprobe analysis to analyze the mobilization of elements, X-ray computed tomography to analyze the volume of total and connected porosity at micron and larger length scales, and small-angle and ultra small-angle neutron scattering to analyze the surface area and porosity at nanometer to micron length scales. In naturally weathered basaltic clasts from Costa Rica, these studies demonstrate a critical porosity of 9% [45], that is reached through calcic-plagioclase dissolution (Fig. 12.3) as the low porosity ($\sim 3\%$) parent material (core) weathers to high porosity saprolite (rind, $\sim 40\%$) prior to any increases in diffusion of reactants to and products from the weathering interface. The expression of this critical porosity is the formation of a sharp weathering front (of only a few millimeters in total thickness) that cannot be simulated with traditional Archies Law dependencies of diffusivity on porosity [46]. The onset of weathering in the basalt clast occurs at the smallest scale, i.e. in nanometer-sized pores, where connectivity first increases through mineral dissolution [64]. In contrast, in a weathering granite in similar tropical climate (Luquillo Experimental Forest, Puerto Rico), the weathering interface is meters in scale and plagioclase weathering appears to be faster, despite the more sodic plagioclase mineralogy, that is less reactive than the more calcic plagioclase in the Costa Rica basalt in laboratory conditions [83]. Weathering in this granitic system is also greatly influenced by porosity generation during weathering. However, the porosity generated is in the form of microcracks that develop during spheroidal weathering processes [84, 85].

The morphology of the porosity created in these two systems (enlargement of pores versus microcracks) explains the apparent, and unexpected, faster weathering rates of granite relative to the basalt. Basalt is comprised of minerals that dissolve faster (calcic-plagioclase and augite) than the minerals that make up the granite (quartz, alkali feldspar, and biotite) [83]. However, the way water gets into the rocks, and the way solutes are transported through the pore space, reverse the effective weathering rates as predicted from the mineral dissolution rates in these two systems. The diffusion limited transport in the basalt leads to saturation of the pore fluids in the reaction front and very slow removal of reaction products, thus weathering proceeds only as fast as solutes are transported away from the weathering interface. In the granite, microcracks allow for advective flow of fluids and fast solute transport away from the weathering interface with a corresponding fast weathering rate. These two examples highlight not only the importance of porosity generation during weathering, but also the impact of the morphology of the porosity created on the overall weathering rate. A better understanding of how porosity is generated as rock weathers to saprolite will lead to more robust methods of predicting rates of weathering in field systems using numerical modeling and laboratory-derived kinetic rate data.

12.4. Geochemical Stability of Shales and Mudstones in Geologic CO_2 Storage

Elevated levels of atmospheric CO_2 have prompted investigation of technologies to mitigate anthropogenic CO_2 emissions. One of the most researched and technically viable options today is geologic carbon capture utilization and storage (CCUS), where CO_2 is captured at point emission sources, compressed, and injected into deep geologic formations as a supercritical fluid. Containment of the supercritical CO_2 in these storage formations relies on a good caprock or seal, which is often a low permeability shale or mudstone [86]. While the seal capacity of formations overlying the injection formation can be evaluated prior to injection, geochemical reactions induced by dissolution of CO_2 into the pore waters and subsequent formation of carbonic acid, change the mineralogy, porosity, and permeability leading to changes in seal capacity. Geochemical interactions between caprock formations and CO_2 rich brine have been experimentally investigated in order to assess the structural long-term stability of caprocks in potential storage sites. The results vary widely with the chemical composition of rocks studied. Precipitation of iron carbonate, illite, and smectite, and dissolution of calcite, quartz, illite/smectite, and chlorite have all been observed [87–89]. Tarkowski and Wdowin [90] demonstrated a decrease in porosity in limestone and dolomite caprocks after exposure to CO_2, and Labus and Bujok [91] linked the impact of reactions between CO_2 and caprock minerals on caprock porosity to initial porosity and hydraulic conductivity of the samples.

Linking geochemical reactions to changes in pore structure in fine-grained rocks such as shales and mudstones is especially difficult due the numerous pores in the micron to nm length scales. In many rocks, a large fraction of the internal surface area is found in narrow pores. Therefore, geochemical reactions in these nm pores may dominate the reaction mechanism, and advanced characterization techniques are required for their characterization and quantification. Our recent study employed such advanced techniques to shale caprock samples from two pilot CO_2 injection sites, Gothic Shale and Marine Tuscaloosa Formation, reacted with CO_2 in hydrothermal autoclave reactors with synthetic formation brines [60, 92]. For each experiment five grams (g) of shale (4 g small chips + 1 g ground) was mixed with 270 g of brine (initial brine/rock ratio of 54) and placed in a 300 milliliter (mL) EZE-seal stirred Hastelloy C-276 benchtop reactor. CO_2 was injected into the headspace of the reactor to a pressure of 150 bar, and a thermal sleeve was used to maintain a temperature of 160°C. Brine samples were extracted at 24 hours and then at 7 day intervals after the start of each experiment for chemical analysis.

High-resolution scanning electron microscope images of the unreacted and reacted samples reveal an increase in the number and sizes of pores in both rocks as a result of reaction with CO_2- rich brines (Fig. 12.4). Gas adsorption isotherm (BET) analysis detected a change in pore structure in the Marine Tuscaloosa sample, but not the Gothic Shale. The Marine Tuscaloosa sample is defined by a large number of nm sized pores and high organic carbon content. Data from small-angle neutron scattering analyses performed on the unreacted and reacted samples made it possible to identify changes to the pore size distribution. A 1.5% increase in porosity in the Gothic Shale (Fig. 12.5) resulted from an increase in pores with diameters of hundreds of nm, but not in pores $< \sim 50$ nm. The increase in porosity at these larger scales is attributed to dissolution of carbonate mineral grains from the rock. A slight increase in porosity in the Marine Tuscaloosa ($< 0.5\%$, Fig. 12.5) resulted from an increase in pores from a few to thousands of nm diameter with no preferential increase in any pore size. In the Marine Tuscaloosa sample, the decrease in connected porosity is attributed to mineral precipitation (Fig. 12.6). Mineral precipitation likely also contributed to blocking of pores with diameters > 100 nm in the Gothic Shale. In both samples, the median pore diameter is slightly larger in the CO_2-reacted samples (~ 4 nm) compared to the unreacted samples (~ 3 nm).

The results from these batch experiments provide information on the dissolution processes that are likely to occur when CO_2 rich brines come into contact with shale caprock formations under CO_2 sequestration conditions. The fluid velocity expected in shale formations in the deep subsurface is very slow, thus in this instance batch reactions may be more applicable to the field scale system than in other types of environments. The brine in the experiments reached chemical equilibrium with respect to dissolving carbonate minerals relatively quickly, on the order of timescales that are likely faster than fluid moves through shales. We, thus, might expect equilibrium control on carbonate mineral dissolution to be the leading factor in pore network evolution immediately after CO_2-rich brines enter the shale pore network. As these brines dissolve carbonate minerals, pH and inorganic carbon concentrations will rise in the mobile pore fluid, leading to the potential for precipitation of carbonate minerals away from the a zone of low pH fluids near the caprock/injection formation interface. This phenomenon is predicted in numerical models of CO_2 intrusion into shale caprocks [93], and carbonate and sulfate mineral precipitate formation on the outer surfaces of the shale chips in the experiments (Fig. 12.6) also suggests that secondary mineral precipitation may be an important process to consider when evaluating how shale pore networks will evolve under CO_2 sequestration conditions.

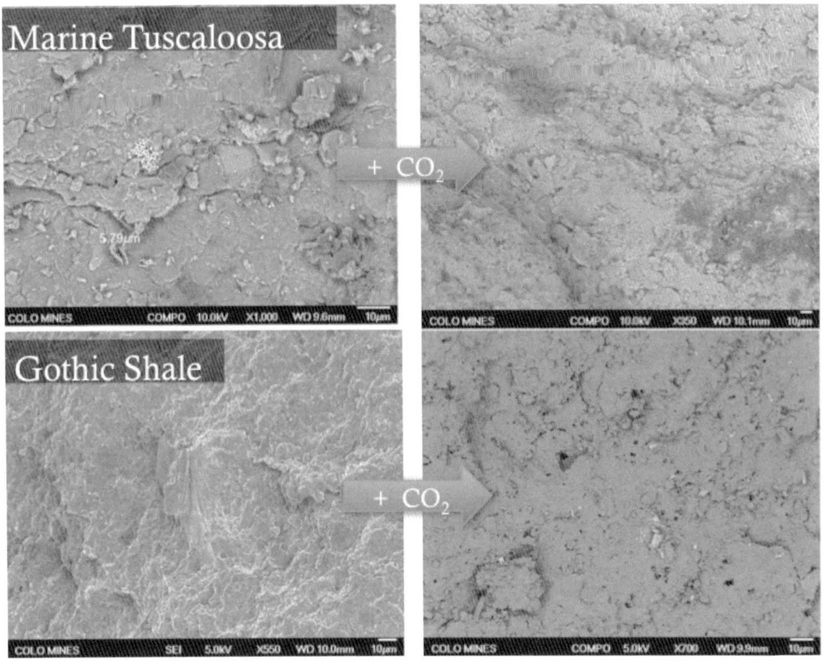

Fig. 12.4. Field emission SEM images of the shale samples before (left) and after (right) reaction with carbon dioxide. The after reaction samples of the Gothic Shale exhibit more and larger pores than the before reaction samples. After reaction samples of the Marine Tuscaloosa do not show significantly more pores than the before reacted samples.

All of the data presented in this study suggest that the Gothic Shale would experience more mineral dissolution under CO_2 sequestration conditions, facilitated by its carbonate content. The carbonate minerals (i.e. calcite) are highly reactive and have fast dissolution kinetics, which contributed to the fast pH buffering in the Gothic Shale experiments. The same degree of pH buffering was not observed the carbonate mineral poor Marine Tuscaloosa. Based on the experimental evidence, the Marine Tuscaloosa appears have lower potential for long-term decreases in sealing capacity due to geochemical reactions.

12.5. Summary

We discussed examples of geochemical reactions during which the porosity of the rock, and likely its permeability, change. These changes and

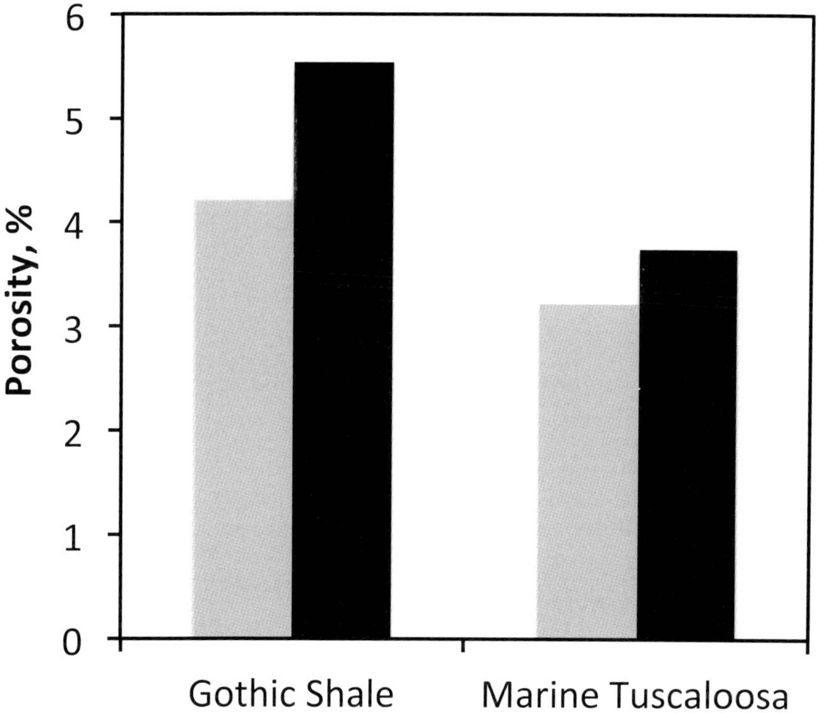

Fig. 12.5. Porosity in unreacted (grey) and O_2-reacted (black) samples of caprock formations.

the nanoscale mechanisms of porosity generation vary with rock type, and greatly impact reaction rates. Small-angle neutron scattering proved to provide the nanoscale-level information necessary to resolve these mechanisms, and is a very valuable tool in addition to the more established techniques like sorption, mercury porosimetry, and imaging. The highlighted examples showcase the need for a detailed understanding of the complex, coupled geochemical reactions often found in nature. Different research fields, for instance surface weathering and subsurface fluid injection, may benefit from the increased use of neutron scattering techniques, for instance in resolving the long-standing discrepancies between field and laboratory scales.

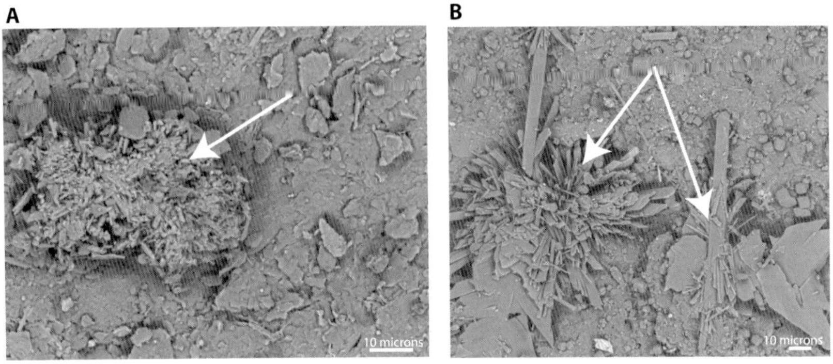

Fig. 12.6. Calcium sulfate (likely gypsum) precipitates on the surface of the CO_2-reacted Marine Tuscaloosa (A) and Gothic Shale (B) samples.

Acknowledgements

A.N-S. and J.K. acknowledge funding for this research from the United States Department of Energy, Grant number DE-FE0000730. Work by G.R. was supported by the US Department of energy (DOE), Office of Science, Basic Energy Sciences (BES) under award ERKCC72. Research conducted at ORNL's High Flux Isotope Reactor was sponsored by the Scientific User Facilities Division, Office of Basic Energy Sciences, US Department of Energy.

References

[1] J. Feth, C. Roberson, and W. Polzer, Sources of mineral constituents in water from granitic rocks, Sierra Nevada, California and Nevada, *US Geological Survey Water-Supply Paper 1535-I.* p. 70 (1964).
[2] A. Taylor and M. Velbel, Geochemical mass balances and weathering rates in forested watersheds of the Southern Blue Ridge II: Effects of botanical uptake terms, *Geoderma.* **51**, 29–50 (1991).
[3] G. J. S. Bluth and L. R. Kump, Lithologic and climatologic controls of river chemistry, *Geochimica et Cosmochimica Acta.* **58**(10), 2341–2359 (1994).
[4] O. A. Chadwick, L. A. Derry, P. M. Vitousek, B. J. Huebert, and L. O. Hedin, Changing sources of nutrients during four million years of ecosystem development, *Nature.* **397**(6719), 491–497 (1999).
[5] E. Hausrath, A. Neaman, and S. L. Brantley, Elemental reslease rates from dissolving basalt and granite with and without organic ligands, *Am. J. Sci.* **309**(8), 633–660 (2009).

[6] R. Graham, A. Rossi, and K. Hubbert, Rock to regolith conversion: Producing hospitable substrates for terrestrial ecosystems, *GSA Today*. **20**(2), 4–9 (2010).

[7] J. Matter, T. Takahashi, and D. Goldberg, Experimental evaluation of in situ co2-water-rock reactions during co2 injection into basaltic rocks: Implications for geological co2 sequestration, *Geochemistry Geophysics Geosystems*. **8** (2007). doi: 10.1029/2007gl031560.

[8] A. Navarre-Sitchler and S. Brantley, Basalt weathering across scales, *Earth and Planetary Science Letters*. **261**(1-2), 321–334 (2007).

[9] K. Pruess, Enhanced geothermal systems (egs) using co2 as working fluida novel approach for generating renewable energy with simultaneous sequestration of carbon, *Geothermics*. **35**(4), 351–367 (2006). doi: 10.1016/j.geothermics.2006.08.002.

[10] J.-P. Nicot and I. J. Duncan, Common attributes of hydraulically fractured oil and gas production and co2 geological sequestration, *Greenhouse Gases: Science and Technology*. **2**(5), 352–368 (2012). doi: 10.1002/ghg.1300.

[11] A. F. White and S. L. Brantley, The effect of time on the weathering of silicate minerals: why do weathering rates differ in the laboratory and field?, *Chemical Geology*. **202**, 479–506 (2003).

[12] A. F. White, *Natural Weathering Rates of Silicate Minerals*, In ed. J. I. Drever, *Treatise on Geochemistry: Surface and ground water, weathering, and soils*, vol. 5, pp. 133–168. Elsevier, Amsterdam (2005).

[13] M. S. Beig and A. Luttge, Albite dissolution kinetics as a function of distance from equilibrium: Implications for natural feldspar weathering, *Geochimica et Cosmochimica Acta*. **70**(6), 1402–1420 (2006).

[14] K. Maher, C. Steefel, D. DePaolo, and B. Biani, The mineral dissolution rate conundrum: Insights from reactive transport modeling of u isotopes and pore fluid chemistry in marine sediments, *Geochimica Et Cosmochimica Acta*. **70**, 337–363 (2006).

[15] F. A. L. Pacheco and A. M. P. Alencoao, Role of fractures in weathering of solid rocks: narrowing the gap between laboratory and field weathering rates, *Journal of Hydrology*. **316**(1-4), 248–265 (2006).

[16] S. Brantley, J. Kubicki, and A. F. White, *Kinetics of Water-Rock Interaction*. Springer, New York (2008).

[17] J. Drever and D. Clow, *Weathering Rates in Catchments*, In eds. A. F. White and S. Brantley, *Chemical Weathering Rates of Silicate Minerals: Reviews in Mineralogy*, vol. 31, pp. 463–481. Mineralogical Society of America, Washington D.C. (1995).

[18] M. E. Malmstrom, G. Destouni, S. A. Banwart, and B. H. E. Stromberg, Resolving the scale-dependence of mineral weathering rates, *Environ. Sci. Technol.* **34**(7), 1375–1378 (2000). doi: 10.1021/es990682u.

[19] E. Busenburg and C. Clemency, The dissolution kinetics of feldspars at 25c and 1 atm co2 partial pressure, *Geochimica et Cosmochimica Acta*. **40**, 41–49 (1976).

[20] L. Chou and R. Wollast, Study of the weathering of albite at room temperature and pressure with a fluidized bed reactor, *Geochimica et Cosmochimica*

Acta. **48**(11), 2205–2217 (1984).

[21] L. Chou and R. Wollast, Steady-state kinetics and dissolution mechanisms of albite, *Am. J. Sci.* **285**(10), 963–993 (1985).

[22] P. V. Brady, Kinetics of quartz dissolution at low temperatures, *Chem. Geo.* **82**(3-4), 253 (1990). 0009-2541.

[23] C. Amrhein and D. L. Suarez, Some factors affecting the dissolution kinetics of anorthite at 25c, *Geochimica Et Cosmochimica Acta.* **56**(5), 1815–1826 (1992). doi: 10.1016/0016-7037(92)90312-7.

[24] C. Anbeek, N. van Breeman, E. L. Meijer, and L. van der Plas, The dissolution of naturally weathered feldspar and quartz, *Geochimica et Cosmochimica Acta.* **58**(21), 4601–4613 (1994).

[25] E. Oelkers, J. Schott, and J. Devidal, The effect of aluminum, ph and chemical affinity on the rates of alumino-silicate dissolution reactions, *Geochimica et Cosmochimica Acta.* **58**, 2011–2024 (1994).

[26] Y. Chen and S. L. Brantley, Temperature- and ph-dependence of albite dissolution rate at acid ph, *Chem. Geol.* **135**(3-4), 275–290 (1997). TY - JOUR.

[27] A. Berg and S. A. Banwart, Carbon dioxide mediated dissolution of ca-feldspar: implications for silicate weathering, *Chem. Geol.* **163**, 25–42 (2000).

[28] J. Cama, V. Metz, and J. Ganor, The effect of ph and temperature on kaolinte dissolution rate under acidic conditions, *Geochimica et Cosmochimica Acta.* **66**(22), 3913–3926 (2002).

[29] S. A. Carroll and K. Knauss, Dependence of labradorite dissolution kinetics on co2 (aq), al(aq), and temperature, *Chem. Geol.* **217**, 213–225 (2005).

[30] A. Luttge, R. E. Arvidson, and C. Fischer, A stochastic treatment of crystal dissolution kinetics, *Elements.* **9**(3), 183–188 (2013).

[31] A. F. White and M. L. Peterson, *Role of reactive surface area characterization in geochemical models*, In eds. D. Melchoir and R. Bassett, *Chemical Modeling of Aqueous Systems II*, vol. 416, pp. 461–475. American Chemical Society Symposium Series (1990).

[32] A. F. White, *Chemical weathering rates of silicate minerals in soils*, In eds. A. F. White and S. L. Brantley, *Chemical weathering rates of silicate minerals*. Revies in Mineralogy. Mineralogical Society of America, Washington D.C. (1995).

[33] L. Kump, S. Brantley, and M. Arthur, Chemical weathering, atmospheric co2, and climate, *Annual Review of Earth and Planetary Science.* **28**, 611–667 (2000).

[34] T. E. Burch, K. L. Nagy, and A. C. Lasaga, Free energy dependence of albite dissolution kinetics at 80c and ph 8.8, *Chem. Geol.* **105**(13), 137–162 (1993). doi: 10.1016/0009-2541(93)90123-Z.

[35] R. Stallard, *Tectonic processes, continental freeboard, and the rate-controlling step for continental denudation*, In ed. S. Butcher, *Global biogeochemical cycles*. Academic Press, London (1992).

[36] C. S. Riebe, J. W. Kirchner, and R. C. Finkel, Erosional and climatic effects on long-term chemical weathering rates in granitic landscapes spanning diverse climate regimes, *Earth Planet. Sci. Lett.* **224**(3-4), 547–562 (2004).

[37] A. J. West, A. Galy, and M. Bickle, Tectonic and climatic controls on silicate weathering, *Earth Planet. Sci. Lett.* **235**(1-2), 211–228 (2005).
[38] C. Zhu and P. Lu, Alkali feldspar dissolution and secondary mineral precipitation in batch systems: 3. saturation states of product minerals and reaction paths, *Geochimica Et Cosmochimica Acta.* **73**, 3171–3200 (2009).
[39] J. I. Drever, The effect of land plants on weathering rates of silicate minerals, *Geochimica et Cosmochimica Acta.* **58**, 2325–2332 (1994).
[40] J. I. Drever and L. Stillings, The role of organic acids in mineral weathering, *Colloids Surf. A: Physicochemical and Engineering Aspects.* **120**, 167–181 (1997).
[41] Y. Lucas, The role of plants in controlling rates and products of weathering: importance of biological pumping, *Annual Review of Earth and Planetary Science.* **29**(135-163) (2001).
[42] R. Millot, J. Gaillardet, B. Dupre, and C. J. Allegre, The global control of silicate weathering rates and the coupling with physical erosion: new insights from rivers of the canadian shield, *Earth Planet. Sci. Lett.* **196**, 83–98 (2002).
[43] C. Dessert, B. Dupre, J. Gaillardet, L. Francois, and C. Allegre, Basalt weathering laws and the impact of basalt weathering on the global carbon cycle, *Chem. Geol.* **202**, 257–273 (2003).
[44] J. Gaillardet, R. Millot, and B. Dupre, Chemical denudation rates of the western canadian orogenic belt: the stikine terrane, *Chem. Geol.* **201**, 257–279 (2003).
[45] A. Navarre-Sitchler, C. Steefel, L. Yang, and L. Tomutsa, Evolution of porosity and diffusivity associated with chemical weathering of a basalt clast, *J. Geophys. Res. - Earth Surface* (2009). doi: 10.1029/2008JF001060.
[46] A. Navarre-Sitchler, C. Steefel, P. Sak, and S. Brantley, A reactive-transport model for weathering rind formation on basalt, *Geochimica Et Cosmochimica Acta.* **75**, 7644–7667 (2011).
[47] L. Li, C. Steefel, and L. Yang, Scale dependence of mineral dissolution rates within single pores and fractures, *Geochimica Et Cosmochimica Acta.* **72**, 360–377 (2008).
[48] A. M. Tartakovsky, G. Redden, P. C. Lichtner, T. D. Scheibe, and P. Meakin, Mixing-induced precipitation: Experimental study and multiscale numerical analysis, *Water Resour. Res.* **44**(6), W06S04 (2008). doi: 10.1029/2006wr005725.
[49] M. Dentz, P. Gouze, and J. Carrera, Effective non-local reaction kinetics for transport in physically and chemically heterogeneous media, *J. Contam. Hydrol.* **120121**(0), 222–236 (2011). doi: 10.1016/j.jconhyd.2010.06.002.
[50] S. Mollins, D. Trebotich, C. Steefel, and C. Shen, An investigation of the effect of pore scale flow on average geochemical reaction rates using direct numerical simulation, *Water Resour. Res.* **48**, W03527 (2012).
[51] I. Battiato, D. M. Tartakovsky, A. M. Tartakovsky, and T. Scheibe, On breakdown of macroscopic models of mixing-controlled heterogeneous reactions in porous media, *Adv. Water Res.* **32**(11), 1664–1673 (2009). doi: 10.1016/j.advwatres.2009.08.008.

[52] A. Putnis, Mineral replacement reactions: from macroscopic observations to microscopic mechanisms, *Mineral. Mag.* **66**, 689–708 (2002).

[53] B. Jamtveit and . Hammer, Sculpting of rocks by reactive fluids, *Geochemical Perspectives.* **1**(3), 341–481 (2012).

[54] W. Ball, C. Bouhler, T. Harmon, D. MacKay, and P. Robers, Characterization of a sandy aquifer material at the grain scale, *J. Contam. Hydrol.* **5**, 253–295 (1990).

[55] E. Rufe and M. Hochella, Quantitative assessment of reactive surfacea rea of phlogopite during acid dissolution, *Science.* **285**, 874–876 (1999).

[56] S. Brantley and N. P. Mellott, Surface area and porosity of primary silicate minerals, *Am. Mineral.* **85**, 1767–1783 (2000).

[57] C. Fischer and R. Gaupp, Multi-scale rock surface area quantification - a systematic method to evaluate the reactive surface area of rocks, *Chemie der Erde - Geochemistry.* **64**, 241–256 (2004).

[58] L. M. Anovitz, G. W. Lynn, D. R. Cole, G. Rother, L. F. Allard, W. A. Hamilton, L. Porcar, and M. H. Kim, A new approach to quantification of metamorphism using ultra-small and small angle neutron scattering, *Geochimica Et Cosmochimica Acta.* **73**(24), 7303–7324 (2009). doi: 10.1016/j.gca.2009.07.040.

[59] L. M. Anovitz, D. R. Cole, G. Rother, L. F. Allard, A. J. Jackson, and K. C. Littrell, Diagenetic changes in macro- to nano-scale porosity in the st. peter sandstone: An (ultra) small angle neutron scattering and backscattered electron imaging analysis, *Geochimica et Cosmochimica Acta.* **102**(0), 280–305 (2013). doi: 10.1016/j.gca.2012.07.035.

[60] A. Navarre-Sitchler, K. Mouzakis, G. Rother, J. Banuelos, X. Wang, J. Kaszuba, and J. E. McCray. Invited: Co2 induced geochemical reactions at the pore scale. (2013).

[61] F. Mildner, David, R. ERezvani, P. Hall, and R. Borst, Small-angle scattering of shaly rocks with fractal pore interfaces, *Appl. Phys. Lett.* **48**(19), 1314–1316 (1986).

[62] A. P. Radlinski, *Small-angle neutron scattering and the microstructure of rocks*, In ed. H. Wenk, *Neutron Scattering in Earth Sciences*, vol. 63, *Reviews in Mineralogy & Geochemistry*, pp. 363–397. Mineralogical Society of America (2006). ISBN 1529-6466.

[63] L. Jin, G. Rother, D. R. Cole, F. Mildner, David, C. J. Duffy, and S. L. Brantley, Characterization of deep weathering and nanoporosity development in shale - a neutron study, *Am. Mineral.* **96**, 498–512 (2011).

[64] A. K. Navarre-Sitchler, D. Cole, G. Rother, L. Jin, H. L. Buss, and S. L. Brantley, Porosity and surface area evolution during weathering of two igneous rocks, *Geochimica Et Cosmochimica Acta* (2013).

[65] A. P. Radlinski, C. J. Boreham, G. D. Wignall, and J.-S. Jin, Microstructural evolution of source rocks during hydrocarbon generation: A small-angle-scattering study, *Phys. Rev. B.* **53**(21), 14 153 (1996).

[66] E. Bazilevskaya, M. Lebedeva, M. Pavich, G. Rother, D. Parkinson, D. R. Cole, and S. L. Brantley, Where fast weathering creates thin regolith and slow weathering creates thick reagolith, *Earth Surf. Process. Landf.* **38**(8),

847–858 (2013).
[67] A. L. Hinde, Prinsas — a windows-based computer program for the processing and interpretation of small-angle scattering data tailored to the analysis of sedimentary rocks, *J. Appl. Crystallogr.* **37**, 1020–1024 (2004).
[68] S. Kline, Reduction and analysis of sans and usans data using igor pro, *Journal of Applied Crystallography.* **39**, 895–900 (2006).
[69] L. DeBeer-Schmidt and K. Bailey. Sans data reduction guide (2011).
[70] P. Hall, F. Mildner, David, and R. Borst, Pore size distribution of shaly rocks by small angle neutron scattering, *Appl. Phys. Lett.* **43**, 252–254 (1983).
[71] A. Navarre-Sitchler, D. Cole, G. Rother, and S. Brantley, Evolution of microporosity during weathering of basalt, *Geochimica Et Cosmochimica Acta.* **72** (12), A673–A673 (2008). Suppl. 1.
[72] A. Kahle, B. Winkler, A. Radulescu, and Schreuer, Small-angle neutron scattering study of volcanic rocks, *Eur. J. Mineral.* **16**, 407–417 (2004).
[73] L. Hedin, P. Vitousek, and P. Matson, Nutrient losses over four million years of tropical forest development, *Ecology.* **84**(9), 2231–2255 (2003).
[74] N. Johnson, C. Driscoll, J. Eaton, G. Liken, and W. McDowell, 'acid rain', dissolved aluminum and chemical weathering at the hubbard brook experimental forest, new hampshire, *Geochimica Et Cosmochimica Acta.* **45**, 1421–1437 (1981).
[75] J. L. Dixon, A. M. Heimsath, and R. Amundson, The critical role of climate and saprolite weatherign in landscape evolution, *Earth Surf. Process. Landf.* **34**(11), 1507–1521 (2009).
[76] J. Walker, P. Hays, and J. Kasting, A negative feedback mechanism for the long-term stabilization of earth's surface temperature, *J. Geophys. Res.* **86** (C10), 9776–9782 (1981).
[77] R. Berner, A. Lasaga, and R. M. Garrels, The carbonate-silicate geochemical cycle and its effect on atmospheric carbon dioxide over the past 100 million years, *Am. J. Sci.* **283**, 641–683 (1983).
[78] K. Maher, C. Steefel, A. White, and D. Stonestrom, The role of reaction affinity and secondary minerals in regulating chemical weathering rates at the santa cruz soil chronosequence, california, *Geochimica Et Cosmochimica Acta.* **73**, 2804–2831 (2009).
[79] Y. Li and S. Gregory, Diffusion of ions in sea water and in deep-sea sediments, *Geochimica et Cosmochimica Acta.* **38**, 703–714 (1974).
[80] C. E. Koltermann and S. M. Gorelick, Heterogeneity in sedimentary deposits: A review of structure-imitating, process-imitating, and descriptive approaches, *Water Resour. Res.* **32**(9), 2617–2658 (1996). doi: 10.1029/96wr00025.
[81] S.-Y. Lee, S. F. Carle, and G. E. Fogg, Geologic heterogeneity and a comparison of two geostatistical models: Sequential gaussian and transition probability-based geostatistical simulation, *Adv. in Water Resour.* **30**(9), 1914–1932 (2007). doi: 10.1016/j.advwatres.2007.03.005.
[82] P. Sak, D. Fisher, T. Gardner, K. Murphy, and S. Brantley, Rates of weathering rind formation on costa rican basalt, *Geochimica et Cosmochimica Acta.* **68**, 1453–1472 (2004).

[83] J. Bandstra, H. L. Buss, R. Campen, L. Liermann, J. Moore, E. Hausrath, A. K. Navarre-Sitchler, J.-H. Jang, and S. L. Brantley, *Appendix: Compilation of Mineral Dissolution Rates*, In eds. S. L. Brantley, J. Kubicki, and A. F. White, *Kinetics of Water-Rock Interaction*. Springer, New York (2008).

[84] R. C. Fletcher, H. L. Buss, and S. L. Brantley, A spheroidal weathering model coupling porewater chemistry to soil thicknesses during steady-state denudation, *Earth Planet. Sci. Lett.* **244**(1-2), 444–457 (2006).

[85] H. L. Buss, P. Sak, R. M. Webb, and S. Brantley, Weathering of the rio blanco quartz diorite, luquillo mountains, puerto rico: Coupling oxidation, dissolution and fracturing, *Geochimica et Cosmochimica Acta.* **72**, 4488–4507 (2008).

[86] J. Heath, T. Dewers, B. McPherson, R. Petrusak, T. Chidsey, A. Rinehart, and P. Mozley, Pore networks in continental and marine mudstones: Characteristics and controls on sealing behavior, *Geofluids.* **7**, 429–454 (2011).

[87] J. P. Kaszuba, D. R. Janecky, and M. G. Snow, Experimental evaluation of mixed fluid reactions between supercritical carbon dioxide and nacl brine: Relevance to the integrity of a geologic carbon repository, *Chemi. Geol.* **217** (3-4), 277–293 (2005).

[88] M. Andreani, P. Gouze, L. Luquot, and P. Jouanna, Changes in seal capacity of fractured claystone caprocks induced by dissolved and gaseous co2 seepage, *Geophys. Res. Lett.* **35**(14), L14404 (2008). doi: 10.1029/2008gl034467.

[89] V. d. Lima, S. Einloft, J. M. Ketzer, M. Jullien, O. Bildstein, and J.-C. Petronin, Co2 geological storage in saline aquifers: Paran basin caprock and reservoir chemical reactivity, *Energy Procedia.* **4**(0), 5377–5384 (2011). doi: 10.1016/j.egypro.2011.02.521.

[90] R. Tarkowski and M. Wdowin, Petrophysical and mineralogical research on the influence of co2 injection on mesozoic reservoir and caprocks from the polish lowlands, *Oil Gas Sci. Technol.* **66**, 137–150 (2011).

[91] K. Labus and P. Bujok, co_2 mineral sequestration mechanisms and capacity of saline aquifers of the upper silesian coal basin (central europe) - modeling and experimental verificaation, *Energy.* **36**, 4974–4982 (2011).

[92] K. Mouzakis. *Porosity and pore networks of caprocks: a nanoscale experimental approach investigating the Gothic shale and Marine Tuscaloosa shale - implications to Carbon Sequestration.* Thesis (2011).

[93] F. Gherardi, T. F. Xu, and K. Pruess, Numerical modeling of self-limiting and self-enhancing caprock alteration induced by co2 storage in a depleted gas reservoir, *Chemi. Geol.* **244**(1-2), 103–129 (2007). doi: 10.1016/j.chemgeo.2007.06.009.

Chapter 13

Quantifying Hydrogeological Heterogeneity of Rocks using Core-Floods

Ronny Pini

Department of Petroleum Engineering,
Colorado School of Mines, Golden, CO, USA
rpini@mines.edu

Sally M. Benson

Department of Energy Resources Engineering,
Stanford University, Stanford, CA, USA
smbenson@stanford.edu

Making accurate predictions of fluid movement and distribution in rocks or soils requires making measurements at the full range of spatial scales that characterize these complex natural systems. Hereby, new knowledge on the small-scale physics is needed to refine upscaling approaches that can be integrated in large-scale models. Core-flooding experiments combined with the simultaneous imaging of flows represent a powerful tool to assess the role of fine-scale heterogeneity on fluid displacement and have the potential to create the necessary link between the pore- and the continuum-scale. In this chapter, we review and discuss the effects of sub-core scale heterogeneities on multiphase flows by considering both experimental and modeling studies. Novel experimental methodologies are presented that allow non-destructive measurements of core- and sub-core-scale hydrogeological properties during conventional core-flooding tests, thus including porosity, permeability and capillary pressure curves. It is shown that commonly applied scaling relationships break down at the sub-core scale and that a quantitative agreement with high-resolution numerical simulations can be achieved only when these properties are evaluated within a unique framework that includes the spatial variability in the capillary pressure curve. Such an approach represents a significant improvement from conventional core analysis and represents a firm-point of departure to investigate complex hydrogeological flows in the subsurface.

13.1. Introduction

As attempts are made to characterize geological formations across various scales, their intrinsic complexity becomes readily apparent. It comes as no surprise that the latter represents a major challenge for predicting the movement of fluids in the subsurface for a variety of applications [1]. The description of groundwater flow following the spill of a petroleum product, the estimation of the amount of recoverable oil upon application of enhancement techniques or the prediction of the spreading of a CO_2 plume in a CO_2 sequestration project are a few practical examples highlighting the importance of this issue. On the one hand, most of the basic processes that control fluid flow within a porous rock take place at the pore-scale, and include interfacial phenomena such as dissolution, capillarity, adsorption and chemical reactions. On the other hand, dynamics at the scale of a reservoir are affected by geological structures such as those that define its stratigraphy. Interestingly enough, mm- to cm-scale features that are commonly observed in sedimentary rocks [2, 3] have been shown to greatly influence fluid transport over much larger observational scales (tens to hundreds of meters) [4]. Such small-scale heterogeneities are often represented by a spatial association of so-called lamina and beds and they become relevant as soon as their presence significantly increases the overall variability in the distribution of permeability of the medium. Permeability variations within length scales < 20 cm have indeed been observed for many depositional environments [5]. Most importantly, these features give rise to capillary phenomena, which in turn control the behavior of multiphase flows and affect process-relevant parameters, such as sweep [6] or trapping efficiencies [7]. In fact, under those circumstances, fluid doesn't move as a uniform front, but penetrates the rock as one ore more fingers, thus giving rise to lateral spreading, preferential flow paths and pooling [8].

There have been a number of both experimental [10–14] and modeling studies [15–18] that have highlighted the importance of small-scale heterogeneities on multiphase flows, but only few of them address the problem in a quantitative way. As an example of general validity, Figure 1 presents results from a core-flooding experiment performed to investigate the impact of a natural capillary barrier in a sandstone rock on the trapping of a nonwetting phase (CO_2) [9]. The rock sample considered has two regions of distinct capillarity (Fig. 13.1(A)) characterized by a high (upstream) and low (downstream) permeability, respectively. The latter is represented by sub-core scale sand bedding planes that cut diagonally across the core

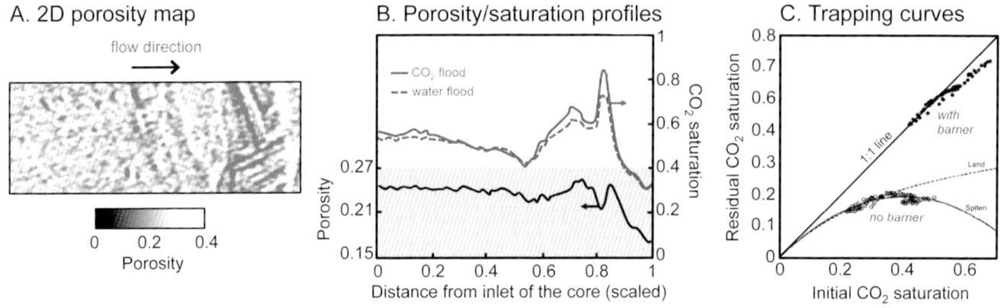

Fig. 13.1. Capillary heterogeneity trapping of CO_2 in a sandstone core. (A) 2D porosity map that shows significant cross-bedding at the downstream end of the core. (B) Slice-averaged porosity and saturation profiles along the length of the core during 100% CO_2 and 100% water flooding. (C) Residual CO_2 saturation (after water flooding) as a function of the initial CO_2 saturation (after CO_2 flooding) for experiments carried out with and without capillary barrier. (Figure modified from Ref. [9].)

and that give rise to a capillary barrier. In fact, during the drainage process of CO_2 displacing water (Fig. 13.1(B)), CO_2 builds up upstream of the capillary barrier, and once in place it cannot be displaced during the subsequent water flooding. Most importantly, this leads to trapped saturations (Fig. 13.1(C)) that are a factor 2–5 times higher than what would be expected from residual trapping alone (and obtained from an experiment performed upon removal of the downstream section of the core). This practical example demonstrates that a nonwetting phase can be immobilized behind capillary barriers as a continuous phase at saturations higher than would be possible as isolated ganglia; we can refer to the former as a trapped *mobile* saturation, rather than a trapped *residual* saturation. At the large (field) scale the implication of this phenomenon can be significant, because it might dramatically affect the storage capacity of a reservoir.

The ability to quantify the processes just described at a scale where flow is described by continuum-scale physics and where these small-scale heterogeneities are taken into account represents a critical step in the effort of creating a link between the pore- and the reservoir-scale. Ideally, process upscaling begins at the size of a Representative Elementary Volume (REV) [19], which, for many depositional environments, is at the mm to cm scale (also called the lamina scale [5]). The lack of access to information about rock property heterogeneity at this scale has restricted the ability to fully take advantage of so-called multistage upscaling approaches, which have been specifically developed to link the core-scale to the size of major reservoir units [1] and whose inputs are typically derived from well-logs and core samples. In fact, properties derived from measurements on reservoir cores are inherently effective, their spatial resolution being limited to a minimum of several centimeters by the measurement or sampling technique. However, making accurate predictions of multiphase flows requires making measurements at the full range of relevant spatial scales, thus referring to the internal structure of the sample and including the effects of the laminations described above [6, 18]. Essential components in this description include continuum properties that are related to the rock (porosity and permeability), to the fluids (saturation) and to both of them (capillary pressure-saturation relationship); the ability to create a link among all these properties is key to a physically-sound description of these naturally complex systems. As described in the following, one way to accomplish this is by adopting an integrated approach that combines displacement experiments in naturally heterogeneous core-samples with the simultaneous imaging of flow as well as with the support of detailed numerical simulations. By applying visualization, laboratory core-floods offer more reliable results because the experiments give spatial and temporal information on local saturation development [20], while numerical simulations represent an efficient tool for the simultaneous investigation of a variety of experimental conditions [18].

Throughout this chapter we are going to intentionally interpose the use of the term sub-core scale with mesoscale, while referring to the same type of heterogeneity. The former implies that most of the analysis presented here has been carried out on reservoir rock cores and that the spatial variation in hydrogeological properties is quantified for sub-elements of the domain. The term mesoscale highlights the intermediate position of this type of heterogeneity: while representing the collective behavior of microscopic features observed at the pore-scale, they form the building blocks that control and influence behavior at the macroscopic scale.

13.2. Core Flooding Experiments and Imaging of Flow

Dynamic displacements (or core-flooding) tests are the most common experimental method for studying multiphase flow in porous media. Very briefly, a typical experiment consists of injecting one or more fluids into a core initially saturated with the fluid to be displaced. The displacement is carried out at either constant rate or constant pressure, while continuously monitoring the pressure differential across the core together with cumulative injected and effluent volumes. The advent of imaging techniques such as X-ray Computed Tomography (CT) has provided a new level of observational detail not previously available to early investigators into the effects of heterogeneity on saturation distribution within the core sample. Hereby, porosity and fluid-phase saturations can be observed at a resolution of a few mm^3 (a volume that exceeds the size of a porosity-based REV for most sandstones) at any time during a core-flooding experiment [21]. Figure 13.2 shows snapshots of steady-state saturation distributions that have been obtained by injecting nitrogen (N_2) at various flow rates in a sandstone core initially saturated with water; the corresponding porosity map is also shown in the figure and the qualitative differences are readily apparent ([22]). Although not visible in the dry or fully saturated state, features in the form of stratifications that are oriented sub-parallel to the axis of the core appear as soon as N_2 is injected to displace the resident water. This confirms that, unlike single-phase flow properties, the effect of micro-structural heterogeneity on multiphase flow is large [20, 23]. The latter are very likely associated to finer-textured strata that are often observed in sandstones [24] and are particularly evident at the lowest to moderate flow rates, i.e. when that portion of the capillary pressure curve is covered where saturation is very sensitive to the value of capillary pressure. Accordingly, the histograms plots of the saturation (shown below each 3D map) show a bi-modal saturation distribution that turns into a skewed distribution upon increasing the flow rate. Statistical measures of the distribution such as the mean (μ_S) and the mode (m_S) can be used to quantify the effects related to the presence of heterogeneity: the former gives the actual average saturation in the core, while the latter represents its most prominent value, thus providing an indication for the saturation of a perfectly homogeneous core. As reported in the figure, the difference between these two measures is significant (3-10%abs.); for horizontal flows, such as those examined here, the effect of neglecting the presence of fine-scale heterogeneities would be to significantly overestimate displacement efficiency, particularly at low flow rates. As explained in the next sections, these saturation distributions can be understood by considering spatial variations in the capillary pressure characteristic curve and they can be directly related to a corresponding distribution of permeabilities.

13.3. Numerical Simulation of Core-floods

The core-flooding experiments described above are very challenging and time consuming to conduct, and numerical simulations represent a very

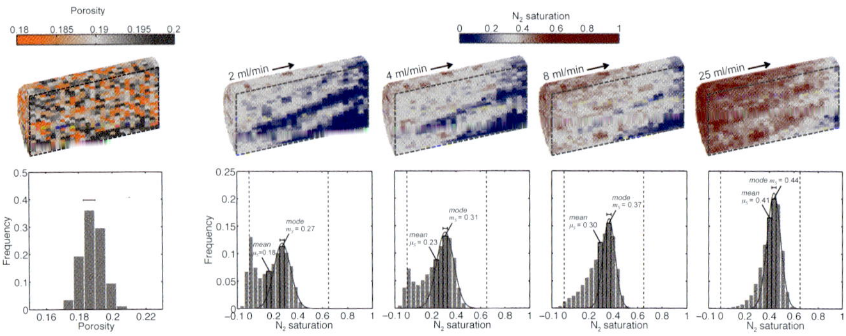

Fig. 13.2. 3D porosity and steady-state saturation maps (cross-section along the core length) of a Berea Sandstone core. The latter have been obtained upon injection of N2 at four different flow rates in the core initially saturated with water. Voxel size $2 \times 2 \times 3$ mm^3; porosity and saturation values are obtained with an accuracy of 0.4 and 1.8%abs., respectively. Below each map, histograms are shown of the corresponding distribution. For comparison, a normal distribution curve (black line) is plotted on top of each histogram that is centered on the mode of the distribution. (Figure taken from Ref. [22].)

valuable complementary tool to investigate the dynamics of core-floods under different flow and thermophysical conditions. To name a few and with relevance to the present discussion, core-flooding experiments have been numerically simulated by using TOUGH2 from the Lawrence Berkeley National Lab (Berkeley, California, USA) [17, 18, 25, 26], UTCHEM from UT Austin (Austin, Texas, USA) [27], METSIM2 from the Imperial College London (London, UK) [14] and GPRS from Stanford University (Stanford, California, USA) [28]. Very briefly, the mathematical formulation of the problem include mass balance equations for multiphase flow that can be solved in a 3D domain; upon application of appropriate boundary and initial conditions, together with so-called characteristic functions (capillary pressure and relative permeability curves), the solution of these equations provides information about fluid distributions and saturation values within the rock sample. The key point here is that in order to reproduce by numerical simulations the details of the fluids saturation distributions shown in Fig. 13.2, the core sample has to be replicated accurately with a simulation grid, which is populated with properties of the rock, such as porosity and permeability values. Subcore-scale porosity distributions can be measured using a CT scanner [21] and the corresponding permeabilities can be calculated at the same scale using common porosity-based techniques (see Ref. [18] for an overview of available models). While it might sound

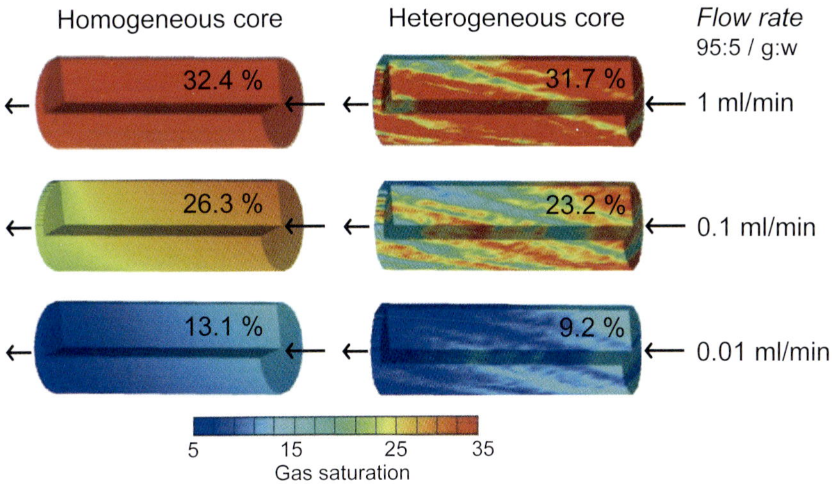

Fig. 13.3. Numerical simulation of a core-flood: 3D steady-state saturation maps of a Berea Sandstone core obtained upon co-injection of CO_2 and brine (95:5 volume ratio) at 12.4 MPa and 50°C and at various total flow rates. The core (5.08 cm diameter and 15.24 cm length) has a porosity of 20.3% and a permeability to water of 430 mD; each grid element has a volume of $\sim$16 mm^3 and the simulations have been performed by using TOUGH2 MP with ECO2N module. (Figure modified from Ref. [17].)

relatively simple, such an exercise provides very useful insights on the influence of various parameters on fluid displacement, such as sub-core scale heterogeneities, gravity or flow rates, and has allowed identifying some of the key mechanisms controlling fluid distribution within a core sample.

Laboratory experiments on cores regarded as homogeneous may be obscured by local heterogeneities, which can dramatically affect residual saturations, recovery efficiency, and relative permeability [18, 20]. Figure 13.3 shows snapshots of steady-state saturation distributions that have been obtained by co-injecting CO_2 and brine (95:5 volume ratio) at various total flow rates in a sandstone core initially saturated with brine [17]; simulation results are compared between an homogeneous and an heterogeneous core. In the former case, each grid element possesses the same porosity and permeability values as the core's mean values, while for the latter, porosity values have been assigned from sub-core scale measurements made using a X-ray CT scanner and the corresponding permeabilities are calculated using the Kozeny-Carman relationship. Hereby, a scaling parameter is applied to ensure that the core average permeability as calculated in the simulation grid is the same as the experimentally measured value. For the homogeneous case, both the average CO_2 saturation *and* the CO_2 saturation distribution change with injection flow rate: these changes are mostly driven by the density difference between CO_2 and brine, and become more apparent as the flow rate is decreased. This picture is further complicated once heterogeneities are included in the model: the stronger variations in saturation observed at the fine-scale are now the result of the complex interplay between all the forces involved in the displacement process, namely viscous, gravity and capillary forces. Moreover, changes in brine displacement efficiency with flow rate (as reflected by the core-averaged CO_2 saturation values) are stronger as compared to the homogeneous case, as suggested by other studies [20]. It is worth noting that the saturation distributions for the heterogeneous case resemble more to those observed during an actual experiment and shown in Fig. 13.2. However, despite the similarity in the general spatial pattern, the degree of saturation contrast observed in the simulations is typically not high enough and significant discrepancies have been reported between measured and modeled sub-core scale saturation distributions [18]. As a matter of fact, it has been shown that the degree of saturation contrast is closely correlated to the degree of permeability contrast, thus suggesting that commonly applied porosity-based relationships are not reliable for predicting permeability variations at the sub-core scale [18]. This last aspect is somewhat intuitive because porosity is not necessarily an indicator of pore connectivity or of the ability of a rock to transmit fluid. As discussed in the section that follows, other methods have been developed to estimate sub-core scale permeability variations, in the effort of improving the quantitative agreement between measured and simulated saturation distributions.

13.4. The Concept of Capillary Heterogeneity

All rock samples possess an inherent heterogeneity that is associated to its microscopic structure. This is related, for instance, to grain shape and orientation that in turn affect the configuration of the pore space (tortuosity), as given by the size and distribution of pore throats and bodies. In a macroscopic sense, this heterogeneity translates into a variation of the permeability to a fluid [12]. Intuitively, rock heterogeneities have significant effects on fluid displacement processes, independently of the scale considered. For instance, the initial distribution of hydrocarbons in an oil reservoir is controlled by a capillary-gravity equilibrium, which in the presence of heterogeneities yields to saturation discontinuities. Moreover, since the factors that affect this process are also effective during the production of the reservoir, any simulation study of fluid flow in heterogeneous media that use single relative permeability and capillary pressure curves would automatically be inadequate [29]. On the contrary, capillary pressure and relative permeability relations should be parameterized, so as to assign the correct fluid saturation distribution in the reservoir. The same concept holds true also when attempting to represent the details of the small-scale (mm-to-cm) fluid behavior during core-floods [19] and we refer to these spatial variations of local capillary pressure curves as capillary heterogeneity.

As anticipated in the previous section, the challenging task lies in the identification of accurate techniques to characterize rock (and in particular permeability) heterogeneities at the sub-core scale; in fact, if those would be known, a scaling function such as the Leverett J-Function could be used to represent the corresponding variations in the capillary pressure-saturation relationship (see Section 13.5). Some investigators have used sand columns packed with alternating layers of fine and coarse sands with well-defined permeability and porosity characteristics [20], while others working with consolidated rock samples have used a so-called mini-permeameter to quantify local permeability heterogeneities [19, 27]. While all these works support the general idea of the importance of capillary heterogeneity to understand fluid distribution in a porous medium, only recently efforts have been directed into the quantification of these phenomenon by combining experimental observations and predictions from numerical simulations. In this context, an effective experimental method to identify permeability heterogeneity works with capillary dominated displacements and exploits the sensitivity of the saturation response to permeability variations [12]. By applying a conceptually similar approach, Krause and co-workers have developed a technique to determine sub-core scale permeability distributions, which, combined with measured capillary pressure data, allowed reproducing via core-flood simulations measured sub-core scale saturation distributions on various rock cores [18, 26].

Besides the development of techniques to obtain reliable porosity and permeability distributions within a rock sample, experimental protocols have been developed that independently validate the existence of these spatially varying capillary pressure curves at the sub-core scale [31]. Some of the results from these studies are shown in Fig. 13.4 for two distinct fluid pairs, namely CO_2/brine and N_2/water, in a Berea Sandstone core. Capillary pressure curves are shown for four distinct positions (voxels) in

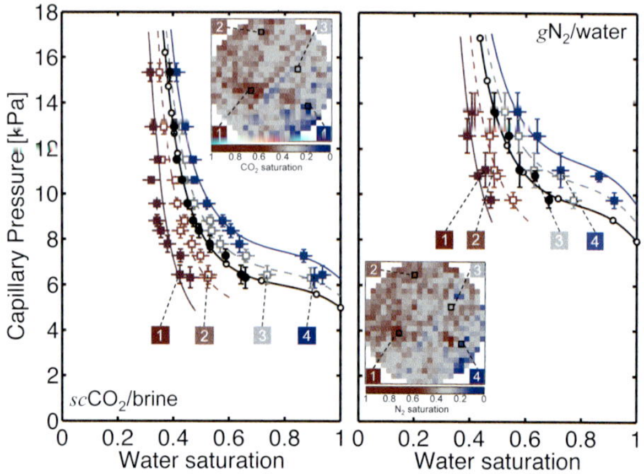

Fig. 13.4. Experimental evidence of the existence of spatially varying capillary pressure curves at the sub-core scale. Results are shown for the CO_2 /brine and N_2/water system in a Berea Sandstone core; $P_c(S)$ curves obtained using these fluid-pairs have been measured at the core- ($\sim$1.9 cm^3, black filled circles) and voxel-scale ($\sim$4 mm^3, empty and filled gray squares), and are compared to converted mercury intrusion data (empty circles). (Figure taken from Ref. [30].)

the core (as indicated by the 2D saturation map shown in the inset of the figure that represent the inlet slice of the core), which are characterized by an early or late invasion of the nonwetting phase as compared to the average slice saturation. These are representative of $\sim$4 mm^3 sub-sets of the rock-sample and are compared to the core-average capillary pressure curve (circles), as obtained from both core-flooding and mercury intrusion techniques. Interestingly enough, at a given capillary pressure level, the four mm-scale capillary pressure curves are shifted towards higher or lower saturation values, but retain a similar shape as the mean curve. This is evidenced by the colored lines in the figure that are drawn primarily to guide the eye, but that result merely from a shift of the mercury data toward higher or lower capillary pressure levels.

13.5. Linking Multiphase and Rock Properties

With the aim of simplifying the description of the statistical variability of a spatially distributed property such as the capillary pressure, hydrologists have often adopted a scaling approach [32, 33]. The latter makes use of so-called scaling factors α_j to relate the capillary pressure at a given location,

$P_{c,j}$, to a representative mean, $\hat{P}_c$, i.e.

$$P_{c,j}(S) = \frac{1}{\alpha_j}\hat{P}_c(S), \qquad j = 1, ..., N \tag{13.1}$$

where N is the number of distinct locations in the dataset. The Leverett J-Function belongs to this category and it explicitly relates the variation in the capillary properties of a given rock to variations in porosity, permeability and fluid-pair [34]:

$$J(S) = \frac{P_c(S)}{\sigma_{12}\cos\theta_{12}}\sqrt{\frac{\hat{k}}{\hat{\varepsilon}}} \tag{13.2}$$

where $\hat{k}$ and $\hat{\varepsilon}$ are the core's average permeability and porosity, σ_{12} is the interfacial tension of the given fluid pair and θ_{12} is the contact angle. While in its original form, this relationship was developed to scale distinct core samples with similar lithologies, a number of investigators has applied it at the sub-core scale [18, 35, 36]. In fact, upon assumption that each subset j of the core sample follows the same scaling function $J(S)$, a relationship between the scaling factor α_j and the properties of the rock can be readily found, i.e.

$$\alpha_j = \sqrt{\frac{k_j/\varepsilon_j}{\hat{k}/\hat{\varepsilon}}}, \qquad j = 1, ..., N \tag{13.3}$$

where k_j and ε_j are the location-dependent permeability and porosity. In a core-flooding experiment, the data set consists of N locations (voxels) within the rock sample and the scaling factors in Eq. 13.1 can be found through a simple minimization problem, when both the representative and local capillary pressure curves are known from measurements. With reference to the data presented in the previous section (and shown in Fig. 13.4), Fig. 13.5 (left) shows the whole set of unscaled capillary pressure values, $P_{c,j}$, from 11,637 locations within the rock sample. In the figure, the solid line represents the characteristic capillary pressure curve, while the plus symbols are the (voxel) unscaled data. As expected, a large spread of saturation values is observed: at $P_c = 10$ kPa, S_w ranges from about 0.4 down to 0.05. The same data, now scaled, are also shown in Fig. 13.5 (right): it can be seen that the originally scattered data now coalesce nicely into a relatively narrow band around the characteristic capillary pressure curve, as

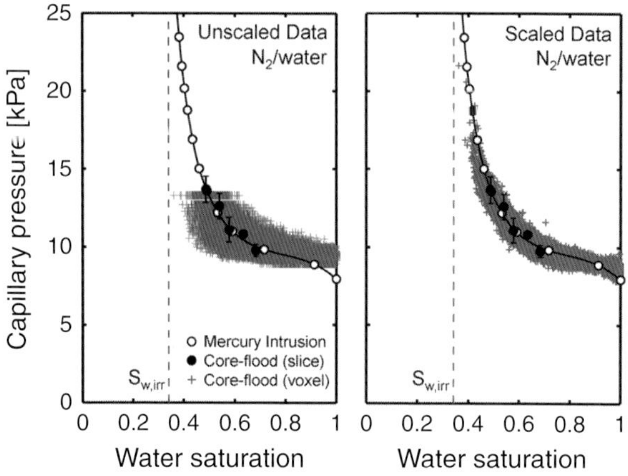

Fig. 13.5. Scaling of sub-core scale capillary pressure curves : unscaled (left) versus scaled (right) N_2/water capillary pressure curves measured in a Berea Sandstone core. Circles are results from the mercury intrusion (empty black symbols) and core-flooding (filled black symbols) experiments, and represent the mean curve against which observations at the voxel scale (plus gray symbols, voxel size ~12 mm^3) are scaled. (Figure modified from Ref. [22].)

confirmed by value of the residual sum of squares, which is reduced to only 15% of the original value upon solving the minimization problem on the unscaled data. In other words, an approach based on constant (location-dependent) scaling factors is able to capture the variability of the P_c curve at the sub-core scale, further implying that the description of capillary heterogeneity can be simplified considerably.

The heterogeneity of a rock sample can be easily visualized by considering 3D maps similar to those shown in Fig. 13.6, which depict the spatial distribution of the scaling factors and permeabilities obtained for the Berea Sandstone core considered in Figs. 13.2, 13.4 and 13.5. In both cases, two scenarios are shown that refer to results obtained from the commonly adopted (and less effective) Kozeny-Carman approach (top) and to those obtained from a capillary pressure based technique (bottom). In the first case, measured permeability values are predicted from the corresponding sub-core scale porosity values and capillary pressure curves can thus be obtained (Eqs. 13.2 and 13.3). In the latter case, both the porosity values and the sub-core scale capillary pressure curves are independently measured, and the corresponding permeability values are obtained from

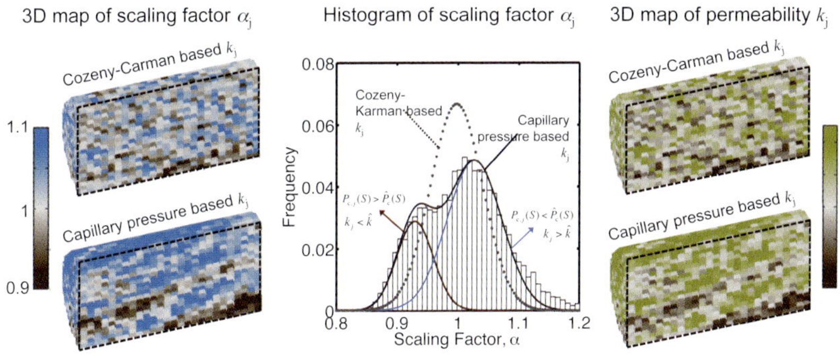

Fig. 13.6. Linking multiphase- and rock properties: scaling factors (capillary pressure) and permeability are estimated within a unique framework. The distribution of scaling factor (center) describe the spatial variability of the capillary pressure curve within a sandstone sample (left) and can be related to a corresponding distribution of permeabilities (right). Voxel size 2×2×3 mm^3. (Figure modified from Ref. [22].)

Eqs. 13.1–13.3. The differences are significant and readily apparent, while the similarity between the results from the capillary pressure-based technique and the saturation maps shown in Fig. 13.2 is indeed as anticipated: the same strata that remain filled with water during the drainage process are now characterized by small scaling factors (and permeabilities), i.e. at these particular locations a larger capillary pressure is required to reach the same saturation as the core-averaged value (see also Eq. 13.1). The graphical representation of the same distribution in a histogram (Fig. 13.6, center) shows that the majority of the scaling factors α lie within a $\pm$ 15% range from the mean and that they follow a bi-modal distribution, as observed in the saturation maps. As expected, when the sub-core scale permeability distribution is predicted from the Kozeny-Carman model, a completely different distribution is obtained (dotted line). It follows that in the latter case high-resolution numerical simulations (such as those described in Section 13.3) would fail to replicate the saturation distributions shown in Fig. 13.2. Only an integrated approach that combines experimentally observed saturations with spatially distributed capillary pressure curves leads to an accurate representation of the sub-core scale permeability distribution within the rock sample [18]. Such an approach evaluates multiphase and rock properties within a unique and consistent framework and represents a significant improvement from conventional core analysis.

As discussed in the previous sections, the accurate representation of the relationship between multiphase- and rock properties is paramount for making quantitative conclusions about the parameters and processes controlling subcore-scale saturation distributions. When applied to core-flooding tests, the results and observations of key multiphase flow properties, such as relative permeability curves, can be directly referred to the internal structure of the sample, thus providing a new level of diagnostic information into the effects of small- scale heterogeneities on fluid displacement. Such insights are in turns key to feed and validate multistage upscaling approaches; in this context, and with reference to the results just described, two main families of voxels can be identified within the rock sample (given by the brown and blue curves plotted on top of the histogram in Fig. 13.6), whose statistics are normally distributed around two mean scaling factor (or permeability) values, which are respectively smaller and larger than the corresponding mean properties of the core. A direct practical implication of this type of analysis would be the definition of a minimum scale at which heterogeneities have to be resolved, so that mathematical models would adequately capture the observed displacement patterns.

13.6. Importance of Mesoscale Heterogeneity

Only recently have the quantitative experimental, theoretical, and simulation tools become available to observe and predict the implications of mesoscale geological heterogeneity on the fate and transport of fluids in the subsurface. Mesoscale heterogeneity has the potential to influence flow and transport properties in a number of important ways, for example: increasing dispersivity for contaminant transport, affecting multiphase displacement efficiency for resource recovery, NAPL removal, and CO_2 storage; and affecting the kinetics of subsurface geochemical reactions. Here we provide just one example illustrating how mesoscale heterogeneity influences multiphase displacements in buoyancy dominated flows.

Figure 13.7 illustrates the numerical simulation grid for a 2-dimensional system used to investigate the influence of mesoscale heterogeneity on CO_2 injection into a brine-filled aquifer. A random permeability field, based on the geostatistical representation of a Berea Sandstone, was used to populate a 250 by 250 (4 mm $\times$ 4 mm) simulation. Leverett J-Scaling (Eq. 13.2) was applied to assign different capillary pressure curves for each cell. CO_2 was injected along the left-hand boundary. For comparison, simulations were also carried out for a homogeneous system with the average properties used in the heterogeneous system. The simulations, using Stanford's General

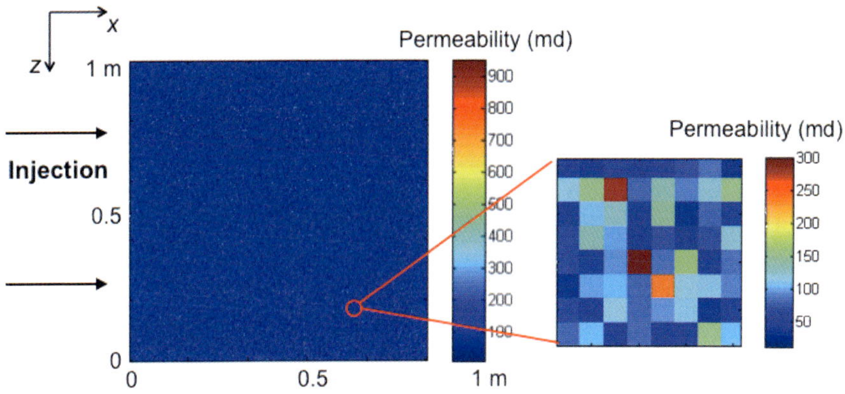

Fig. 13.7. Illustration of the simulation grid used to study the influence of mesoscale capillary heterogeneity on multiphase flow of CO_2 displacing brine. The simulations were conducted using a version of GPRS modified to study capillary heterogeneity [28].(Image from B. Li, Stanford University.)

Purpose Reservoir Simulator (GPRS) include rigorous treatment of viscous, capillary and gravity forces [28].

Figure 13.8 provides a comparison between the two simulations, showing that while for the heterogeneous rock, the CO_2 plume is compact, and only slightly influenced by buoyancy forces (supercritical CO_2 has about 2/3's of the density of brine), CO_2 migration in the homogeneous rock is largely dominated by gravity forces, with the CO_2 rising quickly to the top of the system and migrating laterally out of the simulation grid. Clearly mesoscale heterogeneity plays the important role of stabilizing the CO_2

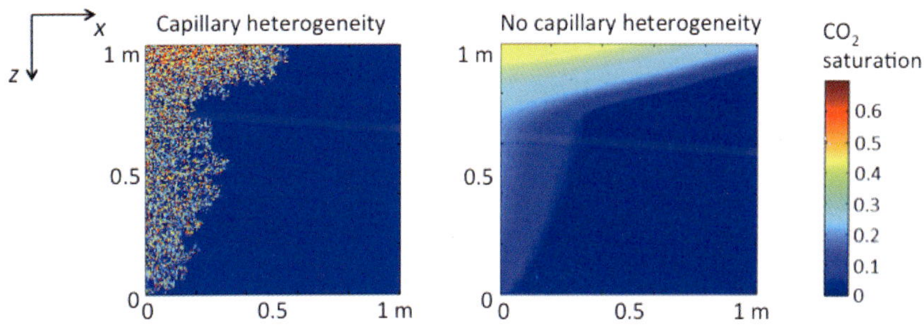

Fig. 13.8. Comparison between CO_2 plume movement with and without mesoscale capillary heterogeneity. (Image from B. Li, Stanford University.)

front and counteracting the influence of gravity at this scale. This provides just one example of the influence of mesoscale heterogeneity on subsurface flow and transport. Over the coming years, as the tools available to study these processes improve, other interesting and important examples can be expected.

13.7. Concluding Remarks

The application of core analysis to provide data for reservoir characterization is a research field under continuous development as new challenges are faced and novel technologies are developed. One of the main challenges is to account for the heterogeneous nature of reservoir rocks; not only is it important to make measurements for the full range of rock types present in a reservoir, but also to quantify the degree and nature of the heterogeneity of the rock samples themselves. Of particular interest are small-scales (mm-to-cm) features whose combined effects can be substantial on the flow behavior over much larger scales. Whether we are dealing with the remediation of a shallow aquifer or with improved/enhanced oil recovery techniques, there is an increasing need to incorporate in current large-scale models new knowledge on the physics at the fine-scale. In fact, an accurate representation of hydrogeological properties across various scales is paramount for making quantitative conclusions about the parameters and processes controlling saturation distributions in the subsurface.

Because core-flooding systems can be studied in such detail, they are very useful for understanding how fluid properties and geological parameters govern multiphase flow under different conditions and flow regimes. In this context, multidimensional scanning techniques, such as X-ray CT, are increasingly being used to obtain detailed information about rock heterogeneity and saturation distribution. Such level of observational and characterization detail has the potential to significantly improve the contribution of core analysis to the characterization of a reservoir and to the prediction of its performance. Given that the opportunity to recover reservoir rock core samples arises only once in the life-time of a well and that the so-called special core-analysis (i.e. the one involving the determination of capillary pressure and relative permeability curves) is expensive and time-consuming, it is indeed in the best interest of reservoir engineers to obtain the most exhaustive and valuable information out of core-flooding tests.

The effects of heterogeneities of sub-core scale properties on multiphase flows have been discussed in this chapter by considering both experimental and modeling studies. Experimental methodologies have been presented that allow to simultaneously and non-destructively measure core- and subcore-scale hydrogeological properties during conventional core-flooding tests, thus including porosity, permeability and capillary pressure curves. Quantitative agreement with high-resolution numerical simulations can be achieved when these properties are evaluated within a unique framework that includes spatial variability in the capillary pressure curve, as reflected by the corresponding variability in the fluids' saturation distribution. Furthermore, the effect of heterogeneities can be quantified through the analysis of statistical measures of the saturation distribution and representative populations of voxels within the core sample can be identified. Besides their exploitation within an upscaling process, the latter define terms for comparison among different rock types.

In this chapter we have restricted the discussion to the flow of two immiscible phases within a porous medium, thus neglecting other processes that contribute to the inherent complexity of natural environments. The tools and techniques presented above can be extended to the study of other interfacial processes that might be also affected by the heterogeneity of these systems and include among others chemical reactions and adsorption phenomena. Efforts in this direction are in fact driven by the need to better understand complex hydrogeological flows, thus increasing our ability to successfully predict the distribution of fluids in the subsurface.

References

[1] A. Lohne, G. Virnovsky, and L. Durlofsky, Two-stage upscaling of two-phase flow: from core to simulation scale, *SPE J.* **11**(3), 304–316 (2006).
[2] W. Murphy, J. Roberts, D. Yale, and K. Winkler, Centimeter scale heterogeneities and microstratifaction in sedimentary rocks, *Geophys. Res. Lett.* **11**(8), 697–700 (1984).
[3] K. A. Klise, V. C. Tidwell, and S. A. McKenna, Comparison of laboratory-scale solute transport visualization experiments with numerical simulation using cross-bedded sandstone, *Adv. Water Res.* **31**(12), 1731–1741 (2008).
[4] M. J. Ronayne, S. M. Gorelick, and C. Zheng, Geological modeling of submeter scale heterogeneity and its influence on tracer transport in a fluvial aquifer, *Water Resour. Res.* **46**(10), W10519 (2010).
[5] P. Corbett, P. Ringrose, J. Jensen, and K. Sorbie. Laminated clastic reservoirs: the interplay of capillary pressure and sedimentary architecture. In *SPE Annual Technical Conference and Exhibition*, Washington, D.C., USA, 4-7 October (1992).
[6] P. Ringrose, K. Sorbie, P. Corbett, and J. Jensen, Immiscible flow behaviour

in laminated and cross-bedded sandstones, *J. Pet. Sci. Eng.* **9**(2), 103–124 (1993).

[7] P. P. van Lingen, J. Bruining, and C. P. J. W. van Kruijsdijk, Capillary entrapment caused by small-scale wettability heterogeneities, *SPE Reservoir Eng.* **11**(2), 93–100 (1996).

[8] T. H. Illangasekare, J. L. Ramsey, K. H. Jensen, and M. B. Butts, Experimental study of movement and distribution of dense organic contaminants in heterogeneous aquifers, *J. Contam. Hydrol.* **20**, 1–25 (1995).

[9] S. Krevor, R. Pini, B. Li, and S. Benson, Capillary heterogeneity trapping of CO_2 in a sandstone rock at reservoir conditions, *Geophys. Res. Lett.* **38**, L15401 (2011).

[10] N. Wardlaw and J. Cassan, Oil recovery efficiency and the rock-pore properties of some sandstone reservoirs, *B. of Can. Petrol. Geol.* **27**(2), 117–138 (1979).

[11] B. H. Kueper, W. Abbott, and G. Farquhar, Experimental observations of multiphase flow in heterogeneous porous media, *J. Contam. Hydrol.* **5**(1), 83–95 (1989).

[12] M. Chaouche, N. Rakotomalala, D. Salin, B. Xu, and Y. Yortsos, Capillary effects in drainage in heterogeneous porous media: continuum modelling, experiments and pore network simulations, *Chem. Eng. Sci.* **49**(15), 2447–2466 (1994).

[13] J.-C. Perrin and S. Benson, An experimental study on the influence of sub-core scale heterogeneities on CO_2 distribution in reservoir rocks, *Transport Porous Med.* **82**(1), 93–109 (2010).

[14] J.-Q. Shi, Z. Xue, and S. Durucan, Supercritical CO_2 core flooding and imbibition in Tako sandstone-influence of sub-core scale heterogeneity, *Int. J. Greenhouse Gas Control.* **5**(1), 75–87 (2011).

[15] B. Ataie-Ashtiani, S. M. Hassanizadeh, and M. A. Celia, Effects of heterogeneities on capillary pressure-saturation-relative permeability relationships, *J. Contam. Hydrol.* **56**(3-4), 175–192 (2002).

[16] E. Saadatpoor, S. Bryant, and K. Sepehrnoori, New trapping mechanism in carbon sequestration, *Transport Porous Med.* **82**(1), 3–17 (2010).

[17] C.-W. Kuo, J.-C. Perrin, and S. M. Benson. Effect of gravity, flow rate, and small scale heterogeneity on multiphase flow of co_2 and brine (spe 132607). In *SPE Western Regional Meeting*, Anaheim, California, USA, 27-29 May (2010).

[18] M. Krause, J.-C. Perrin, and S. Benson, Modeling permeability distributions in a sandstone core for history matching coreflood experiments, *SPE J.* **16** (4), 768–777 (2011).

[19] M. Honarpour, A. Cullick, N. Saad, and N. Humphreys, Effect of rock heterogeneity on relative permeability: implications for scale-up, *J. Petrol. Technol.* **47**(11), 980–986 (1995).

[20] A. Graue, Imaging the effects of capillary heterogeneities on local saturation development in long corefloods, *SPE Drilling & Completion.* **9**(1), 57–64 (1994).

[21] S. Akin and A. R. Kovscek. Computed tomography in petroleum engineer-

ing research. In eds. F. Mees, R. Swennen, M. Van Geet, and P. Jacobs, *Applications of X-ray Computed Tomography in the Geosciences*, vol. 215, pp. 23–38. Geological Society, London (2003).
[22] R. Pini and S. M. Benson, Characterization and scaling of mesoscale heterogeneities in sandstones, *Geophys. Res. Lett.* **40**(15), 3903–3908 (2013).
[23] J.-Y. Arns, C. H. Arns, A. P. Sheppard, R. M. Sok, M. A. Knackstedt, and W. V. Pinczewski, Relative permeability from tomographic images; effect of correlated heterogeneity, *J. Pet. Sci. Eng.* **39**(3–4), 247–259 (2003).
[24] R. H. Brooks and A. T. Corey, Hydraulic properties of porous media, *Hydrology Paper No. 3, Colorado State University.* pp. 1–27 (1964).
[25] M. Krause. Modeling and investigation of the influence of capillary heterogeneity on relative permeability. In *SPE Annual Technical Conference and Exhibition*, San Antonio, Texas, USA, 8-10 October (2012).
[26] M. Krause, S. Krevor, and S. M. Benson, A procedure for the accurate determination of sub-core scale permeability distributions with error quantification, *Transport Porous Med.* **98**(3), 565–588 (2013).
[27] S. Ganapathy, D. G. Wreath, M. T. Lim, B. A. Rouse, G. A. Pope, and K. Sepehrnoori, Simulation of heterogeneous sandstone experiments characterized using ct scanning, *SPE Formation Evaluation.* **8**(4), 273–279 (1993).
[28] B. Li, S. M. Benson, and H. A. Tchelepi. Modeling fine-scale capillary heterogeneity in multiphase flow of CO_2 and brine in sedimentary rocks. In *XIX International Conference on Water Resources*, University of Illinois at Urbana-Champaign, Illinois, USA, 12-22 June (2012).
[29] S. Kocberber and R. Collins. Impact of reservoir heterogeneity on initial distributions of hydrocarbons. In *SPE Annual Technical Conference and Exhibition*, New Orleans, Louisiana, 23-26 September (1990).
[30] R. Pini and S. M. Benson. Characterizing mesoscale heterogeneities in reservoir rocks. In *International Symposium of the Society on Core Analysts*, Napa Valley, California, USA, 16-19 September (2013).
[31] R. Pini, S. C. M. Krevor, and S. M. Benson, Capillary pressure and heterogeneity for the CO_2/water system in sandstone rocks at reservoir conditions, *Adv. Water Res.* **38**, 48–59 (2012).
[32] A. Warrick, G. Mullen, and D. Nielsen, Scaling fieldmeasured soil hydraulic properties using a similar media concept, *Water Resour. Res.* **13**(2), 355–362 (1977).
[33] J. W. Hopmans, A comparison of various methods to scale soil hydraulic properties, *J. Hydrol.* **93**(3), 241–256 (1987).
[34] M. C. Leverett, Capillary behavior in porous solids, *T. Soc. Petrol. En. AIME.* **142**, 152–169 (1941).
[35] J. Chang and Y. Yortsos, Effect of capillary heterogeneity on buckley-leverett displacement, *SPE Reservoir Engineering.* **7**(2), 285–293 (1992).
[36] Y. Huang, P. Ringrose, and K. Sorbie, The effects of heterogeneity and wettability on oil recovery from laminated sedimentary structures, *SPE Journal.* **1**(4), 451–462 (1996).

Chapter 14

Coarse-Grained Monte Carlo Simulation of the Distribution and Capillarity of Multiple Fluids in Porous Media

Benjamin D. Zeidman, Ning Lu and David Wu

Department of Chemistry and Geochemistry
Department of Civil and Environmental Engineering
Department of Chemical and Biological Engineering
Colorado School of Mines, Golden, CO, USA
dwu@mines.edu

The pore-scale distribution of fluids and gasses and the associated capillary phenomena control key macroscale constitutive properties of interest in any porous medium, including systems of water or oil in soils and rocks. However, due to the geometric complexity of the pore space as well as that of interfaces between phases, the distribution of fluids and forces is difficult to obtain. An attractive approach for efficient determination of the spatial distribution of various fluid components–such as air, water, oil, or gas hydrate–in a porous medium is Coarse-Grained Monte Carlo simulation. The method is well-suited to handle complicated pore space geometries, and can exploit advances in imaging technologies such as X-ray micro CT for obtaining high-resolution maps of pore space. Coarse-graining allows for computational efficiency, and can be applied at the inherent size scale of the pores, from nano to macro scale. Monte Carlo simulation foregoes expensive, explicit simulation of hydrodynamic transport, to allow rapid determination of local equilibrium phase distributions.

While Coarse-Grained Monte Carlo has been used before to simulate adsorption in porous media, we make parallel measurements on an experimental system (Ottawa sand) and simulations based on the experimentally determined porespace of the same system, and demonstrate excellent agreement in key constitutive properties including capillary pressure, soil-water retention, specific surface areas, as well as wetting/drying hysteresis. To do this, we introduce a method for using two-dimensional slices of the 3D porespace for efficiency. Our results support the hypothesis that Coarse-Grained Monte Carlo generates metastable configurations consistent with experimental hysteresis, justifying our elimination of Simulated Annealing used in prior approaches.

14.1. Introduction

Knowledge of where fluids are situated in a porous medium is fundamental to an understanding of how heterogeneity at the pore-scale leads to macroscale properties. This is applicable not only in hydrogeological contexts, but also across the whole range of porous systems of technological and scientific interest such as nanomaterials [1, 2] and biological systems [3, 4]. Many of the properties of interest are related to the complex interplay of liquid and gas within the porespace, and the resulting equilibrium distribution of phases contained therein [5, 6].

Given the ubiquity of the problem, several approaches have been used to calculate the phase distribution within a porous medium [7, 8], and we provide here only a brief overview. Exact analytical solutions for capillary shape exist, but for only the simplest of systems (e.g., disks in two-dimensions) [9–16]. As the geometric complexity of the porespace increases, numerical approaches are needed, including continuum hydrodynamic modeling, lattice-Boltzmann method, molecular scale Monte Carlo simulation, and the Coarse-Grained Monte Carlo simulation methods discussed here.

Continuum hydrodynamic modeling of capillarity in a porous medium involves solving the Navier-Stokes equations subject to the geometrical boundary constraints of free interfaces and fixed surfaces, as well as the capillary forces in the system [17, 18]. Systems are thus described by a second order partial differential boundary value problem, but the interfaces can suffer catastrophic events such as rupture or coalescence that need separate treatment [19]. As the size and geometric complexity of the system increases, so does the difficulty of numerically solving the associated equations.

Since interfaces arise from interactions between molecules, it is possible to use direct molecular simulation to capture phase distribution and interfacial geometry in a porous media. Molecular dynamics (MD) simulation would be most fundamental, and would at a larger scale, produce fluid flow consistent with continuum hydrodynamics. However, explicit evaluation of molecular motion for large complex porespaces is prohibitive, both because of the cost of high spatial resolution and to capture realistic dynamics. A standard strategy for overcoming the cost of simulating exact molecular trajectories is to use the Monte Carlo method for molecular simulation, which can more directly and efficiently find local equilibrium configurations of molecules. It also naturally allows for simulating equilibrium adsorption/desorption from a reservoir via Grand Canonical Monte Carlo, which is a standard method in molecular simulation to generate adsorption isotherms in porous materials such as zeolites [20]. As an additional step, fictitious but physically-motivated dynamics can be introduced via kinetic Monte Carlo [21] to mimic time evolution of the fluid system. However, atomistic Monte Carlo simulation is generally computationally limited to relatively simple pore geometries (e.g., Refs. [17, 22]), though advances in computing power and algorithms will continue to push the limits of feasibility [23].

Intermediate between the continuum and molecular models are coarse-grained models that sacrifice atomistic realism for computational efficiency while retaining the physically relevant pore-scale physics. Such methods become increasingly compelling as they present an opportunity to exploit the high-resolution porespace maps made available from advances in imaging, such as in X-ray micro computed tomography.

The lattice Boltzmann method (Ref. [24], and also see Chapter 9), is such a coarse-grained model that retains hydrodynamic fluid transport, but via discrete Boltzmann equations representing particle collisions on a lattice. It has the advantage over continuum hydrodynamic modeling of not requiring special treatment of interfaces, which arise naturally from the interactions between particles in the bulk. This method has been used to compute properties associated with porous media [25], such as hydraulic conductivity. This methodology can be used with systems with complex porespace geometries, but carries a computational cost to capture the hydrodynamic transport, relative to Coarse-Grained Monte Carlo simulations.

Like the lattice Boltzmann method, Coarse-Grained Monte Carlo simulation coarse-grains many individual molecules into effective groups that reside on lattice sites [26–29]. However, these lattice site groups contain information only of which phase is most prevalent at that location. No momentum information is retained, since to speed up determination of local phase equilibria, hydrodynamic behavior (which originates from momentum balance) is replaced by a more efficient stochastic dynamics. These lattice sites do interact through neighboring free-energetic interactions, and thus are capable of naturally capturing interfacial behavior. Just as molecular Monte Carlo simulation is an efficient alternative to molecular dynamics for molecular simulations, Coarse-Grained Monte Carlo is an efficient alternative to the lattice Boltzmann method for coarse-grained simulations of local equilibrium phase distributions.

It should be noted that the application of Coarse-Grained Monte Carlo to fluids in porous media involves a hypothesis that the local, metastable equilibria produced by the stochastic dynamics are representative of the experimental system. In many practical situations, the pore-scale capillary number, $\mathrm{Ca} = \mu \mathrm{v}/\gamma$, (where μ is the viscosity, v is the pore-scale velocity, and γ is the fluid/fluid interfacial tension) is small (e.g., oil trapped in pores of water-wet rock can typically have $\mathrm{Ca} \sim 10^{-6} - 10^{-4}$) [30]. This corresponds to dominance of interfacial capillary stresses over viscous stresses (and usually inertial stresses as well given the low Reynolds numbers at these small length scales), and lends some argument for the quasistatic neglect of hydrodynamic transport in determination of local phase equilibria. There is also some support and analysis on the molecular scale for this hypothesis, where molecular Monte Carlo simulation has been verified to reproduce the hysteresis associated with such metastable states when compared with molecular dynamics simulations [31–33]. We present below direct validation of this hypothesis by comparison with experiment on natural porous media, which has been previously lacking. In the following section, however, we first describe the Coarse-Grained Monte Carlo simulation algorithm, before reviewing and discussing its applications.

14.2. Coarse-Grained Monte Carlo Simulation Algorithm

The simulated system consists of a fixed solid porous medium, inside which several phases can exist. The system is represented by a lattice, with a resolution Δ represents the inter-lattice spacing chosen to adequately represent the pore-scale features of interest. Each lattice point (or cell) is assigned an identity corresponding to the majority phase at that point. To be concrete in the following discussion, we consider the case of a soil partially saturated with water, where there will be a total of three phases: solid (soil), water, and air (vapor). Each lattice point then is assigned to be one of these three phases, with the solid cells, representing the pore space geometry, fixed throughout the simulation, and the remaining cells corresponding to porespace that can be occupied with either water or air as the simulation proceeds. Following the convention for lattice gas models, we mathematically represent the phase of a lattice cell, i, by three occupation numbers, n_i^α, where α can take on the values of s (soil), w (water), or a (air). The occupation number n_i^α takes on the value 1 if lattice cell i is phase α, and 0 otherwise. The set of all occupation numbers for each lattice cell and each phase fully defines a configuration of the system.

The stochastic evolution of the system is governed by its total free energy

$$E = \sum_{i,j} \sum_{\alpha,\beta} J_{\alpha\beta} n_i^\alpha n_j^\beta + \mu \sum_i n_i^w \tag{14.1}$$

where the first term on the right represents the sum of interaction free energies $J_{\alpha\beta}$ between neighboring lattice cells of phases α and β in contact, and the second term on the right is the sum of the chemical potential contributions μ for each cell occupied by water. (We take the zero of chemical potential to be that of the vapor phase; the chemical potential contribution of the solid is neglected since it is constant.) Note the first sum is over all nearest neighbor lattice cells i and j, while the second sum is over all phases for α and β, and the third sum is over all lattice cells i. The values of $J_{\alpha\beta}$ can be assigned based on experimental values for interfacial free energy according to the relation [34]),

$$\gamma\Delta = 2J - T \ln \left[\coth \left(\frac{J}{T} \right) \right] \tag{14.2}$$

where T is the simulation temperature. This allows simulation parameters to be set in a manner that reflect the intrinsic properties concerning interfacial tension and/or contact angles (controlled by the ratio of interfacial tensions) from the interaction of real world materials.

The system is then updated from one configuration to the next by proposing a phase change in one or more cells, and then according to a specific criterion involving a random number, deciding whether to accept the proposed change or to reject it. Coarse-Grained Monte Carlo methods differ in the choice of which changes to propose. Algorithms that propose changes that conserves the total amount of each phase in the system are called "canonical", while those that regulate the total quantity of each phase by its chemical potential are called "grand canonical". Note that in canonical algorithms, the chemical potential term in the total free energy is conventionally neglected, since it is constant as the total water content is fixed.

Coarse-Grained Monte Carlo methods also differ in the choice of the criterion for acceptance of a proposed update in configuration. Here we highlight the standard Metropolis algorithm, which serves as a common basis for several methods [35, 36]: accept the proposed change if $\exp(-\Delta E/k_B T)$ is greater than a random number chosen uniformly between 0 and 1, where ΔE is the change in total system energy for the proposed change in configuration and k_B is Boltzmann's constant. This algorithm is designed to

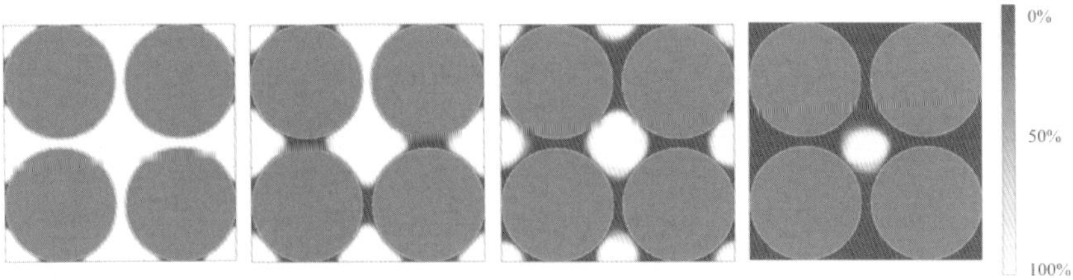

Fig. 14.1. Canonical simulation result for four grains on a periodic 250 × 250 lattice grid at liquid water saturations from left to right of 10%, 25%, 50%, and 90%. The lowest saturation begins with the formation of the wetting layer and as saturation is increased capillary bridges are formed. At higher saturations the grains become fully wet with remaining air bubbles. The color bar at right shows the average density of the liquid phase.

drive the system towards thermodynamic equilibrium, at which point the probability of a given configuration is proportional to the Boltzmann factor, $\exp(-\Delta E/k_BT)$.

Figure 14.1 shows the result of an illustrative example of a Coarse-Grained Monte Carlo simulation on a simple periodic system of 4 circular grains with varying water content. The canonical algorithm used here proposes new configurations with a liquid phase cell and a gas phase cell selected at random and swapped, and then accepts or rejects according to the Metropolis criterion. The values for $J\alpha\beta$ are assigned as -1 for liquid/liquid, -2 for liquid/solid, and 1 for liquid/gas interactions, where the unit of energy is taken to be k_BT. All other values are set to zero. These values correspond to perfect wetting and favor water adsorption onto the solid phase. The simulation is run for approximately 8,000 Monte Carlo time steps, where a time step is defined as N^2 attempted configuration changes (or moves), where N is the total number of lattice cells in the system. Equilibration to a point of less than 1% change in energy fluctuations typically occurs in about 1,500 time steps, with the remaining simulation time used for sampling.

14.3. Applications and Experimental Tests

An early application of Coarse-Grained Monte Carlo simulation to solving for phase equilibria in porous materials of relatively simple geometries was the development of a Monte Carlo annealing method [26]. This approach sought to obtain the global energy minimum using three methods. The first and second methods for obtaining the energy minimum involved starting with either a fully wet or fully dry system and attempting to change individual lattice sites to a different phase (e.g. liquid to gas, or gas to liquid) one at a time until a global energy minimum was reached. The third approach used a Metropolis algorithm starting the system with a given level of liquid saturation and swapping liquid and gas phase cells. The system was evolved using Simulating Annealing, where the temperature was slowly lowered to approach the global minimum.

This method was adapted by Ref. [27] to applications in porous geologic material with the intent of using it to calculate thermal conductivity. The method was used to model fluid location within a complex geologic material. An example is shown below as Fig. 14.2.

While the sandstone geometry was taken from a thin section, there was no experimental image including fluid location to validate against the simulation. After the multiphase equilibrium was established using the Monte Carlo annealing method, Fourier's law was used to calculate the thermal conductivity, showing qualitatively similar behavior as experiments at various moisture content. Interestingly they noticed a percolation behavior at the 40% water saturation level which is best explain by the water making its own thermally conductive pathway in addition to the rock phase to aid in the overall thermal conductivity.

Silverstein and Fort applied the Monte Carlo annealing method to random distributions of model particles [28, 37, 38]. The focus of the work was to calculate interfacial surface areas and hydraulic conductivity. To mimic continuous transport of the liquid phase, a canonical algorithm was employed where swaps are only attempted between liquid and gas cells a short distance apart. For interfacial surface area calculations the systems were started as fully saturated with liquid cells removed that would provide the lowest interfacial free energy. Hydraulic conductivity functions and surface area calculations were carried out in both two and three dimensions. There were discrepancies at saturation levels below 50–60% when comparing the simulation results for interfacial area to experimental data. These were attributed to the experimental system not reaching the global energy minimum at low saturations, which is that generated by the Monte Carlo annealing method. Berkowitz and Hansen applied the Monte Carlo annealing method using porespace geometry data obtained from X-ray micro computed tomography of sandstone [39], but did not present experimental data to compare against.

Due to the importance of metastability, Lu et al. [29] employed a Coarse-

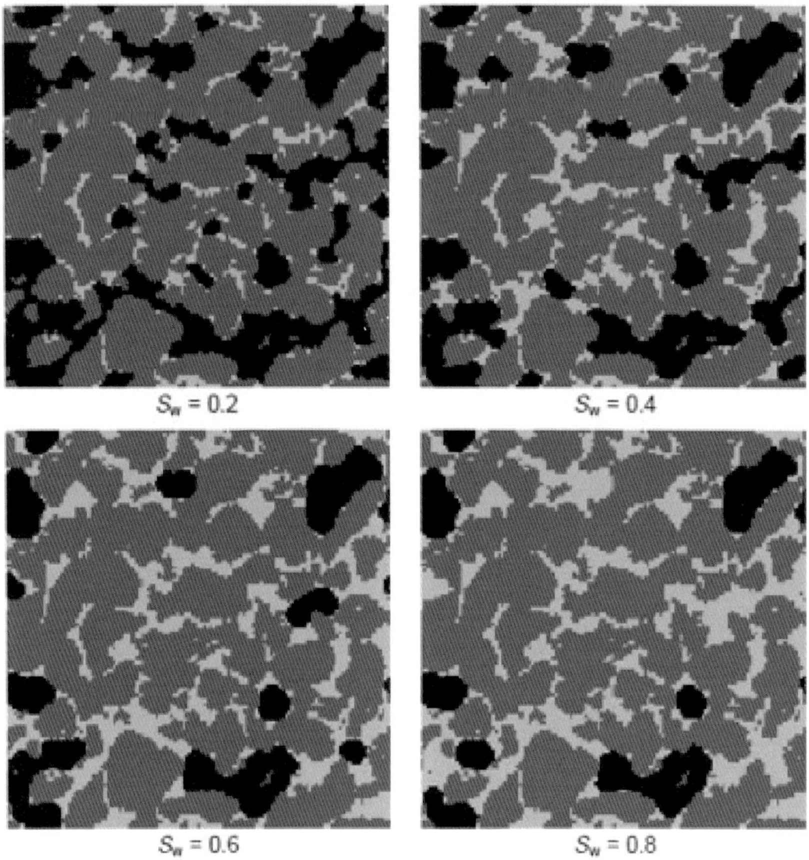

Fig. 14.2. Fluid distribution result in a thin section of Berea sandstone. Dark gray represents rock, light gray is water, and black is air. The S_W values correspond to the level of saturation [27].

Grained Monte Carlo simulation using the Metropolis algorithm directly at fixed simulation temperature, without simulated annealing. Because the simulation is coarse-grained in nature, interfacial energy is significantly greater than k_BT if the experimental temperature is used, and thus configurations are dominated by local energy minima. To address this, the simulation temperature T is assigned to a higher fictitious value that allows for sampling of all possible configurations while maintaining simulation efficiency. This requires the assumption that the distribution of stable and metastable energy minima is representative of that contained in a macroscopic sample. As the simulation proceeds, the system evolves through a series of metastable minima, trending towards equilibrium.

This modified method was applied to two-dimensional sections of an Ottawa sand system whose porespace was determined by X-ray micro computed tomography [29]. The results were directly compared with experiments on the same systems. Results comparing the phase distribution as determined by experiment (top) and simulation (bottom) for various water saturations are shown in Fig. 14.3.

The samples were prepared in small columns each containing a different amount of liquid saturation and then imaged at the Advanced Photon Source 13-BMD at Argonne National Laboratory. Two dimensional (2D) cross-sectional images were analyzed to extract the grain and pore space data from samples containing 15%, 25%, 70%, and 90% liquid saturation. These images are represented in 14.3(a)–14.3(d). The grain and pore space data were discretized into a lattice and MC simulation was carried out with similar saturation conditions. These results are given in Figs. 3(e)–(h). The comparison between experiment (14.3(a)–(d)) and simulation (14.3(e)–(h)) shows similar distributions of adsorbed water in the pore space. Under low saturation conditions, liquid is condensed within many of the smaller pores in the system. At higher saturations the water is also adsorbed in the same locations but perhaps what is more remarkable is the similarity in the shape of the gas "bubble", particularly evident at the 90% saturation level.

Lu et al. also presented a method for calculating the capillary pressure, ΔP, as a function of water content, an important quantity in soil mechanics referred to as the soil-water retention curve (SWRC). Capillary pressure is calculated by measuring the radius of curvature, r, of individual interfaces found in the simulation, and using the Young-Laplace equation for the pressure difference across the liquid-vapor interface, $\Delta P = \dfrac{2\gamma}{r}$. As a result of running simulations in two dimensions, an assumption is made that the two principle radii of curvature are equal to the one observed in the two-dimensional configuration. The result is given in Fig. 14.3(i). The simulation SWRC agrees with the experimental SWRC (prepared by the conventional Tempe cell method [6]) with deviations of less than 8%. Similarly, the specific surface area of a three-dimensional interface can be corrected when deduced from the interfacial length of a two-dimensional slice of that interface. For isotropically distributed three-dimensional orientations of the interface, this results in a correction factor of $(4/\pi)$, a classical result in stereology. While this does not strictly apply to the two-dimensional simu-

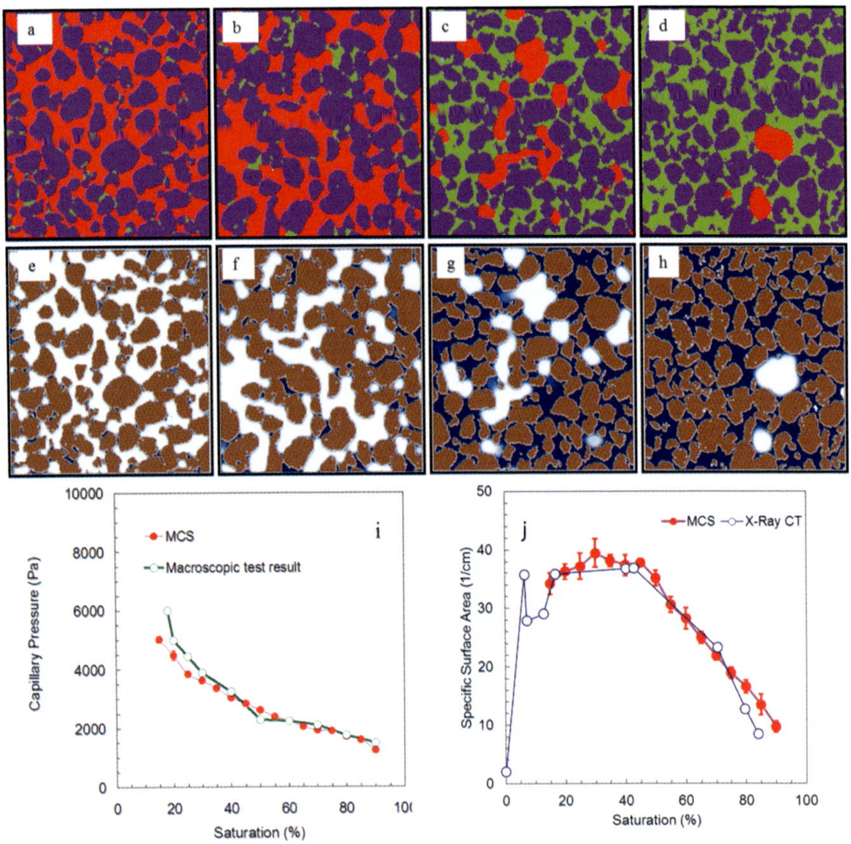

Fig. 14.3. Results for a three phase study of capillarity in Ottawa sand. (a)–(d) show microXCT images of partially wet Ottawa sand with solid represented as purple, gas as red, and liquid as green. Image size is 3.521 mm × 3.521 mm. From left to right, the liquid saturation is 15%, 25%, 70%, 90%. (e)–(h) show Monte Carlo simulation (MCS) output for identical pore spaces with the same liquid saturations, with solid as brown, gas as white, and liquid as blue. Simulation resolution is 250 lattice units × 250 lattice units with an interlace unit spacing of 10.06 microns. (i) shows the soil-water retention curve comparison between simulation and macroscopic experiment. (j) shows the capillary specific surface area between simulation and experiment (with new correction from two-dimensions to three-dimensions) [29].

lations here (as they are not slices of a true three-dimensional surface), we nonetheless hypothesize that the curvatures observed in two-dimensions are representative, and would be isotropically distributed in orientation, and apply the correction factor of $(4/\pi)$ to our simulation results in Fig. 14.3(j).

We note this correction factor was not applied in Ref. [29], but can be seen to bring the experimental and simulation results into close agreement. Thus the simulation methodology can accurately capture quantitative properties of complex porous materials.

14.4. Grand Canonical Monte Carlo Simulation for Wetting/Drying Hysteresis

As porous materials are exposed to either wetting or drying conditions, liquid flows into or out of the material. Properties thus follow a hysteretic, i.e. path-dependent, trajectory, and depend on whether conditions are wetting or drying. The primary reason we see hysteresis in simulation is because of the metastability of local equilibrium interfacial configurations in the porespace. This series of metastable configurations is difficult to characterize or predict in a complex porespace. To isolate the effects and principles behind the hysteresis, many studies on adsorption hysteresis have been carried out on simple pore geometries, and not generally on experimentally determined porespaces. Woo et al. did carry out a Coarse-Grained Grand Canonical Monte Carlo simulation in a random porespace with statistical properties similar to Vycor glass (Woo and Monson, 2003). This was mainly to compare the adsorption hysteresis to a mean-field treatment, concluding that the hysteresis behavior was controlled largely by metastable configurations. Below, we present grand-canonical Coarse-Grained Monte Carlo simulation performed on an experimentally determined porespace.

In grand-canonical Coarse-Grained Monte Carlo, the algorithm for proposing a change in system configuration consists simply of picking a porespace cell at random and changing the phase from water to air or vice versa. The rest of the algorithm proceeds without modification. A higher chemical potential drives the system to lose water, while a lower chemical potential drives the system to gain water. The result is that the simulation functions similarly to experimental adsorption/desorption experiments where adsorbate is driven to adsorb/desorb on a material using a pressure gradient. The chemical potential here functions analogously to the pressure gradient in experiment, and can be related via the equation of state. For a simulation under wetting conditions, the cells in the pore space are all initially assigned as air, i.e., 0% liquid saturation. For drying conditions, the pore space cells are initially all set to liquid, corresponding to 100% saturation.

Figure 14.4 shows the wetting/drying hysteresis in simulations carried out on a system of randomly distributed circular disks. The distribution of these circular disks corresponds to a two-dimensional slice of an X-ray micro computed tomography image taken of a sample consisting of 400–600 μm glass beads. As before, we use a simulation temperature corresponding to $k_BT = 1$. Simulation output for chemical potentials ranging from 2.1-2.45 (also in units of k_BT) is shown in 14.4. 14.4(a)–(b) shows the phase distribution under drying conditions while 14.4(c)–(d) shows that corresponding to wetting conditions. The degree of hysteresis in water content in these figures is generally consistent with that seen in experiment.

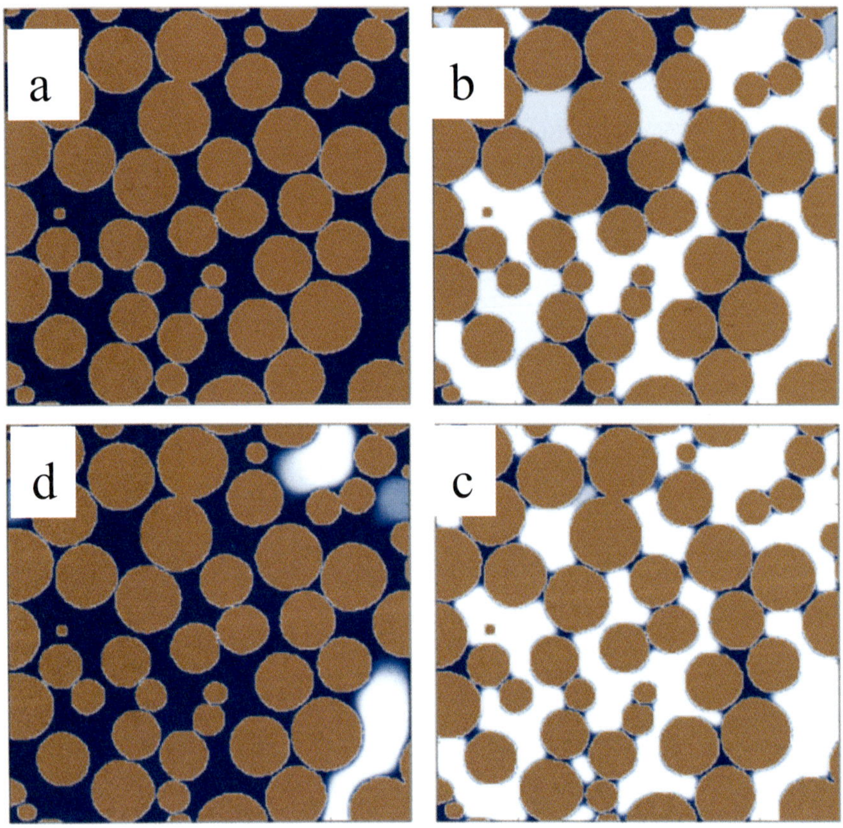

Fig. 14.4. Simulation output showing water distributions at chemical potential values of 2.1 and 2.45 (from left to right). Drying is shown in panels (a) → (b), and wetting in panels (c) → (d). Notice that despite having the same chemical potential value, the resulting capillary distributions are quite different between the wetting and drying cases.

14.5. Conclusions

Coarse-Grained Monte Carlo simulation is an effective tool for the study of phase distribution in porous materials. By coarse graining the system to a lattice and eliminating the computational overhead involved in calculating hydrodynamic flows, the simulation methodology can be used to calculate constitutive properties of these materials in a highly efficient manner. Approaches in the literature are diverse and in the case of adsorption phenomena tend to focus on molecular simulation and simplified model single-pore systems. The Coarse-Grained Monte Carlo simulation can make direct correspondence with experimental systems, by using porespace geometries from experimental imaging techniques (e.g. X-ray micro computed tomography) and interfacial coupling parameters from experimentally determined surface tensions.

By performing simulations and experiments on the same porespaces, we have presented here direct validation of the Coarse-Grained Monte Carlo simulation methodology. Simulation results conform well with experimental data, particularly in the case of capillarity in soils in the equilibrium phase distribution, soil-water retention, and capillary specific surface area. We introduced a method to use a two-dimensional slice of the three-dimensional porespace to model constitutive properties, at a great savings in computational cost. We furthermore have shown that Coarse-Grained Monte Carlo generates metastable configurations that are representative of those seen in experiment, as previously hypothesized based on molecular simulation results and supported by quasistatic arguments about the dominance at the pore scale of capillary forces over hydrodynamic forces for typical conditions. This supports our approach of using a fixed simulation temperature in Coarse-Grained Monte Carlo rather than using Simulated Annealing as in prior methods.

Acknowledgment

We are grateful for support from the Unconventional Natural Gas and Oil Institute (UNGI) CIMMM (Coupled Integrated Multiscale Measurements and Modeling) Consortium.

References

[1] A. Striolo, Water self-diffusion through narrow oxygenated carbon nanotubes, *Nanotechnology.* **18**(47), 475704 (nov, 2007).

[2] G. M. Torrie, G. Lakatos, and G. N. Patey, Structure and adsorption of water in nonuniform cylindrical nanopores., *J. Chem. Phys.* **133**(22), 224703 (dec, 2010).

[3] S. J. Pogorzelski, Z. Berezowski, P. Rochowski, and J. Szurkowski, A novel methodology based on contact angle hysteresis approach for surface changes monitoring in model PMMA-Corega Tabs system, *AppL. Surf. Sci.* **258**(8), 3652–3658 (feb, 2012).

[4] M. R. Urist, Bone: formation by autoinduction, *Science.* **150**(3698), 893–899

[5] P.-G. de Gennes, F. Brochard-Wyart, and D. Quere, *Capillarity and Wetting Phenomena: Drops, Bubbles, Pearls, Waves (Google eBook).* Springer (2004).

[6] N. Lu and W. J. Likos, *Unsaturated Soil Mechanics.* Wiley (2004).

[7] P. Meakin and A. M. Tartakovsky, Modeling and simulation of pore-scale multiphase fluid flow and reactive transport in fractured and porous media, *Rev. Geophys.* **47** (2009).

[8] L. D. Gelb, Modeling amorphous porous materials and confined fluids, *MRS Bulletin.* **34**(8), 592–601 (2009).

[9] R. Fisher, On the capillary forces in an ideal soil; correction of formulae given by W. B. Haines, *J. Agric. Sci.* **16**, 492 (1926).

[10] Z. Grof, C. J. Lawrence, and F. Štěpánek, The strength of liquid bridges in random granular materials, *J. Colloid Interf. Sci.* **319**(1), 182–192 (2008).

[11] T. Gröger, U. Tüzün, and D. M. Heyes, Modelling and measuring of cohesion in wet granular materials, *Powder Technol.* **133**(1), 203–215 (2003).

[12] J. Lechman and N. Lu, Capillary force and water retention between two uneven-sized particles, *J. Eng. Mech.* **134**(5), 374–384 (May, 2008).

[13] G. Mason and W. Clark, Liquid bridges between spheres, *Chem. Eng. Sci.* **20**(10), 859–866 (1965).

[14] F. M. Orr, L. E. Scriven, and A. P. Rivas, Pendular rings between solids: meniscus properties and capillary force, *J. Fluid Mech.* **67**(04), 723 (Mar, 2006).

[15] P. Pierrat and H. S. Caram, Tensile strength of wet granula materials, *Powder Technol.* **91**(2), 83–93 (1997).

[16] V. Richefeu, M. El Youssoufi, and F. Radja, Shear strength properties of wet granular materials, *Physi. Rev. E.* **73**(5), 051304 (May, 2006).

[17] C. Cuvelier and R. M. S. M. Schulkes, Some numerical methods for the computation of capillary free boundaries governed by the Navier Stokes equations, *SIAM Review.* **32**(3), 355–423 (1990).

[18] Y. D. Shikhmurzaev, *Capillary Flows with Forming Interfaces.* Chapman and Hall/CRC, Boca Raton, FL (2008).

[19] A. Q. Raeini, M. J. Blunt, and B. Bijeljic, Modelling two-phase flow in porous media at the pore scale using the volume-of-fluid method, *J. Comput. Phys.* **231**(17), 5653–5668 (2012).

[20] D. Frenkel and B. Smit, *Understanding Molecular Simulation : From Algorithms to Applications*, 2nd edn. Computational Science Series, Academic Press, San Diego (2002).

[21] R. M. Ziff, E. Gulari, and Y. Barshad, Kinetic phase transitions in an irreversible surface-reaction model, *Physi. Rev. Lett.* **56**(24), 2553 (1986).

[22] J. C. Liu and P. A. Monson, Molecular modeling of adsorption in activated carbon: Comparison of monte carlo simulations with experiment, *Adsorption* **11**(1), 5–13 (2005).

[23] J. Kim and B. Smit, Efficient monte carlo simulations of gas molecules inside porous materials, *J. Chemi. Theory Comput.* **8**(7), 2336–2343 (2012).

[24] S. Chen and G. D. Doolen, Lattice Boltzmann method for fluid flows, *Annual Review of Fluid Mechanics*. **30**(1), 329–364 (jan, 1998).

[25] M. E. Kutay, A. H. Aydilek, E. Masad, and T. Harman, Computational and experimental evaluation of hydraulic conductivity anisotropy in hot-mix asphalt, *Int. J. Pavement Eng.* **8**(1), 29–43 (mar, 2007).

[26] R. Knight, A. Chapman, and M. Knoll, Numerical modeling of microscopic fluid distribution in porous media, *J. Appl. Phys.* **68**(3), 994–1001 (1990).

[27] S. Mohanty, Effect of multiphase fluid saturation on the thermal conductivity of geologic media, *J. Phys. D: Appl. Phys.* **30**(24), L80–L84 (dec, 1997).

[28] D. L. Silverstein and T. Fort, Incorporating low hydraulic conductivity in a numerical model for predicting airwater interfacial area in wet unsaturated particulate porous media, *Langmuir.* **16**(2), 835–838 (jan, 2000).

[29] N. Lu, B. D. Zeidman, M. T. Lusk, C. S. Willson, and D. T. Wu, A monte carlo paradigm for capillarity in porous media, *Geophys. Res. Lett.* **37**(23), 1944-8007 (2010).

[30] A. W. Cense and S. Berg. The viscous-capillary paradox in 2-phase flow in porous media. In *International Symposium of the Society of Core Analysts held in Noordwijk, The Netherlands*, pp. 27–30 (2009).

[31] L. Sarkisov and P. A. Monson, Hysteresis in monte carlo and molecular dynamics simulations of adsorption in porous materials, *Langmuir.* **16**(25), 9857–9860 (2000).

[32] L. Sarkisov and P. A. Monson, Modeling of adsorption and desorption in pores of simple geometry using molecular dynamics, *Langmuir.* **17**(24), 7600–7604 (2001).

[33] P. Monson, Understanding adsorption/desorption hysteresis for fluids in mesoporous materials using simple molecular models and classical density functional theory, *Microporous Mesoporous Mater.* **160**, 47–66 (sep, 2012).

[34] L. Onsager, Crystal Statistics. I. A Two-Dimensional Model with an Order-Disorder Transition, *Physi. Rev.* **65**(3-4), 117–149 (1944).

[35] N. Metropolis, A. W. Rosenbluth, M. N. Rosenbluth, A. H. Teller, and E. Teller, Equation of State Calculations by Fast Computing Machines, *J. Chem. Phys.* **21**(6), 1087 (1953).

[36] D. P. Laundau and K. Binder, *A Guide To Monte Carlo Simulations in Statistical Physics.* Cambridge University Press (2005).

[37] D. L. Silverstein and T. Fort, Prediction of airwater interfacial area in wet unsaturated porous media, *Langmuir.* **16**(2), 829–834 (2000).

[38] D. L. Silverstein and T. Fort, Prediction of water configuration in wet unsaturated porous media, *Langmuir.* **16**(2), 839–844 (2000).
[39] B. Berkowitz and D. Hansen, A numerical study of the distribution of water in partially saturated porous rock, *Transport Porous Med.* **45**(2), 301–317 (2001).

Chapter 15

Coupled Thermo-Hydrological Processes in Enhanced Geothermal Systems

Yu-Shu Wu, Yi Xiong and Hossein Kazemi

Department of Petroleum Engineering, Colorado School of Mines, Golden, CO, USA

ywu@mines.edu

Geothermal energy from enhanced geothermal systems (EGS) is highly favored among sustainable energy resources because of its abundance and cleanness. The EGS concept has received worldwide attention and undergone intensive study in the last decade. In comparison, hydrothermal energy, the 'low-hanging fruit' of geothermal energy, is small while energy from EGS is extensive and has great potential to supply the world with needed energy. EGS, in essence, is an engineered subsurface heat mining method where water, or another suitable heat exchange fluid, is injected into hot formations to extract heat from the hot formation. Specifically, EGS relies on the principle that injected water, or another working fluid, penetrates deep into the reservoir through formation fractures to adsorb large quantities of heat by contact with hot rock. Finally, the heated working fluid is produced through production wells for electricity generation or other usage.

Even with significant progress in the laboratory, modeling, and field demonstrations, EGS energy cannot compete with other commonly used forms of energy, such as oil and gas. This is primarily because of technical and economic reasons. There is a great need for further research and development, before EGS energy emerges as a viable energy resource.

Heat mining from fractured EGS reservoirs is subject to complex interactions within the reservoir rock pores involving high temperature heat exchange, multiphase flow, rock deformation, and chemical reactions. We must thoroughly understand these thermal-hydrological-mechanical-chemical (THMC) interactions before any successful field operation is undertaken. In this paper we present the formulation of a transport model, which underlies the THMC processes. Then, we use the model to simulate an enhanced geothermal reservoir.

15.1. Introduction

Geothermal energy spans a wide range of heat sources, from shallow low-temperature formations to deep hot water saturated rocks a few miles beneath the earth surface, and to the deeper extremely high temperatures magma. Thus, geothermal energy includes not only the easily accessible hydrothermal resources; it also includes the earth's deepest thermal energy present anywhere. The grade of a specific geothermal resource depends on its geothermal gradient, reservoir rock permeability, porosity, fluid saturation, and recharge capability [1]. The coincidence of naturally occurring fluid-saturated permeable porous reservoirs containing a substantial amount of heat is not commonplace in earth; thus, high grade geothermal resources are limited [2]. An alternative to naturally occurring high-grade hydrothermal reservoirs is human intervention to create engineered reservoirs in hot rocks, known as Engineered Geothermal Systems (EGS). Tester et al. [1] claim that the accessible US EGS resource base is over 130,000 times the current US annual consumption of primary energy, which has drawn the attention of the government, academia and industry worldwide. In spite of progress in EGS technology, there is a great need for future investment on research and development to make geothermal energy a primary source of energy.

An EGS reservoir can be created by drilling two wells into a hot dry basement rock as depicted by Fig. 15.1. The next step is to fracture the basement rock (via hydraulic fracturing technology) to create flow paths for the injected fluid to travel between the injection and production well. The injected fluid, during its travel to the production well, picks up in-situ heat and exits the reservoir via the production well. At the surface, the heated fluid is directed to a power plant where electricity is generated. Upon leaving the power plant, the fluid is returned to the reservoir via the injection well to complete a closed circulation loop.

Geomechanical stress plays a major role on the flow properties of EGS reservoirs [3]. For instance, Ref. [4] correlated the change in porosity, permeability and capillary pressure magnitude to the reservoir in situ stress conditions. In addition, the geomechanical stress influences the environment, such as land subsidence and seismic events. Specifically, the EGS environment is amenable to strong thermo-poro-elastic deformations and property alterations.

The effect of chemical reactions in EGS reservoirs have been observed in several commercial EGS fields. Ref. [5] modeled the reactive chemical pro-

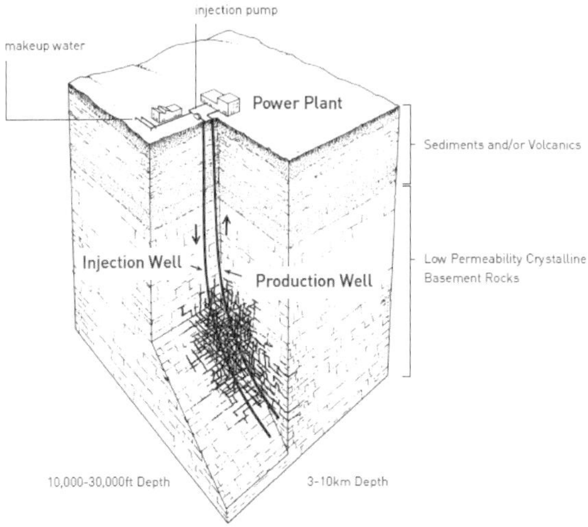

Fig. 15.1. Schematic of a conceptual two-well EGS reservoir [1].

cesses based on the field data from tens of geothermal fields in Kamchatka (Russia) and Japan. In addition, Ref. [6] presented the reactive transport model of injection well scaling and acidizing at Tiwi field in Philippines. Ref. [7] studied the calcite and silica scaling problems with an exploratory model for Ahuachapán and Berlín geothermal fields in El Salvador. Jung et al. [8] performed reactive transport modeling to study fluid-rock interactions for a CO2-based geothermal system related to EGS development. These papers indicate the importance of chemical reactions between fluids and rock minerals and mineral dissolution and precipitation when developing geothermal energy systems.

In this paper, we present a mathematical formulation, numerical solution, and an illustrative example for an EGS reservoir. The formulation couples thermal, hydrological, mechanical, chemical (THMC) processes. We first introduce the mathematical model and a discussion of the governing mass, energy, stress, and solute reactive transport equations. Then, we briefly discuss the numerical discretization used in the model. Finally, we present an example of the THMC process simulation for a geothermal injection well. The simulation results indicate strong thermo-mechanical rock deformation effects at short and intermediate times and chemical effects at longer times. We also observe that the permeability enhancement by thermo-mechanical effects is diminished by chemical precipitation of minerals in the injected water. Finally, we conclude that temperature of the injected water is a major controlling parameter for enhancing reservoir permeability and improving injection well performance.

15.2. Mathematical Model

15.2.1. *Fluid and Heat Flow Equations*

The thermo-hydrological (TH) processes in the EGS reservoirs is governed by the mass and energy balance, which we describe by a set of multiphase, multi-component flow equations coupled to geomechanic and reactive geochemistry equations. Specifically, the differential equation describing mass and heat balance is Eq. (15.1). Integrating this equation in a subdomain n, with volume V_n and boundary Γ_n, and applying Gauss divergence theorem, we obtain Eq. (15.2), which was also used by Ref. [9]:

$$\frac{\partial M^k}{\partial t} = -\nabla \cdot F^k + q^k \tag{15.1}$$

$$\frac{d}{dt}\int_{V_n} M^k dV_n = -\int_{\Gamma_n} F^k \cdot n\, d\Gamma_n + \int_{V_n} q^k dV_n \tag{15.2}$$

In the above equations, M^k is either the mass accumulation term for component k in the total system Eq. (15.3) or the heat accumulation term Eq. (15.4), F^k is the advective mass velocity vector for component k in each fluid phase β Eq. (15.6) or the heat flux vector Eq. (15.7), as described below:

$$M^k = \sum_\beta \phi S_\beta \rho_\beta C_{\beta k} \tag{15.3}$$

In subsurface unsaturated soil modeling of fluid and heat flow $k = 1, 2, h$ has been used to represent water, air, and heat.

The heat accumulation term (per unit volume) includes contributions from the rock matrix, aqueous and gaseous phases:

$$M^h = (1-\phi)\rho_R C_R T + \phi \sum_\beta S_\beta \rho_\beta u_\beta \tag{15.4}$$

The mass fluxes of aqueous and gaseous phases are determined by a multiphase version of Darcy's law, written in the form:

$$F_\beta = -k_0\left(1 + \frac{b_\beta}{P_\beta}\right)\frac{k_{r\beta}\rho_\beta}{\mu_\beta}(\nabla P_\beta - \rho_\beta g) \tag{15.5}$$

The flux for component k, $F_\beta^k$is given by Eq. (15.6):

$$F_\beta^k = \rho_\beta C_\beta^k F_\beta = -\rho_\beta C_\beta^k \left[k_0 (1 + \frac{b_\beta}{P_\beta}) \frac{k_{r\beta}\rho_\beta}{\mu_\beta} (\nabla P_\beta - \rho_\beta g) \right] \tag{15.6}$$

The heat flux includes conduction, advection and radiation heat transfer, given by Eq. (15.7) below:

$$F^h = -[(1-\phi)C_R + \phi \sum_\beta S_\beta C_\beta^h]\ K_\beta \nabla T + \sum_\beta h_\beta F_\beta + f_\sigma \sigma_0 \nabla T^4 \tag{15.7}$$

Where, $-K_\beta \nabla T$ is heat conduction term, h_β is specific enthalpy of phase β, C_β^h heat capacity for phase β, f_σ surface emissivity, and σ_0 Stefan constant for radiation heat transfer.

15.2.2. *Geomechanics Equations*

The geomechanics formulation is based on the classical theory of elasticity extended to porous and non-isothermal media. In the theory of elasticity, the stress-strain behavior of an isothermal elastic material is described by Hooke's law, which when extended to porous media with pore pressure and non-isothermal conditions, is given by the following equation [10]:

$$\sigma_{kk} - (\alpha P + 3\beta K (T - T_{ref})) = \lambda (\varepsilon_{xx} + \varepsilon_{yy} + \varepsilon_{zz}) + 2G\varepsilon_{kk}\ ; k = x, y, z \tag{15.8}$$

$$\sigma_{kl} = 2G\varepsilon_{kl}\ ; k, l = x, y, z \tag{15.9}$$

For poroelastic media, Eq. (15.8) has three principal components in x, y and z directions. After summing, one obtains the following equation in terms of mean normal stress and volumetric strain [11].

$$\sigma_m - \alpha P - 3\beta K (T - T_{ref}) = \left(\lambda + \frac{2}{3} G \right) \varepsilon_v \tag{15.10}$$

Two other fundamental relations in the theory of linear elasticity are the relation between strain tensor and the displacement vector:

$$\bar{\bar{\varepsilon}} = \frac{1}{2} \left[\nabla \bar{u} + (\nabla \bar{u})^T \right] \tag{15.11}$$

In addition, the static equilibrium equation is:

$$\nabla \cdot \bar{\bar{\sigma}} + \bar{F}_b = 0 \tag{15.12}$$

We combine Eq. (15.10), Eq. (15.11), and Eq. (15.12) to obtain the non-isothermal thermo-poroelastic Navier equation:

$$\alpha \nabla P + 3\beta K \nabla T + (\lambda + G) \nabla (\nabla \cdot \bar{u}) + G \nabla^2 \bar{u} + F_b = 0 \tag{15.13}$$

Equation (15.13) contains the displacement vector. Taking divergence of Eq. (15.13) results in an equation containing the divergence of the displacement vector:

$$\alpha \nabla^2 P + 3\beta K \nabla^2 T + (\lambda + 2G) \nabla^2 (\nabla \cdot \bar{u}) + \nabla \cdot \bar{F}_b = 0 \tag{15.14}$$

The divergence of the displacement vector is the sum of the normal strain components, which is the volumetric strain, as shown:

$$\nabla \cdot \bar{u} = \frac{\partial u_x}{\partial x} + \frac{\partial u_y}{\partial y} + \frac{\partial u_z}{\partial z} = \varepsilon_{xx} + \varepsilon_{yy} + \varepsilon_{zz} = \varepsilon_v \tag{15.15}$$

Combining Eq. (15.14), Eq. (15.15), and Eq. (15.10) yields an equation relating mean normal stress, pore pressure, temperature, and body force:

$$\frac{3(1-\nu)}{(1+\nu)} \nabla^2 \sigma_m + \nabla \cdot \bar{F}_b = \frac{2(1-2\nu)}{(1+\nu)} \left(\alpha \nabla^2 P + 3\beta K \nabla^2 T \right) \tag{15.16}$$

Equation (15.16) is the governing geomechanical equation for our THMC model. In this equation ν is Poisson ratio; λ and G, the Lam parameter and shear modulus, were replaced by appropriate elastic properties.

15.2.3. *Chemical Reaction Equations*

The chemical reaction model solves solute transport and chemical reactions, coupled with fluid and heat flow and rock deformation. The solute transport model provides the chemical concentration for each species, and the effects of chemical reactions on the change of concentrations are fed back to the solute transport model through a sink/source term.

15.2.3.1. *Solute Transport*

The solute transport in phase l follows the mass balance Eq. (15.1), where accumulation and flux terms for each phase are given as:

$$M_l^k = \phi S_l \rho_l C_l^k \tag{15.17}$$

$$F_l^k = v_l \rho_l C_l^k - \left(\tau \phi S_l \rho_l D_l^k\right) \nabla C_l^k \ ; k = 1 \ldots N_k \tag{15.18}$$

where, N_k is the total number of chemical components (species) in phase l; C_l^k is the concentration of kth species in phase l; v_l is Darcy velocity; and D_l^k is diffusion coefficient of kth species in phase l.

For efficiency, a subset of aqueous species could be selected as the primary species. All other species, the secondary species, including aqueous complexes and precipitated species, can be represented as a linear combination of primary species as indicated:

$$C_i^l = \sum_{j=1}^{N_P} v_{ij}^l C_j^l \ ; i = 1 \ldots N_S \tag{15.19}$$

where, C_i^l is the concentration of the secondary species i in phase l; j is the primary species index for concentration in phase l; N_P is the number of the primary species; N_S is the number of the secondary species; and v_{ij}^l is the stoichiometric coefficient of jth primary species in the ith reaction.

Molecular diffusion is insignificant compared to hydrodynamic and numerical dispersion. Thus, for the majority of applications molecular diffusion can be ignored.

15.2.3.2. *Mineral Dissolution and Precipitation*

The mineral dissolution and precipitation due to the reactions between the injected water and the reservoir rock or fluid is the main interest with respect to reactive geochemistry for EGS reservoir modeling. The mineral saturation ratio can be expressed as:

$$\Omega_m = X_m^{-1} \lambda_m^{-1} K_m^{-1} \prod_{j=1}^{N_C} c_j^{v_{mj}} \gamma_j^{mj} \ m = 1 \ldots N_P \tag{15.20}$$

where, N_P is the number of minerals at equilibrium conditions; m is he equilibrium mineral index; Ω_m is mineral saturation ratio; X_m is the mole fraction of the mth mineral phase; λ_m is its thermodynamic activity coefficient (X_m and λ_m are taken to be one for pure mineral phases); and K_m is the corresponding equilibrium constant. At equilibrium condition

$$\mathsf{SI}_m = \log \ \Omega_m = 0 \tag{15.21}$$

where, SI_m is the mineral saturation index.

In mineral dissolution and precipitation, the kinetic rate can be a function of non-primary species. We usually consider the species appearing in rate laws as the primary species. In our model, we use the rate expression given by [12]:

$$r_n = f(c_1, c_2, ..., c_{Nc}) = \pm k_n A_n \left|1 - \Omega_n^\theta\right|^\eta \quad n = 1...N_q \tag{15.22}$$

where, N_q is the number of minerals at kinetic conditions; the positive values of r_n indicate dissolution and negative values for precipitation; k_n is the rate constant (moles per unit mineral surface area and unit time) which depends on the temperature; A_n is the specific reactive surface area per kg H_2O. Ω_n is the kinetic mineral saturation ratio defined as Eq. (15.20). The exponent θ and η can be determined from experiments, but they are usually assumed as unity. The reaction rate constant k_n is dependent on temperature and usually rate constants are reported at 25°C.

15.2.4. *Time and Space Discretization*

The mass and energy balance, Eq. (15.1), and the mean stress, Eq. (15.16), are solved simultaneously for the primary variables. The mass and energy balance, Eq. (15.1), is discretized in space using Gauss divergence theorem, resulting in the integral finite-difference (IFD) method presented by Ref. [13]. The time discretization is implicit in time using a first order backward-difference approximation. Applying an average volume to each control volume, V_n, and a discrete sum of surface elements to the surface integrals, Eq. (15.1) will end up with the following residual form:

$$R_n^{k,t+1} = M_n^{k,t+1} - M_n^{k,t} + \frac{\Delta t}{V_n}\left[\sum_m A_{nm} F_{nm}^{k,t+1} - V_n q_n^{k,t+1}\right] = 0 \tag{15.23}$$

where, $R_n^{k,t+1}$ is the mass and energy balance residual for component k at grid cell n and the current time level $t+1$. Index m donates the neighboring grid cells of n, A_{nm} is the interface area between grids n and m, F_{nm} is the flux between the neighboring grid cells. Moreover, $k = 1$ denotes the water component, $k = 2$ the air component, and $k = h$ is the heat component, and $k = M$ is the moment component as Eq. (15.24). Nonetheless, k could also represent the primary species in the solute transport for grid cell n.

The mean normal stress Equation (15.16) is discretized using the Gauss divergence theorem and the same procedure used in obtaining Eq. (15.23) to obtain:

$$R_n^{M,t+1} = \sum_l \left[\frac{3(1-\nu)}{(1+\nu)} \nabla \sigma_m + \bar{F}_b - \frac{2(1-2\nu)}{(1+\nu)} (\alpha \nabla P + 3\beta K \nabla T) \right] A_{ln} = 0 \tag{15.24}$$

where, l is a neighboring grid block of n and A_{ln} is the interface area between l and n grid cells. After solving Eq. (15.23) and Eq. (15.24) for the primary variables for the thermal-hydrological-mechanical (THM) process, the resulting Darcy velocity and phase saturations are used to compute solute transport and chemical reactions in a similar discretization method.

15.2.5. *THMC Process Effects on Reservoir Properties*

Rock deformation and chemical reactions alter the flow properties of the porous media such as permeability, porosity, capillary pressure, and pore surface wettability, which are addressed in the following section.

15.2.5.1. *Porosity and Permeability*

The relationship between mean stress, σ_m, and porosity, ϕ, has been investigated extensively [4, 14–16] which we have incorporated in our mechanical model formulation. We begin with mechanical changes in porosity:

$$\phi_M = \phi(\sigma_M) \tag{15.25}$$

where, subscript M indicates mechanical effects on porosity and σ_M is the current mean stress.

The chemical reaction model incorporates porosity changes due to mineral dissolution and precipitation as follows:

$$\phi_c = 1 - \sum_{m=1}^{Nm} fr^m - fr^u \tag{15.26}$$

where N_m is the number of reactive minerals, fr^m is the total mineral fraction ($V_{\text{mineral}}/V_{\text{medium}}$), and fr^u is the non-reactive fraction.

In the THMC processes, the porosity change within each time step $\Delta t = t^{n+1} - t^n$ is given in terms of the changes in mechanical and chemical properties $\Delta\phi_M = \phi_M^{n+1} - \phi_M^n$ and $\Delta\phi_c = \phi_c^{n+1} - \phi_c^n$:

$$\phi^{t+1} = \phi^t + \Delta\phi_M + \Delta\phi_c \tag{15.27}$$

The permeability change is related to the porosity:

$$k^{t+1} = k\left(\phi^{t+1}\right) \tag{15.28}$$

15.2.5.2. *Mass Accumulation*

The total mass density of a component k in a grid cell or REV changes because of rock deformation. Thus, a volumetric strain term $(1 - \varepsilon_v)$ is included in the mass accumulation term as shown:

$$M^k = \sum_{\beta} (1 - \varepsilon_v) \, \phi S_\beta \rho_\beta C_\beta^k \tag{15.29}$$

15.2.5.3. *Capillary Pressure*

A change in permeability and porosity leads to a change of pore size and capillary pressure. The following J-function is included in our model to correct for capillary pressure change assuming no change in wettability of the pore surface.

$$p_c = p_{c,o} \sqrt{\frac{k_o \phi}{k \ \phi_o}} \tag{15.30}$$

15.2.5.4. *Wettability Change*

Wettability change is also a major issue, which can be included in the above equation by including the change in contact angles.

15.3. Application and Discussion

In the following application example, we present a prototype EGS reservoir, similar to Fig. 15.1, to simulate the THMC process at the vicinity of the injection well for two years of continuous water injection. The EGS reservoir, Fig. 15.2, is overlain by an ideal caprock and the injection well is in the middle of the simulated area. The initial reservoir pressure and temperature are 6.35 MPa and 240°C with a mean normal stress of 15.6 MPa. The computation grid is a 2-D R-Z model, with logarithmic grid distribution in radial direction to capture the near-wellbore details. We simulated two scenarios for 150°C and 100°C injected water.

The reservoir hydraulic properties, including porosity and permeability, change dynamically because of geomechanical and chemical effects. Our simulation incorporates various correlations for stress-dependent porosity and permeability, but we can select an appropriate one for the specific reservoir. We used Ref. [16] correlation for porosity given by Eq. (15.31).

The total porosity change combines the stress and mineral reaction induced effects using Eq. (15.27) indicated earlier. For permeability, we used Carman-Kozeny correlation to arrive at Eq. (15.32).

$$\phi(\sigma'_m) = \phi_r + (\phi_0 - \phi_r)e^{-c\sigma'_m} \tag{15.31}$$

$$k(\sigma'_m) = k_i \frac{(1-\phi_i)^2}{(1-\phi)^2}\left(\frac{\phi}{\phi_i}\right)^3 \tag{15.32}$$

The definition of the subscripted parameters in Eq. (15.31) and Eq. (15.32) are:

- $\sigma'_m = \sigma_m - \alpha P$
- $\alpha =$ Biot coefficient
- $\phi(\sigma'_m) =$ Porosity at the current state of mean net stress σ'_m
- $\phi_r =$ Residual porosity under infinite net stress or minimum porosity
- $\phi_0 =$ Porosity at zero mean net stress (i.e., core sitting on the bench top)
- $\phi_i =$ Porosity at initial reservoir net stress
- $k(\sigma'_m) =$ Permeability at net stress σ'_m
- $k_i =$ Permeability at initial reservoir net stress

The exponent c in Eq. (15.31) is related to reservoir rock compressibility, which is determined experimentally. For the example problem in this paper, $c = 2 \times 10^{-7}\ Pa^{-1}$.

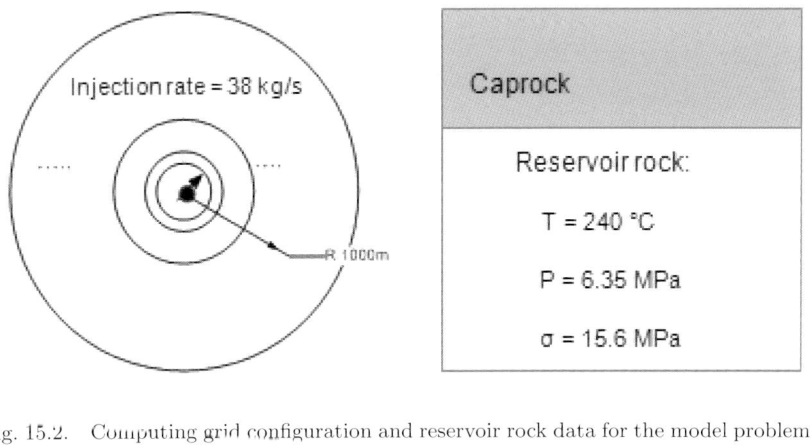

Fig. 15.2. Computing grid configuration and reservoir rock data for the model problem.

To represent a real EGS reservoir, we used reservoir mechanical and chemical properties from the published data of Refs. [4] and [6], presented in Table 15.1 and Table 15.2. Table 15.1 gives the reservoir's initial hydraulic and mechanical properties. The hydraulic properties, such as porosity and permeability, are subject to change because of THMC effects. The mechanical properties, such as Young modulus, Poisson ratio, and thermal expansion coefficient, are required to solve the governing mechanical equations, Eq. (15.23) and Eq. (15.24). The porosity and exponent parameter in Eq. (15.31) are required for the numerical simulation.

Table 15.1. Reservoir and material properties.

Properties	Values
Young Modulus (GPa)	14.4
Poisson ratio (dimensionless)	0.20
Permeability (m^2)	5.37×10^{-14}
Porosity (dimensionless)	0.1
Pore compressibility (Pa^{-1})	5×10^{-7}
Thermal expansion coefficient $(°C)^{-1}$	4.14×10^{-6}
Rock grain specific heat $J/(kg°C)$	1000
Rock grain density (kg/m^3)	2750
Formation thermal conductivity $(W/(m°C))$	2.4
Biot coefficient α (dimensionless)	1.0
High stress residual porosity	0.08
Exponent parameter c of Eq. (15.31) (Pa^{-1})	2×10^{-7}

Table 15.2 lists the primary chemical species in the initial reservoir aqueous system and the injection water at the beginning of the simulation. The chemical reactions occur not only between the injected fluid and reservoir fluid, but also between the injected fluid and reservoir rock minerals; thus, the chemical composition of the reservoir fluid is subject to change because of dissolution of rock minerals and subsequent precipitation. Figure 15.3 presents the initial mineral volume fraction of reservoir rock. The un-reactive minerals do not undergo chemical reactions. In this simulation, 68% of reservoir rock is composed of mineral grains, which are chemically reactive, and 32% of reservoir rock is un-reactive.

Injection temperature is an important parameter, which can be controlled in the EGS field operations. To analyze the injection temperature impact on reservoir mechanical and chemical reactions, two simulation runs were performed with the same input data but with different injection temperatures: 150°C and 100°C.

Table 15.2. Chemical components for reservoir initial water and injection water.

Chemical Species	Initial Water Concentration (mol/L)	Injection Water Concentration (mol/L)
Ca^{2+}	3.32×10^{-3}	1.0327×10^{-3}
Mg^{2+}	8.62×10^{-6}	1.6609×10^{-6}
Na^{+}	1.285×10^{-1}	1.2734×10^{-1}
Cl^{-}	1.418×10^{-1}	1.418×10^{-1}
SiO_2(aq)	1.218×10^{-2}	1.1734×10^{-2}
HCO_3^{-}	1.0423×10^{-3}	1.0423×10^{-3}
SO_4^{2-}	2.6272×10^{-4}	2.6272×10^{-4}
K^{+}	1.5852×10^{-2}	1.5852×10^{-2}
AlO_2^{-}	0.059×10^{-4}	0.0205×10^{-4}

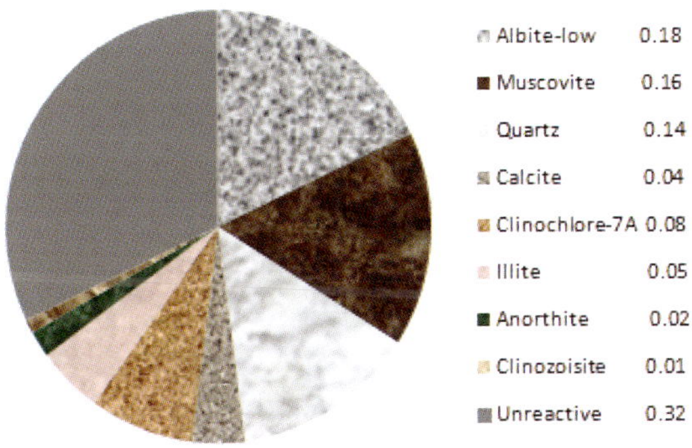

Fig. 15.3. Initial mineral fraction of reservoir rock.

Figure 15.4 gives the distributions of temperature and pressure after two years of injection. Figure 15.5 shows the evolution of effective stress because of geomechanical perturbations, and Fig. 15.6 shows the change of mineral volume fraction because of mineral dissolution (negative change) and precipitation (positive change). The horizontal axis represents the distance from the injection well.

As shown on the temperature profile of Fig. 15.4, the low temperature from injected water has propagated to the 150 meter from the injection well. Since the injection rate is same for both cases, the reservoir pressures

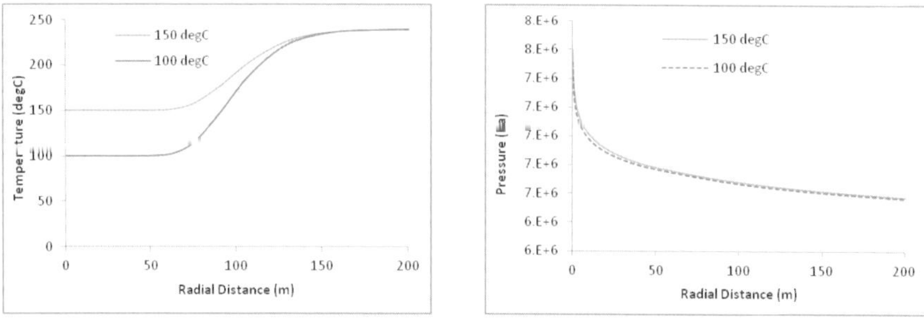

(a) (b)

Fig. 15.4. Temperature (a) and pressure (b) distribution after 2 years of injection.

are very close as shown in the right plot of Fig. 15.4. The small difference between them is due to the different temperature impact on reservoir properties. We also observe that the temperature transition zone from lower injection temperature to higher reservoir temperature is the same for both cases, from about 60 to 150 m. A close look at the pressure profile plot reveals a significant increase in pressure very close to the injector, approximately 10 m radial distance from the injection well.

Figure 15.5 shows the geomechanical effects around the injection well. Results demonstrate that the low temperature injected water results in a significant decrease in effective stress, from about 9MPa to less than 2MPa. In comparison, in the higher injection temperature case, the effective stress only reduces to above 4MPa. This coincides with the geomechanical governing equation that the lower temperature has much higher effects on effective stress decrease. The two plots in Fig. 15.5 also show that the stress profile has the same trend as the temperature profile of Fig. 15.4 for two years injection curves, both propagating to around 150m radial distance, which implies the stress evolution is mainly due to the low temperature injection. It is also observed that the change of stress propagates faster at early time, reaching about 70 meters from the injection well in half year and finally 150 meters in two years.

Figure 15.6 presents the change of mineral fraction of reservoir rock due to chemical reactions. The positive change means the chemical precipitation, over the dissolution, dominates the chemical effects, and vice versa. The two plots have the same trend that the change of mineral fraction increases steadily from half year to two years, and mainly concentrates within the 10 meters area from the injection well. In addition, the lower injection

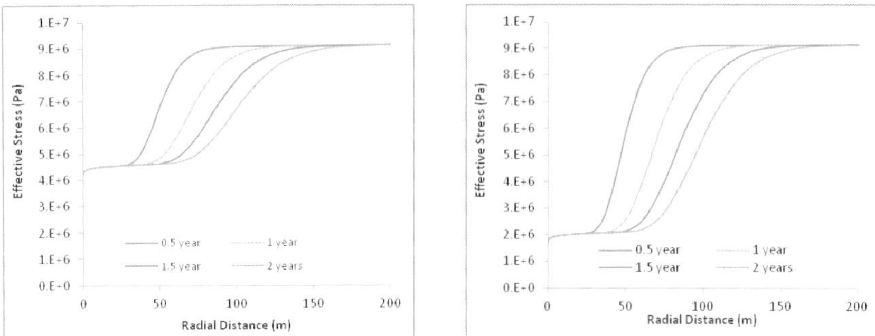

Fig. 15.5. Evolution of effective mean stress for 150°C (left) and 100°C (right) injected water.

temperature results in much stronger precipitation effects from Fig. 15.6. For example, the change of mineral fraction under 100°C injected temperature is over five times more than that under 150°C.

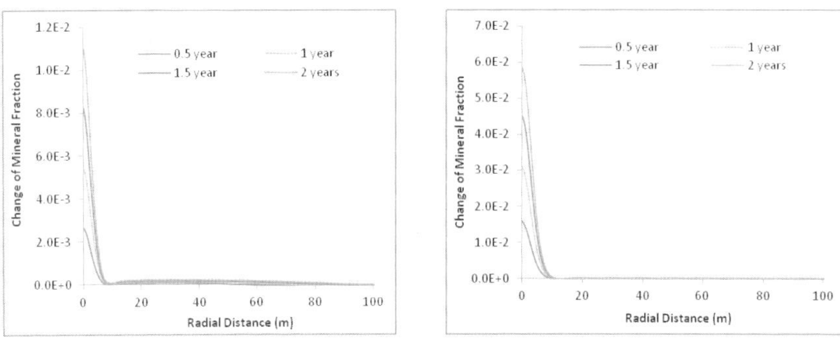

Fig. 15.6. Evolution of the change in mineral fraction for 150°C (left) and 100°C (right) injected water.

The geomechanical and geochemical effects are fed back to flow effects through the change of reservoir properties, such as porosity and permeability. Eqs. (15.31) and (15.32) are used in the simulation to correlate between the reservoir porosity and the stress, as well as between the permeability and porosity. Equations (15.31) and (15.32) show that the lower effective stress, the higher stress enhanced porosity and permeability. On the other hand, Fig. 15.6 reveals that the lower injection temperature introduces stronger chemical precipitation effects, which undermines the reservoir porosity and permeability. Therefore the reduced injection tem-

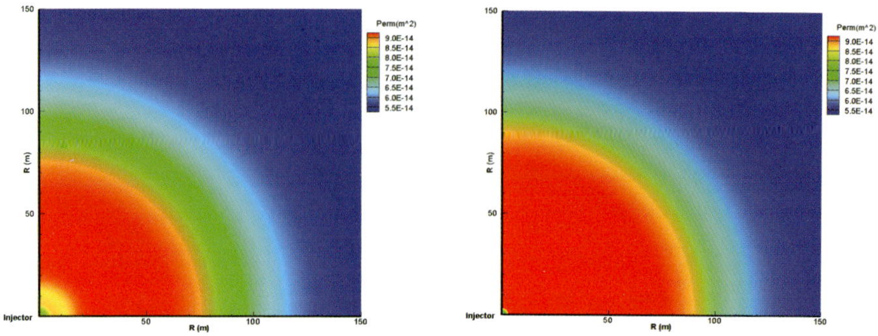

Fig. 15.7. Permeability profile alterations near water injection well because of THMC effects for 150°C (left) and 100°C (right).

perature has opposite effects on mechanics and chemical reaction induced reservoir quality change.

Figure 15.7 shows permeability changes around the injector after two years of water injection at 150°C and 100°C. The area near the injector well, on the left, has a lower permeability over the geomechanical influence due to mineral precipitation effects, which agrees with Fig. 15.6 that the chemical precipitation only concentrate around the injection well. The lower injection temperature on the right plot shows even lower permeability than the left plot because the reduced injection temperature results in larger precipitation effects. Beyond the near-well area, the permeability is enhanced due to geomechanical effects over the chemical influence, which is in agreement with Fig. 15.5 that the reduced effective stress has propagated to about 150 meter radial distance from the injector in both wells. But, in this area, the reduced injection temperature on the right plot leads to larger permeability enhancement because of the geomechanics induced effects.

15.4. Summary and Conclusions

We have presented a mathematical model for simulating four coupled thermo-hydrological processes in Enhanced Geothermal Systems. We used a control-volume finite-difference solution to model a prototype EGS reservoir as an engineering management tool with the following goals: (1) Assess efficacy of coupled fluid and heat flow to include geomechanical and reactive geochemical effects. (2) Determine reservoir property changes around the injection well.

The geomechanical model solution was verified against (1) the analytical solutions of separate one-dimensional consolidation and heat conduction problems, and (2) against the two-dimensional Mandel's problem (Mandel, 1953). We also verified the chemical reaction model against the TOUGHREACT [17]. Furthermore, we verified the simulation results of fluid and heat flow against CMG:STARS [18] commercial simulator.

The conclusions of this study are: (1) Numerical model results indicate that the thermal-hydrological-mechanical-chemical (THMC) processes interact consistent with EGS reservoir observations in practice. (2) The temperature difference between injection fluid and reservoir can enhance reservoir porosity and permeability because of geomechanical effects, but such improvement can be diminished by mineral precipitation. (3) Reducing injection temperature enhances the geomechanical effects to improve formation permeability, but promotes chemical precipitation causing formation damage. (4) Proper design of the chemical composition of injection water with respect to its interaction with reservoir rock minerals is necessary to minimize and eliminate formation damage. (5) In addition to concerns regarding formation damage by injection water incompatibility, reducing the injection temperature as low as possible could be an effective way to enhance reservoir permeability for long-term reservoir performance.

Compared with other geomechanical models for geotechnical analysis of soil, rock, and structural support, such as FLAC3D [19], our geomechanical model only calculates mean normal stress instead of the total stress tensor. This simplification reduces computation workload, but it cannot analyze any shear-dependence phenomena. We have presented the shear effects in another paper by Ref. [20]. Overall, our THMC model is rigorous in handling the physical processes involving coupled flow, rock deformation, and reactive geochemistry. It can be applied to stress-sensible or geochemistry-sensible reservoirs for analyzing multiphase fluid, heat flow, rock deformation and chemical reactions.

Acknowledgement

This paper is the result of the research supported by the US Department of Energy (DOE) on the Development of Advanced Thermal-Hydrological-Mechanical-Chemical (THMC) Modeling Capabilities for Enhanced Geothermal Systems (Contract No. DE-EE0002762). Therefore, the authors are grateful to the US DOE and the technical support from the students in the Energy Modeling Group (EMG) and Foundation CMG in the Department of Petroleum Engineering at Colorado School of Mines.

Nomenclature

A_{ln}	cross section area between grid blocks l and n, m^2
c	exponent parameter of Eq. (15.31), Pa^{-1}
$C_{\beta,k}$	the concentration of component k in liquid phase β, mol $\times L^{-1}$
C_β^h	specific heat of phase β, $J \times$ kg$^{-1} \times K^{-1}$
C_i	the total concentration of the ith species, mol $\times L^{-1}$
C_{kl}	the concentration of the kth species in liquid phase, mol $\times L^{-1}$
C_R	specific heat of reservoir rock, $J \times$ kg$^{-1} \times K^{-1}$
D_l	the diffusion coefficient of liquid phase, m$^2 \times$ s^{-1}
F_b	body force per area, Pa
F_β^k	mass or energy flux of component k in phase β due to advection
g	gravitational acceleration, m $\times$ s^{-2}
G	shear modulus, Pa
h_β	specific enthalpy in phase β, J $\times$ kg^{-1}
k	absolute permeability, m^2
K	heat conduction coefficient, $W \times$ m$^{-1} \times K^{-1}$
K	bulk modulus, Pa
K_R	heat conductivity of rock, $W \times$ m$^{-1} \times K^{-1}$
K_β	heat conductivity of phase β, $W \times$ m$^{-1} \times K^{-1}$
$k_{r\beta}$	relative permeability to phase β
M^k	accumulation of component k for mass or energy: specific mass or heat content, Kg or J
n	normal vector outwards to subdomain surface
P_c	capillary pressure, Pa
q_k	sink or source term of mass or energy component, Kg/s or J/s
R_n^k	residual of mass component or energy component
S_β	saturation of phase β
t	time, s
T	temperature, °C or K
$\bar{u}$	displacement vector, m
u_β	specific internal energy of phase β, J $\times$ kg^{-1}
V_n	volume of grid block n, m^3

Greek

α	Biot coefficient
β	linear thermal expansion coefficient, $(^\circ\text{C})^{-1}$
μ_β	viscosity of phase β, $Pa \times \text{s}$
ϕ	porosity
λ	Lame's constant, Pa
ε	strain
$\bar{\bar{\varepsilon}}$	strain tensor
ν	Poisson's ratio
σ	normal stress, Pa
σ_m	mean stress, Pa
$\bar{\bar{\sigma}}$	stress tensor, Pa
σ'	effective stress, Pa
ρ_R	density of reservoir rock, $\text{kg} \times \text{m}^{-3}$
ρ_β	density of phase β , $\text{kg} \times \text{m}^{-3}$
Γ_n	area of closed surface, m^2

References

[1] J. W. Tester, B. J. Anderson, A. S. Batchelor, D. D. Blackwell, R. DiPippo, E. M. Drake, and J. Garnish. The future of geothermal energy. Report 358, Massachusetts Institute of Technology (2006).

[2] US DOE. An evaluation of enhanced geothermal systems technology. In *Geothermal Technology Program*. US Department of Energy, Energy Efficiency and Renewable Energy (2008).

[3] C.-F. Tsang, Linking thermal, hydrological, and mechanical processes in fractured rocks 1, *Annual Review of Earth and Planetary Sciences*. **27**(1), 359–384 (1999).

[4] J. Rutqvist, Y.-S. Wu, C.-F. Tsang, and G. Bodvarsson, A modeling approach for analysis of coupled multiphase fluid flow, heat transfer, and deformation in fractured porous rock, *Int. J. Rock Mech. Min. Sci.* **39**(4), 429–442 (2002).

[5] A. Kiryukhin, T. Xu, K. Pruess, J. Apps, and I. Slovtsov, Thermal- hydrodynamic chemical (thc) modeling based on geothermal field data, *Geothermics*. **33**(3), 349–381 (2004).

[6] T. Xu, Y. Ontoy, P. Molling, N. Spycher, M. Parini, and K. Pruess, Reactive transport modeling of injection well scaling and acidizing at Tiwi field, Philippines, *Geothermics*. **33**(4), 477–491 (2004).

[7] F. Montalvo, T. Xu, and K. Pruess. TOUGHREACT code applications to problems of reactive chemistry in geothermal production-injection wells.

First exploratory model for Ahuachapán and Berlín geothermal fields. In *Proceedings of World Geothermal Congress 2005*, Antalya, Turkey (2005).

[8] Y. Jung, T. Xu, P. F. Dobson, N. Chang, and M. Petro. Experiment-based modeling of geochemical interactions in CO_2 based geothermal systems. In *PROCEEDINGS of Thirty-Eighth Workshop on Geothermal Reservoir Engineering*, Stanford University, Stanford, California (2013).

[9] K. Pruess, C. Oldenburg, and G. Moridis. TOUGH2 user's guide, version 2.0. Report LBL-43134, Lawrence Berkeley Laboratory, Berkeley, CA (1999).

[10] J. C. Jaeger, N. G. W. Cook, and R. W. Zimmerman, *Fundamentals of Rock Mechanics*, 4th edn. (2007).

[11] Y. Xiong, L. Hu, and Y.-S. Wu. Coupled geomechanical and reactive geochemical simulations for fluid and heat flow in enhanced geothermal reservoirs. In *PROCEEDINGS of Thirty-Eighth Workshop on Geothermal Reservoir Engineering*, Stanford University, Stanford, California (2013).

[12] A. C. Lasaga, J. M. Soler, J. Ganor, T. Burch, and K. L. Nagy, Chemical weathering rate laws and global geochemical cycles, *Geochimica et Cosmochimica Acta*. **58**(10), 2361–2386 (1994).

[13] T. N. Narasimhan and P. A. Witherspoon, An integrated finite difference method for analyzing fluid flow in porous media, *Water Resour. Res.* **12**(1), 57–64 (1976).

[14] R. W. Ostensen, The effect of stress-dependent permeability on gas production and well testing, *SPE Formation Evaluation*. **1**(3), 227–235 (1986).

[15] C. R. McKee, A. C. Bumb, and R. A. Koenig, Stress-dependent permeability and porosity of coal and other geologic formations, *SPE Formation Evaluation*. **3**(1), 81–91 (1988).

[16] J. P. Davies and D. K. Davies, Stress-dependent permeability: characterization and modeling, *SPE Journal*. **6**(2), 224–235 (2001).

[17] T. Xu, E. Sonnenthal, N. Spycher, and K. Pruess. TOUGHREACT user's guide: A simulation program for non-isothermal multiphase reactive geochemical transport in variably saturated geologic media. v1.2.1. Report LBNL-55460-2004, Lawrence Berkeley National Laboratory, Berkeley, CA, USA (2004).

[18] CMG. *STARS Users Guide, Advanced Process and Thermal Reservoir Simulator Version 2010*. Calgary, Alberta, Canada (2010).

[19] Itasca. *FLAC3D Manual. Fast Lagrangian Analysis of Continua in Three Dimensions Version 2.0*. Itasca Consulting Group, Minneapolis (1997).

[20] P. Fakcharoenphol, S. Charoenwongsa, H. Kazemi, and Y.-S. Wu, The effect of water-induced stress to enhance hydrocarbon recovery in shale reservoirs, *SPE Journal*. pp. 897–909 (Oct., 2013).

PART 3

Materials Science, Energy, Devices & Biology

Preamble

Robert J. Kee
Mechanical Engineering Department
Colorado School of Mines, Golden, CO, USA

Section 3 on "Materials Science, Energy, Devices, and Biology" consists of seven chapters. Compared with the first two sections, Section 3 covers a rather diverse set of topics. Although the chapters are certainly all connected through the use of porous materials, the technical applications and the roles of the porous media are quite different. Nevertheless, there are interesting synergies in the science and technology associated with the variety of applications.

Chapter 16 *Pore-scale phenomena and challenges in energy research and technology (Robert Kee, Huayang Zhu, Graham Goldin, and Scott Barnett).* This chapter focuses on electrochemical devices, such as fuel cell and batteries. In such technologies, the composite electrodes are fabricated as porous materials. In addition to diffuse fluid transport in the pore spaces, chemical and electrochemical reactions are active on the pore walls. Moreover, there can be ionic and electronic transport within both the pore and solid volumes.

Chapter 17 *Structural characterization of porous materials (Brian Gorman).* Although porous materials are often characterized as assemblies of spherical particles, the reality can be very different. This chapter concentrates on experimental techniques to measure and characterize the actual microstructures. Methods such as Focused-Ion-Beam–Scanning Electron Microscopy (FIB-SEM) X-ray tomography can be used reconstruct the three-dimensional pore structure. As discussed in Chapter 4, three-dimensional reconstructions provide great quantitative insight abut the actual structure of geological formations. As discussed in Chapters 16, computational models can be developed to model fluid, chemical, and electrochemical phenomena within the porous microstructures.

Chapter 18 *Homogenization approaches for the elastic properties of porous materials (John Berger, Steven Geer, and William Parnell).* Whether by design (e.g., electrodes) or by nature (e.g., geological structure), actual porous materials can be geometrically complex at the microscale. In practice, however, the microstructural details cannot be directly implemented into large-scale simulations. Instead, *effective* macroscopic properties need to be derived, but should be informed by the microstructure. This chapter discusses mathematical homogenization techniques to establish effective properties for application at large scales.

Chapter 19 *The role of pore characteristics on novel membrane processes for water treatment and desalinization (Tzahi Cath).* This chapter discusses the role of porous media in technologically important separation applications such as water purification and desalination. The synthetic membrane materials achieve performance by functionally designing and fabricating pore-scale attributes, including porosity, pore diameter, tortuosity, and hydrophobicity. The chapter discusses alternative separation strategies and the roles played by pore microstructure in realizing particular separation objectives.

Chapter 20 *Microporous crystalline membranes and their application for CO_2 separations (Moises Carreon and Tracy Gardner).* There are a variety of reasons to separate carbon dioxide from gas mixtures. Applications range from the control of green-house gases to removing CO_2 from natural gas as it emerges from a well head. A broad class of crystalline materials known as metal-organic frameworks (MOF) have nanoscale channels or pores that can be highly selective for the transport or rejection of particular molecules. Effective separation membranes can by developed by applying thin films of MOFs or zeolites onto larger-scale porous supports. This chapter discusses the roles played by crystal structure and the performance of alternative materials. It also discusses theoretical and modeling approaches for characterizing and predicting performance.

Chapter 21 *Chemistry in confined pore domains (Megan Otting, Yazhou Ji, Ryan Richards, and Brian Trewyn).* This chapter considers chemical reaction within nano-porous materials that are similar to those used for separations in Chapter 20. Chemical applications range from cell-scale biological phenomena to industrial-scale catalysis. The nano-scale pores facilitate chemical processes that are particularly effective in producing high-value products. The chapter discusses materials synthesis methods as well as chemical performance.

Chapter 22 *Fluid and solute transport in the interstitial space of blood* *clots (Adam Wufsus and Keith Neeves).* Biological tissues are porous materials composed of fibrous matrices. Of course, the biological materials are qualitatively quite different from geological formations, engineered electrodes, or separation membranes. Nevertheless, compared to these other materials, there are interesting common aspects to characterizing the porous microstructures and modeling transport with the pore volumes. Understanding and predicting behaviors in these biological materials is valuable in the development of medical interventions and drug delivery technologies.

Chapter 16

Pore-Scale Phenomena and Challenges in Energy Research and Technology

Robert J. Kee and Huayang Zhu
Mechanical Engineering,
Colorado School of Mines, Golden, CO 80401, USA
rjkee@mines.edu

Graham M. Goldin
ANSYS, Inc.,
Lebanon, NH 03766, USA

Scott A. Barnett
Materials Science,
Northwestern University, Evanston, IL 60208, USA

Numerous energy-conversion devices rely on the design and performance of porous-media components. In many cases, the porous media serves several functions. In addition to fluid transport within the pore volume, the solid phase (or phases), can contribute essential functions that include structural support, heterogeneous catalysis, charge-transfer kinetics, ion and electron transport, etc. Li-ion batteries, for example, rely on porous electrodes. In this case, the pore volume is filled with an organic electrolyte solvent that is a lithium-ion conductor (e.g., ethylene carbonate). The solid phases (e.g., graphite for the anode and Li_xCoO_2 for the cathode) are intercalation materials that can store and release lithium. Fuel cells also rely on porous electrodes. The anode of a solid-oxide fuel cell, for example, is typically a composite of yttria-stabilized zirconia (YSZ) and nickel. The YSZ is an oxide-ion (O^{2-}) conductor. The Ni serves as an electron conductor and a heterogeneous reforming catalyst. Nearly all practical heterogeneous catalytic processes rely on porous media, with the catalysts being introduced in packed beds or as porous washcoats on monolith channel walls. Despite the extraordinarily broad range of applications, there is a great deal of scientific and engineering commonality in the design, development, and predictive understanding of the porous media.

16.1. Introduction

This chapter uses examples drawn from different technology applications to illustrate aspects of porous-media transport and chemistry. Although the focus here is on microstructural behaviors (i.e., micron scale), there is a need to upscale microscale knowledge to the larger length scales that characterize complete devices (e.g., a battery or fuel cell). In all cases of interest, the pore volume is filled with a mixture of chemical species (i.e., a multicomponent mixture). The fluids may be liquid or gas, and in some cases may contain ionic species carrying an electrical charge. In principle, species transport within the pore volume may be via convection (i.e., bulk fluid motion driven by pressure gradients), diffusion (i.e., driven by concentration gradients), migration (i.e., driven by electrostatic-potential gradients). For applications such as fuel cells or Li-ion batteries the bulk fluid velocity, and thus convective transport, is negligible. For applications such as packed-bed catalytic reactors fluid velocities, and hence convective transport, can be important. In most technologies of interest, the fluids within the pore volumes participate in significant chemistry or electrochemistry.

In the study of porous media, one often thinks primarily about the fluid transport. However, for the technologies of interest here, the solid phases play equally important roles. In a solid-oxide fuel cell (SOFC) the solid phases of the porous composite anode serve as oxide-ion and electron conductors. Moreover, the surface of the Ni phase serves as a catalyst that promotes heterogeneous fuel reforming. In a Li-ion battery the solid phases of the porous electrodes intercalate and deintercalate Li. The solid phases are also electronic conductors. Electrochemical charge-transfer reactions proceed at the surfaces between the solid phase and the pore volume, which is filled with a Li^+-conducting electrolyte. Because the charge-transfer chemistry depends upon electrostatic potential differences between the electrode and electrolytes phases, the overall process depends upon the electrostatic-potential fields within the solid and fluid phases of the porous microstructure.

16.2. Solid Oxide Fuel Cells

Solid oxide fuel cells represent a rapidly maturing technology that has numerous potential applications, ranging from large-scale combined-cycle power generation to residential-scale combined heat and power (CHP) [1]. Unlike the more-mature fuel-cell technology based upon polymer electrolyte

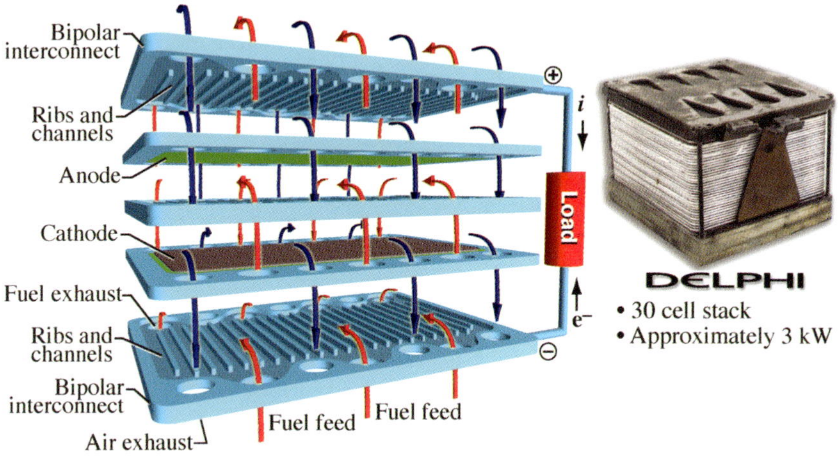

Fig. 16.1. Schematic representation of a planar counter-flow SOFC stack and photograph of a Delphi SOFC stack.

membranes (PEM), SOFC systems operate at elevated temperature (usually around 800°C) and can operate with significant fractions of natural gas (dominantly CH_4) or synthesis gas (H_2-CO mixtures) fuels. The SOFC technology uses relatively low-cost Ni in the anode structure, rather than precious metal catalysts (Pt or Rh) that are required in the PEM technology. The conversion efficiencies for electricity generation can exceed 50%, and can approach 80% for CHP applications.

Although the operating principles are similar, several different fuel cell configurations are being developed, including planar, tubular, and segmented-in-series architectures [2]. The right side of Fig. 16.1 is a photograph of a planar stack developed by Delphi.* The left side of Fig. 16.1 illustrates the salient attributes of a counter-flow planar SOFC stack that is motivated by a system developed in Germany at Forschungszentrum Jülich [3]. The membrane-electrode assembly (MEA) typically is comprised of a porous composite Ni-YSZ anode support (usually on the order of 1 millimeter thick), a dense YSZ electrolyte membrane (on the order of 10 microns), and a composite LSM-YSZ cathode (order 20 microns). Fuel and air are introduced through a manifold of circular passages that carry gases from below. Exhaust is collected on the opposite side of the feed and directed downward via similar circular passages. The metal ribs that form the flow channels also contact the electrodes forming the electrical inter-

*Delphi Powertrain Systems, Fenton, MI

connect between MEA layers. Fuel flows on the anode side of the MEA and oxidizer flows on the cathode side. Under typical conditions each cell layer operates at about 0.7 V. A typical stack may consist of some 50 layers connected electrically in series, producing a stack voltage of around 35 V. Among all the alternatives, planar architectures usually deliver the highest volumetric power density.

16.2.1. *Porous Electrodes*

Porous composite electrodes involve three phases. The ion-conducting phase (yttria-stabilized zirconia, YSZ, is typically used in both the anode and cathode) must be a good oxygen-ion O^{2-} conductor, with very little electronic conduction. The electron-conducting phase (typically nickel, Ni, in the anode and strontium-doped lanthanum manganate, LSM, in the cathode) must have good electronic conduction, with very little ionic conduction. The pore phase must facilitate gas transport. Typically each of these three phases occupies about one third of the overall volume. Charge-transfer reactions proceed at the three-phase boundaries (TPB) that are formed at the intersections of the three phases. The solid-phase particles and the pore spaces are typically on the order of a fraction of a micron.

The composite electrode must accomplish several essential functions. The electron-conducting phase must provide a low-resistance path from the position where the electrons are produced (or consumed) to the current collection components (i.e., bipolar plates in a planar configuration such as illustrated in Fig. 16.1). That is to say the electronic-conducting phase must percolate through the thickness of the composite electrode. Especially within a few tens of microns around the dense electrolyte the ion-conducting phase must percolate, enabling ion-conduction to the three-phase boundaries [4]. When hydrocarbon fuels are used, the Ni phase in the anode also serves as a reforming catalyst.

A porous structure that facilitates one function can sometimes serve to frustrate another function. For example, small particles and pore spaces tend to increase three-phase boundary length, which facilitates the essential charge-transfer chemistry. However, the small pore spaces also serve to increase gas-transport resistance, which diminishes performance. Small particles can also decrease effective ionic and electronic conductivity, which can negatively influence performance. Because there are different functional needs in different regions of the composite electrode, functional grading of the electrode can improve performance. For example, in the few tens of mi-

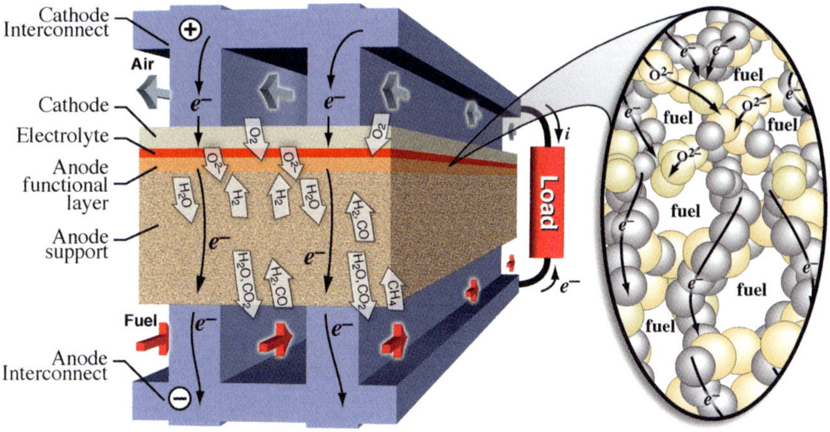

Fig. 16.2. Closeup representation of an SOFC membrane-electrode assembly at the millimeter scale. The balloon shows a representation of the porous anode at the particle scale.

crons near the dense electrolyte layer charge transfer is most important [4]. In this region a functional layer with fine particles and small pore spaces is valuable. However, far from the dense electrolyte there is no charge-transfer chemistry (i.e., all available charge has been transferred nearer to the dense electrolyte). In this region (usually the majority of an anode support structure) performance is improved by using large pores that minimize gas-transport resistance. Finding optimal electrode architectures that balance competing tradeoffs can be challenging. However, the process is facilitated by a predictive quantitative understanding of the competing physical and chemical processes.

Figure 16.2 illustrates a closeup view of an anode-supported membrane-electrode assembly (MEA). The MEA is composed of a relatively thick porous composite anode (e.g., 1 mm of Ni-YSZ), a thin dense electrolyte membrane (e.g., 20 μm YSZ), and a thin porous composite cathode (e.g., 30 μm LSM-YSZ). The dense electrolyte transports only oxygen ions and is impermeable to any gas or electron transport. The MEA is sandwiched between bipolar interconnect plates[†] that also form fuel and air channels. The flow channels typically have height and width dimensions on the order of a millimeter, but may be tens of centimeters long.

[†]When multiple cell are layered atop one another (e.g., Fig. 16.1), the positive electrode of one layer becomes the negative electrode of the adjoining cell. Hence, the designation bipolar.

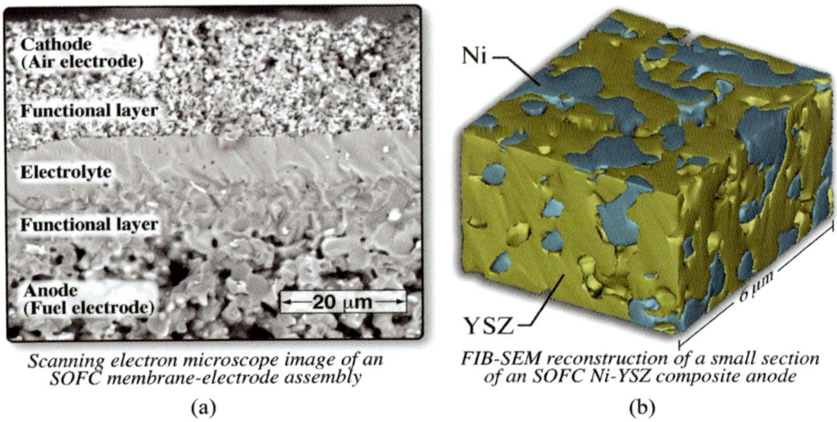

Fig. 16.3. (a) Scanning electron microscope image of an SOFC membrane electrode assembly (MEA). Only a small portion of the full anode thickness is shown. (b) FIB-SEM reconstruction of a microscopic segment of a different SOFC anode.

As illustrated in Fig. 16.2 the electrode microscale is represented as ideal spherical particles. This representation is helpful in visualizing the microstructure, and has practical utility. However, Fig. 16.3 shows clearly that actual electrodes are not simply assemblies of spheres.

16.2.2. *Gas-phase Transport*

Mass-continuity equations for the K_{g} gas-phase species and the overall mass-continuity equations in the electrode pore phase can be written in the transient conservative form as

$$\frac{\partial \left(\phi_{\mathrm{g}} \rho Y_k\right)}{\partial t} + \nabla \cdot \mathbf{j}_k = W_k \dot{s}_k, \tag{16.1}$$

$$\frac{\partial \left(\phi_{\mathrm{g}} \rho\right)}{\partial t} + \sum_{k=1}^{K_{\mathrm{g}}} \nabla \cdot \mathbf{j}_k = \sum_{k=1}^{K_{\mathrm{g}}} W_k \dot{s}_k. \tag{16.2}$$

In these equations ρ is the gas-phase density, ϕ_{g} is the porosity, and Y_k, W_k, $\mathbf{j}_k$ and $\dot{s}_k$ are the gas-phase mass fractions, molecular weights, mass fluxes, and molar production rates from the thermal and electrochemical reactions, respectively.

Local gas-phase mass fluxes $\mathbf{j}_k$ within the porous structure can be evaluated using a Dusty-Gas Model (DGM) [5, 6], which is written as an implicit relationship among the gas-phase molar fluxes $\mathbf{J}_k$, molar concentrations $[X_k]$, concentration gradients, and the pressure p gradient as

$$\sum_{\ell \neq k} \frac{[X_\ell]\mathbf{J}_k - [X_k]\mathbf{J}_\ell}{[X_\mathrm{T}]D^\mathrm{e}_{k\ell}} + \frac{\mathbf{J}_k}{D^\mathrm{e}_{k,\mathrm{Kn}}} = -\boldsymbol{\nabla}[X_k] - \frac{[X_k]}{D^\mathrm{e}_{k,\mathrm{Kn}}} \frac{B_\mathrm{g}}{\mu} \boldsymbol{\nabla} p, \tag{16.3}$$

where $[X_\mathrm{T}] = p/RT$ is the total molar concentration, B_g is the permeability, and μ is the mixture viscosity. The mass fluxes $\mathbf{j}_k$ are related simply to the molar fluxes $\mathbf{J}_k$ as $\mathbf{j}_k = W_k \mathbf{J}_k$. Knudsen diffusion represents mass transport assisted by gas-surface collisions. The Knudsen diffusion coefficients depend upon the porous-media microstructure, including porosity, average pore radius r_p, and tortuosity τ_g. The *effective* binary and Knudsen diffusion coefficients $D^\mathrm{e}_{k\ell}$ and $D^\mathrm{e}_{k,\mathrm{Kn}}$ can be evaluated as

$$D^\mathrm{e}_{k\ell} = \frac{\phi_\mathrm{g}}{\tau_\mathrm{g}} D_{k\ell}, \qquad D^\mathrm{e}_{k,\mathrm{Kn}} = \frac{2}{3} \frac{r_p \phi_\mathrm{g}}{\tau_\mathrm{g}} \sqrt{\frac{8RT}{\pi W_k}}. \tag{16.4}$$

The binary diffusion coefficients $D_{k\ell}$ and the mixture viscosities μ are determined from kinetic theory [7]. The permeability can be evaluated from the Kozeny-Carman relationship as

$$B_\mathrm{g} = \frac{\phi_\mathrm{g}^3 d_\mathrm{p}^2}{72\tau_\mathrm{g}(1-\phi_\mathrm{g})^2}, \tag{16.5}$$

where d_p is the particle diameter. Further details of the DGM and its numerical implementation can be found in Zhu, et al. [6].

The production rates of gas-phase species $\dot{s}_k$ are due to the thermal catalytic chemistry and electrochemical charge-transfer reactions. The pore spaces are typically sufficiently small that the most likely collisions are between gas molecules and surfaces. Thus, gas-phase homogeneous kinetics is usually negligible. In general, the production rates $\dot{s}_k$ are functions of temperature, gas composition, surface-species coverages, and electric-potential differences between the electrode and electrolyte phases. In addition to gas-phase species, the chemistry also depends on surface-adsorbed species. The temporal variations of site coverages θ_k of surface-adsorbed species are represented as

$$\frac{d\theta_k}{dt} = \frac{\dot{s}_k}{\Gamma}, \quad k = 1, \cdots, K_\mathrm{s} \tag{16.6}$$

where Γ is available site density, and K_s is the number of the surface-adsorbed species [7].

16.2.3. *Electrostatic Potential and Charge Conservation*

Because the charge-transfer chemistry depends upon the local electrostatic potential between phases $\Delta\Phi = \Phi_{\rm ed} - \Phi_{\rm el}$, the electrostatic potential variations within the electrode and electrolyte phases that comprise the porous media play central roles. These variations are governed by Poisson equations as

$$\nabla \cdot \left(\sigma_{\rm ed}^{\rm eff} \nabla \Phi_{\rm ed}\right) = -\dot{s}_{\rm e}, \tag{16.7}$$

$$\nabla \cdot \left(\sigma_{\rm el}^{\rm eff} \nabla \Phi_{\rm el}\right) = \dot{s}_{\rm e}. \tag{16.8}$$

The electrode phase may be either an anode or a cathode phase. The effective conductivities $\sigma_{\rm ed}^{\rm eff}$ and $\sigma_{\rm el}^{\rm eff}$ can be quite different from the intrinsic material properties because they must accommodate the porous-media microstructure. For example, the geometrical "necking" where particles are joined decreases the effective conductivity. These relationships are discussed subsequently in the context of battery electrodes. The source terms reaction rates $\dot{s}_{\rm e}$ represent the charge that is transferred across the phase boundaries via charge-transfer reactions.

16.2.4. *Boundary Conditions*

Boundary conditions are needed to solve Eqs. 16.1, 16.2, 16.7, and 16.8 within the porous electrode structures. At the interfaces between the fuel and air compartments the following boundary conditions apply:

- The gas-phase composition is assumed to be that within either the bulk fuel or air flow.
- The ion flux within the ionic-conducting phase must vanish.
- On the air side (cathode) the electrostatic potential in the electron-conducting phase is set to a reference potential (e.g., $\Phi_{\rm c} = 0$ V). On the fuel side (anode), the electrostatic potential in the electron-conducting phase is set to the cell operating potential (e.g., $E_{\rm cell} = \Phi_{\rm c} - \Phi_{\rm a} = 0.7$ V).

At the interfaces between the porous electrodes and the dense electrolyte membrane, the following boundary conditions apply:

- The gas-phase fluxes must vanish.
- The oxide-ion flux within the ionic-conducting phase of the electrode must be equal to the oxide-ion flux through the dense electrolyte membrane.

- The electron flux within the electron-conducting phase of the porous electrodes must vanish.

Zhu and Kee discuss details about the boundary conditions and computational implementation [4].

16.2.5. *Chemistry and Electrochemistry*

16.2.5.1. *Global Charge Transfer Reactions*

At the anode three-phase boundaries, fuel is oxidized with oxygen ions from the electrolyte. In global terms a charge-transfer reaction for hydrogen electrochemical oxidation may be written as

$$\mathrm{H_2(g) + O^{2-}(el) \rightleftharpoons H_2O(g) + 2e^-(a)}. \tag{16.9}$$

In words, gas-phase hydrogen $H_2(g)$ reacts with an oxygen ion emerging from the electrolyte $O^{2-}(el)$ to form steam in the gas phase $H_2O(g)$ and deliver electrons to the anode $e^-(a)$. At the cathode, molecular oxygen is reduced, replenishing the electrolyte with oxygen ions,

$$\frac{1}{2}\mathrm{O_2(g) + 2e^-(c) \rightleftharpoons O^{2-}(e)}. \tag{16.10}$$

The parentheses associated with the species names indicates the phase in which the species resides. Because these charge-transfer reactions proceed at three-phase boundaries, an important characteristic of the porous-media microstructure is the TPB length per unit volume of the electrode. As discussed subsequently, these global reactions are in fact the result of many elementary reactions.

16.2.5.2. *Global Reforming Reactions*

Solid-oxide fuel cells can operate with mixtures of H_2, CO, and CH_4 as the fuel. However, it is usually the H_2 that dominates the charge-transfer chemistry [6, 8, 9]. Thus, the Ni in the anode structure plays an important role in catalytically transforming the fuel to H_2. In global terms, the following steam-reforming and water-gas-shift processes are important:

$$\mathrm{CH_4 + H_2O \rightleftharpoons CO + 3H_2}, \qquad \Delta H \approx 206\ \mathrm{kJ\ mol^{-1}}, \tag{16.11}$$

$$\mathrm{CO + H_2O \rightleftharpoons CO_2 + H_2} \qquad \Delta H \approx -41\ \mathrm{kJ\ mol^{-1}}. \tag{16.12}$$

As the electrochemically produced steam from Reaction 16.9 diffuses outward toward the flow channel, it encounters small hydrocarbons and CO diffusing inward toward the anode functional layer. At high temperature and in the presence of an appropriate catalyst (e.g., Ni), the heterogeneous chemistry produces H_2. This benefits SOFC performance because H_2 is much more electrochemically active than CH_4 or CO. The highly endothermic steam-reforming process requires high temperature (above 800°C for CH_4 but less for higher hydrocarbons), but the water-gas-shift can proceed at relatively lower temperatures (below 600°C) on a Ni catalyst. When the fuel stream contains small hydrocarbons such as methane endothermic reforming can be used to assist controlling the cell temperature.

16.2.5.3. *Detailed Chemical and Electrochemical Kinetics*

The SOFC literature is dominated by the global chemistry as discussed above. Nevertheless, there is compelling evidence that the actual chemistry is considerably more complex. Two groups are leading the development of detailed charge-transfer reaction mechanisms, providing much better physical and chemist representations than are possible with the relatively more straightforward global reactions. The results from Caltech and Colorado School of Mines [6, 9, 10] and the Bessler and Deutschmann groups in Germany [11–15] lead generally to similar mechanisms as shown in Fig. 16.4.

When using fuel mixtures of H_2, CO, CO_2, and CH_4 the catalytic chemistry is more complex that can be accurately represented by the global reactions 16.11 and 16.12. As represented in Fig. 16.4, Deutschmann and colleagues have developed detailed reactions mechanisms for catalytic reforming on Ni surfaces [6, 12, 16]. Recent research shows compelling evidence the detailed catalytic chemistry influences measured electrochemical impedance spectra (EIS) in ways that cannot be explained with global chemistry [17]. There is evidence that incorporating low levels of O_2 into the SOFC fuel stream can improve performance [18, 19]. The important effects of catalytic partial oxidation within the porous anode structure cannot be represented in terms of the global reactions.

Chemical and electrochemical reactions can be written generally as

$$\sum_{k=1}^{K} \nu'_{ki} \Lambda_k^{z_k} \rightleftharpoons \sum_{k=1}^{K} \nu''_{ki} \Lambda_k^{z_k}, \tag{16.13}$$

where $\Lambda_k^{z_k}$ is the chemical symbol of the kth species with charge z_k. The forward and backward stoichiometric coefficients for the kth species in the ith

Ni surface

1. $H_2(g) + 2(Ni) \rightleftharpoons H(Ni) + H(Ni)$
2. $O_2(g) + (Ni) + (Ni) \rightleftharpoons O(Ni) + O(Ni)$
3. $CH_4(g) + (Ni) \rightleftharpoons CH_4(Ni)$
4. $H_2O(g) + (Ni) \rightleftharpoons H_2O(Ni)$
5. $CO_2(g) + (Ni) \rightleftharpoons CO_2(Ni)$
6. $CO(g) + (Ni) \rightleftharpoons CO(Ni)$
7. $O(Ni) + H(Ni) \rightleftharpoons OH(Ni) + (Ni)$
8. $OH(Ni) + H(Ni) \rightleftharpoons H_2O(Ni) + (Ni)$
9. $OH(Ni) + OH(Ni) \rightleftharpoons O(Ni) + H_2O(Ni)$
10. $O(Ni) + C(Ni) \rightleftharpoons CO(Ni) + (Ni)$
11. $O(Ni) + CO(Ni) \rightleftharpoons CO_2(Ni) + (Ni)$
12. $HCO(Ni) + (Ni) \rightleftharpoons CO(Ni) + H(Ni)$
13. $HCO(Ni) + (Ni) \rightleftharpoons O(Ni) + CH(Ni)$
14. $CH_4(Ni) + (Ni) \rightleftharpoons CH_3(Ni) + H(Ni)$
15. $CH_3(Ni) + (Ni) \rightleftharpoons CH_2(Ni) + H(Ni)$
16. $CH_2(Ni) + (Ni) \rightleftharpoons CH(Ni) + H(Ni)$
17. $CH(Ni) + (Ni) \rightleftharpoons C(Ni) + H(Ni)$
18. $O(Ni) + CH_4(Ni) \rightleftharpoons CH_3(Ni) + OH(Ni)$
19. $O(Ni) + CH_3(Ni) \rightleftharpoons CH_2(Ni) + OH(Ni)$
20. $O(Ni) + CH_2(Ni) \rightleftharpoons CH(Ni) + OH(Ni)$
21. $O(Ni) + CH(Ni) \rightleftharpoons C(Ni) + OH(Ni)$

YSZ surface

22. $O_O^{2-} + (\chi) \rightleftharpoons O^{2-}(\chi) + V_O$
23. $H_2O(g) + (Zr) + O^{2-}(\chi) \rightleftharpoons OH^-(Zr) + OH^-(\chi)$
24. $OH^-(\chi) + (Zr) \rightleftharpoons OH^-(Zr) + (\chi)$
25. $H_2(g) + (Zr) + O^{2-}(\chi) \rightleftharpoons H^-(Zr) + OH^-(\chi)$

Three-phase boundary

26. $H(Ni) + O^{2-}(\chi) \rightleftharpoons OH^-(\chi) + (Ni) + e^-$
27. $H(Ni) + OH^-(\chi) \rightleftharpoons H_2O(Ni) + (\chi) + e^-$

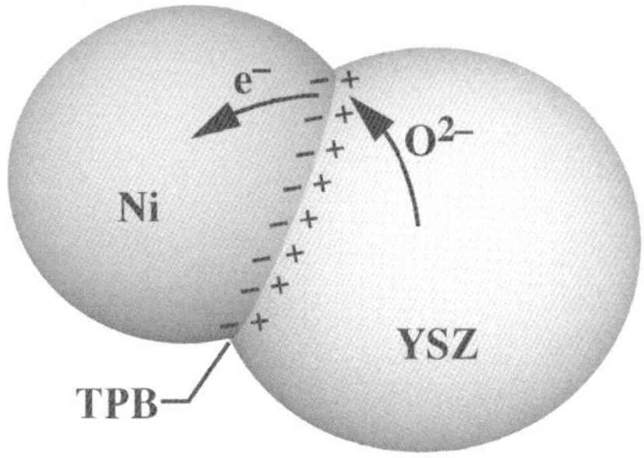

Fig. 16.4. Detailed reactions on the Ni surface, the YSZ surface, and at the three-phase boundary.

reaction are written as ν'_{ki} and ν''_{ki}, respectively. Each participating species is associated with a phase (e.g., electrolyte phase such as YSZ, or electrode phase such as Ni). Gas-phase species are presumed to be uncharged (i.e., $z_k = 0$). When a reaction involves transferring charge between phases, the reaction rates depend on the electrostatic-potential differences between phases. When all the participating species are uncharged or in the same phase, the reaction rate depends only on the temperature and species activities. The source terms $\dot{s}_k$ that appear in Eqs. 16.1, 16.2, and 16.6 are the result of chemical and electrochemical reactions that proceed on the surfaces of the porous media.

As an example, consider the following reversible charge-transfer reaction,

$$H(Ni) + OH^-(\chi) \rightleftharpoons H_2O(Ni) + (\chi) + e^-(Ni), \qquad (16.14)$$

where two surface sites are involved. One is the Ni surface and the other is a χ site on the YSZ surface [10]. The anodic (forward) and cathodic (backward) stoichiometric coefficients are:

$$\nu'_{H(Ni)} = 1,\ \nu'_{OH^-(\chi)} = 1,\ \nu''_{H_2O(Ni)} = 1,\ \nu''_{e^-} = 1,\ \nu''_{(\chi)} = 1. \qquad (16.15)$$

The H(Ni), H_2O(Ni), and (χ) are assumed to be charge-neutral (i.e., $z = 0$) and the hydroxyl $OH^-(\chi)$ and the electron both carry a single negative charge (i.e., $z = -1$).

The net production rate for kth species resulting from the ith reaction $\dot{s}_{ki}$ can be represented in terms of the reaction rate of progress q_i as

$$\dot{s}_{ki} = (\nu''_{ki} - \nu'_{ki})\, q_i = \nu_{ki} q_i, \tag{16.16}$$

where the net stoichiometric coefficients are defined as $\nu_{ki} \equiv \nu''_{ki} - \nu'_{ki}$. The rate of progress for each reaction can be evaluated as

$$q_i = k_{\mathrm{a}i} \prod_{k=1}^{K} C_k^{\nu'_{ki}} - k_{\mathrm{c}i} \prod_{k=1}^{K} C_k^{\nu''_{ki}}, \tag{16.17}$$

where C_k are the activities of the participating species. The activity of gas-phases species is its molar concentration $[X_k]$. The activity of surface-adsorbed species is the molar surface coverage $\Gamma_m \theta_{k,m}$, where $\theta_{k,m}$ is the site fraction for species k on the surface of site-type m and Γ_m is the total available surface-site density for site-type m [7, 20].

Charge-transfer reactions are said to be anodic when producing electrons and cathodic when consuming electrons. By the convention adopted here, reactions are written such that the forward direction is anodic. The anodic (i.e., producing electrons) and cathodic (i.e., consuming electrons) rate expressions for ith reaction are written as

$$k_{\mathrm{a}i} = k^{\mathrm{t}}_{\mathrm{a}i} \exp\left[-\beta_{\mathrm{a}i} \frac{F}{RT} \sum_{k=1}^{K} \nu_{ki} z_k \Phi_k\right], \tag{16.18}$$

$$k_{\mathrm{c}i} = k^{\mathrm{t}}_{\mathrm{c}i} \exp\left[+\beta_{\mathrm{c}i} \frac{F}{RT} \sum_{k=1}^{K} \nu_{ki} z_k \Phi_k\right]. \tag{16.19}$$

The anodic and cathodic symmetry factors are $\beta_{\mathrm{a}i}$ and $\beta_{\mathrm{c}i}$, respectively. For elementary charge-transfer reactions (i.e., those transferring a single electron) $\beta_{\mathrm{a}i} + \beta_{\mathrm{c}i} = 1$. The Faraday constant and gas constant are represented as F and R, respectively, and T is the temperature.

Each species is assumed to have an associated charge z_k; charge-neutral species have $z_k = 0$. Each phase m (e.g., the Ni and YSZ in Eq. 16.14) is assumed to be at an electrostatic potential Φ_m. The rate expressions in Eqs. 16.18 and 16.19 are written with the assumption that each charged species k is at the electrostatic potential of the phase in which it resides. As a convenience in computational implementation Eqs. 16.18 and 16.19 associate the electrostatic potential Φ_k with the species, not the phase directly.

The thermal component of the anodic (forward) rate expression is written in modified Arrhenius form as

$$k^{\mathrm{t}}_{\mathrm{a}i} = A_i T^{n_i} \exp\left(-\frac{E_i}{RT}\right), \tag{16.20}$$

where E_i represents an activation energy, A_i a pre-exponential factor, and n_i a temperature exponent. To satisfy microscopic reversibility and maintain thermodynamic consistency, the thermal component of the cathodic (reverse) rate is related to the forward rate via the reaction's equilibrium constant K_i as

$$K_i = \frac{k^{\mathrm{t}}_{\mathrm{a}i}}{k^{\mathrm{t}}_{ci}} = \exp\left(\frac{-\Delta G^\circ}{RT}\right), \tag{16.21}$$

where ΔG° is the change in standard-state free energy for the reaction. Evaluating ΔG°, and hence the equilibrium constant, depends upon quantitative thermochemical properties for all participating species. For gas-phase species such information is readily available. However, finding or estimating properties for surface adsorbates and ionic species can be difficult.

When all species in a reaction are at the same electrostatic potential, or when all participating species are charge-neutral, the exponential factors in Eqs. 16.18 and 16.19 become unity. In this case, the rate expressions revert to the purely thermal rate expressions.

Despite clear evidence that predictive models benefit from higher-fidelity representations of the chemistry, there are impediments to widespread use. The porous-media models become more complex, thus increasing complexity of computational implementation. In addition to including more gas-phases species, the detailed reaction mechanisms must also represent local surface adsorbates throughout the porous electrode structure. Although no problems in principle, there is a need for sophisticated models and implementation software that can handle significantly increased chemical complexity. Appropriate software packages include CHEMKIN (developed initially by R. J. Kee and colleagues [7]), CANTERA (developed initially by D. G. Goodwin, https://code.google.com/p/cantera), DETCHEM (developed initially by O. Deutschmann and colleagues, www.detchem.com), and DENIS (developed by W. Bessler and colleagues [11]).

16.2.6. *Separated Anode Experiment*

Models depend on numerous parameters that characterize the porous-media transport and chemistry. Thus, there is a need for experiments that can

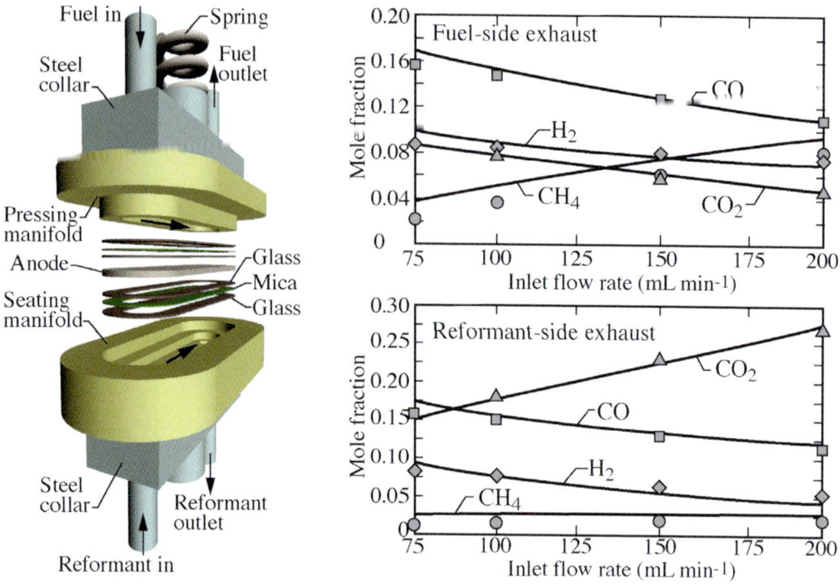

Fig. 16.5. Cartoon image of the separated-anode experiment. The graphs show measured and modeled exhaust mole fractions from CH_4-CO_2 dry-reforming experiments with a Ni-YSZ anode support. The upper graph shows the fuel-side exhaust and the lower graph shows the reformant-side exhaust stream. Measurements are represented by symbols and the model is shown as lines.

assist establishing the parameters and validating the models. Some parameters, such as the porosity, are relatively straightforward to measure. However, other parameters, such as the tortuosity are not easily measured. The catalytic reforming chemistry within a composite electrode must also be validated.

The so-called Separated Anode Experiment (Fig. 16.5) can be used to assist parameter estimation and model development for SOFC anode structures [21, 22]. A porous composite anode (typically on the order of 1 mm thick) is sandwiched between two small flow channels. As illustrated in Fig. 16.5, fuel (e.g., H_2, CO, CH_4) enters the upper channel and reforming agents (e.g., H_2O, CO_2) enter the lower channel. In an actual fuel cell, the reforming agents are the products of electrochemical fuel oxidation. The fuel and reformant gases cross-diffuse through the porous anode structure, which replicates the configuration within an operating fuel cell. Such cross-flow transport is different from more-typical catalyst experiments in which fuel and reforming agents are premixed and introduced onto a packed bed.

As illustrated in the graphs on the right side of Fig. 16.5, the exhaust-gas composition of both channels is measured as a function of inlet flow rates. Maintaining relatively high channel flow rates keeps the catalytic chemistry in the kinetic regime. To validate chemical kinetics reactions mechanisms, the catalytic process should not be allowed to reach equilibrium. As discussed in previous sections, the reactive porous-media transport situation can also be modeled. Thus, the comparison of model and measurement can validate models and submodels, which are then used in complete fuel-cell models. Before introducing reactive gases, the experiment is typically used to characterize non-reacting aspects of the gas-phase transport within the porous structure. For example, two different inert gases can be introduced in the upper and lower channels. In this case, measuring the non-reactive cross diffusion as a function inlet flow rate, important parameters such as the tortuosity (Eq. 16.4) can be determined.

16.2.7. *Percolation Theory*

The conservation equations (e.g., Eqs. 16.1, 16.2, 16.7, and 16.8) are written as continuum partial differential equations, assuming that *effective* material properties can be established. In fact, as illustrated in Figs. 16.2 and 16.3, at the microscopic scale the electrodes are not continuous media, but rather are composed of discrete interconnected phases. Historically, percolation theory has been used to estimate the needed effective properties [23–27].

Percolation theory provides an analytic approach to predict important porous-media parameters, including effective conductivities and three-phase boundary lengths. The approach is based upon large numbers of randomly packed particles, where an important parameter is the coordination number, representing the number of contacts that a certain particle makes with neighboring particles [23]. The percolation probability represents the probability that clusters of certain particle types (e.g., electrode or electrolyte) form connected networks that span the porous media. As illustrated by the cartoon in Fig. 16.6, the clusters labeled "A" percolate through the porous media. The clusters labeled "B" and "C" do not percolate through the structure. The percolation probabilities depend upon numerous parameters, such as particle sizes, packing density, and overlap between particles.

The right-hand side of Fig. 16.6 shows predicted results for "intra-particle" conductivities of a composite SOFC electrode [27]. The abscissa is the electrode volume fraction, meaning the volume of electrode parti-

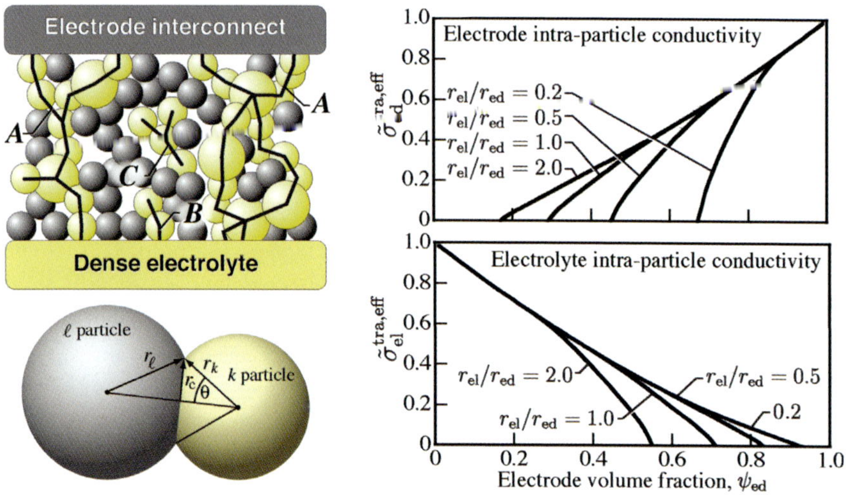

Fig. 16.6. The cartoon illustrates aspects of percolating and non-percolating particle clusters within a porous composite electrode. The graphs show predicted dimensionless effective intra-particle conductivity for binary mixtures as functions of electrode volume loading ψ_{ed} and the particle-radii ratio r_{el}/r_{ed}. The upper panel shows the electrode conductivity and the lower panel shows electrolyte conductivity.

cles (e.g., Ni) relative to the total solid phases (e.g., Ni plus YSZ). The family of curves represent the ratios of electrode and electrolyte particle radii. The intra-particle conductivity considers geometry effects, such as area constrictions at particle-particle intersections. The percolation theory is also capable of including inter-particle conductivity, such as, for example, grain-boundary resistance at the particle intersections [27].

16.2.8. *Illustrative SOFC Anode Example*

Figure 16.7 shows simulated results for a typical SOFC membrane-electrode assembly, considering distributed charge transfer throughout the composite electrodes. Electrons returning from an external circuit enter the electron-conducting phase of the cathode. Charge-transfer reactions (e.g., Eq. 16.10) at the three-phase boundaries use the electrons to reduce gas-phase O_2 to oxide ions O^{2-} within the electrolyte phase. The dense electrolyte membrane is a pure oxide-ion conductor. On the anode side, charge-transfer reactions at the three-phase boundaries use the oxide ions to electro-oxidize the gas-phase fuel (e.g., Eq. 16.9) and deliver electrons into the electron-conducting phase, which is connected to an external circuit.

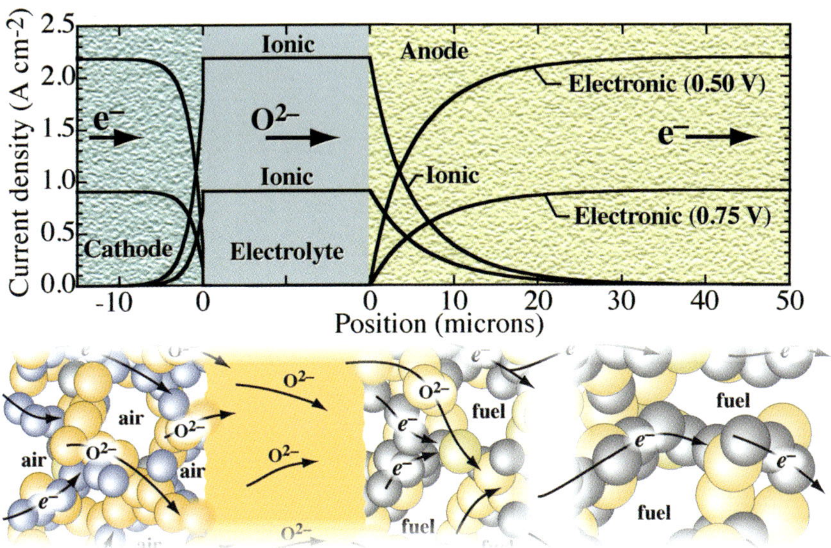

Fig. 16.7. Predicted ionic and electronic fluxes in the membrane-electrode assembly of a solid-oxide fuel cell. The cartoon at the bottom illustrates an expanded view of a porous composite electrode structure. The ion-conducting phase is shown as lighter yellow particles and the electron-conducting phase is shown as darker blue-grey particles.

Overall, the chemical potential associated with an overall reaction (e.g., $H_2 + \frac{1}{2}O_2 \rightleftharpoons H_2O$) is used to produce net electricity. The charge-transfer regions within the porous electrodes typically extend a few tens of microns from the dense ion-conducting membrane.

16.3. Li-ion Batteries

Figure 16.8 illustrates the cathode end-cap region of a Li-ion battery in a cylindrical, jelly-roll, 18650 format (the designation "18650" means 18 mm in diameter and 65 mm high). In such batteries, the lower portion of the external case is the anode (negative terminal). Typical batteries include safety features that interrupt electrical current under abnormal operating environments. As a result of overcharging, or other accidental operation, the electrolyte solvent can decompose leading to high internal gas pressure. In this case, the burst disk ruptures, relieving the gas pressure and severing the current path to the cathode terminal. The positive-temperature-coefficient (PTC) layer is a material that is electrically conductive at nor-

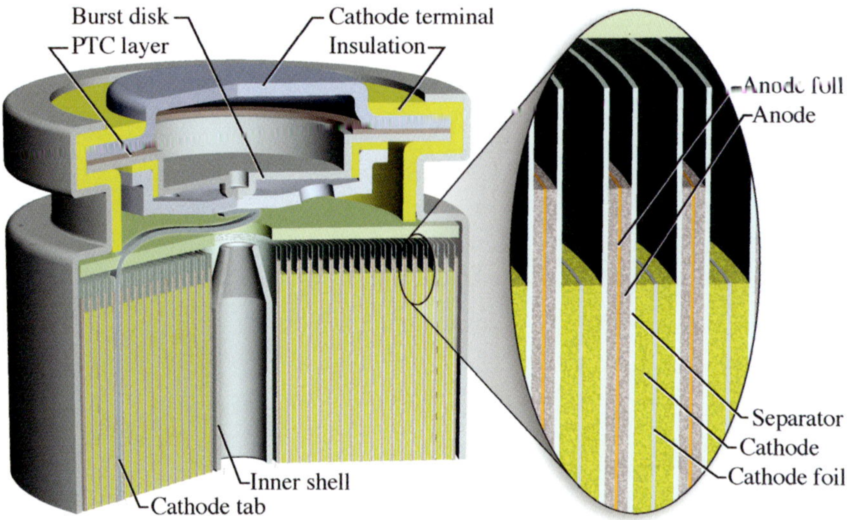

Fig. 16.8. Illustration of the cathode (positive electrode) end of a cylindrical Li-ion battery.

mal operating temperature, but becomes an electrical insulator at elevated temperature. Thus, when a certain battery temperature is exceeded the electrical pathway is broken and the cell cannot continue to function.

As illustrated in the balloon in the right-hand side of Fig. 16.8, the anode and cathodes are thin porous composite structures. The solid phases intercalate Li and the pore volume is filled with a Li-ion-conducting electrolyte solvent, such as ethylene carbonate. The electrode structures are typically in the order of 50 to 100 μm thick. The solid phase of the anode is typically graphite and the solid phase of the cathode is typically a material such as Li_xCoO_2. The anodes and cathodes are formed on both sides of metal current-collection foils that are typically around 20 μm thick. Separator sheets are typically porous polymers that electrically isolate the anode and cathode.

Figure 16.9 illustrates a typical configuration, with the intercalating electrodes being represented as ideal spherical particles. Lithium ions are mobile in the electrolyte, which is typically a Li-ion-conducting organic solvent. At the electrode particle surfaces charge-transfer reactions move Li across the electrode-electrolyte interfaces as

$$\mathrm{Li[ed]} \rightleftharpoons \mathrm{Li^+[el]} + \mathrm{e^-[ed]}. \tag{16.22}$$

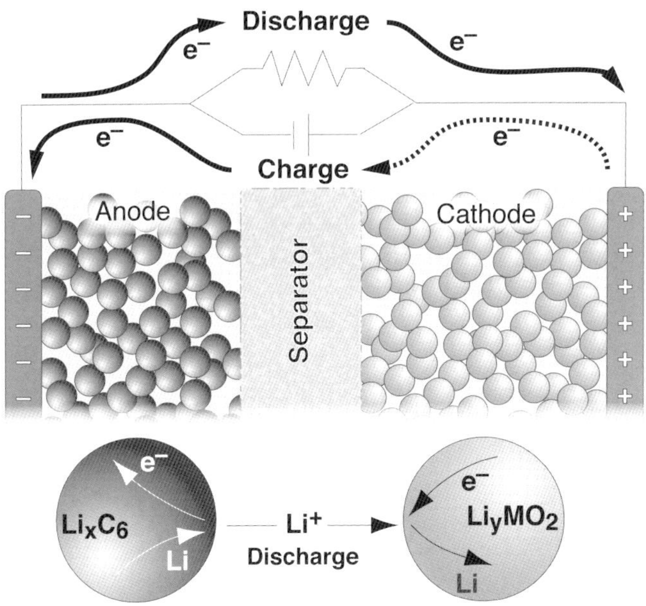

Fig. 16.9. Illustration of a Li-ion battery electrode layout with the electrodes being represented as spherical particles.

Note that the forward direction of reaction 16.22 is written to be anodic (i.e., producing electrons), which is relevant in the subsequent evaluation of the rate of progress using the Butler–Volmer formulation.

Following pioneering research of Newman and colleagues [28–31], numerous researchers have developed models based upon spherical-particle electrodes. This approach is based upon solving mass- and charge-conservation equations within the electrolyte and within the electrode particles. The electrolyte is modeled as a porous media using a finite-volume discretization on a mesh network that spans both porous electrodes and the separator. Within each finite volume an intercalation-diffusion problem is solved for a single representative electrode particle. The approach is summarized in the following sections.

16.3.1. *Electrolyte Charge Conservation*

Charge conservation within the electrolyte phase can be represented as [32]

$$\nabla \cdot \left(\kappa^{\mathrm{eff}} \nabla \Phi_{\mathrm{el}}\right) + \nabla \cdot \left(\kappa_{\mathrm{D}}^{\mathrm{eff}} \nabla \ln C_{\mathrm{Li^+(el)}}\right) + A_{\mathrm{s}} \dot{s}_{\mathrm{Li^+(el)}} F = 0, \qquad (16.23)$$

where the dependent variables are the electric potential of the electrolyte phase $\Phi_{\rm el}$ and the lithium-ion concentration $C_{\rm Li^+(el)}$. The specific surface area $A_{\rm s}$ represents the surface area of active electrode particles per unit volume of composite electrode. The net production rate of lithium ions via charge-transfer reactions is represented as $\dot{s}_{\rm Li^+(el)}$ and F is the Faraday constant. Because bulk velocity of the electrolyte is neglected there is no convective contribution.

Charge enters and leaves the electrolyte phase via surface reactions (third term in Eq. 16.23). Within composite electrodes an electric double layer is formed at the interface between the electrode particles and electrolyte. The charge-transfer rates depend upon the electrostatic-potential differences between the electrode and electrolyte phases. Because the characteristic time for charging and discharging the double layer is on the order of milliseconds or shorter [28], double layer charging can be neglected on typical charging/discharging timescales. Within the separator there are no electrode phases, thus eliminating any charge-transfer reactions.

The effective ionic conductivity of the electrolyte phase $\kappa^{\rm eff}$ and the effective diffusional conductivity $\kappa_{\rm D}^{\rm eff}$ can be evaluated as

$$\kappa^{\rm eff} = \kappa \phi_{\rm el}^{p}, \tag{16.24}$$

$$\kappa_{\rm D}^{\rm eff} = \frac{2RT\kappa^{\rm eff}}{F}\left(t^{\circ}_{\rm Li^+(el)} - 1\right). \tag{16.25}$$

Ion conductivity of the pure electrolyte phase is represented as κ, the volume fraction (porosity) of the electrolyte phase is $\phi_{\rm el}$, and p the Bruggeman porosity exponent (typically, $p \approx 1.5$). Both positively and negatively charged species can produce ion current. The transference number t°_{k} is the fraction of current carried by species k via migration in a electrolyte solution with uniform composition [33]. For example, if the lithium-ion transference number $t^{\circ}_{\rm Li^+(el)} = 1$, charge movement is entirely the result of electrostatic-potential gradients.

16.3.2. *Electrolyte Species Conservation*

Lithium-ion concentration within the binary electrolyte solution is represented as

$$\frac{\partial \phi_{\rm el} C_{\rm Li^+(el)}}{\partial t} = \nabla \cdot \left(D_{\rm el}^{\rm eff} \nabla C_{\rm Li^+(el)}\right) + \left(1 - t^{0}_{\rm Li^+(el)}\right) A_{\rm s} \dot{s}_{\rm Li^+(el)}. \tag{16.26}$$

The effective diffusion coefficient is evaluated using a Bruggeman expression

$$D_{\rm el}^{\rm eff} = D_{\rm el} \phi_{\rm el}^{p}, \tag{16.27}$$

which approximates the effect of porosity $\phi_{\rm el}$. The electrolyte diffusion coefficient $D_{\rm el}$ is a function of local concentrations, species charges, and binary diffusion coefficients within the electrolyte phase. The last term represents the effects of lithium ions entering the electrolyte via heterogeneous charge-transfer chemistry at the electrode surfaces. In the limit of the lithium-ion transference being unity, the charge-transfer chemistry does not affect the electrolyte concentration. Hence the electrolyte composition would be spatially uniform. Conversely, if the transference number is zero, then the charge-transfer chemistry causes significant lithium-ion variations throughout the electrolyte [28]. Unlike the charge-conservation equation (Eq. 16.23), the species-conservation equation (Eq. 16.26) must retain the transient term. Species transport through the electrolyte proceeds on time scales that are commensurate with charge and discharge times.

16.3.3. *Electrode Charge Conservation*

Charge conservation within the solid-phase electrodes may be represented as

$$\nabla \cdot \left(\sigma_{\rm ed}^{\rm eff} \nabla \Phi_{\rm ed}\right) - A_{\rm s} \dot{s}_{{\rm Li}^+({\rm el})} F = 0, \tag{16.28}$$

where $\Phi_{\rm ed}$ is the bulk electrode-phase electric potential and $\sigma_{\rm ed}^{\rm eff}$ is the effective solid-phase conductivity. The first term in Eq. 16.28 represents charge transport via electronic conduction. The second term represents a source (sink) of charge associated with electrons entering (leaving) the solid electrode phase via charge-transfer reactions. The effective solid-phase conductivity may be approximated as $\sigma_{\rm ed}^{\rm eff} = \sigma_{\rm ed} \phi_{\rm ed}^{p}$, where $\phi_{\rm ed}$ is the solid-phase volume fraction. However, alternatives such as percolation theory may provide the basis for a more accurate approximation [27].

16.3.4. *Particle Intercalation*

For the purposes of modeling lithium intercalation within electrode particles, lithium transport between particles may neglected [30]. That is, intercalated lithium diffuses within a spherical particle, but not from particle to particle. Additionally the electrode particles are presumed to be spherical. With these assumptions, the intercalated lithium transport within an electrode particle can be represented as a one-dimensional spherical diffusion problem,

$$\frac{\partial C_{\rm Li}}{\partial t} = \frac{1}{r^2} \frac{\partial}{\partial r} \left(r^2 D_{\rm Li} \frac{\partial C_{\rm Li^I}}{\partial r} \right), \tag{16.29}$$

where C_{Li} is the concentration of intercalated lithium. The diffusion coefficient of intercalated lithium D_{Li} can be a strong function of the concentration of intercalated lithium [34], which renders the diffusion problem to be nonlinear. The transient term is retained because the intercalation time scales are commensurate with the charge/discharge times. In fact, intercalation is often the rate-limiting process.

Solution of Eq. 16.29 requires an initial condition (concentration profile) and two boundary conditions. At the center of a spherical particle the radial gradient vanishes. At the particle surface the internal diffusion is balanced by the net production rate of intercalated lithium via surface reaction,

$$D_{\mathrm{Li}}\frac{\partial C_{\mathrm{Li}}}{\partial r} = \dot{s}_{\mathrm{Li}}. \tag{16.30}$$

Newman-type models solve the spherical intercalation problem for a single representative particle within each mesh cell of the full problem that considers long-range transport of charge and species (i.e., Eqs. 16.23, 16.24, and 16.28.

16.3.5. *Charge-transfer Chemistry*

The charge-transfer chemistry at the electrolyte-electrode interfaces is usually modeled using a Butler–Volmer formulation, where Faradaic current density i is represented as

$$i_{\mathrm{BV}} = i_{\mathrm{o}}\left[\exp\left(\frac{\alpha_{\mathrm{a}}F\eta}{RT}\right) - \exp\left(-\frac{\alpha_{\mathrm{c}}F\eta}{RT}\right)\right]. \tag{16.31}$$

In this expression i_{o} is the exchange current density, and α_{a} and α_{c} are the anodic and cathodic symmetry factors. The activation overpotential η is defined as

$$\eta = \Phi_{\mathrm{ed}} - \Phi_{\mathrm{el}} - E^{\mathrm{eq}}, \tag{16.32}$$

where Φ_{ed} and Φ_{el} represent electrostatic potentials of the electrode and electrolyte phases at the electrode-electrolyte interfaces, respectively, and E^{eq} is the equilibrium electric-potential difference. The two terms in the Butler–Volmer equation represent the difference in the anodic and cathode rates of a reversible charge-transfer reactions. The signs of the activation over potentials are consistent with the forward reaction being written as anodic. The equilibrium electric-potential differences, which depend upon the electrode material and the extent of intercalation, are often represented by empirical expressions [32, 34]. The exchange current density for Li-ion batteries is often modeled as

$$i_o = k_r F C_{Li^+}^{\alpha_a} (C_{Li}^{max} - C_{Li})^{\alpha_a} C_{Li}^{\alpha_c}, \tag{16.33}$$

where k_r is a rate constant, C_{Li^+} is the local Li-ion concentration in the electrolyte solution adjacent to the electrode interface, C_{Li} is the intercalated Li concentration at the outer edge of the electrode phase, and C_{Li}^{max} is the saturated (maximum) intercalated Li concentration within the electrode phase.

Molar production rates $\dot{s}_k$ (mol m^{-2} s^{-1}) appear in Eqs. 16.23, 16.26, and 16.30, not the current density i_{BV} (A m^{-2}) that comes directly from the Butler–Volmer expression (Eq. 16.31). The two are related by the Faraday constant as $\dot{s}_k = i_{BV}/F$. Assuming that the Li charge transfer chemistry is represented by reaction 16.22, $\dot{s}_{Li^+} = i_{BV}/F$, $\dot{s}_e = i_{BV}/F$, and $\dot{s}_{Li} = -i_{BV}/F$.

Although batteries are usually modeled using a global Butler–Volmer formalism, there can be benefits in developing multi-step elementary reaction mechanisms [32]. In addition to the primary charge-transfer chemistry, describing the growth of solid-electrolyte-interface (SEI) layers requires multiple heterogeneous and charge-transfer reactions [35, 36].

16.3.6. *Battery Electrode Microstructure*

Although spherical-particle models are widely used and are usually a reasonable approximation, it is now possible to reconstruct actual electrode microstructures. A number of groups are using focused-ion-beam–scanning-electron-microscopy (FIB-SEM) and X-ray tomography to develop three-dimensional electrode reconstructions [37–42]. Figure 16.10(a) shows an ion-milled face of a Li_xCoO_2 battery cathode from an FIB-SEM experiment [39]. Figure 16.10(b) shows a typical three-dimensional reconstruction from FIB-SEM data [42]. It is evident from these images that actual electrode microstructures can be quite different from spherical particles. The chemistry and transport within such actual microstructures can be modeled, providing significant new quantitative insight about battery performance [42]. Moreover, such models can assist in developing the empiricism and upscaling strategies that are needed in modeling complete cells.

In many ways, the governing equations used in modeling actual microstructures are similar to those used in spherical-particle models. However, there are also significant differences. Within the electrolyte phase,

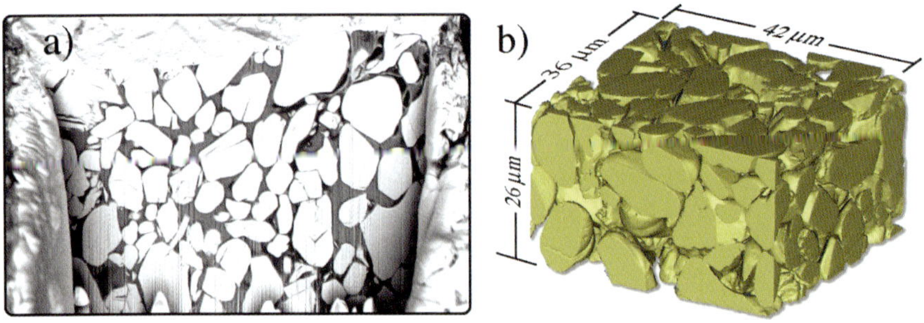

Fig. 16.10. (a) FIB-SEM trench from a commercial Li-ion battery cathode (Prof. Scott Barnett, Northwestern University. (b) Three-dimensional reconstruction of a commercial Li-ion battery cathode from approximately 200 SEM images from FIB-SEM experiments.

the governing equations for Li-ion conservation and charge conservation are represented using concentrated solution theory as [31]

$$\frac{\partial C_{\mathrm{Li}^+}}{\partial t} = \boldsymbol{\nabla}\cdot\left[\left(D_{\mathrm{el}} - \frac{2\kappa(1-t_{\mathrm{Li}^+})t_{\mathrm{Li}^+}RT}{C_{\mathrm{Li}^+}F^2}\right)\boldsymbol{\nabla}C_{\mathrm{Li}^+}\right] + \boldsymbol{\nabla}\cdot(\kappa_{\mathrm{C}}\boldsymbol{\nabla}\boldsymbol{\Phi}_{\mathrm{el}}), \tag{16.34}$$

$$\boldsymbol{\nabla}\cdot(\kappa\boldsymbol{\nabla}\boldsymbol{\Phi}_{\mathrm{el}}) + \boldsymbol{\nabla}\cdot(\kappa_{\mathrm{D}}\boldsymbol{\nabla}\ln C_{\mathrm{Li}^+}) = 0, \tag{16.35}$$

where C_{Li^+} is the Li-ion concentration, D_{el} is the experimentally measured electrolyte diffusion coefficient, $\kappa_{\mathrm{C}} = \kappa t_{\mathrm{Li}^+}/F$, Φ_{el} is the electrostatic potential of the electrolyte phase, and κ is the ionic conductivity. The diffusional conductivity, κ_{D}, is evaluated as,

$$\kappa_{\mathrm{D}} = \frac{2RT\kappa}{F}(1 - t_{\mathrm{Li}^+}), \tag{16.36}$$

where t_{Li^+} is the lithium ion transference number (defined as the fraction of current carried by lithium ions relative to the total current carried by all ionic species in an electrolyte solution with uniform composition) [28].

Some of the differences between modeling approaches are apparent when comparing Eq. 16.23 with Eq. 16.34. Because the pore and particle volumes of the actual microstructure are fully resolved, the three dimensional models do not involve parameters such as porosity. Similarly, the diffusion coefficient that appears in Eq. 16.34 is the actual Li^+ diffusion coefficient in the electrolyte solvent, and is not an effective diffusion coefficient that is modified by the porosity as it is in Eq. 16.23. In the three-dimensional microstructural modes the charge-transfer chemistry enters as a boundary (or interface) condition, and does not appear as a source term as it does on Eq. 16.23.

Within the electrode particles, Li concentration and electronic charge conservation may be represented as

$$\frac{\partial C_{\mathrm{Li}}}{\partial t} = \boldsymbol{\nabla}\cdot(D_{\mathrm{ed}}\boldsymbol{\nabla} C_{\mathrm{Li}}), \tag{16.37}$$

$$\boldsymbol{\nabla}\cdot(\sigma_{\mathrm{ed}}\boldsymbol{\nabla}\boldsymbol{\Phi}_{\mathrm{ed}}) = 0, \tag{16.38}$$

where C_{Li} is the intercalated Li concentration, D_{ed} is the diffusivity of the Li within the electrode, Φ_{ed} is the electrostatic potential within the electrode and σ_{ed} is the electronic conductivity of the electrode phase.

Equation 16.37 is the same as Eq. 16.29. However, in the three-dimensional model the diffusion Li diffusion is modeled within the actual particle geometry, not within an idealized one-dimensional spherical particle. Additionally, Li is free to be transported between particles. Using Eq. 16.38 the three-dimensional electrostatic potential is modeled within the electrode particles. In the Newman-type models that use Eq. 16.28 the electrostatic potential is not resolved within electrode particles. Rather, the electrostatic potential is modeled across the porous electrode structure, but assumed to be spatially uniform within individual particles.

16.3.7. *Illustrative Examples*

Figure 16.11 shows predicted Li contour maps within a reconstructed $\mathrm{Li}_x\mathrm{CoO}_2$ cathode at four times during a 1C discharge[‡]. The planes shown are from the mid-plane of a three-dimensional reconstructed electrode. When fully charged, the cathode has low levels of intercalated lithium (i.e., the Li is mostly in the anode). During the discharge, the Li deintercalates from the anode particles, Li^+ is transported through the electrolyte solvent, and Li intercalates into the cathode particles. The left-hand columns show clearly that small particles, or small segments of particles, intercalate faster than the interior regions of larger particles. The right-hand columns show the Li-ion concentrations within the electrolyte solution. The Li^+ concentrations certainly vary within the pore volume. However, because the Li-ion diffusion coefficients are relatively high, the Li^+ variations are relatively small.

Understanding physical and chemical behaviors at the electrode microscale contributes significant new insight. For example, consider the results shown in Fig. 16.12. During the discharge process, the electrostatic

[‡]The "C" rate is a nomenclature used to specify charge of discharge rates. A rate of 1C means that the battery charges or discharges in one hour. A rate of 2C means that the battery charges/discharges in 1/2 hour, etc.

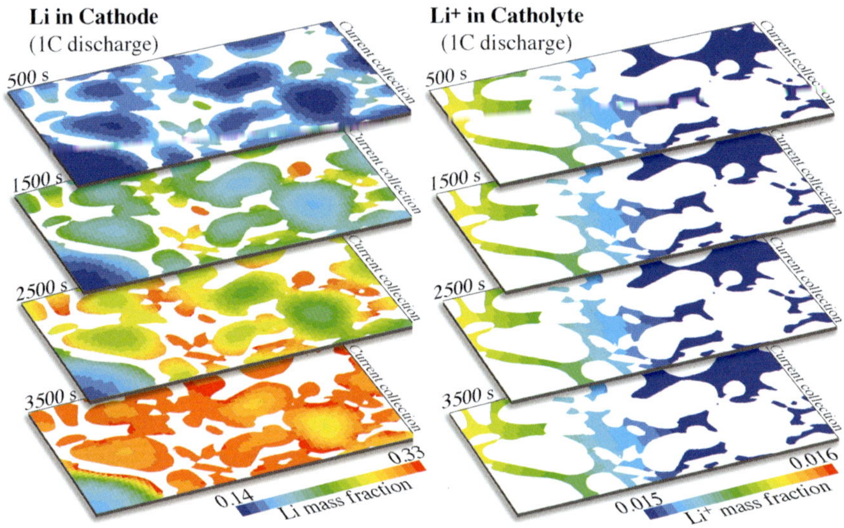

Fig. 16.11. Illustrative results from a three-dimensional reconstructed microstructural model of a Li-ion battery cathode during a 1C discharge. The color contours are from mid-plane of a three-dimensional model. The left column shows Li concentration (mass fraction) within the Li_xCoO_2 electrode particles. The right column shows Li^+ mass fractions within the electrolyte solvent.

potential within the cathode decreases. The contour maps on the left side of Fig. 16.12 show variations within the particles as well as across the width of the reconstructed electrode volume. The graph on the right shows the volume-weighted distribution of electrostatic potentials. The simulation contains on the order of one million finite volumes within the electrode particles, each of which has an electrostatic potential that varies in time during the discharge. Especially at high discharge rates (here 5C), It is interesting to note that local electrostatic potentials at the microscale can differ on the order of 100 mV from the battery terminal voltage. These microscopic variations can have significant influence on the design of charge and discharge control strategies for a complete cell.

16.3.8. *Upscaling*

Given available, or even anticipated, computational resources it is clear that complete battery models cannot resolve the complex microstructural phenomena. Thus, an important aspect of the detailed microstructural models is to derive empiricism needed in larger-scale models. For example,

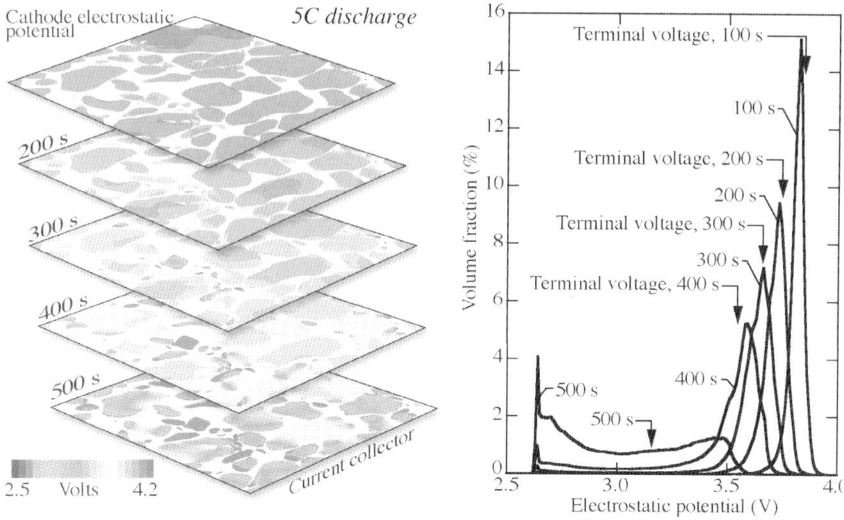

Fig. 16.12. Local electrostatic potential variations on a mid plane within the microstructure of a reconstructed Li_xCoO_2 cathode. The graph shows the terminal voltage as well the volume-weighted electrostatic-potential distribution throughout the microstructure.

the Newman-type battery models depend upon *effective* properties of the porous electrodes. Parameters such as porosity (i.e., relative pore volumes) are easily extracted from the reconstructed electrode microstructures where the geometrical details are known. Effective physical properties, such as average electrode density or heat capacity, are also easily evaluated using a volume-weighted average of the contributing phases (e.g., solid electrode phase and electrolyte solvent within the pore volume).

Effective transport properties, such as electrical or thermal conductivity, are relatively straightforward to evaluate, but require some computation. Figure 16.13 shows the results of a simulation in which fixed temperatures are imposed on two opposing faces of a three-dimensional microstructural cube. The remaining four faces are assumed to be symmetry boundaries. Although the geometry is complex, solving the linear heat equation $\nabla \cdot (\lambda \nabla T) = 0$ is relatively straightforward, assuming of course that software such as FLUENT[§] is available to handle the geometric complexity. In this computation, each finite-volume cell is assigned the intrinsic conductivity of the material from which it is composed (i.e., solid electrode or electrolyte solvent). The net heat flux q'' through the composite material

[§]ANSYS, Inc., Lebanon, NH

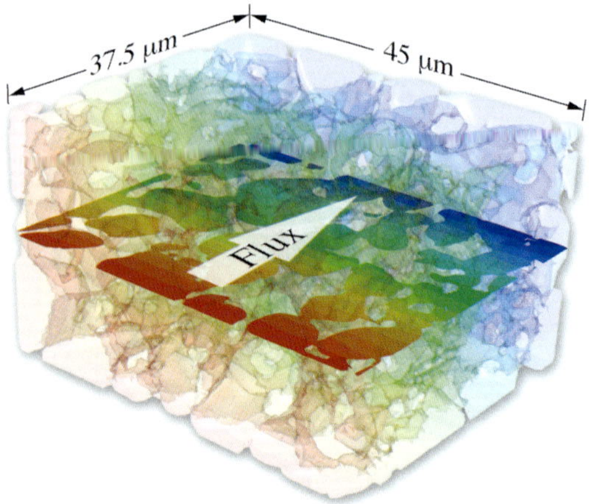

Fig. 16.13. Simulated diffusion through a three-dimensional electrode microstructure.

is easily derived from the three-dimensional solution. Then, the effective conductivity is evaluated as

$$\overline{\lambda} = q'' \frac{L}{\Delta T}, \tag{16.39}$$

where ΔT is the imposed temperature difference on the opposing cube faces and L is the distance between the opposing faces. For any reconstructed cube (e.g., Fig. 16.13), the procedure can be followed three times, in the x, y, and z directions. Because the battery electrodes are reasonably isotropic, the three effective conductivities are usually close to each other. A simple average of the three can be used as the effective conductivity. A similar approach can be used to evaluate the effective electrical conductivities or diffusion coefficients through individual phase of the microstructure.

In addition to the electrochemical function, battery safety and possible thermal runaway are important considerations in Li-ion batteries. Figure 16.14 illustrates volumetric heat-release and temperature contours within a cathode microstructure at a very high (120°C) discharge rate, such as might occur under accidental short-circuit conditions. Two physical and chemical behaviors contribute to the initial stages of heating. One is Faradaic, which is the result of inefficiencies (polarization) associated with irreversibilities in the charge-transfer chemistry at the electrode-electrolyte interfaces. The other is Ohmic, which is associated with conducting electric current through the electrode phase (i.e., "i^2R" heating). The high heating rates around the edges of the particles is primarily Faradaic and the high

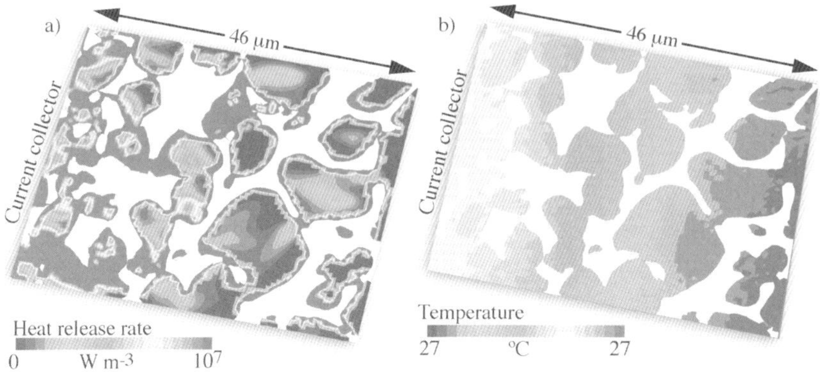

Fig. 16.14. Two-dimensional contour maps at the mid plane of a three-dimensional cathode microstructure at 1 s into an internal short, causing a very high discharge rate of 120°C. (a) local volumetric heat generation rate $\dot{q}$. (b) local temperature field.

heating rates in the particle interior near the current collector is primarily Ohmic. Because the heat-release rates depend upon temperature, state-of-charge, and discharge rate, evaluating effective heat-release rates in general terms is difficult. A potentially viable approach is based upon precomputed tables that can be used to retrieve effective local heating rates in complete battery models. However, at present, this approach is a work-in-progress.

Figure 16.14(b) shows local temperature distributions in the microstructure. Even in this very aggressive short-circuit condition, the temperature differences within the microstructural cube are less than 1°C. The small electrode domain and the relatively high thermal conductivities serve to spread the temperature rapidly. Of course, within a full battery the temperatures can vary greatly, especially in abnormal environments. Thus, an important aspect of the microstructural models is to predict the effective heat-release rates that can be used in complete cell models to predict the macroscopic temperature variations and the overall cell performance.

16.4. Catalysis

Catalytic reactors are used very widely in industrial chemical processing [43]. Every automobile has a catalytic converter to assist environmental pollution control. Although specific implementations can vary greatly, essentially every catalytic reactor introduces the catalyst in a porous media. An important function of the porous structure is to expose the reacting fluid to a high catalyst surface area. However, there are usually design and

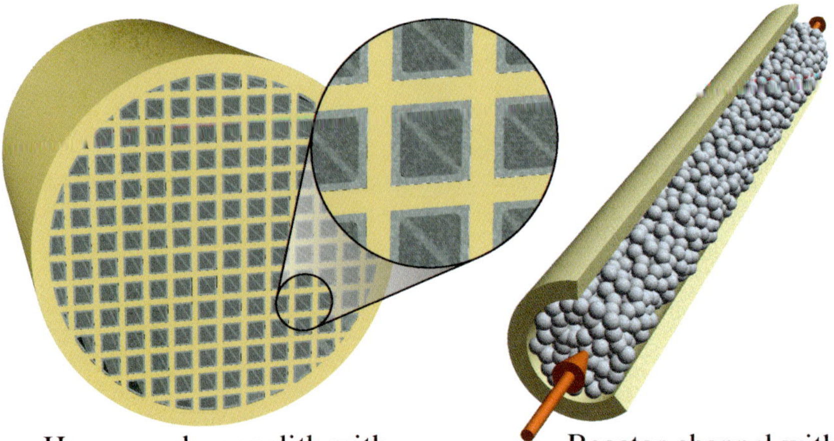

Fig. 16.15. Two common catalytic reactor configurations. The washcoated monolith is typical of automotive catalytic converters. The packed-bed configuration is common for a broad range of industrial processes.

operation tradeoffs between high surface area and gas-transport restrictions within the porous media.

In the monolith configuration the individual channels are typically on the order of a millimeter in effective diameter. The washcoats on the channel walls are typically a few tens of microns thick. In the packed-bed configuration, gases flow in the pore volume between catalyst pellets. The relatively large catalyst pellets can also have microscopic porosity within the pellets. Thus, there is relevant porous-media behavior on a cascade of multiple length scales.

The fundamental principles of representing the porous media chemistry and transport is essentially the same as described earlier in the context of SOFC anodes. However, because the macroscopic reactor configurations can vary greatly, the interactions between the porous-media and the macroscopic flow and heat transfer can vary significantly.

16.5. Conclusions and Outlook

Porous media and reacting flow play central roles in a very wide range of practical technologies. New capabilities to represent and understand the physical, chemical, and electrochemical processes at the microscale are playing increasingly valuable roles. In addition to improving fundamental understanding, new knowledge can be upscaled to assist the optimal design and control of complete devices and components at the macroscale.

Experimental techniques such as FIB-SEM are rapidly improving, enabling high-resolution measurement of microstructural geometric details. Techniques such as X-ray computed tomography (CT) using synchrotron X-rays are enabling real-time, in-situ, measurements of battery functions. The rapidly developing capabilities to model physics and chemistry in complex microstructural detail amplifies the value of new experimental imaging techniques. The modeling will also benefit from independent measurements of fundamental physical and chemical properties that are needed as inputs to models. Moreover, macroscale measurements in complete systems or components will be valuable in validating models and upscaling strategies.

The discussion and examples in this chapter are focused on only a few technologies and applications. However, the general approaches are far more general, with potential applications ranging from small power sources to geological-scale challenges.

16.6. Acknowledgement

We gratefully acknowledge support from the Office of Naval Research, enabling fundamental and applied research in a broad range of electrochemical power technologies.

References

[1] R. Braun, T. Vincent, H. Zhu, and R. Kee, Analysis, optimization, and control of solid-oxide fuel cell systems, *Adv. Chem. Eng.* **41**, 383–446 (2012).

[2] R. Kee, H. Zhu, R. Braun, and T. Vincent, Modeling the steady-state and dynamic characteristics of solid-oxide fuel cells, *Adv. Chem. Eng.* **41**, 331–381 (2012).

[3] A. Gubner, T. Nguyen-Xuan, M. Bram, J. Remmel, and L. de Harrt, Lightweight cassette type SOFC stacks for automotive applications, *European Fuel Cell Forum, Luzern, Switzerland, Paper B042* (2006).

[4] H. Zhu and R. Kee, Modeling distributed charge-transfer processes in SOFC membrane electrode assemblies, *J. Electrochem. Soc.* **155**, B175–B729 (2008).

[5] E. Mason and A. Malinauskas, *Gas Transport in Porous Media: The Dusty-Gas Model.* American Elsevier, New York (1983).

[6] H. Zhu, R. Kee, V. Janardhanan, O. Deutschmann, and D. Goodwin, Model-

ing elementary heterogeneous chemistry and electrochemistry in solid-oxide fuel cells, *J. Electrochem. Soc.* **152**, A2427–A2440 (2005).

[7] R. Kee, M. Coltrin, and P. Glarborg, *Chemically Reacting Flow: Theory and Practice*. John Wiley, Hoboken, NJ (2003).

[8] Y. Jiang and A. Virkar, Fuel composition and diluent effect on gas transport and performance of anode-supported SOFCs, *J. Electrochem. Soc.* **150**, A942–A951 (2003).

[9] R. Kee, H. Zhu, and D. Goodwin, Solid-oxide fuel cells with hydrocarbon fuels, *Proc. Combust. Inst.* **30**, 2379–2404 (2005).

[10] D. Goodwin, H. Zhu, A. Colclasure, and R. Kee, Electrochemical charge-transfer processes for hydrogen-steam mixtures on Ni-YSZ pattern anodes, *J. Electrochem. Soc.* **156**, B1004–B1021 (2009).

[11] W. Bessler, A new computational approach for SOFC impedance from detailed electrochemical reaction-diffusion models, *Solid State Ionics.* **176**, 997–1011 (2005).

[12] V. Janardhanan and O. Deutschmann, CFD analysis of a solid oxide fuel cell with internal reforming: Coupled interactions of transport, heterogeneous catalysis and electrochemical processes, *J. Power Sources.* **162**, 1192–1202 (2006).

[13] W. Bessler, J. Warnatz, and D. Goodwin, The influence of equilibrium potential on the hydrogen oxidation kinetics of SOFC anodes, *Solid State Ionics.* **177**, 3371–3383 (2007).

[14] M. Vogler, A. Bieberle-Hütter, L. Gauckler, J. Warnatz, and W. Bessler, Modelling study of surface reactions, diffusion, and spillover at a Ni/YSZ patterned anode, *J. Electrochem. Soc.* **156**, B663–B672 (2009).

[15] V. Yurkiv, D. Starukhin, H.-R. Volpp, and W. G. Bessler, Elementary reaction kinetics of the CO/CO_2/Ni/YSZ electrode, *J. Electrochem. Soc.* **158**, B15–B10 (2011).

[16] E. Hecht, G. Gupta, H. Zhu, A. Dean, R. Kee, L. Maier, and O. Deutschmann, Methane reforming kinetics within a Ni-YSZ SOFC anode, *Appl. Catal. A.* **295**, 40–51 (2005).

[17] H. Zhu, A. Kromp, A. Leonide, E. Ivers-Tiffée, O. Deutschmann, and R. Kee, Model-based interpretation of the influence of anode surface chemistry on solid oxide fuel cell electrochemical impedance spectra, *J. Electrochem. Soc.* **159**, F255–F266 (2012).

[18] Y. Lin, Z. Zhan, J. Liu, and S. Barnett, Direct operation of solid oxide fuel cells with methane fuel, *Solid State Ionics.* **176**, 1827–1835 (2005).

[19] M. Pillai, Y. Lin, H. Zhu, R. Kee, and S. Barnett, Stability and coking of direct-methane solid oxide fuel cells: Effect of CO2 and air additions, *J. Power Sources.* **195**, 271–279 (2010).

[20] M. Coltrin, R. Kee, and F. Rupley, Surface Chemkin: A generalized formalism and interface for analyzing heterogeneous chemical kinetics at a gas-surface interface, *Int. J. Chem. Kinetics.* **23**, 1111–1128 (1991).

[21] G. Gupta, E. Hecht, H. Zhu, A. Dean, and R. Kee, Gas-phase reactions of methane and natural-gas with air and steam in non-catalytic regions of a solid-oxide fuel cell, *J. Power Sources.* **156**, 434–447 (2006).

[22] A. Richards, M. McNeeley, R. Kee, and N. Sullivan, Gas transport and internal-reforming chemistry in Ni-YSZ and ferritic-steel supports for solid-oxide fuel cells, *J. Power Sources.* **196**, 10010–10018 (2011).

[23] D. Bouvard and F. Lange, Relation between percolation and particle co-ordination in binary powder mixtures, *Acta Metal. Mater.* **39**, 3083–3090 (1991).

[24] P. Costamagna, P. Costa, and V. Antonucci, Micro-modelling of solid oxide fuel cell electrodes, *Electrochimica Acta.* **43**, 375–394 (1998).

[25] M. Suzuki and T. Oshima, Comparison between the computer-simulated results and the model for estimating the co-ordination number in a three-component random mixture of spheres, *Powder Technol.* **43**, 19–25 (1985).

[26] D. Jeon, J. Nam, and C. Kim, Microstructural optimization of anode-supported solid oxide fuel cells by a comprehensive microscale model, *J. Electrochem. Soc.* **153(2)**, A406–A417 (2006).

[27] D. Chen, Z. Lin, H. Zhu, and R. Kee, Percolation theory to predict effective properties of solid oxide fuel-cell composite electrodes, *J. Power Sources.* **191**, 240–252 (2009).

[28] M. Doyle, T. Fuller, and J. Newman, The importance of the lithium ion transference number in lithium/polymer cells, *Electrochimica Acta.* **39**, 2073–2081 (1994).

[29] M. Doyle, T. Fuller, and J. Newman, The use of mathematical modeling in the design of lithium/polymer battery systems, *Electrochem. Acta.* **40**, 2191–2196 (1995).

[30] M. Doyle and J. Newman, Modeling of galvanostatic charge and discharge of the lithium/polymer/insertion cell, *J. Electrochem. Soc.* **140**, 1526–1533 (1993).

[31] J. Newman and K. Thomas-Alyea, *Electrochemical Systems.* John Wiley and Sons (2004).

[32] A. Colclasure and R. Kee, Thermodynamically consistent modeling of elementary electrochemistry in lithium-ion batteries, *Electrochimica Acta.* **55**, 8960–8973 (2010).

[33] A. Bard and L. Faulkner, *Electrochemical Methods Fundamentals and Applications.* Wiley (2001).

[34] D. Karthikeyan, G. Sikha, and R. White, Thermodynamic model development for lithium intercalation electrodes, *J. Power Sources.* **185**, 1398–1407 (2008).

[35] D. Aurbach, Review of selected electrode-solution interactions which determine the performance of Li and Li-ion batteries, *J. Power Sources.* **89**, 206–218 (2000).

[36] A. Colclasure, K. Smith, and R. Kee, Modeling detailed chemistry and transport for solid-electrolyte-interface (SEI) films in Li-ion batteries, *Electrochimica Acta.* **58**, 33–43 (2011).

[37] P. Shearing, L. Howard, P. Jørgensen, N. Brandon, and S. Harris, Characterization of the 3-dimensional microstructure of a graphite negative electrode from a Li-ion battery, *Electrochem. Comm.* **12**, 374–377 (2010).

[38] D. Kehrwald, P. Shearing, N. Brandon, P. Sinha, and S. Harris, Local tor-

tuosity inhomogeneities in a lithium battery composite electrode, *J. Electrochem. Soc.* **158**, A1393–A1399 (2011).

[39] J. Wilson, J. Cronin, S. Barnett, and S. Harris, Measurements of 3-dimensional microstructure in a $LiCoO_2$ positive electrode, *J. Power Sources.* **196**, 3443–3447 (2011).

[40] M. Ender, J. Joos, T. Carraro, and E. Ivers-Tiffée, Three-dimensional reconstruction of a composite cathode for lithium-ion cells, *Electrochem. Comm.* **13**, 166–168 (2011).

[41] T. Hutzenlaub, S. Thiele, R. Zengerle, and C. Ziegler, Three-dimensional reconstruction of a $LiCoO_2$ Li-ion battery cathode, *Electrochem. Solid-State Let.* **15**, A33–A36 (2012).

[42] G. Goldin, A. Colclasure, A. Wiedemann, and R. Kee, Three-dimensional particle-resolved models of Li-ion batteries to assist the evaluation of empirical parameters in one-dimensional models, *Electrochimica Acta.* **64**, 118–129 (2012).

[43] O. Deutschmann, *Modeling and Simulation of Heterogeneous Catalytic Reactions*. Wiley-VCH, Weinhiem (2012).

Chapter 17

Structural Characterization of Porous Materials

Brian P. Gorman

Department of Metallurgical and Materials Engineering, Colorado School of Mines, Golden, CO, 80401, USA

bgorman@mines.edu

Understanding the microstructure – property relationships in porous materials presents unique experimental challenges. Both engineered and natural materials can have a wide range of pore size distributions and interconnectivities that define their mechanical, electrochemical, and fluid transport properties. Disparate changes in these pore sizes leads to the requirement of many different characterization techniques needed in order to fully understand their structures. Both real and reciprocal space techniques can be useful in understanding porous material structures over 108 orders of magnitude in size. Scattering techniques using UV, Visible, IR, X-rays, or neutrons can give a large sampling pore sizes and interconnectivities. 3-dimensional real space characterization using serial sectioning or tilt series imaging can enable quantitative understanding of pore interconnectivity and tortuosity, and when combined with finite element modeling can lead directly to understanding the material properties. Combining imaging with analytical techniques will enable a full understanding of the 3-D pore structure along with the chemistry of pore / solid phase boundaries.

17.1. Introduction

As noted elsewhere in this volume, a wide variety of porous materials are being studied throughout both the natural and engineering sciences. One common theme runs through all of these topic areas, specifically the need to understand the pore size, geometry, and interconnectivity. These structural attributes frequently dominate the thermal, electrical, mechanical, and chemical properties.

In general, the main parameters needed in order to understand the properties of a porous material are the density, pore size distribution, solid phase thickness between pores, interconnectivity of the solid phase, and the tortuosity of the pore network [1]. Pore structure characterization can be completed using a wide variety of techniques, the choice of each being dependent upon the scale of the pores. Recently, many traditionally 2-D characterization techniques such as microscopy have evolved to give 3-D information quantitatively [2]. Such information is highly valuable in porous materials as these datasets can be directly input into modeling such as finite element analyses to give insight into their properties and ultimately provide feedback to improved processing in the case of engineered materials [3].

The field of pore structure characterization is currently very rich, with over 11,000 articles published with "porous" and "tomography" in the title from 2013 to 2014 alone according to Google Scholar. Searching additional keywords would certainly open this database further, illustrating the importance of characterizing these structures in 3-D.

Choosing the correct characterization technique depends heavily on the anticipated scale of the pore structure as well as the intended use of the characterization data [4]. If material density and pore sizes are only required, non-destructive bulk averaged techniques such as scattering are most appropriate, with the choice of radiation depending on the scale of the pores. If more information about the material property is desired, 3-D real space characterization techniques such as serial sectioning, tilt series, and analytical microscopy can be applied. Choosing the correct microscopy technique depends on the scale of the pore structure and what analytical information about the pore / solid interface is needed.

This chapter will discuss scattering techniques such as UV-Vis-NIR, X-ray, or neutron scattering for determining the bulk averaged density and pore size on the nanometer scale. Scattering techniques also have the advantage of using mathematical models to produce real scale depictions of nanoscale pore geometries. Serial sectioning techniques such as metallographic polishing coupled with optical microscopy or focused ion beam sectioning coupled with scanning electron microscopy are also being utilized for mm to m scale 3-D real space characterization. Tilt series tomography with X-rays and electrons are also being utilized for similar length scales, albeit somewhat non-destructively. Analytical techniques such as transmission electron microscopy and atom probe tomography are also being utilized for quantifying pore structure at the μm to sub-nm scales, with the added advantage of chemical information at the pore / solid interface. Gas adsorption techniques such as BET [5] will not be covered in this chapter.

17.2. Scattering with Photons and Neutrons

Scattering radiation from a material results from the interaction of the radiation with interfaces in a material and can typically be modeled using different implementations of the Rayleigh criterion [4]. Since porous materials inherently have a large number of solid / gas interfaces, scattering techniques are particularly insightful into the bulk averaged structure of these materials. The scale at which scattering techniques are useful is inherently defined by the wavelength of the incoming radiation as well as their interaction with either the electrons or nuclei in the solid phase. In this section, examples of scattering using UV-Vis-Near Infrared radiation as well as neutrons are given to demonstrate their applicability to understanding the average pore structure in nanoscale engineered ceramics.

17.2.1. *UV-Vis-Near Infrared Spectroscopy*

Spectroscopy using optical wavelengths is one of the most straightforward measurements to make, but the information gathered from careful analyses gives important information about the material that is otherwise very difficult to obtain [6]. This statement is especially true when analyzing nanoscale porous thin films where nanostructural information about the pores and overall density are nearly impossible to achieve using gas adsorption or gravimetric techniques.

Specifically, transmittance spectra contain information related to scattering and absorption of the light, allowing differentiation between the grain size and porosity effects [7]. Calculation of the refractive index from transmission and reflection spectra can give information about thin film density, and scattering can give information about grain and pore sizes. However, these values are less straightforward to obtain, primarily because proper normalization is needed to obtain quantitative information about reflected or transmitted intensities. If transmission and reflection spectra are normalized correctly, mixing rules provide the connection between the refractive index and the porosity of the films, which is difficult to measure by other techniques. In the example below, UV-Vis-NIR is used to quantify the porosity fraction and particle size in a nanoporous CeO_2 thin film [8].

Colloidal suspensions provide the advantage of being able to form films that have a wide range of porosity and grain size by simply varying the sintering conditions [9]. In this investigation, colloidal suspensions of CeO_2 were used as the precursor for preparation of the film [10]. Powders were prepared using precipitation techniques and had an initial grain size of ~5 nm. Stabilization of an aqueous colloidal suspension was completed using electrostatics by adjusting the pH of the solution to 4 using dilute nitric acid. Films were spin coated on double side polished basal plane oriented sapphire at 500 rpm for 30 sec and subsequently dried at 150°C in air. Final thicknesses of the films were approximately 2 μm. Subsequent anneals at different temperatures were used to investigate the densification processes of these films as well as understanding the applicability of UV-Vis-NIR to studying these processes.

Following air sintering at a variety of temperatures, films were optically characterized using a Cary- 5UV-vis near-infrared (NIR) spectrophotometer. Although the main goal of the investigation was the interpretation of reflection and transmission spectra, other methods were also used to check the accuracy of the spectrophotometry results, including ellipsometry, mechanical profilometry, and scanning electron microscopy (SEM).

Spectral measurements of the transmission and reflection were in the near-IR, visible, and UV region of the spectra 3000200 nm. However, the spectral region of interest was limited to 300700 nm because the refractive index does not change in the IR region and strong absorption took place in the UV region. For calculation of the refractive index, it is necessary to normalize the measured spectra in order to have quantitative values of the transmittance and reflectance. This procedure is well documented for the transmittance and is accomplished by dividing the measured intensity by the measured baseline. Reflectance from the uncoated sapphire substrate was used to calculate the reflection baseline.

Normalized reflectance and transmittance spectra are presented in Fig. 17.1. Three different effects are observed in the spectra: interference by the ceria coating, band gap absorption, and scattering from ceria grains. There are no exact matches between maxima (or minima) of the reflectance and transmittance spectra with the values calculated using standard Fresnel equations [7]. This is connected with the extinction of light in the coating and this extinction should be taken into account in the theoretical model.

Such a model was used to separate these effects and to quantitatively calculate the refractive index, scattering, and absorption in the ceria layer. The complex refractive index of the layer,

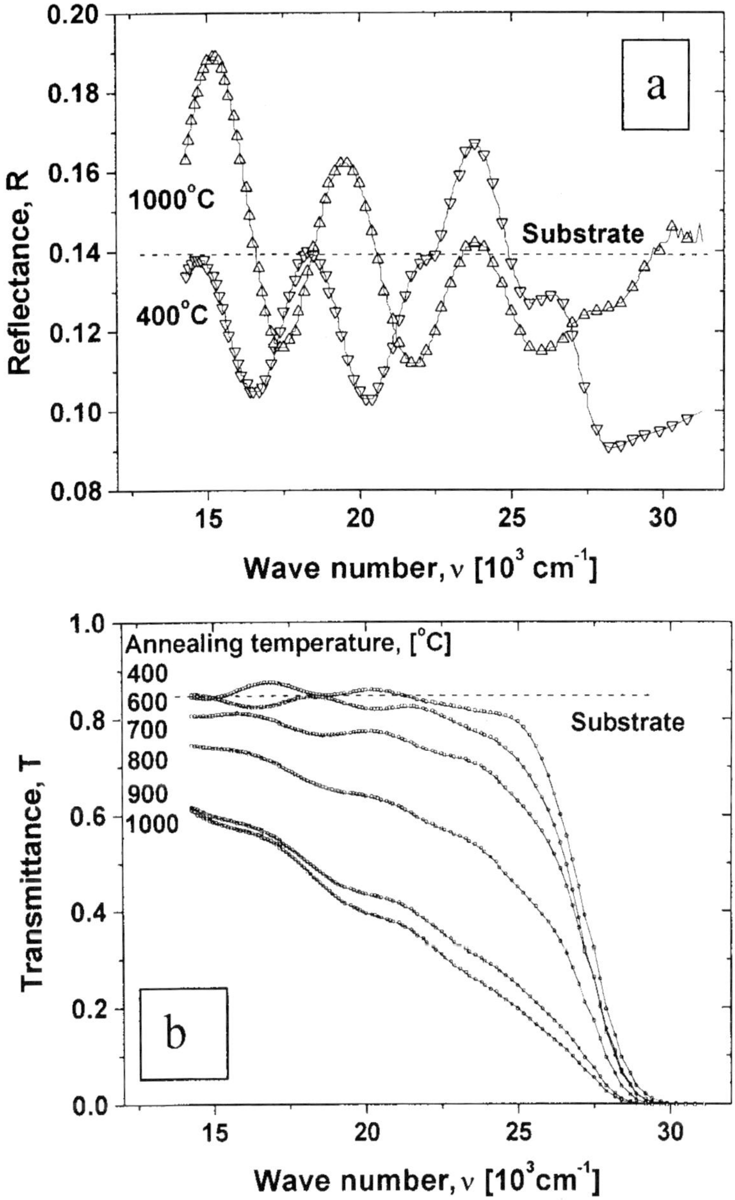

Fig. 17.1. Reflectance and transmission spectra for colloidal CeO_2 films annealed at different temperatures.

$$\hat{n} = n_0 - ik \tag{17.1}$$

contains the real part n_0, which is connected with the refractive index of the dense ceria by the Lorenz equation:

$$\frac{n^2 - 1}{n^2 + 2} = \frac{n_0^2 - 1}{n_0^2 + 2} \star P \tag{17.2}$$

where P is the volume fraction of the oxide in the coating. The extinction coefficient k is the sum of the scattering coefficient k_{scat} and absorption coefficient k_{abs}.

Standard Fresnel equations [9] were used for theoretical spectra calculations for reflectance and transmittance of the substrate. The results of the calculations are shown as lines in Fig. 17.1.

Figure 17.2 illustrates the results of the porosity calculations using Eq. (17.2). The initial volume fraction of ceria in the coating is 50% after heat treatment at 400°C and increases with the sintering temperature up to 85% at 1000°. However, over this range of porosity, the product $(d*P)$ does not change, and thus there is no change in the CeO_2 mass per thickness.

Calculated extinction coefficients versus wavelength are presented in Fig. 17.3. This extinction is connected with the scattering in the wavelength region of 400–700 nm. Strong band gap absorption takes place only at the shorter wavelengths and therefore do not contribute to the calculated extinction coefficient. Optical scattering is therefore the strongest contributor to the extinction coefficient at these wavelengths, from which the CeO_2 particle size can be quantified. As seen in Fig. 17.2, the scattering intensity is dependent upon the sintering temperature (i.e., of the particle size and pore distribution) and increases with increasing wave number. Thus it is possible to investigate geometric parameters of the particles from the scattering and then subtract the scattering from the extinction index to yield the actual absorption by the ceria particles.

It is known from theory that the amplitude and wavelength dependence of the scattering depends strongly on the shape and size of the particles [11]. Rayleigh theory gives λ^{-4} dependence for the scattering intensity on small particles if the size of the particle (a) is much smaller than the wavelength ($a<\lambda$). This means that the scattering coefficient k_{scat} has a $(a/\lambda)^3$ dependence on the wavelength. For one-dimensional defects (such as an infinite cylinder), the scattering coefficient should be proportional to

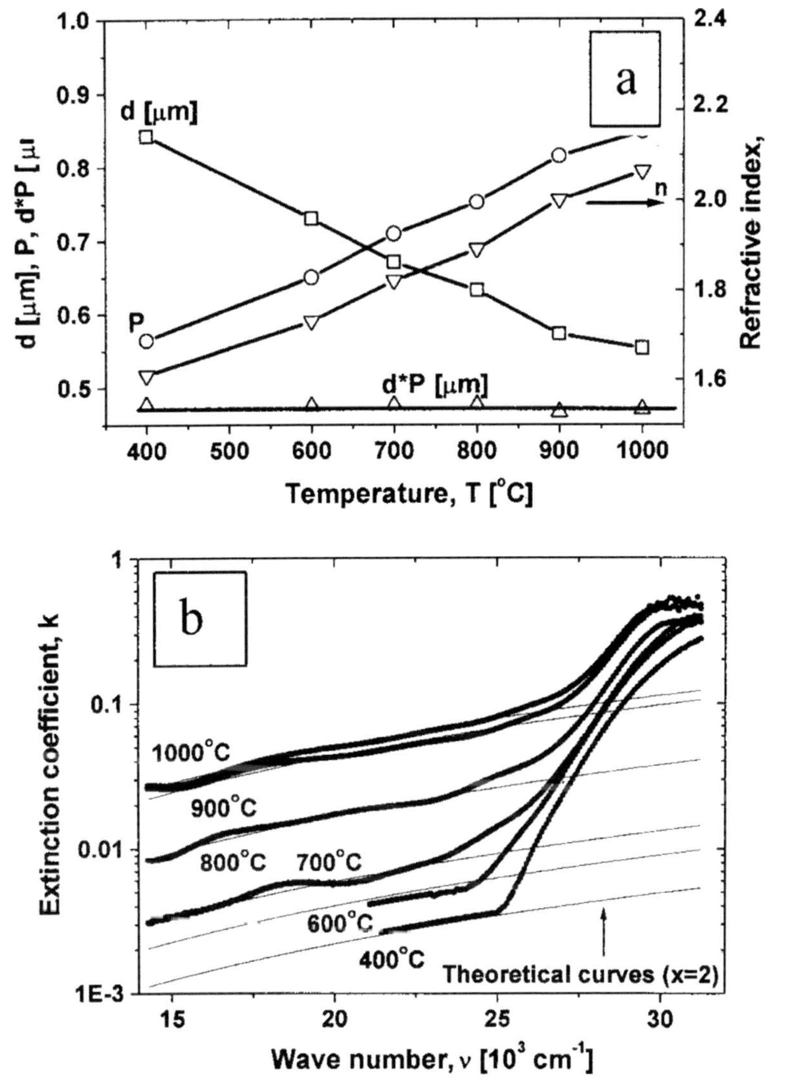

Fig. 17.2. Film porosity calculated from transmission and reflection measurements using Fresnel theory and Lorenz mixing rules.

$(a/\lambda)^2$ if the defect diameter is less than λ. Finally, it will be proportional to (a/λ) for two-dimensional defects with thickness $a<\lambda$.

The physical nature of defects depends on the sample structure. In the present case, the porosity of the coating and the pores are most likely the scattering defects in the dense ceria matrix, because the porosity is in the range of 1550%. Thus, the wavelength dependence of the scattering should be connected with the size and the distribution of the pores [11]:

$$k_{scat} = A\,(n_0)\left(\frac{\alpha}{\lambda}\right)^x (1 - P) \qquad (17.3)$$

where x will be in the range of 13, depending on the pore distribution, and A is a numerical constant which depends on the refractive index of the dense ceria n_0. The theoretical dependence of the extinction coefficient k as function of the wave number is presented in Fig. 17.3 as straight lines. As can be seen, the slope is equal to 2, demonstrating that the pores have a nonuniform distribution and form linear structures. The defect size (a) calculated from Eq. (17.3) correlates to the grain size found from XRD peak broadening and from SEM images.

Ultimately, a nanoporous film density, thickness, particle size, and pore morphology can be quantified as long as the film thickness and particle (pore) size is smaller than the wavelength used in the scattering measurement. This technique enables an easy, non-destructive method for quantifying a porous material's structure.

17.2.2. *Small Angle Neutron Scattering*

Instead of measuring the wavelength dependent response of scattering as in UV-Vis-NIR, neutron and X-ray scattering instead measure the angular dependence using a variety of optics [11]. However, the information achieved is similar quantitative information on bulk averaged material density, particle size, and pore geometry.

In short, these techniques utilize the wave nature of X-rays and neutrons to scatter from internal interfaces. Each scattering event has a characteristic angle relatable to the real space scale of the structure. As these are reciprocal space techniques, smaller scattering angles relate to larger real space structure, and vice versa. X-ray scattering intensity can be linearly related to the atomic density of the solid phase material. X-ray sources range from laboratory scale instruments adapted from X-ray diffraction instruments to highly monochromatic synchrotron light sources.

Neutron scattering intensity, conversely, is dependent on the neutron cross-section of each atom, which is isotope dependent and has no linear trends within the period table. Quantitative SANS analysis thus requires a-priori knowledge of the material composition and isotopic ratios. Neutron sources normally require either reactors or linear accelerators, neither of which are available on a laboratory scale. Using mathematical modeling techniques such as Gaussian random fields (GRFs) [12, 13], it is possible to convert the reciprocal space data into real space 3-D structures. The following example will illustrate the use of small-angle neutron scattering (SANS) to study the microstructure of porous silica aerogels and the applicability of Gaussian random field models for understanding the real space microstructure from SANS reciprocal space data. It should be noted that small angle X-ray scattering (SAXS) is a well-known technique for accomplishing similar studies. Both techniques have been applied to examine a range of length scales and to study changes in network structure. The following example illustrates the applicability of SANS and GRF analysis to understanding the nanostructure of silica based aerogels [14].

Experimentally, silica based aerogels were synthesized from a tetraethoxysilane (TEOS) precursor in ethanol (4:1 ethanol/ TEOS ratio) using a base catalyst (0.1N NH_4OH; pH = 9.6) in a 4:1 catalyst/ TEOS ratio, forming a gel in 1 cm diameter vials at 23°C. The sample was aged in the solvent for 3 days and supercritically dried in CO_2. SANS experiments were conducted on the 30 m SANS instrument on neutron guide NG-7 at the NIST Center for Neutron Research. Detector distances of 1.2, 4, and 15.3 m were used to effectively examine a large range of length scales, and 5 Å neutrons were employed to resolve extremely small length scales. The Q-space intensity curve for these aerogel samples is shown in Fig. 17.3. Gaussian Random Field models were applied to this data in order to extract average chord lengths for the solid and pore phases, the details of which are given in Ref. [14]. Quantifying the GRF model for this material gives nominal pore sizes (chord lengths) of 120 Å.

Figure 17.4 is a transmission electron microscopy (TEM) image of the base-catalyzed TEOS aerogel. A loosely connected microstructure and spheroidal shape of the primary particles are evident. Correlating TEM images to the neutron scattering fitting data demonstrated excellent correlations between the two techniques. This is quite evident when comparing Fig. 17.4 with the realization of the GRF model as presented in Fig. 17.5.

In general, SANS is demonstrated as an effective technique for quantification of the bulk averaged pore scale structure, with features 12 nm easily resolved. Using mathematical treatments such as Gaussian Random Fields for fitting the data results in a 3-D understanding of the nanostructure that is comparable to real space imaging techniques such as TEM. Similar results should be attainable using SAXS, given the proper instrumentation.

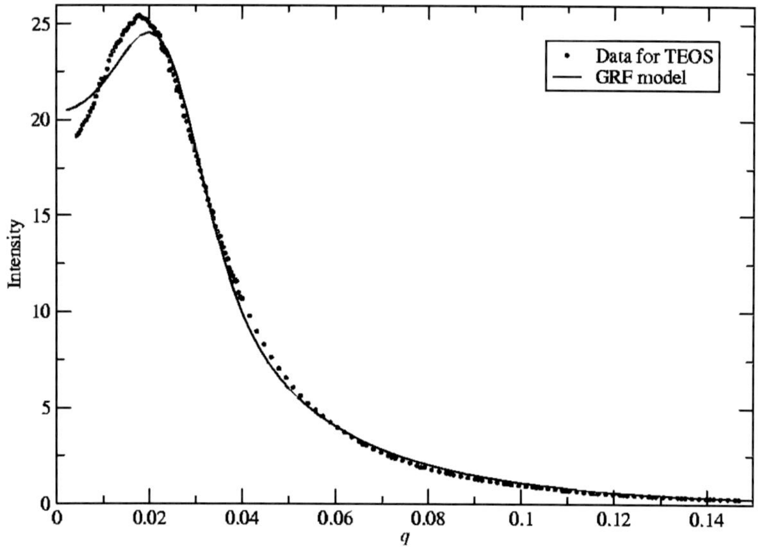

Fig. 17.3. Intensity curve for the intersection model and the experimental scattering data for TEOS.

17.3. Serial Sectioning

One of the most widely used techniques for studying the 3-D microstructure of materials involves serial sectioning. In this technique, a method for removing or sectioning the surface of a material is combined with real space microscopic imaging. Repeating this process over several hundred to several thousand iterations results in a real space 3-D dataset to be reconstructed. Sectioning is most often completed using either metallographic polishing techniques [3] or focused ion beam (FIB) sputtering [15]. Microscopic imaging can be completed using optical or scanning electron microscopy (SEM). Different implementations of these techniques such as polarized light microscopy, backscattered electron imaging, energy dispersive X-ray spectroscopy [16], or electron backscattered diffraction (EBSD) [17] can quantify different phases of material, in addition to their microstructures. Reconstructing a 3-D volume from serial sectioning data is a simple matter of matching fiducial features from one image to the next vertical section and comparing image intensity values in the x-y plane. Several commercial software packages are available for both visualization and quantification of the 3-D microstructures.

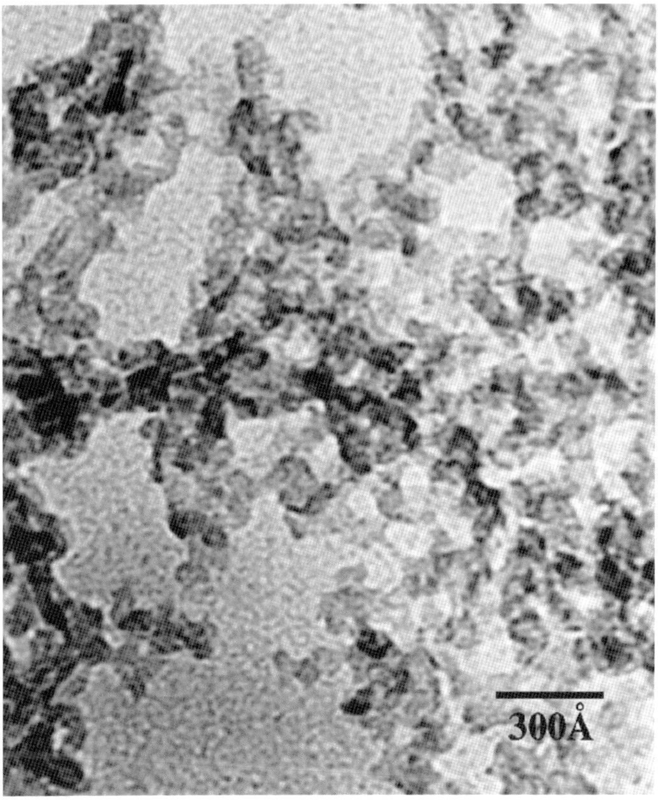

Fig. 17.4. Bright field TEM image of base-catalyzed TEOS aerogel.

17.3.1. *Serial Sectioning with Optical Microscopy*

In the most common implementation of serial sectioning for mm to μm spatial resolutions, optical microscopy is combined with metallographic polishing techniques. Polishing removes a known thickness of material, typically less than one tenth the size of the solid microstructure. After polishing, the surface may be imaged with optical microscopy using a variety of optical microscopy techniques and at a variety of magnifications. Repeating this iteration many times gives a volume that is statistically relevant to the microstructure of the porous material. At the current time, a commercial optical microscopy / metallography system is available.

In the example below from the literature [3], serial metallographic sectioning combined with optical microscopy is utilized for the quantification

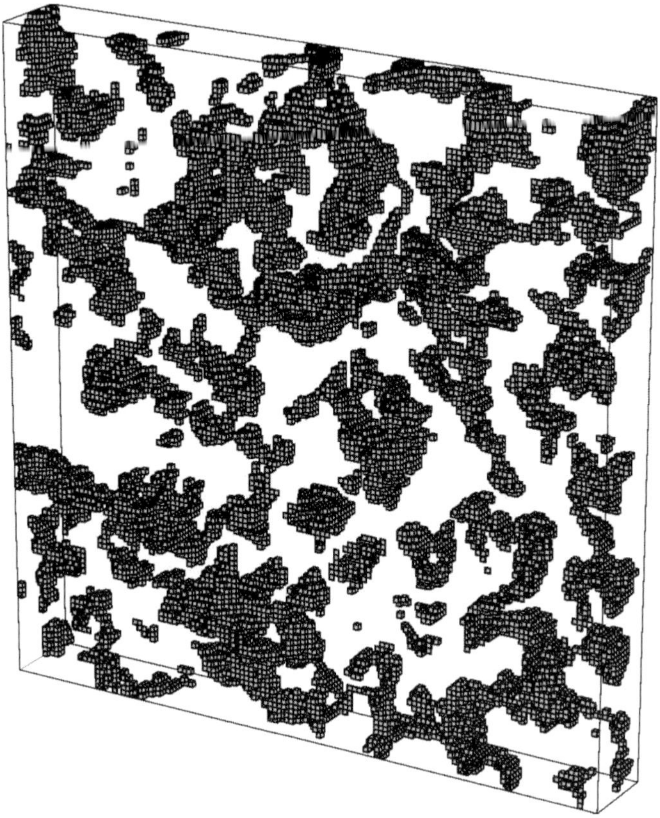

Fig. 17.5. A realization of the intersection model for TEOS. The dimensions of this realization are $1500 \times 1500 \times 187.5$ Å^3. Comparisons with Fig. 17.3 are instructive for understanding the applicability of the mathematical model to SANS data.

of 3-D microstructure in carbon foams. Accurate reconstruction followed by the formation of triangulation matrices allows for not only quantitative information to be obtained; the obtained microstructure may be modeled using finite element methods to determine the mechanical properties as a function of processing and microstructure.

Carbon foams were processed using several different techniques and subsequently vacuum infiltrated with low viscosity epoxy resin. This was completed due to the difficulty in mechanically polishing a brittle, porous solid. Epoxy is used to solidify the structure to allow for polishing while maintaining adequate contrast in the optical microscope. Serial sections were polished using a 1.0 μm diamond lapping film such that the polishing rate

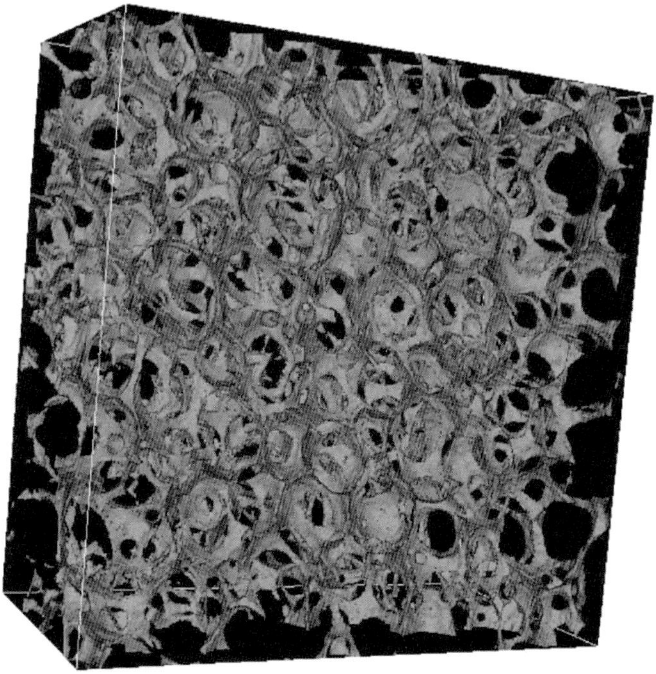

Fig. 17.6. Isosurface image illustrating the porous structure of a carbon foam. Data acquired using metallographic polishing serial sectioning, along with optical microscopy.

was targeted to approximately 3.5 μm per slice. A total of 310 image slices were taken for this sample, extending over a thickness of 780 μm of foam. Thus, the total volume of the sample imaged was 2.7 mm 2.0 mm 0.78 mm, resulting in a volume of 4.2 mm^3.

Following serial sectioning and image acquisition, images were then binarized in order to highlight the 2 different phases (carbon & epoxy). An isosurface rendering of the foam is shown in Fig. 17.6; the foam dimensions are 579 μm $\times$ 606 μm $\times$ 776 μm.

Following binarization, the 3-D data could be converted into a triangular mesh for finite element analysis (Fig. 17.7). From the finite element analysis, the thermal conductivity was determined to be 0.034 for this 90% porous foam, consistent with experimental values. Similarly, stress analysis for the same sub-model gave an overall stiffness of 154 MPa, again consistent with literature values. This consistency between experimental characterization and measured properties illustrates the strength of characterization methods in predicting the properties of pore-dominated microstructures.

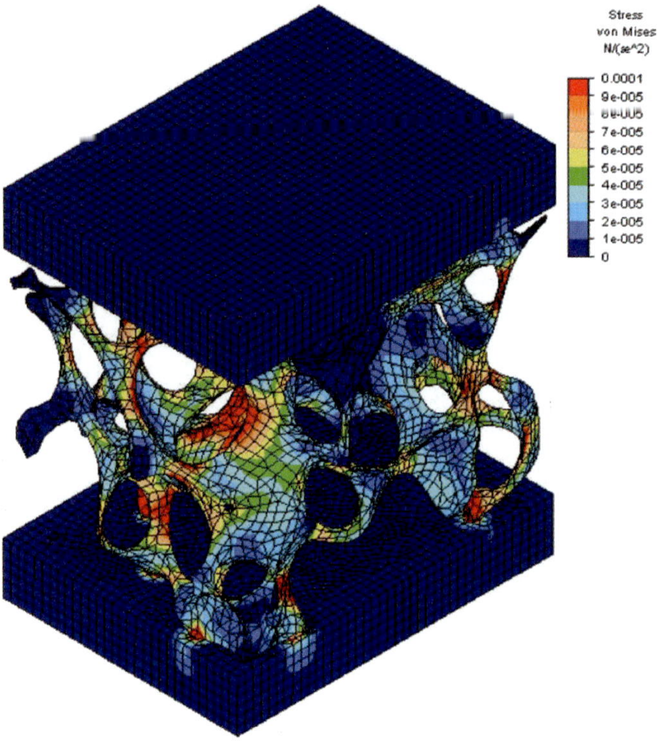

Fig. 17.7. Finite element model of stress states induced in a carbon foam. Matrix calculated from metallographic polishing serial sectioning coupled with optical microscopy.

17.3.2. *Serial Sectioning with Focused Ion Beams and Scanning Electron Microscopy*

Similar to metallography / optical microscopy, serial sectioning using FIB / SEM is also garnering much attention in the 3-D characterization community. Specifically, it has been used for analyzing the 3-D microstructure of fuel cell electrodes [15] and shale deposits [18].

In the most basic implementation, an ion beam is incident on a sample surface, causing surface sputtering and thus sectioning the sample. FIB systems normally use $^{69}Ga^{+}$ ions from a liquid metal ion source accelerated between 0.5 and 30 kV [19]. This kinetic energy allows for collision cascades to occur in the knock-on regime, the most efficient for surface sputtering.

FIB ion columns can now focus these ions to a spot less than 10 nm, allowing highly localized sputtering. In modern FIB systems, an additional SEM column is present within the sample vacuum chamber such that FIB milling can occur immediately prior to SEM imaging, albeit sometimes following a stage rotation and tilt. Throughput in these imaging experiments can approach 30 sections per hour with each slice representing approximately 50 nm. Coupling SEM imaging with EBSD or EDS analyses strongly influences analysis time. SEM imaging using field emission columns can now approach 10 nm spatial resolution, allowing for structural analysis of nanoporous materials.

One of the major drawbacks to this technique is "curtaining" artifacts. When an ion beam is incident on a porous material, the solid phase is initially sputtered away until a free surface (pore) is exposed. At this point, the ion beam is free to sputter the underlying solid phase, albeit after significant beam broadening. The resulting structure (Fig. 17.8) resembles theater curtain creases. These artifacts are problematic as they do not represent real microstructural features and can significantly alter the 3-D microstructural data. Examples of this serial sectioning technique with associate modeling are given in other chapters of this volume [20].

17.4. Tilt Series Tomography and Analytical Techniques

Since serial sectioning techniques are inherently destructive and scattering techniques do not give real space data, tilt series tomography may be another option in the arsenal of 3-D characterization techniques. In these methods, radiation (typically X-rays or electrons) are transmitted through a planar or (preferably) cylindrically shaped specimen. Subsequent images are acquired at various specimen tilt angles up to and including 90°. 3-D datasets are assembled through mathematical treatments, normally weighted back projections or Simultaneous Iterative Reconstruction Technique (SIRT). Fortunately, these techniques are also capable of high resolutions, from 50 nm with X-rays to near atomic with electrons. Cylindrical specimens are typically used in tilt series tomography experiments to facilitate near 180° rotations, and eliminate "missing wedge" artifacts.

17.4.1. *X-ray Tomography (XRT)*

Many previous works in the literature have reviewed the general concepts of XRT2 [2, 21]. A recent review in *Annual Review of Materials Research*

Fig. 17.8. STEM-HAADF image of a FIB milled specimen exhibiting "curtaining" artifacts.

[1] has shown the applicability of XRT to porous materials. In general, XRT constructs a 3D map of the X-ray attenuation in the solid phase of a material. As noted in Sec. 17.3, the attenuation of X-rays can be linearly related to the average atomic number and atomic density of the material. Obviously, pore spaces have significantly lower atomic density than the solid phase so attenuation based contrast is normally easily obtained.

Experimentally, XRT requires an X-ray source, a rotation stage, a detector, and software capable of performing the reconstruction algorithms. In practice, XRT instruments take 2 forms: (1.) Laboratory instruments with resolutions down to about 1 μm, and (2.) Synchrotron based instruments with significantly higher spatial resolutions possible [22]. While phase contrast imaging is also available using phase plates placed between

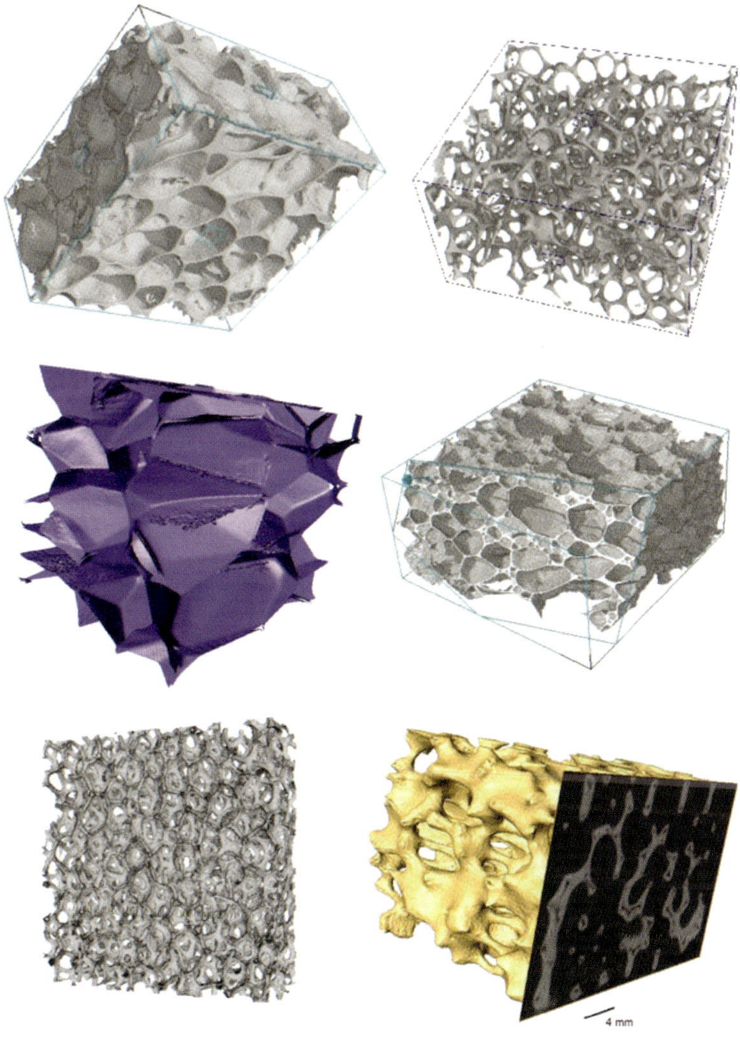

Fig. 17.9. Various examples of the mesostructure of several highly porous solids [1].

the detector and specimen, attenuation contrast is normally sufficient in porous materials where the interfaces between solid and gas phases are of primary interest. Figure 17.9 shows a selection of tomographic images of porous materials [1].

A recent article [23] compared XRT and FIB-SEM serial sectioning on the sample fuel cell electrode. Overall, it was noted that XRT produced more reliable data, which was attributed to the larger volume of bulk material sampled by XRT. Additionally, artifacts such as redeposition of sputtered material may fill in pore spaces during FIB milling, thereby reducing the volume of measured pore spaces. Although not noted by the authors, curtaining artifacts are also a possibility in FIB-SEM, while being absent during XRT experiments. Curiously, the authors noted the high cost and time required to acquired data using the XRT compared with FIB-SEM. However, times required for analyses were not quantified. Normally, images in XRT are acquired every ~30 seconds for each tilt angle, meaning approximately 3 hours for 180° data acquisition. In the FIB-SEM experiment, 180 slices can exceed 6 hours of acquisition, primarily due to time required for stage movements and fiducial alignments. In either case, the value of the acquired data greatly exceeds the time cost associated with data acquisition.

17.4.2. *Transmission Electron Microscopy Tomography*

While commercial XRT systems are available on the laboratory scale, their resolutions are typically limited to 50 nm even in state of the art systems. If the size of the pores is still smaller, then (scanning) transmission electron (STEM) tomography can be applied [24]. Depending upon the specimen geometry, resolutions approaching the atomic scale are possible.

STEM images can be acquired in several different electron optical geometries [25]. In traditional TEM mode that is prevalent in the biological community, contributions from transmitted, Rutherford scattered, Bragg diffracted, and inelastically scattered electrons all contribute to the image contrast. Since the intensity of Bragg scattered electrons varies widely with specimen tilt, contributions from only transmitted or Rutherford scattered electrons are typically used in crystalline solids. This is accomplished through the use of STEM imaging with bright field and high angle annular dark field (HAADF) detectors [25]. When using STEM-HAADF imaging, the contrast is due primarily to Rutherford scattering, and therefore atomic number contrast is prevalent and diffraction contrast is minimized. In porous crystalline materials, STEM-HAADF is now the primary method for STEM imaging.

A primary limitation in STEM tomography is that planar specimen tilt is typically limited by the geometry of the objective lens pole gap. Specialized holders now allow for up to ±70° of tilt, however, the lack of projections for a range of angles results in a "missing wedge" of information in Fourier space and consequently induces artifacts in the 3D reconstruction. To overcome this issue, cylindrical specimens prepared using FIB techniques are now being utilized. While the field of view is subsequently limited, the range of tilts can achieve 90° using specialized hardware.

One example from the literature [26] illustrates the use of STEM tomography at high tilt angles to quantify the porosity in chemical solution derived $La_2Zr_2O_7$(LZO) oxide thin films. Cylindrical specimens were prepared using FIB techniques and data acquired using a stage that allowed for 180° of rotation, eliminating any missing wedge reconstruction errors. Images were collected in STEM-HAADF mode at 200 keV at every 2° from 0° to 180°. Following reconstruction using 20 SIRT iterations, a 3-D visualization of the specimen was produced from which the nanostructure could be quantified. Pores less than 10 nm in diameter could be quantified when missing wedge effects are minimized through large tilt angles.

17.4.3. *Analytical TEM*

While STEM tomography allows for the visualization of nanoscale pore structure, analytical STEM techniques, especially electron energy loss spectroscopy (EELS), can be used to quantify porosity in combination with [27], or even without tilt series imaging. The following example illustrates the use of EELS for quantifying the nanoscale porosity quantitatively.

Figure 17.10 is a STEM-HAADF image of a bulk metallic glass (BMG, composition $Cu_{50}Zr_{45}Ti_5$) that had been implanted with $2x10^{17}$ / cm^2 of 140 keV He^+ ions [28]. The He ions caused crystallization of the structure and also coalesced into "bubbles". In order to evaluate the structure of the pores, STEM imaging, EELS spectroscopy, and atom probe tomography (discussed in the next section) were applied.

STEM-EELS works through the analysis of an electron's energy after inelastically scattering in the specimen [29]. Normally, EELS is used for quantifying the chemistry of the specimen; however, other useful information exists within the energy spectrum. In particular, the ratio of elastically scattered electrons (present within the zero-loss peak, Fig. 17.11) to inelastically scattered electrons gives quantitative information on the inelastic mean free path length of the specimen. Since the inelastic mean free path length is directly proportional to the atomic density of the material, the pore structure can also be quantified as a function of position.

Figure 17.12 is a line profile of the zero loss peak intensity as a function

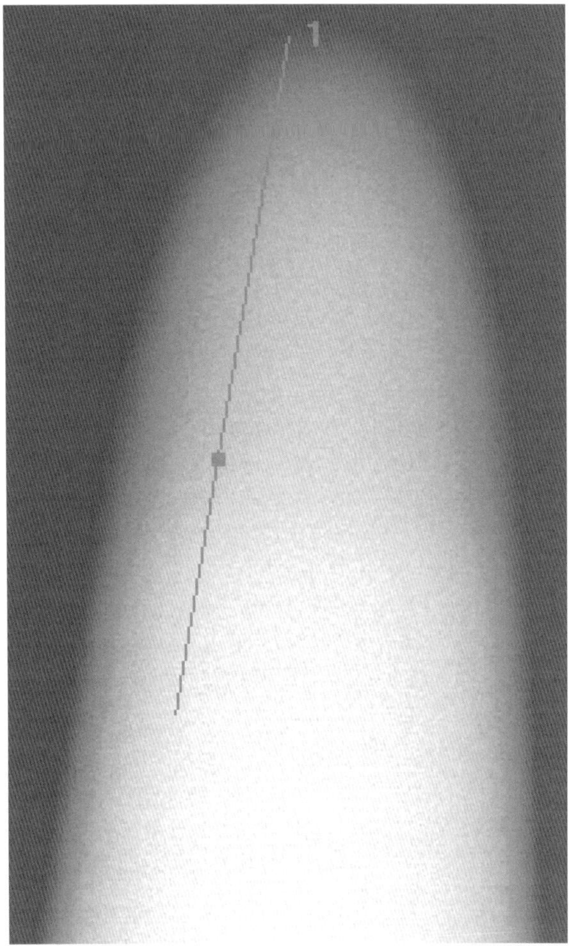

Fig. 17.10. STEM-HAADF image of a bulk metallic glass implanted with energetic He ions. A change in contrast illustrates a change in atomic density. Field of view = 150 nm.

of position (blue trace is background subtracted), with the x-axis corresponding to the line in Fig. 17.10. The dashed box highlights a significant change in the zero loss peak intensity not associated with a change in specimen geometry. Specimen porosity can be directly quantified from this profile using the log-ratio method as outlined by Egerton [29]. Assuming a constant thickness, changes in the zero loss intensity are due instead to a change in mean free path length of the specimen, which in this case is due to

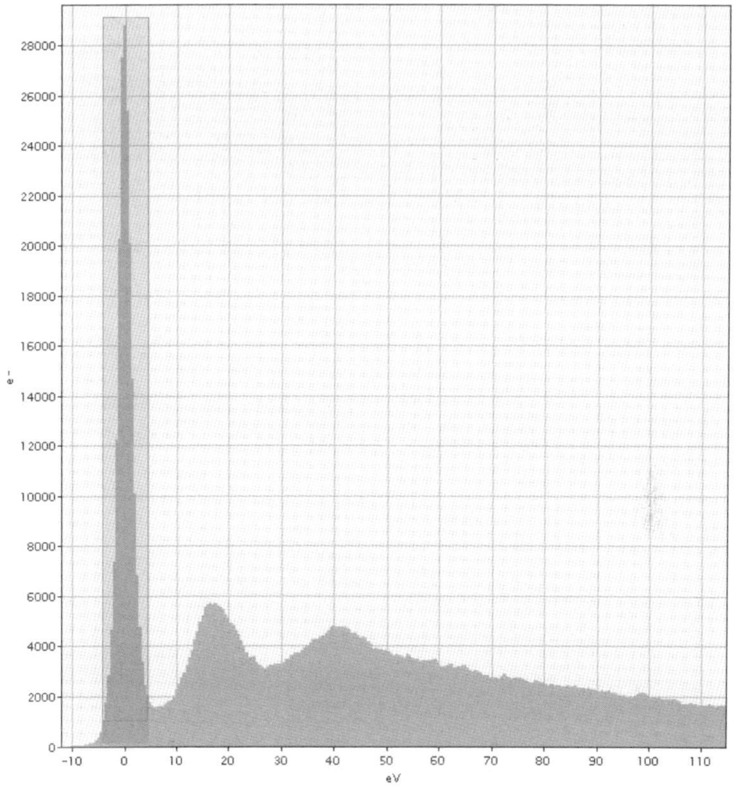

Fig. 17.11. Low-loss EELS spectrum of the BMG in Fig. 17.10. The zero loss peak is highlighted for clarity.

the presence of porosity in the specimen due to a constant chemistry. The porosity was quantified as 33 ± 12% in this specimen using these methods, in good agreement with TEM imaging and atom probe tomography (next section).

17.4.4. *Atom Probe Tomography*

The holy grail of porous materials characterization would include not only 3-D information about the solid and pore phases size distribution and tortuosity, but also include chemical information about the solid / pore interface. Currently, atom probe tomography (APT) is positioned to accomplish this in the <1 to 200 nm scales, but some challenges remain.

APT works on the principle of controlled field ion evaporation from

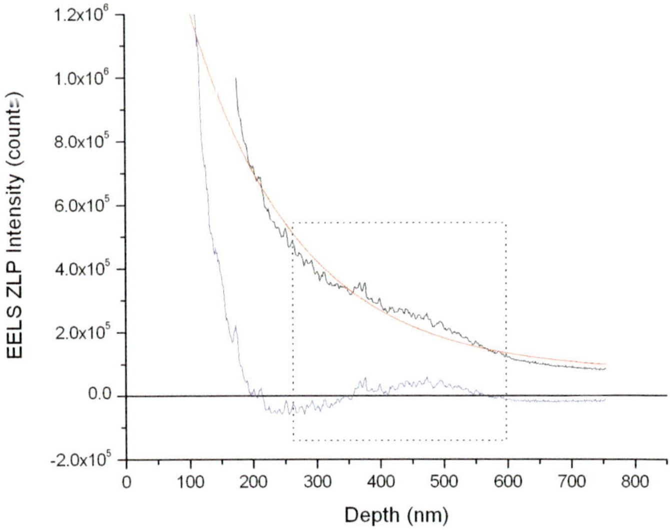

Fig. 17.12. Line profile of the zero loss peak (ZLP) intensity as a function of position. The x-axis corresponds to the red line in Fig. 17.10. The change in ZLP intensity is a direct measurement of the inelastic scattering, and therefore the pore density.

a <200 nm diameter needle shaped specimen [30]. Field evaporation can be completed using pulsed electric fields or thermally using focused laser pulses. Measuring the time of flight from the specimen to a microchannel plate detector, the mass to charge ratio of each atom or molecule can be determined, usually with <0.1 Da mass resolution. Since the magnification from the specimen to the detector is on the order of 10^7, spatial resolutions <1 nm are possible in x-y, with <0.2 nm demonstrated in the z-direction.

Applying APT to porous materials presents significant challenges. The high fields (< 1 V/Å) required for field evaporation frequently lead to dielectric breakdown inside the pores, leading to specimen failure. Similarly, the high fields also lead to ~3% electrostriction strains within the specimens, leading to mechanical failures at pores. When atom probe is successfully completed, trajectory aberrations at the solid / pore interface can lead to inaccurate atom positions. As a solid / pore boundary is exposed at the surface apex, the field can be localized around the solid phase, causing the evaporated ions to follow the local field lines rather than those dominated by the full specimen radius. Despite these challenges, APT has indeed been

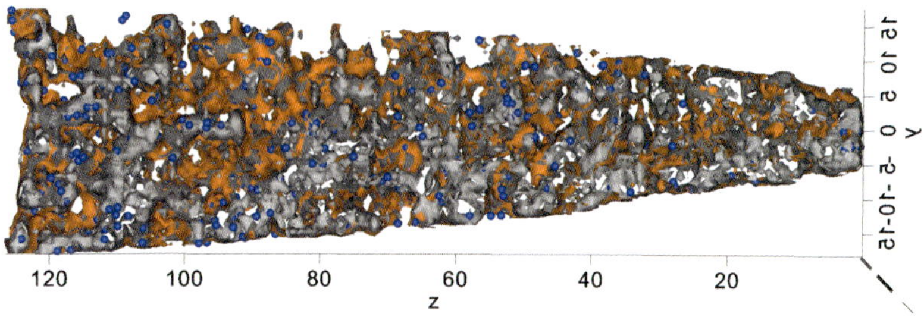

Fig. 17.13. Atom probe reconstruction of a bulk metallic glass following He ion irradiation. Pore spaces are evident throughout the field of view. Not only can the 3-D structure of the pores be quantified, but the chemistry of the solid / pore interface can also be determined on the atomic scale.

successful in some cases, especially with pore diameters <5 nm [28, 31].

APT was applied to the same BMG material illustrated in the previous section. As seen in Fig. 17.12, APT is able to detect changes in the Zr / Ti / Cu ratios (presented as black / gray / orange isosurfaces, respectively). Additionally, the 3-D pore structure is also observed, albeit with the limitations of trajectory aberrations discussed earlier. Each pore surface can be identified as a local atom density minima, and the solid phase surrounding the pore quantified. Extending these studies to, e.g., water / solid phases in geochemistry or gas / solid phases in fuel cell electrodes will guide future understanding of these materials' reactivities.

17.5. Summary and Future Work

Several different techniques for quantifying the bulk density, pore size distribution, solid phase thickness, 3-D pore tortuosity, and pore / solid interface chemistry were presented. These values can be quantified on scales from mm to sub-nm using a suite of different techniques from scattering to serial sectioning to tilt series tomography. Significant challenges exist in each experimental method, but the scientific understanding gained through combined characterization and modeling are well worth the effort to overcome these challenges.

The next frontier for pore scale characterization is a combined understanding of pore structure and the interface chemistry of the solid phase. Combining tilt series and serial sectioning tomography with analytical techniques such as X-ray fluorescence spectroscopy and electron energy loss spectroscopy are on the verge of being well established. Utilizing APT would improve the spatial resolution of these techniques to a true fundamental understanding of the surface chemistry in 3-D.

References

[1] E. Maire, X-ray tomography applied to the characterization of highly porous materials, *Ann. Rev. Mat. Res.* **42**, 163–178 (2012).

[2] G. Möbus and B. J. Inkson, Nanoscale tomography in materials science, *Materials Today.* **10**(12), 18–25 (2007).

[3] B. Maruyama, J. E. Spowart, D. J. Hooper, H. M. Mullens, A. M. Druma, C. Druma, and M. K. Alam, A new technique for obtaining three-dimensional structures in pitch-based carbon foams, *Scripta Materialia.* **54** (9), 1709–1713 (2006).

[4] J. Wachtman and Z. Kalman, *Characterization of Materials.* Butterworth-Heinemann (1993).

[5] S. Lowell, J. E. Shields, M. Thomas, and M. A. Thommes, *Characterization of Porous Solids and Powders: Surface Area, Pore Size, and Density.* Kluwer (2004).

[6] G. D. Brabson, *Ultraviolet / Visible Absorption Spectroscopy*, In *Handbook of Materials Characterization*, pp. 61–71. ASM International.

[7] O. S. Heavens, *Optical Properties of Thin Solid Films.* Courier Dover Publications (1991).

[8] B. Gorman, V. Petrovsky, H. U. Anderson, and T. Petrovsky, Optical characterization of ceramic thin films: Applications in low-temperature solid oxide fuel-cell materials research, *J. Mat. Res.* **19**(02), 573–578 (2004).

[9] B. P. Gorman and H. U. Anderson, Microstructure development in unsupported thin films, *J. Am. Ceram. Soc.* **85**(4), 981–985 (2002).

[10] V. Petrovsky, B. Gorman, H. Anderson, and T. Petrovsky, Optical properties of ceo2 films prepared from colloidal suspension, *J. Appl. Phys.* **90**(5), 2517–2521 (2001).

[11] D. G. LeGrand, *Small-Angle X-ray and Neutron Scattering*, In *Handbook of Materials Characterization*, pp. 402–406. ASM International.

[12] A. P. Roberts, Morphology and thermal conductivity of model organic aerogels, *Phys. Rev. E.* **55**(2), R1286 (1997).

[13] A. P. Roberts and S. Torquato, Chord-distribution functions of three-dimensional random media: approximate first-passage times of gaussian processes, *Phys. Rev. E.* **59**(5), 4953 (1999).

[14] J. Quintanilla, R. F. Reidy, B. P. Gorman, and D. W. Mueller, Gaussian random field models of aerogels, *J. Appl. Phys.* **93**(8), 4584–4589 (2003).

[15] J. R. Wilson, W. Kobsiriphat, R. Mendoza, H.-Y. Chen, J. M. Hiller, D. J. Miller, K. Thornton, P. W. Voorhees, S. B. Adler, and S. A. Barnett, Three-dimensional reconstruction of a solid-oxide fuel-cell anode, *Nature Materials.* **5**(7), 541–544 (2006).

[16] K. Scott, J. M. Davis, and E. P. Vicenzi, Three-dimensional microanalysis

using fib sem: Variations in technique, *Microscopy and Microanalysis*. **15** (S2), 476–477 (2009).
[17] S. J. Dillon and G. S. Rohrer, Characterization of the grain-boundary character and energy distributions of yttria using automated serial sectioning and ebsd in the fib, *J. Am. Ceram. Soc.* **92**(7), 1580–1585 (2009).
[18] B. Bera, N. S. K. Gunda, S. K. Mitra, and D. Vick, Characterization of nanometer-scale porosity in reservoir carbonate rock by focused ion beam–scanning electron microscopy, *Microscopy and Microanalysis*. **18**(01), 171–178 (2012).
[19] L. A. Giannuzzi and F. A. Stevie, *Introduction to Focused Ion Beams: Instrumentation, Theory, Techniques and Practice*. Springer (2005).
[20] R. J. Kee, H. Zhu, G. M. Goldin, and S. A. Barnett. ibid.
[21] P. J. Withers, X-ray nanotomography, *Materials Today*. **10**(12), 26–34 (2007).
[22] J. Baruchel, J.-Y. Buffiere, P. Cloetens, M. Di Michiel, E. Ferrie, W. Ludwig, E. Maire, and L. Salvo, Advances in synchrotron radiation microtomography, *Scripta Materialia*. **55**(1), 41 – 46 (2006). ISSN 1359-6462.
[23] E. A. Wargo, T. Kotaka, Y. Tabuchi, and E. C. Kumbur, Comparison of focused ion beam versus nano-scale x-ray computed tomography for resolving 3-d microstructures of porous fuel cell materials, *J. Power Sources*. **241**, 608–618 (2013).
[24] P. A. Midgley and M. Weyland, 3d electron microscopy in the physical sciences: the development of z-contrast and eftem tomography, *Ultramicroscopy*. **96**(3), 413–431 (2003).
[25] D. B. Williams and C. B. Carter. The transmission electron microscope. In *Transmission Electron Microscopy*, pp. 3–22. Springer (2009).
[26] E. Biermans, L. Molina, K. J. Batenburg, S. Bals, and G. Van Tendeloo, Measuring porosity at the nanoscale by quantitative electron tomography, *Nano Letters*. **10**(12), 5014–5019 (2010).
[27] G. Möbus, R. C. Doole, and B. J. Inkson, Spectroscopic electron tomography, *Ultramicroscopy*. **96**(3), 433–451 (2003).
[28] L. Shao, B. P. Gorman, A. Aitkaliyeva, N. D. Theodore, and G. Xie, Nanometer-scale tunnel formation in metallic glass by helium ion irradiation, *App. Phys. Lett.* **101**(4), 041901 (2012).
[29] R. Egerton, *Electron Energy-Loss Spectroscopy in the Electron Microscope*. Plenum Press (1996).
[30] M. K. Miller and R. G. Forbes. Atom-probe tomography: The local electrode atom probe. Technical report, Springer (2014).
[31] M. K. Miller, L. Longstreth-Spoor, and K. Kelton, Detecting density variations and nanovoids, *Ultramicroscopy*. **111**(6), 469–472 (2011).

Chapter 18

Homogenization Approaches for the Elastic Properties of Porous Materials

J. R. Berger and S. R. Geer

Department of Mechanical Engineering
Colorado School of Mines
Golden, CO, 80401, USA
jberger@mines.edu

William J. Parnell

Department of Mathematics
University of Manchester
Manchester, UK

Homogenization approaches for determining the effective elastic constants in porous media are reviewed. These homogenized elastic constants relate, by definition, volume averages of stress and strain. This necessitates precisely defining the representative volume element, as well as boundary conditions which provide equivalent definitions of the elastic constants from both a mechanical and energetic viewpoint. The dilute estimate and the self consistent method are detailed, and these methods are applied to calculate effective elastic properties of an oil shale sample. Finally, a predictive method based on the fabric tensor is presented. The fabric tensor provides a description of the shape, orientation, and distribution of the pores in a solid. An example application for the prediction of elastic constants as a function of porosity in a PZT (porous lead zirconate titanate) ceramic demonstrates the usefulness of the fabric tensor approach.

18.1. Introduction

Engineering analysis of macro-scale porous solids typically requires a homogenization approach for the elastic constants in order to perform stress-strain calculations. The homogenized elastic constants are spatially constant, and replace the position-dependent elastic constants of the actual material. The homogenized elastic constants relate volume averages of stress and strain. For stress analysis, the homogenized elastic constants are calculated in such a way as to be representative of the microstructural features that may affect the stress and strain response of the solid. With most porous solids there will be some degree of anisotropy, either structural or material in nature. Even if the matrix material itself may be isotropic, the porous material can display structural anisotropy due to the shape and distribution of the voids in the material.

There are a number of different approaches which may be used to homogenize a porous material, including the early work of Hill [1], Kröner [2], Hashin [3], Hashin and Shtrikman [4], and many others. A recent review article [5] provides an excellent state of the art, as well as a critical review of the homogenization approach. In this chapter, homogenization methods based on the inclusion solution of Eshelby [6] will be reviewed, along with the predictive method of Zysett [7] which is useful in studying the variation of elastic constants with porosity.

In order to proceed with homogenization, it is important to understand the *Representative Volume Element* (RVE) concept, and the role that boundary conditions can play in any calculations performed on a volume element. The volume element must satisfy the ergodicity assumption: all statistical information about the material must be contained in the RVE. As such, the RVE will contain sufficient statistical and microstructural information to be representative of any similar volume taken from any location in the heterogeneous solid. It can then be assumed that elastic properties extracted from the volume averages of stress and strain on the volume element can be taken as the overall properties of the homogenized material. These *effective* elastic constants can be extracted from experimental or numerical experiments on the RVE. However, boundary conditions which satisfy the Hill criterion [1] must be used in order for mechanical and energetic definitions of the elastic constants to be identical. Details of these boundary conditions can be found in Refs. [8] and [9].

In this chapter, the differences between effective and apparent elastic constants are reviewed, which depend on whether the volume element is sufficient to fulfill the ergodicity assumptions. Next, homogenization methods based on the equivalent inclusion approach are reviewed, which include the Dilute Estimate and Self Consistent Methods. These methods are then applied to a geological material, in this case, an oil shale. Next, the concept of a fabric tensor is introduced, which is one method to characterize the orientation distribution function for pore shape, size, and distribution in a porous solid. Finally, an example of a porous PZT ceramic is used to illustrate the compuations necessary to predict homogenized elastic constants as the porosity is varied.

18.2. Effective and Apparent Elastic Stiffness Tensors

Huet [10] presents an elegant discussion on effective and apparent elastic constants, and this was briefly reviewed in Ref. [11]. If a material volume is statistically uniform, then

$$\langle \sigma_{ij} \rangle = C^E_{ijkl} \langle \varepsilon_{kl} \rangle \tag{18.1}$$

where the indices are with respect to a rectangular cartesian coordinate system, σ_{ij} is the stress tensor, C^E_{ijkl} is the *effective* elastic stiffness tensor, ε_{kl} is the strain tensor, and $\langle \cdot \rangle$ represents the volume average of the quantity between the angled brackets, i.e,

$$\langle \sigma_{ij} \rangle = \frac{1}{V} \int_V \sigma_{ij} dV$$

where V is the volume of the element being integrated over.. The loading conditions are such that homogeneous fields would result if the material were homogeneous. Then C^E_{ijkl} is independent of the boundary conditions taken on the material volume.

If the local relationship is $\sigma_{ij} = C_{ijkl}\varepsilon_{kl}$, then Eq. 18.1 can be written as

$$\langle C_{ijkl}\varepsilon_{kl} \rangle = C^E_{ijkl} \langle \varepsilon_{kl} \rangle \tag{18.2}$$

which is a definition of C^E_{ijkl}. We can also write this as

$$\langle C_{ijkl}\varepsilon_{kl} \rangle = \langle C_{ijkl} \rangle \langle \varepsilon_{kl} \rangle + \langle C'_{ijkl}\varepsilon'_{kl} \rangle \tag{18.3}$$

where $\langle C'_{ijkl}\varepsilon'_{kl} \rangle$ represents the fluctuation about the average value. Note here that $\langle C_{ijkl} \rangle$ is *not* equal to C^E_{ijkl}. Hill [1] showed that $\langle C_{ijkl} \rangle$ is an upper bound for C^E_{ijkl}. Huet [10] termed this the Hill-Voigt upper bound on the stiffness tensor.

Alternatively, an effective compliance tensor can be formed,

$$\langle S_{ijkl}\sigma_{kl} \rangle = S^E_{ijkl} \langle \sigma_{kl} \rangle \tag{18.4}$$

which is a definition of S^E_{ijkl}. This can also be written as

$$\langle S_{ijkl}\sigma_{kl} \rangle = \langle S_{ijkl} \rangle \langle \sigma_{kl} \rangle + \langle S'_{ijkl}\sigma'_{kl} \rangle \tag{18.5}$$

Again, $\langle S_{ijkl}\rangle$ is *not* equal to S^E_{ijkl}. Hill [1] showed that $\langle S_{ijkl}\rangle$ is an upper bound for S^E_{ijkl}. Huet [10] termed this the Hill-Reuss upper bound on the compliance tensor, or equivalently the lower bound on the stiffness tensor.

One can also define effective properties from an energetic perspective by computing volume averages of stress and strain in a free-energy formulation. The energetic definition and the mechanical definition of the effective elastic stiffness tensors are the same if the Hill [1] criterion is satisfied,

$$\langle \sigma_{ij}\varepsilon_{ij}\rangle = \langle \sigma_{ij}\rangle\langle \varepsilon_{ij}\rangle. \tag{18.6}$$

Note that the Gauss theorem can be used to recast the Hill condition as

$$\int_\Gamma (t_i - \langle \sigma_{ij}\rangle n_j)(u_i - \langle \varepsilon_{ij}\rangle x_j)d\Gamma = 0 \tag{18.7}$$

where Γ is the boundary of the volume and $t_i = \sigma_{ij}n_j, u_i, n_j$ and x_j are the components of the traction, displacement, normal and position vectors, respectively. This form of the Hill criterion lends itself directly to calculations using boundary element methods as discussed in Ref. [11].

If certain boundary conditions can be identified which satisfy the Hill criterion, then an effective elastic stiffness tensor can be obtained which is identical from the mechanical and energetic definitions. This can then be done without any statistical assumptions concerning the volume element, which is of critical importance in the case of volume elements smaller than the RVE. Recall that for the RVE, the relationship between stress and strain should be independent of the boundary conditions. If the volume size is less than the RVE this is no longer generally true.

In many materials applications, the volume element available for analysis may be less than the RVE. The question then arises as to what one can say concerning the elastic stiffness or compliance tensors for the homogenized material. This question has been investigated in a series of papers [10, 8, 12, 9] where the idea of *apparent* elastic stiffness and compliance tensors was presented. Although, in general, stiffness-tensor results on a volume less than the RVE will be dependent on the specific boundary conditions taken, the boundary conditions can still be specified such that the Hill criterion is satisfied. Typically, these boundary conditions are taken as kinematic uniform boundary conditions (KUBC), static uniform boundary conditions (SUBC), or mixed uniform boundary conditions (MUBC). Hill [1] originally determined the KUBC and SUBC, as these generate uniform strain (KUBC) or uniform stress (SUBC) conditions in the volume. The MUBC were developed by Hazanov [8, 12], who also showed that for the apparent elastic stiffness and compliance tensors,

$$C^{\sigma}_{ijkl} \le C^{m}_{ijkl} \le C^{u}_{ijkl}, \quad S^{u}_{ijkl} \le S^{m}_{ijkl} \le S^{\sigma}_{ijkl} \tag{18.8}$$

where $C^{\sigma}_{ijkl}, C^{m}_{ijkl}$ and C^{u}_{ijkl} are the apparent stiffness tensors obtained from SUBC, MUBC, and KUBC, respectively, with similar notation used for the effective compliance tensors. The MUBC were further refined as periodic mixed uniform boundary conditions (PMUBC) in Ref. [9].

Therefore, if volumes less than the RVE are used, upper and lower bounds on the stiffness or compliance tensor can be calculated by applying SUBC or KUBC, or an intermediate estimate can be obtained using MUBC on the volume element. As the volume element becomes larger, these bounds eventually converge to the effective stiffness or compliance tensor.

18.3. Determination of Effective Elastic Constants: Equivalent Inclusion Methods

In this section two common homogenization methods are reviewed: The Dilute Estimate (DE), and the Self Consistent Method (SCM). There are many other approaches available that build on these techniques to include effects such as viscoelasticity, multiple inclusions, etc., yet these methods provide a basic foundation for most approaches. The Dilute Estimate (DE) is computationally cheap, explicit, and relatively easy to formulate when using the developments based on the seminal work by Eshelby [6]. However, the DE has been found to only have good accuracy when the volume fraction of inclusions or pores in a host material are fairly small. Additionally, as it is assumed that the volume fraction of the inclusions or pores is much less than unity, the interactions between two or more of these are neglected, causing a loss of accuracy at higher volume fractions. However, if the inclusions are very stiff then the dilute estimate performs well up to quite high volume fractions.

The work of Mori and Tanaka [13] provides a method that does take into account the interactions between particles albeit approximately. In a comparison of many well known homogenization methods, Klusemann and Svendson [14] point out that inhomogeneity concentrations upwards of 30% are not suitably analyzed using the Mori-Tanaka (MT) method. Additionally, Castañeda and Willis [15] showed that MT is not suitable for randomly-oriented inhomogeneities. However, the Self Consistent Method (SCM) is an implicit method that solves for the unknown effective material properties, the number of which depends on the assumed general behavior of the material system, e.g., isotropy, transverse isotropy, orthotropy, etc. [14] suggests that SCM is well suited for polycrystal materials.

All the equivalent inclusion-based homogenization methods seek to calculate a strain concentration tensor from which the effective elastic properties can be determined. The effective stiffness tensor of the material can be expressed in terms of the stiffness tensors of the constituent materials, their corresponding volume fraction, and the strain concentration tensors [16]. The strain concentration tensor is a tensor that relates the average strain in a composite material to the average strain in an inclusion or void. If $\langle\epsilon\rangle$ is the average strain of the entire composite material, and $\langle\epsilon_r\rangle$ is the average strain of the rth inhomogeneity, then

$$\langle\epsilon_r\rangle = A_r\langle\epsilon\rangle, \tag{18.9}$$

where A_r is the 4th rank strain concentration concentration tensor of the r-th inhomogeneity. The concentration tensors are dependent on either Eshelby's S-tensor formulation or Hill's P-tensor approach, which are discussed briefly here.

The strain concentration tensor for a single inhomogeneity of material 1, embedded in an assumed infinite medium of material 0, can be represented by the relation

$$A_1 = [I + S_1 C_0^{-1}(C_1 - C_0)]^{-1}, \tag{18.10}$$

where I is the 4th rank identity tensor and S_1 is the 4th rank Eshelby tensor. Eshelby tensors do not contain major symmetries, i.e.$S_{ijkl} \neq S_{klij}$; as such, one may wish to express the strain concentration tensor in terms of Hill's polarization tensor, also known as Hill's P-tensor, which is given by $P = S\ C_0^{-1}$. Equation 18.10 may then be expressed as

$$A_1 = [I + P_1(C_1 - C_0)]^{-1}. \tag{18.11}$$

Details concerning the structure of both the S- and P-tensors can be found in [17].

In order to compute the P-tensor for each inclusion or void in the material, one must first consider the inclusion's geometry and orientation. For example, if a single spherical particle is embedded in a matrix material, the P-tensor can be calculated quite readily. The calculation becomes more complicated for other, non-spherical geometries.

Finally, when dealing with materials that exhibit macroscopic behaviors of a higher complexity than isotropy, it may be convenient to express the

4th rank tensors used in elastostatics in terms of Hill's tensor basis. More information on Hill's tensor basis can be found in Ref. [1] and [18]. For example, using this compact notation, one can write the 4th-rank stiffness tensor for an isotropic material, C_{iso}, in terms of two constants,

$$C_{\text{iso}} = (3K, 2G), \tag{18.12}$$

where K and G are the bulk and shear moduli, respectively. Equation 18.12 can easily be converted to the familiar 6×6 matrix form as

$$C_{\text{iso}} = 3KI^h + 2GI^d \tag{18.13}$$

$$I^h = \frac{1}{3}\delta_{ij}\delta_{kl} \tag{18.14}$$

$$I^d = \frac{1}{2}[(\delta_{ij}\delta_{kl} + \delta_{il}\delta_{jk}) - \frac{2}{3}\delta_{ij}\delta_{kl}]. \tag{18.15}$$

In a transversely isotropic (TI) material, the complexity of expressing the properties in a Hill tensor basis increases significantly. The complexity is such that the properties must be represented by six equations. For example, assuming that the axis of rotational symmetry for the TI material is oriented with $X_A \equiv X_3$ then

$$H^1_{ijkl} = \frac{1}{2}\Theta_{ij}\Theta_{kl} \tag{18.16}$$

$$H^2_{ijkl} = \Theta_{ij}\delta_{k3}\delta_{l3} \tag{18.17}$$

$$H^3_{ijkl} = \Theta_{kl}\delta_{i3}\delta_{j3} \tag{18.18}$$

$$H^4_{ijkl} = \delta_{i3}\delta_{j3}\delta_{k3}\delta_{l3} \tag{18.19}$$

$$H^5_{ijkl} = \frac{1}{2}(\Theta_{ik}\Theta_{lj} + \Theta_{il}\Theta_{kj} - \Theta_{ij}\Theta_{kl}) \tag{18.20}$$

$$H^6_{ijkl} = \frac{1}{2}(\Theta_{ik}\delta_{l3}\delta_{j3} + \Theta_{il}\delta_{k3}\delta_{j3} + \Theta_{jk}\delta_{l3}\delta_{i3} + \Theta_{jl}\delta_{k3}\delta_{i3}), \tag{18.21}$$

$$I_{ijkl} = H^1_{ijkl} + H^4_{ijkl} + H^5_{ijkl} + H^6_{ijkl}, \tag{18.22}$$

where $\Theta_{ij} = \delta_{ij} - \delta_{i3}\delta_{j3}$. The above set of equations are then written in terms of a TI tensor basis and Hill moduli, k,l,n,m,and p as

$$C_{ijkl} = \sum_{n=1}^{6} X_n H^n_{ijkl}, \tag{18.23}$$

where the X_n terms are defined as

$$\begin{aligned} X_1 &= 2k \\ X_2 &= l \\ X_3 &= l \\ X_4 &= n \\ X_5 &= 2m \\ X_6 &= 2p \end{aligned} \tag{18.24}$$

Hill [19] defines the moduli in Eq. 18.24 as follows: k is the plane-strain bulk modulus for lateral dilatation without longitudinal extension; m is the rigidity modulus for shearing in any transverse direction; n is the modulus for longitudinal uniaxial straining, and l is the associated cross-modulus. The terms p and m are associated with shear-shear coupling in constitutive elastostatic relations. The relation of Hill's moduli to the usual engineering constants, (bulk and shear moduli, Young's moduli, Poisson's ratios, etc) depend on the overall behavior of the material in question. For example, for an isotropic material, the Hill moduli can be expressed in terms of bulk and shear moduli, K and G, as

$$\begin{aligned} k &= K + \frac{G}{3} \\ l &= K - \frac{2G}{3} \\ n &= K + \frac{4G}{3} \\ m &= G \\ p &= G \end{aligned} \tag{18.25}$$

The formulations seen related to Eq. 18.23 can also be written in an equivalent contracted form, in terms of Eq. 18.24, as

$$C = (2k, l, l, n, 2m, 2p). \tag{18.26}$$

Conveniently, using this method for a TI composite made of isotropic materials is made easier by the fact that an isotropic material is expressible not only in the form of Eq. 18.12, but also in terms of Eq. 18.26 when the following substitutions are made using the r-th material's bulk and shear moduli, K_r and G_r:

$$\begin{aligned}
X_1^r &= 2K_r + \frac{2}{3}G_r \\
X_2^r &= K_r - \frac{2}{3}G_r \\
X_3^r &= K_r - \frac{2}{3}G_r \\
X_4^r &= K_r + \frac{4}{3}G_r \\
X_5^r &= 2G_r \\
X_6^r &= 2G_r
\end{aligned} \tag{18.27}$$

The strain concentration tensor can also be expressed in a form similar to Eq. 18.23, although the coefficients of H_n take on different definitions based on Hill's tensor formulations for a circular cylinder whose axis of rotational symmetry is $X_A \equiv X_3$:

$$A_{ijkl} = \sum_{n=1}^{6} A_n H_{ijkl}^n \tag{18.28}$$

where

$$\begin{aligned}
A_1 &= 1 + \rho_1 X_1 + 2\rho_2 X_3 \\
A_2 &= \rho_1 X_2 + \rho_2 X_4 \\
A_3 &= \rho_3 X_1 + \rho_4 X_3 \\
A_4 &= 1 + 2\rho_3 X_2 + \rho_4 X_4 \\
A_5 &= 1 + \rho_5 X_5 \\
A_6 &= 1 + \rho_6 X_6
\end{aligned} \tag{18.29}$$

The derivation of the problem-specific multipliers, ρ_n, in Eq. 18.29, can be found in [20] for a circular cylinder in a transversely isotropic material. The multipliers, p_n, are the components of the relevant P-tensor with respect to the Hill basis; here, this is appropriate for spheroidal inclusions which lend themselves to materials with macroscopic transverse isotropy, including the limiting ellipsoidal case of a circular cylinder.

The following relations, which can also be written as $X_n = X_n^{\text{inclusion}} - X_n^{\text{matrix}}$, can be developed between the X_n, the bulk and shear moduli of the inclusion material, K_r and G_r, respectively, and the Hill moduli, k,l,n,m,p:

$$\begin{aligned}
X_1^r &= 2K_r + \frac{2}{3}G_r - 2k \\
X_2^r &= K_r + \frac{2}{3}G_r - l \\
X_3^r &= K_r - \frac{2}{3}G_r - l \\
X_4^r &= K_r + \frac{4}{3}G_r - n \\
X_5^r &= 2G_r - 2m \\
X_6^r &= 2G_r - 2p
\end{aligned} \tag{18.30}$$

Finally, these can also be given in terms of five engineering constants needed for transverse isotropy:

$$\begin{aligned}
k &= \frac{l^2}{n - E_A} \\
l &= \frac{E_T E_A \nu_A}{E_A - E_A \nu_T - 2E_T(\nu_A)^2} \\
n &= \frac{E_A^2(1 - \nu_T)}{E_A - E_A \nu_T - 2E_T(\nu_A)^2} \\
m &= \frac{E_T}{2(1 + \nu_T)} \\
p &= G,
\end{aligned} \tag{18.31}$$

where E and ν are the Young's modulus and Poisson's ratio, the subscripts T and A indicate properties in the transverse and axial directions, and G is the out-of-plane shear modulus. For Poisson's ratio, ν_T refers to contraction in the T direction when pulled in the A direction.

18.3.0.1. *The Dilute Estimate*

The Dilute Estimate (DE), also known as the explicit Eshelby Method, is formulated in a general way for the effective stiffness tensor of a composite composed of a matrix material and N inclusions,

$$C_{\text{eff}} = C_0 + \sum_{r=1}^{N} \phi_r (C_r - C_0) A_r, \tag{18.32}$$

where the subscripts eff, 0, and r denote the identity of the effective, matrix, and the rth inclusion properties, respectively. The volume fraction of the inclusion is ϕ_r, and A_r is the strain concentration tensor for the rth inclusion.

The work related to homogenization in this chapter deals primarily with transverse isotropy, and the Hill/Walpole representations of TI tensors, as portrayed in Eqs. 18.25 and 18.26, eliminate some of the algebraic clutter associated with finding effective properties. The strain concentration tensor, A_r, in Eq. 18.32, while perfectly acceptable in matrix form as seen in Eq. 18.28, can be written in terms of its identity components and tensor products as

$$A_r = [I + P_r(C_r - C_0)]^{-1}, \tag{18.33}$$

where $I = (1, 0, 0, 1, 1, 1)$, C_r and C_0 are defined with their respective Hill moduli as seen in Eq. 18.26, and $P_r = (\rho_1, \rho_2, \rho_3, \rho_4, \rho_5, \rho_6)$. The ρ_n terms are defined by the following relations and are specific to the limiting case of a spheroidal inclusion with aspect ratio tending to infinity (infinite circular cylinder)

$$\begin{aligned}
\rho_1 &= \frac{1}{2(k_0 + m_0)} \\
\rho_2 &= 0 \\
\rho_3 &= 0 \\
\rho_4 &= 0 \\
\rho_5 &= \frac{k_0 + 2m_0}{4m_0(k_0 + m_0)} \\
\rho_6 &= \frac{1}{4p_0}
\end{aligned} \tag{18.34}$$

18.3.0.2. *The Self-Consistent Method*

The self-consistent method (SCM) is an implicit method by which the effective properties of a heterogeneous material can be estimated. Qu and Cherkaoui [16] present a simple example of how to obtain sets of nonlinear algebraic equations whose roots are the effective elastic constants of a two-phase composite material. Furthermore, [16] shows that, in a comparison of the Eshelby Method to SCM, the SCM gives better results for concentrations not limited to the dilute regime but less than $\phi < 0.5$. It should be noted that a Generalized Self-Consistent Method (GSCM) is usually used or particulate media whereas SCM is used for polycrystalline media.

When using SCM, the strain concentration tensor, A_r, is dependent upon the *unknown* effective properties which are being solved for. This method is therefore an implicit method, and the solution usually requires iterative numerical solutions. The SCM, in order to better approximate the interaction between phases, approximates the strain inside an inclusion as that which would be felt if the inclusion were in the effective material. The basic formalism of the SCM appears to be quite similar to that of Eq. 18.32:

$$\overline{C} = C_0 + \sum_{r=1}^{N} \phi_r (C_r - C_0)\overline{A_r}, \tag{18.35}$$

where the overbar indicates that the elements of the unknown tensors must be computed based on the effective elastic constants of the homogenized material [17]. The SCM is therefore an implicit method since the unknowns appear on both sides of Eq. 18.35. The strain concentration tensor can be represented by the relation

$$\overline{A_r} = [I + \overline{S_r}\ \overline{C_0^{-1}}(C_r - \overline{C})]^{-1}, \tag{18.36}$$

where I is the identity tensor and $\overline{S_r}$ is the Eshelby tensor for problem-specific inclusion geometries and material behaviors. Equation 18.36 may be equivalently expressed in terms of Hill's P-tensor as

$$\overline{A_r} = \left(I + \overline{P_r}(C_r - \overline{C})\right)^{-1}. \tag{18.37}$$

The known terms in Eqs. 18.35–18.37 have thus been defined. The unknown portions of these equations must be put in terms of the unknown material properties. For a transversely isotropic (TI) homogenized material, the five unknown properties include the axial and transverse Young's moduli (E_A and E_T), the axial and transverse Poisson's ratios (ν_A and ν_T), and the anti-plane shear modulus, p. Note that the formulations in this chapter followed both Refs. [20] and [21], and are based on the assumption that the axis of material symmetry of the TI composite is aligned as $X_A \equiv X_3$.

The unknown 4th-rank strain concentration tensor, $\overline{A}_{ijkl}$, shown in Eq. 18.37 can now be written as

$$\overline{A}_{ijkl} = \sum_{n=1}^{6} \overline{A}_n H^n_{ijkl} \tag{18.38}$$

where

$$
\begin{aligned}
\overline{A}_1 &= 1 + \overline{\rho}_1 \overline{X}_1 + 2\overline{\rho}_2 \overline{X}_3 \\
\overline{A}_2 &= \overline{\rho}_1 \overline{X}_2 + \overline{\rho}_2 \overline{X}_4 \\
\overline{A}_3 &= \overline{\rho}_3 \overline{X}_1 + \overline{\rho}_4 \overline{X}_3 \\
\overline{A}_4 &= 1 + 2\overline{\rho}_3 \overline{X}_2 + \overline{\rho}_4 \overline{X}_4 \\
\overline{A}_5 &= 1 + \overline{\rho}_5 \overline{X}_5 \\
\overline{A}_6 &= 1 + \overline{\rho}_6 \overline{X}_6 .
\end{aligned}
\tag{18.39}
$$

The formulations shown in Eqs. 18.35–18.37 can also be expressed in terms of Hill/Walpole notation. In Hill/Walpole form, the P-tensor can be written in terms of the unknown effective Hill moduli as

$$P_r = (\overline{\rho_1}, \overline{\rho_2}, \overline{\rho_3}, \overline{\rho_4}, \overline{\rho_5}, \overline{\rho_6})$$

where

$$
\begin{aligned}
\overline{\rho_1} &= \frac{1}{2(\overline{k} + \overline{m})} \\
\overline{\rho_2} &= 0 \\
\overline{\rho_3} &= 0 \\
\overline{\rho_4} &= 0 \\
\overline{\rho_5} &= \frac{\overline{k} + 2\overline{m}}{4\overline{m}(\overline{k} + \overline{m})} \\
\overline{\rho_6} &= \frac{1}{4\overline{p}}
\end{aligned}
\tag{18.40}
$$

The derivation of the problem-specific multipliers, ρ_n, in Eqs. 18.39 and 18.40, can be found in Ref. [22] for a circular cylinder in a transversely isotropic material. The multipliers, ρ_n, are the components of the relevant P-tensor with respect to the Hill basis. The following relation can be developed between the $\overline{X}_n$, the bulk and shear moduli of the inclusion material,

K_r and G_r, and the unknown Hill moduli, $\overline{k}, \overline{l}, \overline{n}, \overline{m}$, and $\overline{p}$:

$$\begin{aligned}
\overline{X}_1 &= 2K_r + \frac{2}{3}G_r - 2\overline{k}\\
\overline{X}_2 &= K_r - \frac{2}{3}G_r - \overline{l}\\
\overline{X}_3 &= K_r - \frac{2}{3}G_r - \overline{l}\\
\overline{X}_4 &= K_r + \frac{4}{3}G_r - \overline{n}\\
\overline{X}_5 &= 2\mu_r - 2\overline{m}\\
\overline{X}_6 &= 2\mu_r - 2\overline{p},
\end{aligned} \tag{18.41}$$

which are finally given in terms of the unknown material properties we are solving for using the SCM as

$$\begin{aligned}
\overline{k} &= \frac{\overline{l}^2}{\overline{n} - \overline{E_A}}\\
\overline{l} &= \frac{\overline{E_T}\,\overline{E_A}\,\overline{\nu_A}}{\overline{E_A} - \overline{E_A}\,\overline{\nu_T} - 2\overline{E_T}(\overline{\nu_A})^2}\\
\overline{n} &= \frac{\overline{E_A}^2(1 - \overline{\nu_T})}{\overline{E_A} - \overline{E_A}\,\overline{\nu_T} - 2\overline{E_T}(\overline{\nu_A})^2}\\
\overline{m} &= \frac{\overline{E_T}}{2(1 + \overline{\nu_T})}\\
\overline{p} &= \overline{G}.
\end{aligned} \tag{18.42}$$

18.4. Application to a Geologic Material

As an example of an application of the homogenization methods described above, consider the sample of a geomaterial shown in Fig. 18.1. The figure shows a backscatter image of an oil shale sample. The material represented by the backscatter image is assumed, in this case, to be composed of two species. While the identity and associated properties of each species can be arbitrarily chosen, the matrix material is assumed to be clay (black) and the inclusions are assumed to be mainly kerogen (white). The bulk and shear moduli for clay are 20.9 GPa and 6.85 GPa, respectively; white regions are assumed to be kerogen with bulk and shear moduli of 2.9 GPa and 2.7 GPa, respectively. The volume fractions of the matrix and inclusion phases are determined using image processing algorithms, and effective

Table 18.1. Effective properties of Clay/Kerogen sample shown in Fig. 18.1 calculated using DE and SCM. Young's and shear moduli are in units of GPa.

	DE	SCM
% Kerogen	19.33	19.33
E_A	16.16	16.21
E_T	14.10	14.78
ν_A	0.318	0.330
ν_T	0.265	0.308
G	5.70	5.76

elastic properties are calculated using the Dilute Estimate (DE). The results of the analyses are given in the first column of Table 18.1. Notice that the kerogen volume fraction is very close to the value for which the DE solution will fail, particularly for the normalized Hill moduli, k and n. Note also that the elastic constants are obtained using the relations shown in Eq. 18.25 after the effective Hill moduli were determined via the use of Eq. 18.32 and its associated formulations and assumptions associated with a transversely isotropic material.

The image shown in Fig. 18.1 is next analyzed following the Self Consistent Method (SCM), Eq. 18.35. Again, to maintain consistency with the previous example, the matrix material is assumed to be clay (black) and inclusions kerogen (white); the bulk and shear moduli for clay are assigned values of 20.9 GPa and 6.85 GPa, respectively, and the bulk and shear moduli for kerogen are 2.9 GPa and 2.7 GPa, respectively. The results of the SCM analysis are shown in the second column of Table 18.1. It is reasonable to consider the results shown in Table 18.1 as a good estimate of the effective properties of the sample due to the fact that the normalized Hill moduli were well within bounds for all values of inclusion volume fractions.

18.5. Fabric Tensor — Elastic Stiffness Relations

In addition to computing homogenized elastic constants for a porous solid, one may be interested in the behavior of these constants as the volume fraction is varied. One example of this situation concerns trabecular bone as the solid volume fraction is decreased due to osteoporosis. Predictions of the elastic stiffness tensor as a function of the solid volume fraction can be made provided a measure of the orientation distribution function of the void structure is available. One method for quantifying this distribution is

Fig. 18.1. Backscatter image of mineral/non-mineral sample of petroleum-bearing rock, approximately 2.5 cm on edge.

the fabric tensor. The fabric tensor is a second-rank tensor used to describe the nature of the size, shape, and distribution of the voids in a porous solid. For porous solids, the fabric tensor may be constructed from mean intercept lengths [23], volume orientation distribution [24], star-volume distribution [25], or the star-length distribution [26]. The mean intercept length (MIL) definition of the fabric tensor is used here, and all methods provide very similar results in applications [27]. The mean intercept length is simply the average distance between two solid/void interfaces in a given direction. As discussed in [23], the distribution of the MIL forms an ellipsoid, and provides a second-rank fabric tensor $\mathbf{H}$.

Three slightly different definitions of the second-rank fabric tensor are encountered in the literature [23], [28], [7]: **H, M** and **G**. All are related via

$$\mathbf{M} = \mathbf{H}^{-\frac{1}{2}} = v_s \mathbf{I} + \mathbf{G} \tag{18.43}$$

where v_s is the solid volume fraction. Also, $\mathbf{M}$ must be normalized such that Tr $\mathbf{M} = 3$ in order for certain constants appearing in the fabric tensor-stiffness tensor relationship to have physical meaning. Following Eq. 18.43, the eigenvalues m_i of the normalized fabric tensor $\mathbf{M}$ are related to the eigenvalues h_i of the fabric tensor $\mathbf{H}$ by

$$m_i = \frac{3}{\sqrt{h_i}} \frac{\sqrt{h_1 h_2 h_3}}{\sqrt{h_1 h_2} + \sqrt{h_1 h_3} + \sqrt{h_2 h_3}} \tag{18.44}$$

Relations between the stiffness and fabric tensors have been established using two different approaches [7, 28, 29]. Here, we will follow the approach of [7]. The relations between the orthotropic elastic constants and the fabric tensor are given by [7]

$$E_i = E_0 v_s^k m_i^{2\ell}$$

$$\nu_{ij} = \nu_0 \left(\frac{m_j}{m_i}\right)^{\ell}$$

$$G_{ij} = G_0 v_s^k m_i^{\ell} m_j^{\ell} \tag{18.45}$$

where k, ℓ are constants and E_0, ν_0 and G_0 are the Young's modulus, the Poisson ratio, and the shear modulus of the matrix material ($v_s = 1$) provided the fabric tensor is properly normalized. Note in Eq. 18.45 that the Poisson ratio is independent of v_s. This is supported by many experimental studies included in [30]. Eqs. 18.45 were used in [31] in the context of boundary element analysis to predict the effect of solid volume fraction on principal stresses in anisotropic bimaterials.

18.6. Prediction of Elastic Constant Variation with Porosity in a Ceramic Specimen

To demonstrate the use of the fabric tensor in predicting the behavior of elastic constants with varying porosity, a micrograph of a porous lead zirconate titanate (PZT 95/5) ferroelectric ceramic is analyzed. This specimen was characterized in [32] and the micrograph is shown in Fig. 18.2. Here,

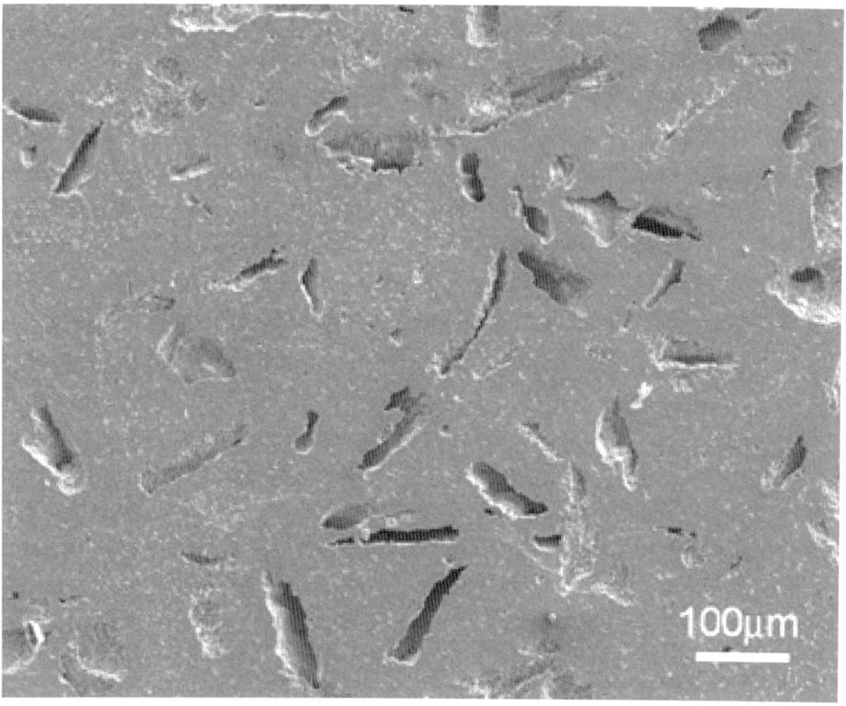

Fig. 18.2. PZT ceramic. Reprinted from Ref. [32] with permission from Elsevier.

the fabric tensor will be constructed, apparent elastic stiffness tensors calculated using a boundary element analysis, and Eqs. 18.45 used to predict the elastic stiffnesses as a function of solid volume fraction.

To construct the fabric tensor, mean intercept length measurements are obtained on the micrograph of the ceramic specimens using commercial computerized tomography software*. The mean intercept lengths are used to construct the fabric tensor $\mathbf{H}$, then the eigenvalues are extracted and properly normalized as m_1 and m_2 for the two-dimensional analysis. For the two-dimensional problem studied here, $m_2 = m_3$ in order for $E_2 = E_3$, $\nu_{12} = \nu_{13}$, and $G_{12} = G_{13}$ from Eq. 18.45. Also, the eigenvectors corresponding to the fabric tensor eigenvalues were extracted from the image analysis, these are needed to know the orientation of the principal material axes relative to the (x, y) coordinates fixed by the boundaries of the micrograph. It is in the principal material axes where the relation-

*Quant3D from the High Resolution X-ray CT Facility at the University of Texas Austin ships given by Eq. 18.45 are valid. It has been shown in previous studies [29, 33, 34] that the principal axes of the fabric tensor are coincident with the principal axes of the material. The image analysis provides values of $m_1 = 1.0612, m_2 = 0.9687$, and a rotation angle for the principal material axes of $\theta = 41.13°$.

Next, the ceramic micrograph shown in Fig. 18.2 is discretized for analysis by a symmetric Galerkin boundary element method as discussed in Ref. [11]. As discussed previously in this chapter, these models were subjected to boundary conditions which satisfy the Hill criterion. The specific boundary conditions used in the present analysis were:

(1) Kinematic Uniform Boundary Conditions (KUBC).
The displacement u_j is imposed at point x_j belonging to the boundary such that

$$u_i = \varepsilon^0_{ij} x_j \quad \forall \mathbf{x} \in \Gamma_0$$

where ε^0_{ij} is a constant strain tensor prescribed *a priori*. Note that this boundary condition satisfies Eq. 18.6 as it produces uniform strain in the volume.

(2) Static Uniform Boundary Conditions (SUBC).
A uniform traction t_j is imposed on the boundary such that:

$$t_i = \sigma^0_{ij} n_j \quad \forall \mathbf{x} \in \Gamma_0$$

where σ^0_{ij} is a constant stress tensor prescribed a priori. This boundary condition produces homogeneous stress in the volume.

(3) Mixed Uniform Boundary Condition (MUBC).
There are several choices for MUBC which will satisfy the form of the Hill criterion given in Eq. 18.7,

$$\left(t_i - \sigma^0_{ij} n_j\right)\left(u_i - \varepsilon^0_{ij} x_j\right) = 0 \quad \forall \mathbf{x} \in \Gamma_0$$

Appropriate choices for MUBC are reviewed in Ref. [9].

The average stresses and strains in the material are computed from the boundary element analysis, subjected to the boundary conditions given above. This provides a system of five equations for the five material properties needed to compute the apparent stiffness tensor. Note that since the apparent stiffness tensor needs to be computed in the principal material coordinates, the stresses and strains used in the volume averaging must be transformed to the principal material coordinates. This is straightforward

Table 18.2. Results for the apparent elastic constants in the PZT specimen. (E and G values are in GPa).

BCs	KUBC	SUBC	MUBC
E_1	101.10	91.20	92.76
E_2	86.90	80.80	85.32
G_{12}	43.11	31.98	35.18
ν_{12}	0.25	0.21	0.22

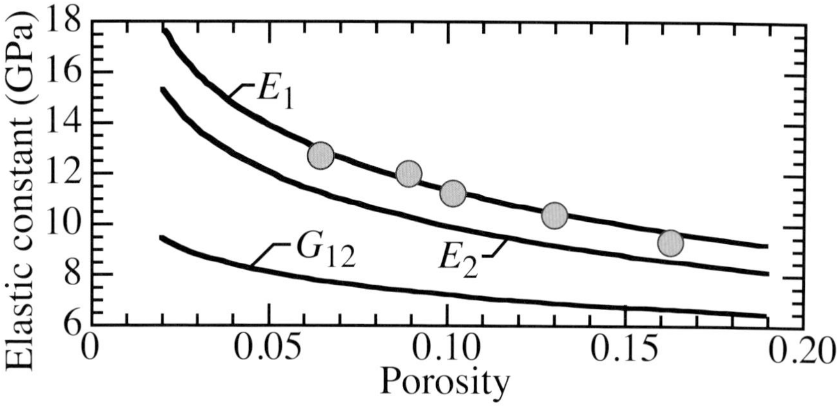

Fig. 18.3. Variation of elastic moduli with porosity for PZT1 (irregular pores). The circles represent the experimentally measured values from Ref. [32].

since the rotation angle is known from the eigenvectors of the fabric tensor. The results for the elastic constants are shown in Table 18.2.

Having estimates of the apparent stiffness tensor as well as the eigenvalues of the fabric tensor, the constants k, ℓ appearing in Eq. 18.45 may now be estimated. For the bulk isotropic material properties E_0, ν_0 values for PZT from [32] of $E_0 = 165$ GPa and $\nu_0 = 0.22$, are used. G_0 can be calculated from E_0 and ν_0 since the bulk material is assumed to be isotropic. The MUBC results from Table 18.2 are used to perform a least-squares estimate which yields $k = 0.35$ and $\ell = 0.48$. Predictions for E_1, E_2 and G_{12} as functions of porosity (1-v_s) using Eq. 18.45 and the estimates for k and ℓ are shown in Fig. 18.3.

18.7. Summary

Homogenization methods can be of great help in performing stress analysis of materials containing voids and multiple constituents. Methods based on elastic inclusion solutions such as the dilute estimate, and the self consistent method, were reviewed. These methods were applied to a geologic specimen, and effective properties were extracted. Additional homogenization methods based on the fabric tensor concept were reviewed, and these methods allow for the variation of elastic constants with porosity to be predicted. An example application of the fabric tensor approach on a porous ceramic demonstrated the usefulness, and accuracy, of the method.

Acknowledgments

The authors gratefully acknowledge helpful discussions with Prof. P. A. Martin in the Applied Mathematics and Statistics Department at the Colorado School of Mines.

References

[1] R. Hill, Elastic properties of reinforced solids: some theoretical principles, *J. Mech. Phys. Solids.* **11**, 357–372 (1963).

[2] E. Kröner, Berechnung der elastischen konstanten des vielkristalls aus den konstanten des einkristalls, *Zeitschrift für Physik.* **151**, 504–518 (1958).

[3] Z. Hashin, On elastic behaviour of fibre reinforced materials of arbitrary transverse phase geometry, *J. Mech. Phys. Solids.* **13**, 119–134 (1965).

[4] Z. Hashin and S. Shtrikman, A variational approach to the theory of the elastic behaviour of multiphase materials, *J. Mech. Phys. Solids.* **11**, 127–140 (1963).

[5] M.-J. Pindera, H. Khatam, A. S. Drago, and Y. Bansal, Micromechanics of spatially uniform heterogeneous media: a critical review, *Composites: Part B.* **40**, 349–378 (2009).

[6] J. Eshelby, The determination of the elastic field of an ellipsoidal inclusion, and related problems, *Proc. Roy. Soc. London A.* **241**, 376–396 (1957).

[7] P. K. Zysset and A. Curnier, An alternative model for anisotropic elasticity based on fabric tensors, *Mech. Mater.* **21**, 243–250 (1995).

[8] S. Hazanov and C. Huet, Order relationships for boundary conditions effect in heterogeneous bodies smaller, than the representative volume, *J. Mech. Phys. Solids.* **42**, 1995–2011 (1994).

[9] D. H. Pahr and P. K. Zysset, Influence of boundary conditions on computed apparent elastic properties of cancellous bone, *Biomech. Model. Mechan.* **7**, 463–476 (2008).

[10] C. Huet, Application of variational concepts to size effects in elastic heterogeneous bodies, *J. Mech. Phys. Solids.* **38**, 813–841 (1990).

[11] B. Elmabrouk, J. Berger, A.-V. Phan, and L. Gray, Apparent stiffness tensors for porous solids using symmetric galerkin boundary elements, *Comput. Mech.* **49**, 1411–419 (2012).

[12] S. Hazanov and M. Amieur, On overall properties of elastic heterogeneous bodies smaller than the representative volume, *Int. J. Eng. Sci.* **33**, 1289–1301 (1995).

[13] T. Mori and K. Tanaka, Average stress in matrix and average elastic energy of materials with misfitting inclusions, *Acta Metall.* **21**, 571–574 (1973).

[14] B. Klusemann and B. Svendsen, Homogenization methods for multi-phase elastic composites: Comparisons and benchmarks, *Technische Mechanik.* **30**, 374–386 (2010).

[15] P. C. neda and J. Willis, The effect of spatial distribution on the effective behavior of composite materials and cracked media, *J. Mech. Phys. Solids.* **43**, 1919–1951 (1995).

[16] J. Qu and M. Cherkaoui, *Fundamentals of Micromechanics of Solids.* John Wiley and Sons, Hoboken New Jersey (2006).

[17] D. G, *Micromechanics of Composite Materials.* Springer Science and Business Media, New York, New York (2013).

[18] L. Walpole, On the overall elastic moduli of composite materials, *J. Mech. Phys. Solids.* **17**, 235–251 (1969).

[19] R. Hill, Theory of mechanical properties of fibre-strengthened materials: I. elastic behavior, *J. Mech. Phys. Solids.* **12**, 199–212 (1964).

[20] W. Parnell and I. Abrahams, *An Introduction to Homogenization and Its Applications.* Cambridge University Press, Cambridge, United Kingdom (2015).

[21] W. Parnell and C. Calvo-Jurado, On the construction of the hashin-shtrikman bounds for transversely isotropic two-phase linearly elastic fibre reinforced composites, *J. Eng. Math.*

[22] K. Markov and L. Preziosi, *Heterogeneous Media: Micromechanics Modeling, Methods and Simulations.* Springer, New York, New York (2000).

[23] T. P. Harrigan and R. W. Mann, Characterization of microstructural anisotropy in orthotropic materials using a second rank tensor, *J. Mat. Sci.* pp. 761–767 (1984).

[24] A. Odgaard, E. B. Jensen, and H. J. G. Gundersen, Estimation of structural anisotropy based on volume orientation: A new concept, *J. Microscopy.* **159**, 335–342 (1990).

[25] L. M. Cruz-Orive, L. M. Karlsson, and S. E. Larsen, Characterizing anisotropy: a new concept, *Micron and Microscopia Acta.* **23**, 75–76 (1992).

[26] T. H. Smit, E. Schneider, and A. Odgaard, Star length distribution: a volume-based concept for the characterization of structural anisotropy, *J. Microscopy.* **191**, 249–257 (1998).

[27] B. Elmabrouk and J. Berger, Boundary element analysis for effective stiffness tensors: effect of fabric tensor determination method, *Comput. Mech.* **51**, 391–398 (2013).

[28] S. C. Cowin, The relationship between the elasticity tensor and the fabric

tensor, *Mech. Mater.* **4**, 137–147 (1985).

[29] C. H. Turner and S. C. Cowin, Dependence of elastic constants of an anisotropic porous material upon porosity and fabric, *J. Mat. Sci.* **22**, 3178–3184 (1987).

[30] L. J. Gibson and M. F. Ashby, *Cellular Solids: Structure and Properties, Second Edition.* Cambridge Solid State Science Series, Cambridge University Press, Cambridge, United Kingdom (1999).

[31] J. R. Berger, Fabric tensor based boundary element analysis of porous solids, *Eng. Anal. Bound. Elem.* **35**, 430–435 (2011).

[32] T. Zeng, X. Dong, C. Mao, Z. Zhou, and H. Yang, Effect of pore shape and porosity on the properties of porous pzt 95/5 ceramics, *J. Eur. Cera. Soc.* **27**, 2025–2029 (2007).

[33] A. Odgaard, J. Kabel, B. van Rietbergen, M. Dalstra, and R. Huiskes, Fabric and elastic principal directions of cancellous bone are closely related, *J. of Biomech.* **30**, 487–495 (1997).

[34] C. H. Turner and S. C. Cowin, Errors induced by off-axis measurements of the elastic properties of bone, *J. Biomech. Eng.* **110**, 213–215 (1988).

Chapter 19

The Critical Role of Pore Characteristics on the Performance of Novel Membrane Processes for Water Treatment and Desalination

Tzahi Y. Cath

Civil & Environmental Engineering
Colorado School of Mines, Golden, CO 80041, USA
tcath@mines.edu

Synthetic membranes for novel separation processes and energy recovery systems require unique characteristics and morphology to facilitate efficient mass transport of both solvents and solutes through the porous support layer of these membranes. The quick advancement of membrane separation processes began with the successful development of thin film composite membranes that enabled the combination of high permeability of solvent (water), high selectivity (solute solvent separation), and high mechanical strength provided by thin, robust, and highly porous polymeric support layer. While in traditional membrane separation processes such as reverse osmosis and nanofiltration the porosity, tortuosity, and thickness play lesser role in mass transport and selectivity, novel thermally driven and osmotically driven separation processes strongly depend on these characteristics of the porous support layer. This chapter introduces these two groups of processes and their potential applications in water treatment and energy recovery from renewable sources. Further development of optimized membranes for osmotically and thermally driven membrane processes, and especially their porous structure, is critical for the broad implementation of these technologies.

19.1. Introduction

Synthetic polymeric and inorganic membranes are becoming integral and critical component of water and wastewater treatment and of energy generation and recovery systems. Some membranes are porous throughout their depth and other membranes use porous material as a substrate for the support of a dense semi-permeable thin layer. While in some applications only the average pore diameter plays a major role in the separation and rejection of specific constituents, in specific novel membrane processes the membrane porosity, thickness, hydrophobicity, and pore geometry play a significant role in the performance and feasibility of the separation process. Thermally driven and osmotically driven membrane processes are being extensively studied at CSM and around the world for their use in a broad range of desalination and industrial water and wastewater treatment applications. This manuscript highlights the significance of pore characteristics of the membranes in these processes.

19.2. Thermally Driven Membrane Processes

Around the world, reverse osmosis (RO) is the dominant commercial process for desalination of brackish water and seawater. RO is also increasingly being utilized for treatment of impaired water and effluents from wastewater treatment processes, facilitating advanced reclamation and reuse of water. However, the two main drawbacks of RO processes are high energy demand due to high operating pressures and fouling and scaling of the membranes by organic and inorganic constituents. Due to the high osmotic pressure of seawater and other saline streams, practical seawater desalination is limited to approximately 50% water recovery (half of the treated volume becomes purified product water), thus generating large volumes of brine. The disposal of large volumes of brine represents inefficiencies associated with pretreatment of seawater before RO desalination (i.e., both physical and chemical processes) and with the loss of chemical energy embedded in the brine (see description on pressure retarded osmosis).

Unlike in pressure driven membrane processes such as RO, osmotic pressure has negligible effects on distillation processes. Of unique interests are distillation processes through membranes, which provide a large surface area to volume ratio and low cost material of construction. In thermally driven membrane processes [1] a microporous, hydrophobic membrane is placed between a heated saline feed stream and a colder fresh water stream (distillate). The vapor pressure difference across the membrane facilitates diffusion of water vapors through the pores, and thus production of ultrapure water from even hypersaline, mildly heated streams (Fig. 19.1). While membrane distillation is still a process under development, many applications have been successfully tested, including high recovery desalination of seawater, recovery of minerals, treatment of blowdown water from cooling towers, and recovery and reuse of water in life support systems (space applications).

In membrane distillation the microporous hydrophobic membrane plays

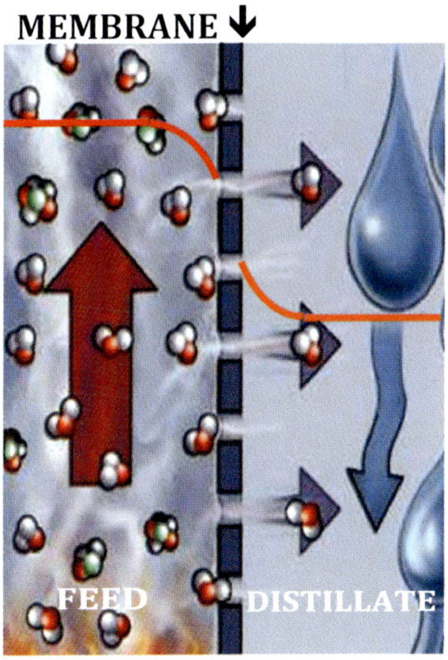

Fig. 19.1. Evaporative mass transport through a membrane distillation microporous hydrophobic membrane.

a critical role in establishing vapor-liquid interfaces, and it has to facilitate efficient heat and mass transport through the membrane. Porosity, pore tortuosity and size, membrane thickness, and the hydrophobicity of the membrane are critical to successful and cost effective desalination. Two of the most important properties of membrane distillation membranes are membrane material and pore diameter.

19.2.1. *Membrane Material*

Microporous membranes for membrane distillation must possess hydrophobic characteristics to prevent wetting of the membrane material and spontaneously wet the pores. The hydrophobicity of the membrane and the high surface tension of the working fluids (feed and distillate) facilitate liquid-vapor interface at the entrance to each pore of the membrane. At the feed-membrane interface, water evaporates and diffuses through the pore in a vapor phase until it reaches the membrane-distillate interface on the distillate side of the membrane. If the membrane loses its hydrophobicity or the surface tension of the water drops, the pore will be flooded and the separation capabilities of the membrane/process will be lost. It is worth noting that membrane distillation is a versatile process and can be used to separate volatile compounds from water. In this case, the more volatile solutes are the constituents that will diffuse more rapidly through the membrane's pores and will be concentrated on the distillate side of the membrane. Traditionally, hydrophobic polymers are used in the fabrication of membrane distillation membranes. These include polytetrafluoroethylene (PTFE/Teflon), polypropylene, and Polyvinylidenefluoride (PVDF). Other novel hydrophobic materials, including carbon nanotubes, might in the future become good building blocks for membrane distillation membranes.

19.2.2. *Pore Size*

The nominal pore size of membrane distillation membranes, in conjunction with the surface tension of the liquids in contact, dictate the entry pressure of water, which is the minimum hydraulic pressure that will break the water-vapor interface at the pore entrance and flood the pores. For most practical purposes with membranes that are currently being developed and evaluated, the nominal diameter of the pore is between 0.1 and 0.4 micron. Membrane distillation membranes having this range of pore diameter will have entry pressure range of 80 to 30 psi, respectively.

19.2.3. *Pore Tortuosity and Membrane Thickness*

Once water vapors start diffusing through the pores, the resistance to vapor diffusion will dictate the efficiency of the process, measured through water throughput or flux across the membrane, in the case of desalination, and conductive heat transport across the membrane from the feed to the distillate. Thus, thin membrane (active layer thickness of less than 10 microns) and pores with low tortuosity (as close as possible to cylindrical pores) will best suit for membrane distillation. Highly porous membrane ($> 70\%$ porosity) will enable both higher water flux and low conductive heat loss from the feed into the distillate.

19.3. Thermally Driven Membrane Processes: Novel Applications and Process Intensification

Increased water demand and the need to improve process efficiency resulted in the exploration of membrane distillation in applications such as concentrate management, mineral recovery, and energy recovery. Recent investigation at CSM focused on membrane distillation desalination of Great Salt

Lake brines for enhancement of mineral recovery. Water flux through unoptimized, currently available membrane distillation membranes was sustained and enabled concentration of Great Salt Lake water from approximately 150,000 mg L^{-1} of total dissolved solids (TDS) to more than 300,000 mg L^{-1} TDS (more than 8 times the concentration of sea water). Preliminary economic evaluation revealed that membrane distillation is more than 150 times faster than natural evaporation (current industry practice) and that an acre of evaporation pond can be replaced with a membrane module encased in a one meter long and 20-cm diameter vessel.

A DOE/ARPA-e funded investigation conducted by CSM, in collaboration with Yale University, focused on membrane distillation in a novel hybrid system that converts low-grade heat from industrial or natural sources into useful energy. The novel closed-loop osmotic heat engine (Fig. 19.2) competes with well-established processes that implement the Organic Rankine Cycle (ORC), using low-grade heat in the membrane distillation process to produce a concentrated brine and a distillate stream from a diluted saline water. The two streams produced by the membrane distillation process are being used in one of two processes (i.e., pressure retarded osmosis or reverse electrodialysis) to produce useful energy and diluted brine. The diluted brine is being used in the membrane distillation system to regenerate the concentrated brine and distillate for the next process cycle. One of the main challenges of the hybrid process is the development of optimized, durable, and robust membrane distillation membranes to enable the operation of the hybrid process.

19.4. Osmotically Driven Membrane Processes

In osmotically driven membrane processes a unique semipermeable membrane is place between a feed stream and a concentrated brine (draw solution), and the osmotic pressure difference between the streams facilitates transport of water molecules by diffusion from the feed into the draw solution [2]. The two main osmotically driven membrane processes are forward osmosis and pressure retarded osmosis (Fig. 19.3). While the two processes use similar physical and chemical principles and process components, they are very different from the application perspective. Forward osmosis uses the osmotic pressure difference across the membrane to extract clean water and to concentrate the feed stream, or to dilute the concentrated draw solution stream for downstream processing with minimal energy consumption (e.g., osmotic dilution). Pressure retarded osmosis, on the other hand,

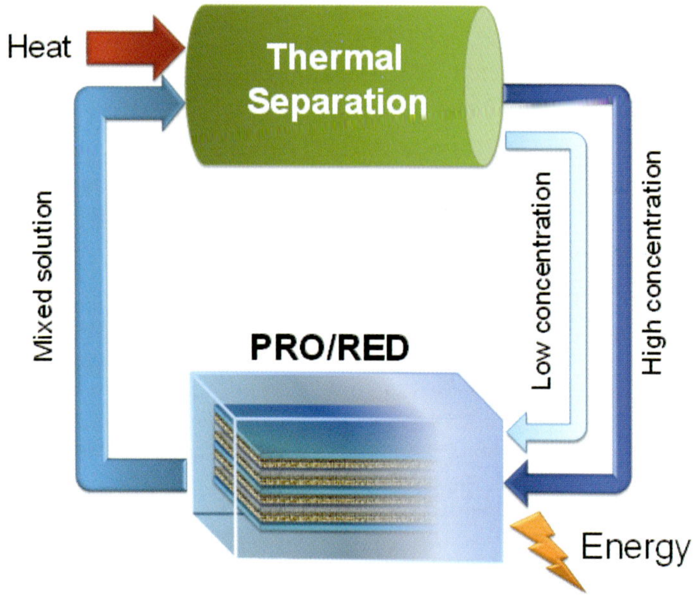

Fig. 19.2. A closed loop osmotic heat engine incorporating membrane distillation and pressure retarded osmosis.

facilitates conversion of osmotic pressure difference to hydraulic pressure difference, thus enabling generation of useful work (e.g., electricity) when releasing the hydraulic pressure through a turbine or other devices. As such, pressure retarded osmosis is a unique membrane process. Its main objective is to convert and recover energy, rather than to perform chemical separation.

Forward osmosis and pressure retarded osmosis use the same type of semipermeable membrane, which is different in structure from thin film composite membranes used for the RO and nanofiltration (NF). While separation by an osmosis membrane is accomplished using a very thin and dense polymeric film, the microporous and nanoporous support layers of these membranes play a significant role in both providing a mechanical support for the thin dense layer and facilitating efficient mass transport through the membrane. Characteristics similar to those of membranes used for thermally driven processes, substantially reduce concentration polarization effects inside and outside osmosis membranes.

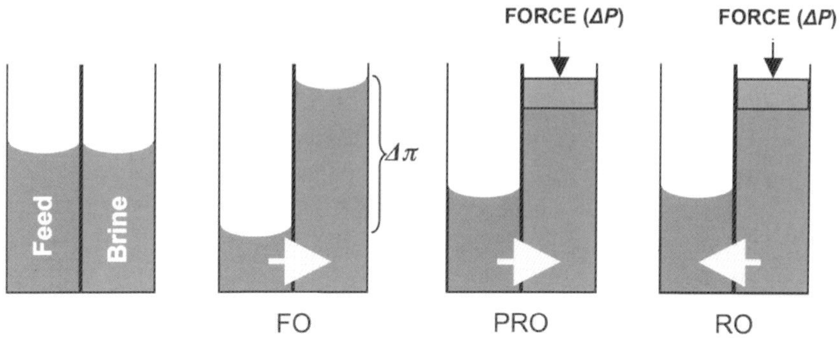

Fig. 19.3. Solvent flows in FO, PRO, and RO. For FO, ΔP is approximately zero and water diffuses to the more saline side of the membrane. For PRO, water diffuses to the more saline liquid that is under positive pressure ($\Delta\pi > \Delta P$). For RO, water diffuses to the less saline side due to hydraulic pressure ($\Delta P > \Delta\pi$).

19.5. The Effect of Membrane Properties on the Performance of Osmotic Membranes

Water flux (J_W) through osmosis semipermeable membranes depends on the permeability (A) of the membrane and the driving force, which is commonly the osmotic pressure (π) difference across the membrane:

$$J_W = A(\pi_D - \pi_F) \tag{19.1}$$

This equation, however, assumes a well-mixed system without the presence of boundary layers, which in forward osmosis and pressure retarded osmosis occur on the feed and draw solution sides of the membrane and inside the porous support layer of the membrane. McCutcheon et al. [3, 4] presented a model for osmotic flux across a dense, symmetric osmosis membrane:

$$J_W = A\left(\pi_{D,b}\exp\left(\frac{-J_W}{k_D}\right) - \pi_{F,b}\exp\left(\frac{J_W}{k_F}\right)\right) \tag{19.2}$$

where $\pi_{D,b}$ and $\pi_{F,b}$ are the bulk osmotic pressures of the draw and feed solutions, respectively, and k_D and k_F are the mass transfer coefficients on the draw and feed solution sides of the membrane, respectively. This implicit flux model incorporates concentration polarization moduli that account for boundary layer phenomenon on both sides of the membrane. The solutes are rejected on the feed side (dense active layer) of the membrane, resulting in an increase in local concentration near the membrane interface on the feed side, and dilution (a negative exponential term) on the draw solution side of the membrane as water that permeates the membrane dilutes the draw solution at the downstream interface.

However, even Eq. 19.2 is incomplete as today's osmosis membranes are asymmetric and comprise a thin selective layer and a thick, non-selective, porous support layer. Because the effective osmotic pressure of a solution is only established at the interface with the selective layer, the asymmetric structure of a membrane ensures that one of the boundary layers occurs within the porous support layer, resulting in internal concentration polarization (ICP) [3, 5, 6]. To account for this change, an effective mass transfer coefficient, k_{eff}, was defined, which takes into account the impact that the porous support layer has on mass transfer:

$$k_{\text{eff}} = \frac{D_S \varepsilon}{\tau \delta} = \frac{D_S \varepsilon}{\tau t} \tag{19.3}$$

Here, D_S is the diffusivity of the draw solution solute, δ is the thickness of the boundary layer (here assumed as the thickness of the porous support layer), and ε, τ, and t are the porosity, tortuosity, and thickness of the porous support layer of the membrane, respectively.

ICP has been investigated in many studies and it is widely considered to be one of the significant obstacles to further development of forward osmosis and pressure retarded osmosis processes. Salt flux during osmosis is also a major factor in the performance of the processes. Recent studies provided thorough experimental and modeling approaches to predict forward and reverse diffusion of electrolytes through semipermeable membranes during forward osmosis and pressure retarded osmosis [7–12]. One useful equation that relates the water flux, J_W, and reverse solute flux, J_S, has been developed by Phillip, et al. [9] and Tang, et al. [13]:

$$\frac{J_W}{J_S} = \frac{A}{B} n R_g T \tag{19.4}$$

where B is the overall solute permeability coefficient of the membrane, n is the number of species that the draw solute dissociates into ($n = 2$ for NaCl), R_g is the gas constant, and T is the absolute temperature. This relationship indicates that the ratio of the water and reverse salt flux is a function of the transport properties of the membrane active layer and is independent of the structure of the support layer.

It can be seen from Eqs. 19.1-19.4 that the membrane transport properties and operating conditions have a major influence on the performance of

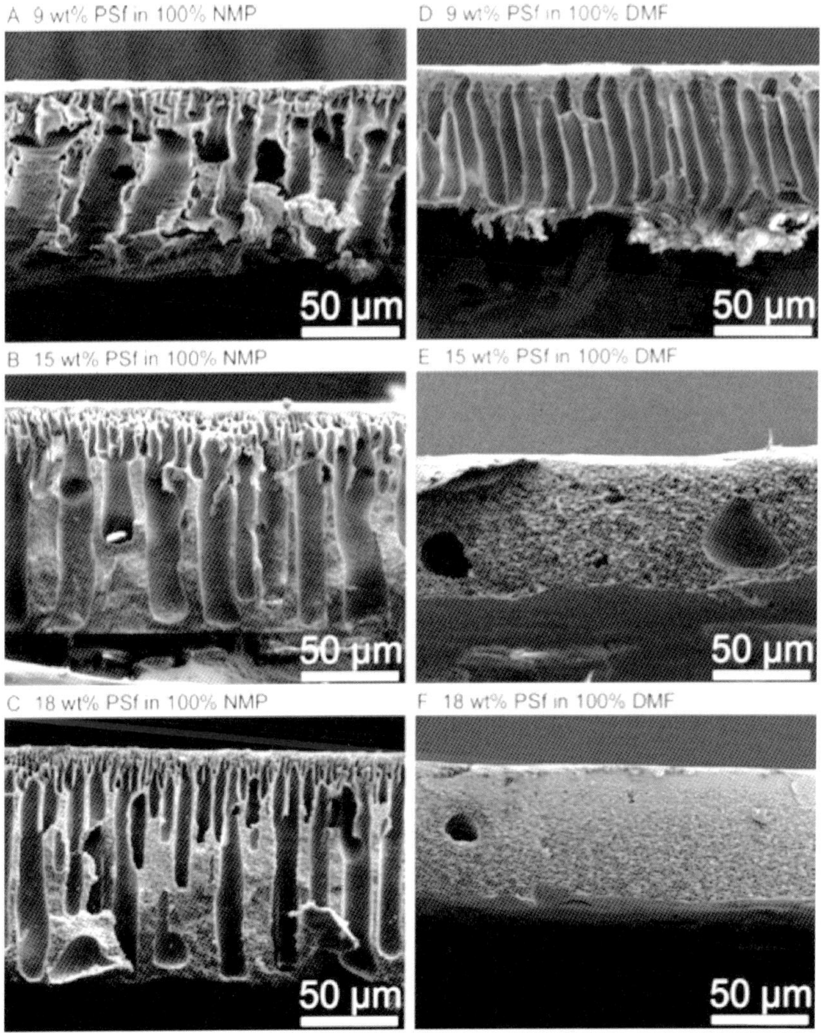

Fig. 19.4. Cross section view of novel porous support layers for osmosis membranes (Elimelech et al., Yale U).

forward osmosis and pressure retarded osmosis. Therefore, accurate characterization of the membrane properties is critical to the prediction of process performance under given operation conditions of forward osmosis and pressure retarded osmosis.

19.6. Novel Membranes for Osmotic Membrane Processes

Novel membranes for osmotically driven membrane processes are being developed around the world. The main goal in current development of these membranes is to produce a membrane with a thin active layer, which is chemically stable (e.g., polyamide), and a very thin and highly porous support layer having a high k_{eff} coefficient, and specifically low structural parameter (D_S/k_{eff}). Examples of currently developed membranes are shown in Fig. 19.4. The membranes are very thin and their pores are cylindrical (finger-shape). This provides tortuosity (τ) of close to unity, high porosity (ε), and low thickness, all of which reduce the structural parameter ($S = t\tau/\varepsilon$) of the novel membranes. Yet, membranes for pressure retarded osmosis also need to be robust and withstand high hydraulic pressure, which conflicts with the need of the membranes to be thin and porous.

19.7. Conclusions

While osmotically and thermally driven membrane processes and the materials used for their membranes are distinctively different, the structure of the membranes used for these processes, and especially the porous layer, is uniquely similar. Currently available membranes for membrane distillation are not porous enough (commonly less than 50% porosity) and their material of construction often loses hydrophobicity over time. Pore tortuosity of current membrane distillation membranes is too high and reduces vapor permeability. These two characters of the porous membrane are critical for further improvement of the process. For engineered osmosis membranes to be successful, simultaneous optimization of porosity, tortuosity, thickness, and robustness are critically needed. In both cases, the ability to control the production of unique pore structures and characteristics, with different materials, will be critical for the advancement of the technologies. Both groups of novel processes will play a significant role in future separation and energy and nutrient recovery applications, roles that current membrane processes in many cases cannot play.

Acknowledgments

We acknowledge the support received from the Advanced Research Projects Agency-Energy (ARPA–E), US Department of Energy, via Grant DE–AR0000306, and from the National Science Foundation (NSF), via Award

CBET–1236846.

References

[1] K. Lawson and D. Lloyd, Review: Membrane distillation, *J. Memb. Sci.* **124** (1), 1–25 (1997).

[2] T. Y. Cath, A. E. Childress, and M. Elimelech, Forward osmosis: Principles, applications, and recent developments, *J. Mem. Sci.* **281**(1–2), 70–87 (2006).

[3] J. R. McCutcheon and M. Elimelech, Influence of concentrative and dilutive internal concentration polarization on flux behavior in forward osmosis, *J. Memb. Sci.* **284**(1–2), 237–247 (2006).

[4] J. R. McCutcheon and M. Elimelech, Modeling water flux in forward osmosis: implications for improved membrane design, *AIChE Journal.* **53**, 1736–1744 (2007).

[5] G. D. Mehta and S. Loeb, Internal polarization in the porous substructure of a semi-permeable membrane under pressure-retarded osmosis, *J. Memb. Sci.* **4**, 261 (1978).

[6] G. T. Gray, J. R. McCutcheon, and M. Elimelech, Internal concentration polarization in forward osmosis: role of membrane orientation, *Desalination.* **197**(1–3), 1–8 (2006).

[7] N. T. Hancock, W. A. Phillip, M. Elimelech, and T. Y. Cath, Bidirectional permeation of electrolytes in osmotically driven membrane processes, *Environmental Science and Technology.* **45**(24), 10642–10651 (2011).

[8] N. T. Hancock and T. Y. Cath, Solute coupled diffusion in osmotically driven membrane processes, *Environ. Sci. Technol.* **43**, 6769–6775 (2009).

[9] W. A. Phillip, J. S. Yong, and M. Elimelech, Reverse draw solute permeation in forward osmosis: modeling and experiments, *Environ. Sci. Technol.* **44**, 5170–5176 (2010).

[10] J. Yong, W. Phillip, and M. Elimelech, Coupled reverse draw solute permeation and water flux in forward osmosis with neutral draw solutes, *J. Memb. Sci.* (2012). 10.1016/j.memsci.2011.11.020.

[11] X. Jin, C. Y. Tang, Y. Gu, Q. She, and S. Qi, Boric acid permeation in forward osmosis membrane processes: modeling, experiments, and implications, *Environ. Sci. Technol.* **45**(6), 2323–2330 (2011).

[12] A. Achilli, T. Y. Cath, and A. E. Childress, Power generation with pressure retarded osmosis: An experimental and theoretical investigation, *J. Memb. Sci.* **343**(1–2), 42–52 (2009).

[13] C. Y. Tang, Q. She, W. C. L. Lay, R. Wang, and A. G. Fane, Coupled effects of internal concentration polarization and fouling on flux behavior of forward osmosis membranes during humic acid filtration, *J. Memb. Sci.* **354** (1–2), 123–133 (2010).

Chapter 20

Microporous Crystalline Membranes and Their Application for CO_2 Separations

Moises A. Carreon and Tracy Q. Gardner

Chemical and Biological Engineering Department
Colorado School of Mines, Golden, CO 80401, USA
tgardner@mines.edu

Crystalline materials such as zeolites and metal organic frameworks (MOFs) have nanometer-scale pores and differing affinities for various gas and liquid molecules. Single crystal or polycrystalline films of these materials, typically on porous supports, form membranes that can separate fluid mixtures based on differences in adsorption and/or diffusion properties. This chapter highlights synthesis methods and key characteristics of zeolite and MOF materials and membranes, transport modeling in these materials, and three important examples of their application for CO_2 separations from flue gas, natural gas, and hydrogen mixtures.

20.1. Introduction

From environmental and energy perspectives, purification and recovery of carbon dioxide from power generation, industrial operation and natural gas drilling are of great interest. Carbon dioxide is the main greenhouse gas, and its accumulation in the environment contributes to global warming issues. The total CO_2 emissions in the US grew to an estimated 6,022 million metric tons (MMT) in 2007, representing more than 80% of the total greenhouse gas emissions in the US that year [1]. World CO_2 emissions in 2005 were estimated at 28,051 MMT and are projected to be 42,325 MMT in 2030 [2]. To alleviate CO_2 accumulation in the atmosphere, it is important to separate and recycle carbon dioxide from flue gases rather than emitting it into the air. From the energy point of view, CO_2 is an undesirable impurity in natural gas wells, with concentrations as high as 70% by volume. About 17% of all domestic natural gas in the US is treated to remove CO_2 before it is sent to the pipeline [3]. Pipeline specifications for natural gas require a CO_2 concentration below 2–3%, whereas raw natural gas can contain varying concentrations of CO_2 above this limit. It is desirable to separate CO_2 from CH_4 because CO_2 reduces the energy content of the gas and because it is acidic and corrosive in the presence of water. Furthermore, pumping CO_2 back down a natural gas well can enhance the recovery of CH_4. CO_2 is the leading enhanced oil recovery agent in the US, and is used to recover CH_4 from coal beds and to increase yield from oil reserves as well. Another relevant technology related to CO_2 capture involves the reaction of CH_4 and other fuels with oxygen or air to produce H_2 and CO (syngas). In the pre-combustion process the formed CO reacts with steam producing a mixture of CO_2 and H_2, which must be separated to obtain pure hydrogen. Membranes that preferentially permeate CO_2 over N_2, CH_4, H_2, and other light gasses at high selectivities can thus significantly reduce negative environmental impacts and improve the profitability of natural gas drilling, syngas production, and other industrial processes [4].

The benchmark technology used for CO_2 removal is amine adsorption, but amine plants require high capital and maintenance costs and are complex and labor intensive to operate [3]. Removing CO_2 from flue gas, natural gas streams, and pre-combustion processes involves treating enormous gas volumes. Membrane technology could play a key role in making CO_2 removal simpler and less expensive compared to amine adsorption [5] since membrane process equipment is simple and easy to operate, control and scale-up and gas separation would be less energy intensive since no phase transformation is required.

Zeolites [6] and metal organic frameworks (MOFs) [7, 8] in membrane form are crystalline microporous materials with uniform micropores, high surface areas, and exceptional thermal and chemical stability, making them ideal candidates for CO_2 gas separations. In this chapter we review the research progress on zeolite and metal organic framework membranes, which have demonstrated ability to separate carbon dioxide from different light gases. Though these materials can be used for a wide variety of gas/liquid separations and the fundamental characteristics and modeling discussed here apply to many different separations, we focus primarily on CO_2/N_2, CO_2/CH_4 and CO_2/H_2 gas separations due to the important industrial applications mentioned above in flue gas treatment, natural gas purification, and hydrogen purification, respectively.

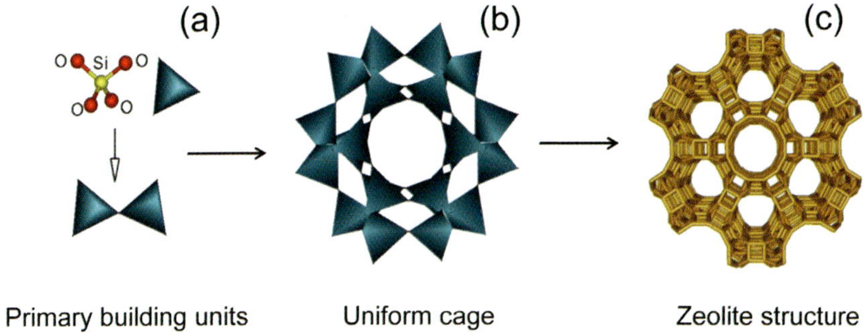

Fig. 20.1. Schematic representation of zeolite primary building units, periodic cages, and zeolite structure. Pore sizes of zeolites range from ~ 0.3–1.3 nm depending on structure type and chemical composition of the zeolite.

20.2. Zeolites and Zeolite Membranes

Zeolites are microporous, crystalline oxide minerals whose structures arise from three-dimensional networks of atoms, typically Si and Al though other metals such as Ga, Ge, B, Zn, Be, etc. can be substituted into the framework as well [9]. These metal atoms are referred to as "T-atoms". If each T-atom in a framework is Silicon (4+ charge) and is tetrahedrally linked to four other T-atoms through oxygen bridges, the framework will be neutral. If some Si^{4+} atoms in the framework are replaced by Al^{3+} atoms, for example, the framework has a net negative charge and can accommodate associated cations (e.g. Na^+, H^+) at each substituted site. The chemical composition of the zeolite framework and any associated cations determines its catalytic and ion-exchange properties, and the number of oxygen atoms in the rings of the crystalline structure determine the pore size, ranging from ~ 0.3–1.3 nm. For a given zeolite composition and structure the pore size is well defined, though it is possible for the zeolite framework to expand and contract slightly depending on the guest molecules within it. These molecular-sized interconnected pores and the electrical charge or polarity make zeolites effective for a wide range of reaction and/or separation applications. Figure 20.1 shows a schematic representation of (a) primary building units – T-atoms tetrahedrally bound to 4 oxygen atoms, forming (b) uniform cages, which connect into a (c) zeolite framework.

20.2.1. *Characteristics of Zeolite Membranes*

Zeolites can occur naturally and they can also be synthesized in either powder or film form in a laboratory. Large single crystals or dense films of zeolite crystals, either self-supported or more commonly grown on porous supports to provide mechanical stability, are called zeolite membranes. Zeolite membranes tend to be more mechanically, thermally, and chemically stable than polymeric membranes for many applications [6]. The molecular sieving ability of zeolite membranes is well known and has been exploited for many relevant gas mixture separations [10–14]. The rapid progress in synthesizing, characterizing, and utilizing zeolite membranes for separations, and in some cases combined separation and reaction, can be seen in these recent reviews [15–22].

As previously mentioned, a zeolite membrane can be prepared as a single crystal [23] or as a self-supported film [24], but typically, it is prepared as a continuous polycrystalline layer on a porous support to ensure mechanical strength [25]. Various porous materials have been used as supports for zeolite membranes – among them porous alumina [25] and stainless steel [26] are most common. Minimizing flow resistance is an important factor in choosing a support, especially for thin zeolite films. Asymmetric supports with good strength and low flow resistance are preferred. The geometry of the support is also an important consideration. Growing a zeolite film may be easier on a disc than on a tubular support, but a tube has a higher surface area to volume ratio and can be easier to incorporate into a flow module, making this geometry more amenable for scale-up.

Zeolite membrane synthesis, and hence the membrane properties, will be influenced by factors such as synthesis gel/solution composition; aging and pH of the gel; synthesis temperature, pressure, and time; chemical and structural nature of the support; position of the support and nature of the autoclave. Zeolite membranes are often prepared under conditions similar to those used to synthesize zeolite powders, which might not be optimal conditions for membrane formation. The goal of membrane synthesis is to form a continuous zeolite layer, and the approach to doing this will affect the membrane thickness, crystal orientation, number and size of defects, and even the chemical composition of the zeolite itself.

Ideally, zeolite membranes would be thin, leading to high flux, and defect-free, leading to high selectivity. However, there tends to be a tradeoff between flux and selectivity, which as mentioned can be affected by how the zeolite membranes are synthesized. The synthesis gel for making zeolites is usually composed of a water-based alkaline solution containing the silicon and aluminum precursors and a structure-directing agent, or template, that determines the structure of zeolite that will be formed. Because of the high alkalinity of the gel and the high reaction temperature required for hydrothermal synthesis, alumina supports can partly dissolve into the reactive solution, affecting the composition of the synthesized zeolite [27]. During in-situ crystallization, zeolite crystals nucleate and grow on the support surface from direct contact between the support and the alkaline precursor gel under hydrothermal conditions. In order to reduce the number and size of inter-crystalline voids to increase selectivity, most zeolite membranes are prepared with more than one hydrothermal synthesis layer. This leads to thicker membranes, and hence lower flux. Some creative synthesis methods to make thinner and/or more selective zeolite membranes have had some success.

Secondary growth of a seed layer is an effective approach for forming supported zeolite membranes and films with good quality and reproducibility [28]. In this "secondary growth" method, crystal seeds are deposited on a support, and then a dense film of crystals is grown through hydrothermal treatment of the seeded support in the presence of the precursor gel [29]. Membranes made by secondary growth can be relatively thin and even have preferentially oriented crystals, thus potentially leading to higher fluxes, compared to zeolite films grown without seeding [30]. One drawback to the secondary growth method is that the films can have more defects or non-zeolitic pores, reducing selectivity of the membranes. In an interesting one-step method (pore-plugging synthesis) zeolite crystals are grown within the pores of a macroporous support, yielding highly robust, defect-free membranes, however with relatively low permeances (hence low flux) due to the thickness of the zeolite layer inside the pores [31]. To obtain higher fluxes, penetration of the gel into the support should be inhibited. This can be done using a masking technique to plug the support pores during zeolite membrane synthesis by coating a poly(methyl methacrylate) (PMMA) layer on the support surface [32].

Zeolite membranes separate mixtures based on differences in molecular size (affecting diffusivity) and/or adsorption strength of the permeating species in the zeolite [17, 33]. Though the term "molecular sieving" makes it sound like they act as size filters, separating solely based on keeping larger molecules out and allowing smaller molecules through, the separation mechanism is more complex. Three separation regimes have been identified for binary gas mixture components through zeolite pores [34]. If both molecules have similar adsorption strengths under given conditions, the smaller molecule will tend to have a higher diffusivity and the membrane will be selective for that species. In the second regime, adsorption differences dominate and the more strongly adsorbing molecule, sometimes the larger molecule, will preferentially permeate. In the third regime, both diffusion and adsorption differences play a role in determining the membrane selectivity. For similar hydrocarbons, for example, there tends to be a tradeoff between adsorption strength and diffusivity – typically larger hydrocarbons adsorb more strongly but have lower diffusivity. For these separations selectivity can switch with operating conditions from favoring the smaller molecule at high temperature and/or low pressure, where adsorption coverages are low, to favoring the larger molecule at low temperature and/or high pressure, where adsorption coverages are high. For some CO_2 separations on the other hand, advantage can be taken of the third regime, as CO_2 can both adsorb more strongly and have a higher diffusivity than its competitor!

20.2.2. *Why Zeolite Membranes for CO_2 Separation?*

Carbon dioxide strongly adsorbs on several zeolites because it has a strong electrostatic quadrupolar moment (4.30×10^{-26} esu·cm^2), which significantly contributes to its attraction to polar surfaces [35]. Carbon dioxide preferentially permeates through some zeolites from CO_2/N_2, CO_2/CH_4, and CO_2/H_2 mixtures at low temperatures, because CO_2 adsorbs more strongly than the other gases. Carbon dioxide is also smaller (0.33 nm kinetic diameter) than CH_4 (0.38 nm) and N_2 (0.36 nm) and is preferentially permeated in the third regime at lower temperatures, where differences in diffusivity and competitive adsorption combine to enhance selectivity. At higher temperatures, the CO_2/N_2 or CO_2/CH_4 mixture can be separated by only differences in diffusivity (first regime). In addition, because CO_2 is smaller than CH_4 or N_2, small-pore zeolites with pore sizes close to the kinetic diameter of CH_4, can separate CO_2/CH_4 mixtures by true molecular sieving. For example, SAPO-34 with pore size $\sim$ 0.38 nm, zeolite Linde type T zeolite with pore size $\sim$ 0.41 nm, DD3R with pore size 0.36×0.44 nm, and AlPO-18 with pore size $\sim$ 0.38 nm (among others) can be used to separate CO_2 from CH_4 or N_2. In particular, these small-pore zeolite membranes have shown high CO_2/CH_4 selectivities due to a combination of differences in diffusivity and competitive adsorption.

20.2.3. *Modeling Transport in Zeolites*

Since zeolite pores are similar in size to the molecules diffusing through them, the molecules interact (often strongly) with the walls and diffusion is activated, with an activation energy that correlates with the heat of adsorption. This can be considered like surface diffusion, molecules moving from one site to the next, though through the material rather than over an exposed surface. Multicomponent diffusion through zeolite pores has been qualitatively and quantitatively well-modeled as Maxwell-Stefan surface diffusion with Langmuir adsorption [36–48]. Adsorption of light gases such as CO_2, N_2, and CH_4 in zeolites typically follows Langmuir type I, or single-site adsorption behavior [41], in which each site can only accommodate one molecule and the available sites are energetically equivalent. With these diffusion and adsorption models, the transient surface occupancies across the thickness of the membrane can be modeled by:

$$\frac{\partial \theta_i}{\partial t} = D_{\mathrm{MS},i} \frac{\partial}{\partial z} \left(\frac{1}{1-\theta_i} \frac{\partial \theta_i}{\partial z} \right), \tag{20.1}$$

where $D_{\mathrm{MS},i}$ is the Maxwell-Stefan diffusion coefficient, θ_i is the fractional coverage of component i and z is the dimension in the direction of the flux. Here θ_i is defined by:

$$\theta_i = \frac{q_i}{q_{\mathrm{sat,i}}} = \frac{b_i P_i}{1 + b_i P_i}, \tag{20.2}$$

where b_i is the adsorption equilibrium constant, P_i is the partial pressure and q_i and $q_{\mathrm{sat},i}$ are the coverage and the saturation coverage, respectively, of component i. The adsorption equilibrium constant varies exponentially with inverse temperature:

$$b_i = b_{0,i} \exp\left(-\frac{\Delta H_{\mathrm{ads},i}}{RT} \right), \tag{20.3}$$

where $\Delta H_{\mathrm{ads},i}$ is the heat of adsorption of component i on the zeolite. The Maxwell-Stefan diffusion coefficient has an Arrhenius-type temperature dependence as well:

$$D_{\mathrm{MS},i} = D^{\circ}_{\mathrm{MS},i} \exp\left(-\frac{E_{\mathrm{a},i}}{RT} \right), \tag{20.4}$$

where $E_{\mathrm{a},i}$ is the activation energy for diffusion of component i through the zeolite. As mentioned above, $E_{\mathrm{a},i}$ and $\Delta H_{\mathrm{ads},i}$ are often correlated since the diffusion mechanism involves adsorbing onto the surface, diffusing into the zeolite pore, diffusing from site to site within the pore, and desorbing back off of the zeolite on the downstream side. The flux through zeolite pores is proportional to the local coverage gradient:

$$J_i = -\rho q_{\text{sat},i} \frac{D_{\text{MS},i}}{1-\theta_i} \frac{\partial \theta_i}{\partial z}\bigg|_z, \tag{20.5}$$

where ρ is the zeolite density. Note that the Maxwell-Stefan diffusivity is assumed to be constant. Hence, at steady state the coverage gradient is steeper where the coverage is lower; that is, the coverage profile across the membrane is non-linear and decreases in slope from the feed side to the permeate side since components' coverages are greater at the feed side than the permeate side.

Equation 20.5 can be applied at $z = 0$ to model the flux into the zeolite and at $z = \delta$ (the membrane thickness) to model the flux out of the zeolite. Using this analysis, a transient permeance response to a step change in the feed concentration to a membrane can be used to determine steady state sorption amounts [40–44]. This method was used to measure adsorption isotherms and Maxwell-Stefan diffusivities of CO_2, N_2, and CH_4 in ZSM-5 zeolite membranes [41]. This technique was used also with a dual-site adsorption model to measure heats of adsorption for butane isomers in both cages and channels of ZSM-5 and ZSM-11 zeolites [42, 44]. Using this combined modeling and experimental technique to quantitatively characterize fundamental diffusion and adsorption properties in the zeolite pores allowed for differences in butane isomer selectivity between the two structure types to be explained [44].

20.2.4. *Modeling Transport in Zeolite Membranes*

While surface diffusion dominates in zeolite pores, zeolite membranes are typically polycrystalline films grown on porous supports. In support pores and in defects between crystals, other types of diffusion and convection can occur in parallel or in series with the zeolite pore diffusion. Depending on the support or defect pore size and operating conditions, bulk diffusion, Knudsen diffusion, and viscous flow can all contribute to the transport. Knudsen diffusion occurs when molecules diffuse due to a partial pressure gradient through small enough pores that they collide with the walls more often than with each other. Knudsen diffusion differs from surface diffusion in that the molecule/wall collisions are assumed to be elastic rather than the molecule adsorbing and hopping. Bulk diffusion occurs when the mean free path of the molecules is greater than the pore diameter and molecules run into each other more often than into the walls. Bulk diffusion and

Knudsen diffusion can occur in parallel, and which dominates depends on operating conditions and pore properties [45]. The combined diffusive flux is related to the partial pressure gradient by:

$$-\frac{\nabla p_i}{RT} = \sum_{j \neq i}^{n} \frac{y_j J_{\mathrm{d},i} - y_i J_{\mathrm{d},j}}{D_{ij}} + \frac{J_{\mathrm{d},i}}{D_{\mathrm{Kn},i}}, \tag{20.6}$$

where y's are mole fractions, D_{ij} is the bulk diffusivity, $D_{\mathrm{Kn},i}$ is the Knudsen diffusivity of component i, and $J_{\mathrm{d},i}$ is the diffusive flux of component i. The diffusivities depend on operating conditions, fluid properties, and pore size, and their relative magnitudes will determine if one mode dominates or if both affect diffusive flux. Bulk diffusion becomes more significant as the system pressure increases since molecule-molecule interactions increase. Bulk diffusion is also likely to be controlling in large pores as the contribution from the molecule-wall interactions accounted for in Knudsen diffusion becomes negligible [36]. In this case Eq. 20.6 simplifies to:

$$-\frac{\nabla p_i}{RT} = \sum_{j \neq i}^{n} \frac{y_j J_{\mathrm{d},i} - y_i J_{\mathrm{d},j}}{D_{ij}}. \tag{20.7}$$

When a system is at low pressure and pores are small, Knudsen diffusion dominates and Eq. 20.6 simplifies to [36]:

$$J_{\mathrm{Kn},i} = -\frac{D_{\mathrm{Kn},i}}{RT} \nabla p_i. \tag{20.8}$$

Ideally viscous flow contributions to the flux of components through zeolite membranes will be minimal, as viscous flow is not selective. However, viscous flow (convection) will occur if there is a total pressure drop in the direction of the flux. The flux contribution from viscous flow depends on the viscosity, η, and the permeability, B $(= d_{\mathrm{pore}}^2/32)$, of the pores in which viscous flow is occurring:

$$J_{\mathrm{visc},i} = -\frac{p_i}{RT} \frac{B}{\eta} \nabla p_i. \tag{20.9}$$

Notice that the flux due to viscous flow is proportional to total pressure squared since the pressure is in both the gradient driving force and in the partial pressure concentration term. Thus whether or not flux through a supported membrane is proportional to the square of the pressure is sometimes used as an indication of whether or not viscous flow is occurring. One must be careful, however, to recognize that which transport mechanisms contribute or dominate can change with operating conditions, and support resistance can affect flux and selectivity of components through membranes even if the flux is not proportional to the square of the pressure. In fact, it can be beneficial to deposit films on both sides of a support in order to decrease the effect of support resistance [45]. In certain circumstances, particularly when the more strongly adsorbing component is the one that preferentially permeates as is the case for the CO_2 separations discussed in this chapter, double-sided zeolite membranes can have higher flux and selectivity than single supported films [45]. This can be true even if the total membrane thickness is greater with the two films as opposed to a single supported film [45].

As mentioned before, these various transport modes can occur in series and in parallel, so proper modeling should consider any and all potentially relevant modes and in what locations relative to each other those modes might be contributing [36, 45].

20.3. Metal Organic Frameworks (MOFs) and MOF Membranes

Metal organic frameworks (MOFs) consist of metal cations or metal-based-clusters linked by organic molecules forming a crystalline network, which after removal of guest species may result in three dimensional structures with permanent porosity [7, 49, 50]. These crystalline materials are relatively easy and inexpensive to synthesize and can be tailored to specific applications through varying the metals, ligands and linkers making up the MOF [51]. The combination of organic and inorganic building blocks offers an almost infinite number of variations, enormous flexibility in pore size (from micro to mesoporous), shape, structure, and functionalization and grafting. Figure 20.2 shows a typical representation of the components and final structure of a MOF.

20.3.1. *Characteristics of MOF Membranes*

The highly accessible porosity of MOFs (inferring high fluxes), their wide range of pore sizes, and their unique and tailorable surface properties make them ideal candidates to tackle classical and relevant industrial separations [52] such as hydrogen or CO_2 removal, alkanes from alkenes, linear from branched alkanes, and aromatic and larger isomers. Synthesizing MOF membranes for molecular gas separations is an emerging research area that is rapidly evolving [53–75], with the main challenges including poor intergrowth at the membrane-support interface, moisture limited stability, and

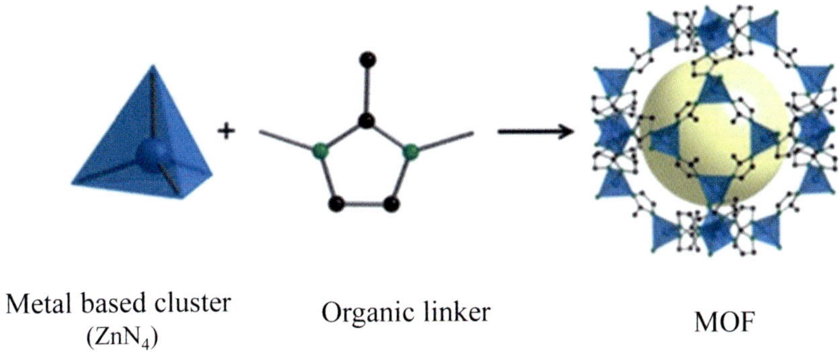

Fig. 20.2. A typical MOF structure composed of a metal-based cluster and an organic linker. The yellow sphere denotes "open space".

the existence of "non-porous pathways" resulting in poor selectivity. In addition, the high framework flexibility due to the presence of the organic linker limits molecular sieving ability. As with zeolite membranes, it is also challenging to reproducibly prepare MOF membranes with consistent separation performance.

Synthesis methods similar to those used to prepare zeolite membranes have been used to prepare MOF membranes. In general these methods involve direct synthesis of the MOF membrane on porous supports (in-situ approach), or seeding the porous support with crystals to promote membrane growth (secondary growth method). The latter is generally preferred since it allows better microstructural control of the resulting membrane.

20.3.2. *Why MOF Membranes for CO_2 Separations?*

Due to the remarkably high affinity of MOFs for CO_2 uptake, their open porous framework structure with large accessible pore volumes, their chemical stability in the presence of hydrocarbons and water (some of them), and limiting pore apertures in the range of the kinetic diameters of several relevant gas molecules, MOFs are appealing materials for CO_2 separations. In particular, zeolitic imidazolate frameworks (ZIFs), a subclass of MOFs, have emerged as a novel type of crystalline porous materials which combine highly desirable properties of zeolites and MOFs microporosity, high surface area, and exceptional thermal and chemical stability making them ideal candidates for gas separation applications [8]. In ZIFs, metal atoms such as Zn, Co and Cu are linked through N atoms by ditopic imidazolate

(Im) or functionalized Im links to form neutral frameworks and to provide tunable nanosized pores formed by 4- and 6-ring ZnN_4, CoN_4, and CuN_4 clusters. The frameworks of ZIF compounds closely resemble the frameworks of zeolites, i.e., the T-O-T bridges (T = Si, Al, P, etc.) in zeolites are replaced by M-Im-M bridges (M = Zn, Co, Cu) and the bond angles in both structures are 145° [76]. Therefore it is not surprising that most of the documented MOF membranes for CO_2 separation have ZIF compositions.

In addition to ZIFs, MOF-5, a crystalline structure made up from $[ZnO_4]^6+$ clusters linked together octahedrally with benzene-1,4-dicarboxylate (BDC) groups, is particularly interesting for H_2 and CO_2 separation and storage. Recently metal-adeninate biometal organic frameworks (Bio-MOFs) [77–82] have also been identified as appealing candidates for CO_2 separation because of their permanent microporosity, high surface area and chemical stability, and due to the presence of basic bio-molecule building units that preferentially adsorb CO_2. A prototypical type of metal-adeninate biometal organic framework is Bio-MOF-1 ($Zn_8(ad)_4(BPDC)_6O \cdot 2Me_2NH_2$; where ad=adeninate, BPDC=biphenyldicarboxylate) – a 3D porous MOF with infinite zinc-adeninate columnar secondary building units (SBUs) which are interconnected via biphenyldicarboxylate linkers [80]. The pore sizes of BioMOFs can be adjusted by choosing appropriate organic linkers [82]. A comprehensive review of CO_2 adsorption and separation with MOFs has been reported elsewhere [83].

20.3.3. *Modeling Transport in MOFs and MOF Membranes*

Transport in MOFs and MOF membranes has also been modeled with the Maxwell-Stefan diffusion model [84–87]. With the vast array of structures, pore sizes, functionalities, etc. possible in MOFs, the generalized transport model encompassing all possible diffusion and convection modes is necessary whether or not support effects are significant. Adsorption and diffusion properties for the specific components in the mixed matrix membranes need to be characterized on individual bases and combined into the generalized model. Molecular simulations have been used to estimate adsorption and diffusion properties of components in ZIFs, and the results used to predict ideal and mixture permeation selectivities for the gases of interest in this chapter [88].

While the detailed modeling described above is preferable for understanding the fundamental behavior of molecules in micro- and meso-porous materials and for predicting fluxes and selectivities of mixtures, many experimentalists follow a simpler approach of estimating transport properties through zeolites and MOFs from permeation experiments using the time-lag method to estimate Fickian diffusivity, D_i ($m^2\ s^{-1}$); gas permeation experiments to measure the permeability, $P_{m,i}$ ($mol\ m^{-1}\ s^{-1}\ Pa^{-1}$), and the definition of permeability to estimate apparent solubility, S ($mol\ m^{-3}\ Pa^{-1}$) [89]:

$$D_i = \frac{\delta^2}{6\tau}, \qquad P_{\mathrm{m},i} = \frac{\delta \dot{n}_i}{A\Delta P_i}, \qquad S_i = \frac{P_{\mathrm{m},i}}{D_i}. \tag{20.10}$$

In these equations δ is the membrane thickness (m), τ is the time lag, $\dot{n}_i$ is the steady state molar flow rate of component i through the membrane, A is the permeation area, and ΔP_i is the transmembrane partial pressure driving force for component i. This analysis assumes Fickian diffusion through the membranes with concentration-independent diffusivity (hence linear concentration profiles across the membrane thickness, δ), and a linear adsorption isotherm. While this level of analysis can help with determining whether solubility or diffusivity differences dominate separations under given conditions, more rigorous modeling of the transport in MOFs accounting for interactions between permeants in a mixture, competitive adsorption outside of the regime where isotherms are linear, and other diffusion and convection transport modes occurring in series and/or in parallel with diffusion through the MOF crystals elucidates much more understanding.

20.4. CO_2 Separation through Microporous Crystalline Membranes

20.4.1. *Microporous Crystalline Membranes for CO_2 Separation from Flue Gas*

As mentioned previously, amine absorption is the current Department of Energy (DOE) and industry benchmark technology for capture of CO_2 from power plant flue gases. DOE/NETL systems analysis studies estimated that using a chemical absorption with an aqueous monoethanolamine system to capture 90% of the CO_2 from flue gas will require an increase in the levelized cost of energy (LCOE) services of 75-85% [90]. These values are well above the 2020 DOE NETL Sequestration Program post-combustion capture goal [90]. Therefore, it is important to develop new advanced CO_2 capture technologies in order to maintain the cost-effectiveness of US coal-fired power generation. Compared to amine absorption, membrane processes require less energy to operate, and do not need chemicals or regenerating absorbents [5]. Also, membranes are compact and can be retrofitted onto the tail end of power-plant flue gas streams without complicated integration schemes. Recently published *systems analysis and feasibility studies demonstrate that membranes are a technically and economically viable option for CO_2 capture from flue gas* exhaust in pulverized coal power plants [91]. For CO_2 capture from flue gases, CO_2 permeance and CO_2/N_2 selectivity in the intended operating environment are the two basic criteria to determine whether a membrane can be effectively utilized.

Polymeric membranes such as cellulose acetate, polyimide and polyaramide can separate CO_2 from N_2 mixtures, but the high CO_2 pressures plasticize them and decrease their separation ability considerably [92]. Polymeric membranes containing poly (ethylene oxide) units have shown moderate to good CO_2/N_2 separation properties [93, 94]. Recently, amine carriers functionalized in cross-linked poly(vinyl alcohol) membranes have shown remarkable separation performance for CO_2/N_2 [95]. Although the current membrane technology is dominated by polymers, these membranes suffer from low CO_2 permeability and selectivity because they are based on a solution-diffusion mechanism that separates mainly based on small size differences of the gas molecules [96, 97]. These polymeric membranes are not suitable for the high temperature and high humidity conditions required for CO_2 capture from flue gas from power plants [96].

As mentioned previously, zeolite and MOF membranes are more thermally, mechanically, and chemically stable than polymeric membranes [6]. They can operate as molecular sieves or separate gases based on adsorption and diffusion differences and so can be more selective than polymeric membranes as well, including for CO_2 separation from N_2. The molecular sieving ability of zeolite membranes is well known and has been exploited for a vast number of relevant gas mixture separations [10–14], including CO_2 separation from N_2. Some zeolite membrane types that have been used for this particular separation include zeolites Y [98], T [99], ZSM-5 [100], SAPO-34 [64], DD3R [101] and AlPO-18 [102]. Typically these zeolite membranes have moderate to high CO_2/N_2 selectivities due to a combination of differences in diffusivity and adsorption strengths, but CO_2 permeability tends to be moderate in these materials. Although the separation ability of several MOFs have been demonstrated for several gas pairs [57–59, 61–64, 66–68, 70, 103] including CO_2/CH_4, H_2/CH_4, H_2/N_2, H_2/O_2, H_2/CO_2,CO_2/CO, only few reports exist on the separation abil-

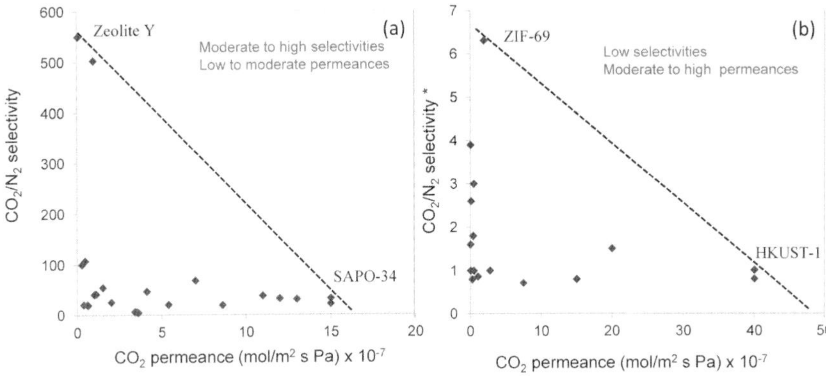

Fig. 20.3. CO_2/N_2 separation selectivity versus CO_2 permeance for (a) zeolite and (b) MOF membranes. Note that in (b) these are ideal selectivities.

ity of MOF membranes for CO_2/N_2 [56, 61, 69, 104]. Low to moderate CO_2/N_2 selectivities were observed for these MOF compositions.

Figure 20.3 summarizes the performance of (a) zeolite membranes and (b) MOF membranes for CO_2/N_2 separation. Zeolite membranes show much higher CO_2/N_2 separation selectivities as compare to MOF membranes. This may be related to the molecular sieving effect of zeolites due to their "rigid crystallographic pore size" as compared to the flexible MOF structure caused by the organic linkers leading to "gate opening" or "breathing" effects which limit the molecular sieving ability of MOFs. On the other hand, at least in part due to the flexible framework nature and higher internal surface area of MOFs, CO_2 permeances are typically higher in MOFs. For clarity, only the zeolite and MOF compositions with the highest CO_2/N_2 selectivity and CO_2 permeance are illustrated in Fig. 20.3. For example in Fig. 20.3(a), zeolite Y shows the highest CO_2/N_2 separation selectivity with low CO_2 permeance [98]. Due to the 0.73 nm pore size of zeolite Y, there is no steric hindrance for CO_2 or N_2 to enter in the pores. However, CO_2 interacts preferentially with the aluminosilicate framework and the extraframework cations, leading to a favorable combination of sorption and mobility for CO_2, leading to the high CO_2/N_2 separation selectivities. In addition, the nanometer-sized zeolite Y reported by the authors [98] may have contributed to the formation of close-packed arrays of crystals which led to a well intergrown membrane. The highest CO_2 permeance reported was through SAPO-34 membranes, but the separation

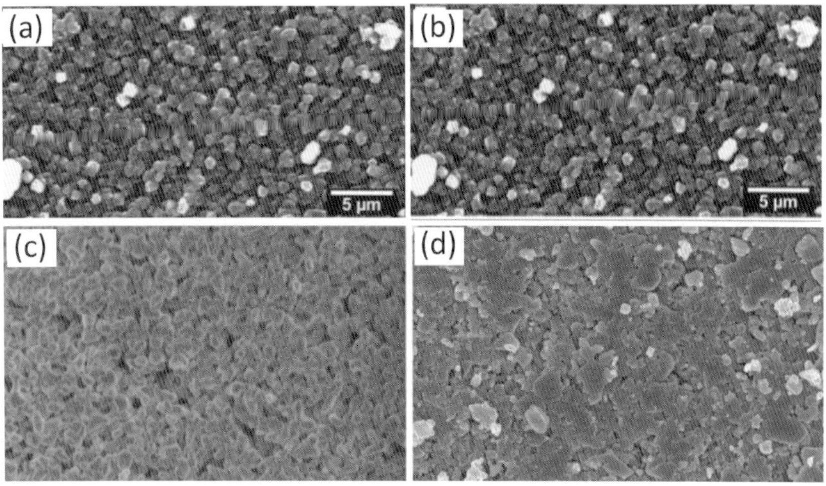

Fig. 20.4. Zeolite Y membrane (a) top view and (b) cross section view [98]. MOF membrane (c) top view and (d) cross section view [60] employed to separate CO_2/N_2 gas mixtures.

selectivity was moderate [105]. SAPO-34 is a silicoalumino phosphate with 0.38 nm pore size which in principle allows for CO_2 molecular sieving over N_2. Authors attributed the high CO_2 permeance to the formation of a high quality (low concentration of non-selective pores) thin membrane [105]. The selectivities shown for the MOF membranes (Fig. 20.3(b)) are ideal selectivities (ratios of pure component permeances), so competitive adsorption effects from a mixture are not taken into account. ZIF-69, a zeolitic imidazolate framework with pore size of 0.78 nm, shows the highest CO_2/N_2 selectivity of the MOF membranes [61]. The moderate selectivity was attributed to the surface diffusion of CO_2 being slower because of the strong adsorption affinity of CO_2 to ZIF-69. HKUST-1, a Cu based MOF with 0.9 nm pores, shows the highest CO_2 permeance of the MOFs [60], which was attributed to the large pore size and large pore volume of the framework.

Figure 20.4 shows representative SEM images of zeolite and MOF membranes that have shown high separation selectivity performance for CO_2/N_2 gas mixtures.

The Robeson plot for CO_2/N_2 separation selectivities versus CO_2 permeabilities in polymeric membranes [106] has been widely used to compare the performance of membranes. Figure 20.5 summarizes the separation performance of polymeric, and zeolite membranes for CO_2/N_2 gas mixtures.

Several zeolites, including Y, T, ZSM-5 and SAPO-34 perform much better (above the upper bound) than any polymeric membrane reported (red points). Although several zeolite membranes perform much better than any other membrane composition (including MOF compositions), still the CO_2 permeability needs to be improved for these membranes to be an economically viable option for CO_2 capture from flue gas. In a recent study, Merkel et al. showed the optimal membrane CO_2/N_2 selectivity for separation of CO_2 from flue gas can be as low as 30-40 [107], but the CO_2 permeance must be much greater than approximately 6.6×10^{-7} mol m^{-2} s^{-1} Pa^{-1} (≈2000 GPU). The commercially available CO_2-selective polymer membranes, however, only have a permeance of 100 GPU [107] which is too low for flue gas CO_2 capture.

20.4.2. *Microporous Crystalline Membranes for CO_2 Separation from Natural Gas*

Processing of natural gas is by far the largest industrial gas separation application [108]. Every year close to 100 trillion standard cubic feet of natural gas are used worldwide. The US consumption of natural gas accounts for approximately 22 trillion standard cubic feet per year. As mentioned previously, CO_2 needs to be removed from natural gas before it enters the pipeline. Polymeric membranes including cellulose acetate [109], polyimide [110], and perfluoro-based [111] polymer membranes have also been used to separate CO_2 from natural gas. Similar to flue gas treatment, polymeric membranes can separate CO_2/CH_4 mixtures, but the high CO_2 pressures can plasticize them and decrease their separation ability [92].

Zeolite membrane types that have been used to separate CO_2/CH_4 mixtures are essentially the same ones employed for CO_2/N_2 – types Y [98], T [99], ZSM-5 [100], SAPO-34 [112–117], DD3R [101] and AlPO-18 [102]. Since the size difference between CO_2 and CH_4 is larger than that between CO2 and N2 zeolite membranes have shown high CO_2/CH_4 selectivities due to a combination of molecular sieving and preferential CO_2 adsorption. However, most of these zeolite types have had only moderate CO_2 permeability. The best separation performance of any zeolite membrane type for this separation has been through SAPO-34 membranes, which have high selectivities even at high feed pressures [115–117].

Several MOF compositions have been successfully prepared in membrane form, and have been evaluated for CO_2/CH_4 gas mixture separation as well. These compositions include: ZIF-8 [57, 58, 62, 68, 70, 118],

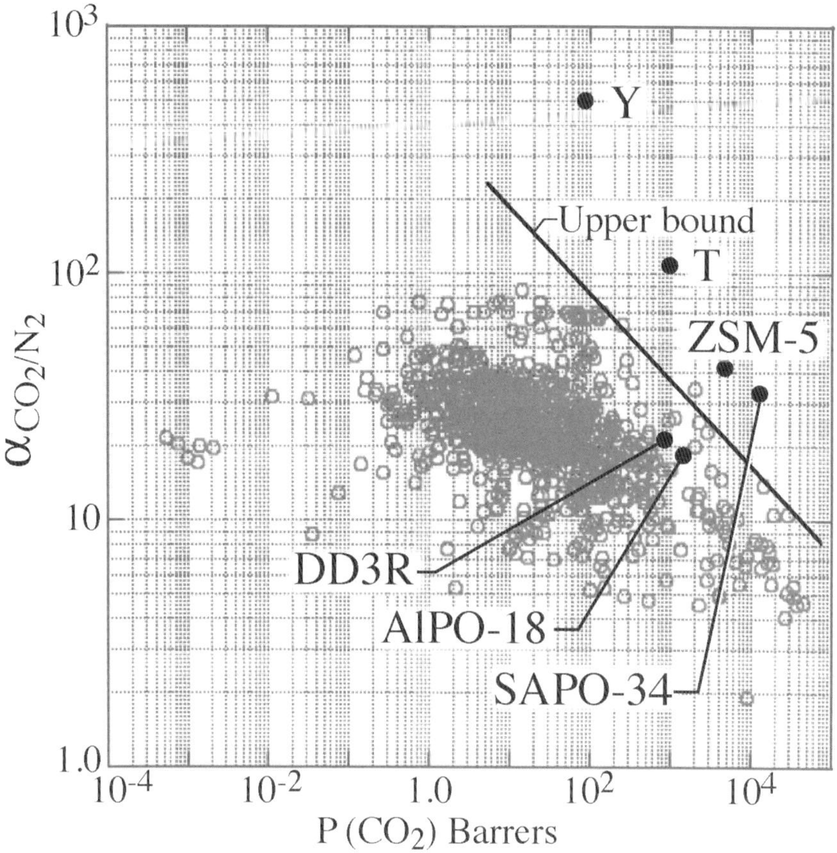

Fig. 20.5. Robeson plot for CO_2/N_2 gas mixtures. Points represent experimental data for polymeric membranes. For comparison zeolite membranes are included.

HKUST-1 [55, 60], MOF-5 [53], MIL-53 [119], $[Cu_2L_2P]_n$ [120], SIM-1 [121], ZIF-7 [59], ZIF-22 [63], ZIF-69 [69], ZIF-95 [122] and ZIF-90 [66, 67, 123, 124]. Some of these MOFs have had low to moderate CO_2/CH_4 selectivities with moderate to high CO_2 permeances. Once again, the flexibility of the MOF framework is in part responsible for the low separation selectivities.

Figure 20.6 summarizes the performance of (a) zeolite membranes and (b) MOF membranes for CO_2/CH_4 separation. Similar to CO_2/N_2 separations, zeolite membranes are more selective for CO_2/CH_4 separation, but have lower CO_2 permeance. Again the rigid small pores of the zeolites compared to the flexibile framework of the MOFs cause the zeolites to be

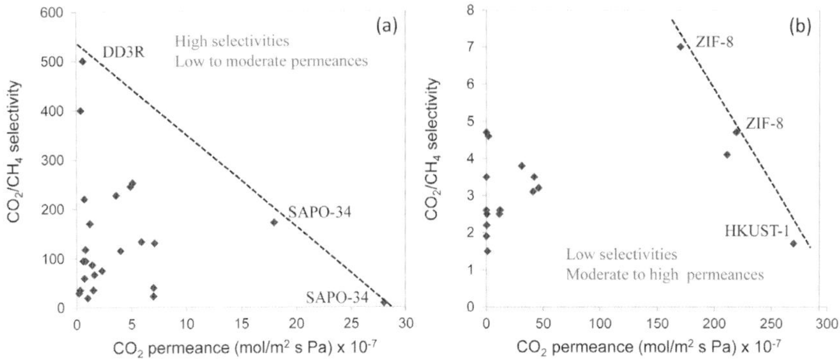

Fig. 20.6. CO_2/CH_4 separation selectivity vs CO_2 permeance for (a) zeolite and (b) MOF membranes.

more selective but less permeable than MOFs. Notice in Fig. 20.6(a) zeolite DD3R has the highest CO_2/CH_4 separation selectivity but low CO_2 permeance [101]. DD3R is a microporous structure consisting of cages, each connected by three windows of 8-membered rings forming a 2-dimensional pore structure with pore sizes of 0.36×0.44 nm. Here too, the authors attributed the high separation selectivities to steric effects introduced by the window opening of DD3R leading to molecular sieving and activated transport, as well as to preferential adsorption effects for CO_2 over CH_4 [101]. Overall, SAPO-34 membranes have shown the best separation performance for CO_2/CH_4 mixtures with high selectivities and moderate to high CO_2 permeance [112–117] (Fig. 20.6(a)), again due to molecular sieving and preferential adsorption of CO_2. The moderate to high CO_2 permeances have been achieved by preparing thin membranes [113, 115–117]. The highest CO_2/CH_4 selectivity for a MOF membrane (Fig. 20.6(b)) was through ZIF-8, a zeolitic imidazolate framework with limiting pore aperture of 0.34 nm [58]. The high observed CO_2 permeances on ZIF-8 membranes were attributed to the small crystals with a narrow size distribution, which led to thin membranes, and to the textural properties of the alumina porous support. Although ZIF-8 is composed of large 1.16 nm pores and small 0.34 nm pores, density functional theory simulation data suggest that the smaller pores are the preferential adsorption sites for CO_2 molecules. Therefore, the small pore aperture of ZIF-8 favored diffusion of CO_2 over CH_4, resulting in moderate separation selectivities. HKUST-1 showed the highest CO_2 permeance [60], but low separation selectivities because of the large pore size of that MOF structure.

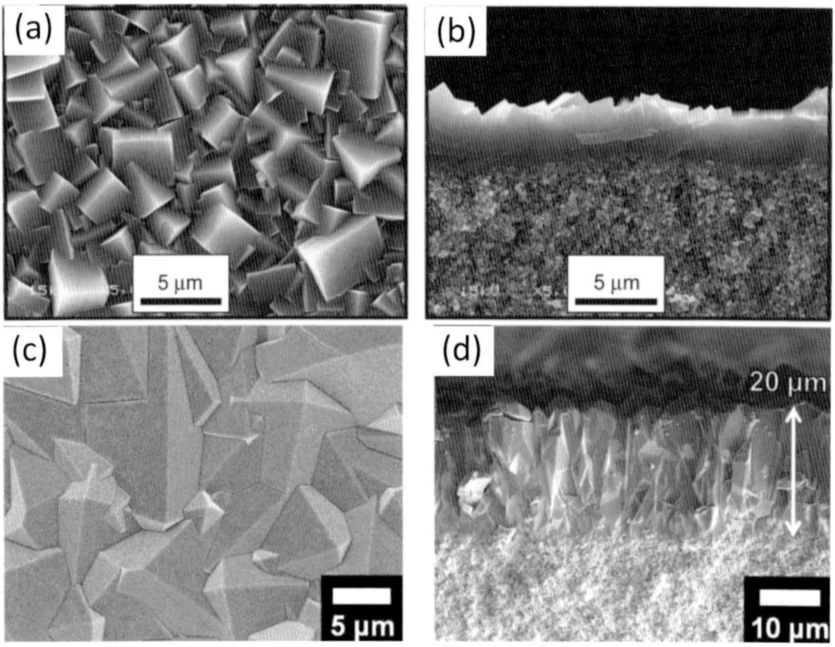

Fig. 20.7. SAPO-34 membrane (a) top view and (b) cross section view [115]. ZIF-8 membrane (c) top view and (d) cross section view [62] employed to separate CO_2/CH_4 gas mixtures.

Figure 20.7 shows representative SEM images of SAPO-34 (zeolite) and ZIF-8 (MOF) membranes that have shown the best separation performance for CO_2/CH_4 gas mixtures.

Figure 8 shows the Robeson plot for CO_2/CH_4 gas mixtures for polymeric membranes, and for comparison the performance of SAPO-34 membranes has been included. The SAPO-34 membranes are significantly above the upper bound of the Robeson plot showing that these zeolite membranes perform better than any polymeric or MOF membrane. Although the separation performance of SAPO-34 membranes may be sufficient for natural gas treatment, as pure inorganic membranes SAPO-34 membrane costs may limit their application at the industrial level. Current efforts to decrease the cost of SAPO-34 membrane synthesis are underway [115, 116].

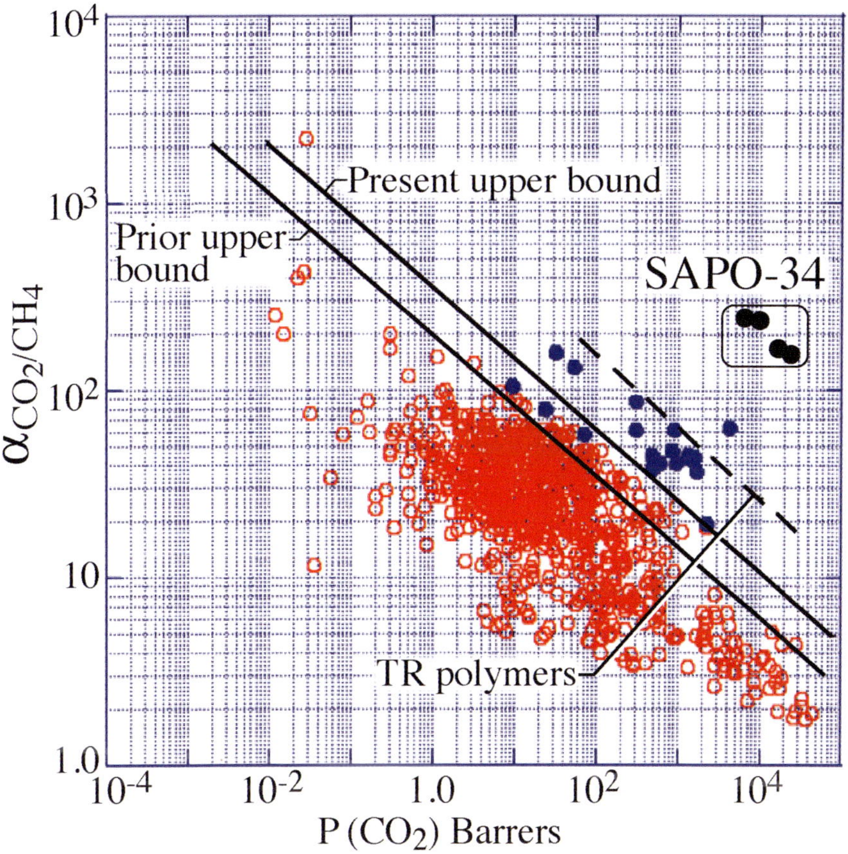

Fig. 20.8. Robeson plot for CO_2/CH_4 gas mixtures. Red and blue points represent experimental points for polymeric membranes For comparison SAPO-34 zeolite membranes are included.

20.4.3. *Microporous Crystalline Membranes for CO_2/H_2 Separations*

Membranes represent an alternative technology for carbon dioxide capture and hydrogen purification at hydrogen production facilities and Integrated Gasification Combined Cycle (IGCC) power plants due to their simple operation mode and potentially low energy consumption [125]. Hydrogen selective [96, 126–130] and CO_2 selective [4, 131–135] membranes of polymeric or dense metal materials have been studied for this separation. Only few examples, discussed below, of porous crystalline membranes for CO_2/H_2

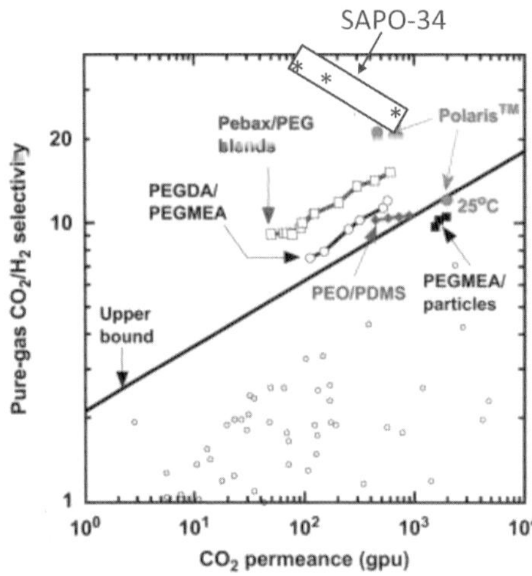

Fig. 20.9. Robeson plot for CO_2/H_2 gas mixtures modified from Lin et al. [135]. For comparison SAPO-34 zeolite membranes are included. The left point on SAPO-34 region corresponds to a separation selectivity of 150.

(CO_2-selective) separation have been reported. Carbon dioxide selective membranes have attracted significant interest because pure hydrogen can be retained at the high pressure side, thus avoiding recompression costs. In particular, these membranes are of great relevance for CO_2/H_2 separations in steam reforming of methane, where it is desirable to have pure hydrogen at high pressure by permeating CO_2 from syngas [4, 130–135], and in air-blown gasification processes, where CO_2 needs to be removed from syngas containing H_2 and N_2 [135].

SAPO-34 membranes can selectively permeate CO_2 from CO_2/H_2 mixtures [13, 114] with CO_2 permeances as high as 2.5 $\times10^{-7}$ mol m^{-2} s^{-1} Pa^{-1} and a separation selectivity of 22 [114], and permeance of 0.3×10^{-7} mol m^{-2} s^{-1} Pa^{-1} with separation selectivity higher than 100 [13] at room temperature [114]. Although hydrogen has a higher diffusivity than most molecules within zeolite crystals, it adsorbs weakly, leading to preferential permeation from mixtures of other molecules such as CO_2. As shown in Figure 20.9, SAPO-34 membranes perform better than any polymeric membrane for this separation as well. Other zeolite membranes that have been employed for CO_2/H_2 separations include MFI [136, 137] (of which ZSM-5 is an example) and silicalite-1 [138]. These membranes had separation factors from 0.7 to 12 [136–138].

20.5. Concluding Remarks and Outlook

Both zeolite and MOF membranes have demonstrated ability to separate CO_2 from other light gases, such as CH_4, N_2 and H_2. Each family of membranes shows distinctive features that make it appealing for CO_2 separation. In the case of zeolite membranes, molecular sieving due to the rigid pore structure in addition to preferential adsorption has led to high CO_2 separation selectivities. Typically, zeolite membranes are chemically, thermally, and mechanically stable due to their structure and chemical composition. From the technical point of view, zeolite membrane synthesis is a mature field, in which relatively well developed synthesis approaches (in-situ growth, secondary seeded growth) typically lead to reproducible high quality membranes. In addition, low synthesis costs due to relatively cheap structure directing agents employed for zeolite growth and water make zeolite synthesis an attractive approach for potential scale up of these membranes for industrial applications. In fact, the scalability of SAPO-34 membranes for natural gas treatment has been demonstrated [139]. Due to their superior separation performance, SAPO-34 membranes for the separation of CO_2 from natural gas are close to being commercialized.

On the other hand, MOF membranes for CO_2 separation is an emerging research field that is evolving rapidly. The extremely high porosity and high surface areas of several MOF membranes has led to high CO_2 permeances. Another unique feature of MOF membranes is related to its unprecedented high CO_2 uptakes which translates into high CO_2 ideal selectivities over other gases. As compared to zeolites, MOFs (if prepared in membrane form) offer a wide range of pore sizes leading to a nearly limitless number of structures with flexible pore sizes and structures. Finally, due to the presence of organic linkers in MOF structures, proper functionalization of MOFs may result in pore size control and control of adsorption properties. This in principle would allow one to rationally design "tailored MOF membranes" for CO_2 separation.

Although both types of microporous crystalline membranes are promising for CO_2 separation from key gases, we need to consider practical challenges that may impede the future application of these membranes for CO_2 related industrial separations. First of all it is important to note that even if a material exhibits unique and remarkable separation properties as prepared in powder or particle form, the same material may not be suitable for membrane preparation due to limited adhesion of the membrane to the support, considerable adsorption induced stresses, poor crystal intergrowth, large crystal size leading to non-continuous membranes, poor reproducibility from the synthesis method, etc.

Membrane synthesis approaches aimed at improving membrane intergrowth, controlling thickness, and reducing the concentration of "non-selective pores" may result in high quality membranes able to effectively separate CO_2 from gas mixtures. These synthetic approaches should attempt to (a) produce small microporous crystals with a narrow size distribution allowing for better "crystal packing" during membrane growth; (b) improve membrane intergrowth through chemical, physical, thermal or other seeding techniques, and (c) promote effective membrane-support interaction. Novel MOF membrane synthesis methods potentially allowing for thickness, crystal size, and intergrowth control include electrochemical synthesis [140, 141] and evaporation induced self-assembly [142–144].

Finally, important challenges to consider for zeolite and MOF membranes for practical industrial applications include:

Reproducibility Poor membrane reproducibility is a common problem found in particular for MOF membranes.

Long term and chemical stability Limited studies have explored the stability of membranes under real flue gas and natural gas conditions (presence of impurities, effect of temperature, and pressure).

Economics In general, zeolite membranes do not show good economics for CO_2 capture from flue gas or natural gas. The cost of the membrane module is at least one order of magnitude more expensive than the state-of the art polymeric membranes used for this particular separation [3, 107].

Scalability To the best of our knowledge only one successful documented example demonstrating zeolite or MOF membrane scale-up has been reported (SAPO-34 membranes for natural gas treatment).

If these challenges are properly addressed, one may see in the near future some of these microporous membranes operating effectively at industrial conditions for relevant CO_2 separations.

References

[1] DOE/EIA. *EIA Estimates on Global Warming Potentials: Intergovernmental Panel on Climate Changes, Climate Change: The Physical Science Basis,* Cambridge UK, Cambridge University Press. (2007).

[2] Annual Energy Outlook, 0383 (June, 2008).

[3] R. W. Baker, Future directions of membrane gas separation technology, *Ind. Eng. Chem. Res.* **41**(6), 1393–1411 (Mar., 2002). ISSN 0888-5885.

[4] H. Lin, E. Van Wagner, B. D. Freeman, L. G. Toy, and R. P. Gupta, Plasticization-enhanced hydrogen purification using polymeric membranes., *Science (New York, N.Y.)* **311**(5761), 639–42 (Mar., 2006). ISSN 1095-9203.

[5] S. Stern, Polymers for gas separations: The next decade, *J. Memb. Sci.* **94**, 1–65 (1994).

[6] E. Thompson, R.W. (H.G. Karge, J. Weitkamp, *Recent Advances in the Understanding of Zeolite Synthesis, Molecular Sieves Vol. 1,* Springer, Berlin Heidelberg, (1998).

[7] O. M. Yaghi, M. O'Keeffe, N. W. Ockwig, H. K. Chae, M. Eddaoudi, and J. Kim, Reticular synthesis and the design of new materials., *Nature.* **423** (6941), 705–14 (June, 2003). ISSN 0028-0836.

[8] K. S. Park, Z. Ni, A. P. Côté, J. Y. Choi, R. Huang, F. J. Uribe-Romo, H. K. Chae, M. O'Keeffe, and O. M. Yaghi, Exceptional chemical and thermal stability of zeolitic imidazolate frameworks., *Proceedings of the National Academy of Sciences of the United States of America.* **103**(27), 10186–91 (July, 2006). ISSN 0027-8424.

[9] J. Čejka, *Introduction to Zeolite Molecular Sieves, 3rd Edition,* Elsevier, (2007).

[10] T. C. Bowen, R. D. Noble, and J. L. Falconer, Fundamentals and applications of pervaporation through zeolite membranes, *J. Memb. Sci.* **245**(1-2), 1–33 (Dec., 2004). ISSN 03767388.

[11] K. Aoki, K. Kusakabe, and S. Morooka, Separation of Gases with an A-Type Zeolite Membrane, *Ind. Eng. Chem. Res.* **39**(7), 2245–2251 (July, 2000). ISSN 0888-5885.

[12] M. A. Snyder and M. Tsapatsis, Hierarchical nanomanufacturing: from shaped zeolite nanoparticles to high-performance separation membranes., *Angewandte Chemie* (International ed. in English). **46**(40), 7560–73 (Jan., 2007). ISSN 1433-7851.

[13] M. Hong, S. Li, J. L. Falconer, and R. D. Noble, Hydrogen purification using a SAPO-34 membrane, *J. Memb. Sci.* **307**(2), 277–283 (Jan., 2008). ISSN 03767388.

[14] C. D. Baertsch, H. H. Funke, J. L. Falconer, and R. D. Noble, Permeation of Aromatic Hydrocarbon Vapors through SilicaliteZeolite Membranes, *J. Phys. Chem.* **100**(18), 7676–7679 (Jan., 1996). ISSN 0022-3654.

[15] A. Tavolaro and E. Drioli, Zeolite Membranes, *Adv. Mat.* **11**(12), 975–996 (Aug., 1999). ISSN 0935-9648.

[16] T. Bein, Synthesis and Applications of Molecular Sieve Layers and Membranes, *Chem. Mater.* **8**(8), 1636–1653 (Jan., 1996). ISSN 0897-4756.

[17] J. Caro, M. Noack, P. Kölsch, and R. Schäfer, Zeolite membranes – State of their development and perspective, *Microporous Mesoporous Mater.* **38** (1), 3–24 (July, 2000). ISSN 13871811.

[18] E. McLeary, J. Jansen, and F. Kapteijn, Zeolite based films, membranes and membrane reactors: Progress and prospects, *Microporous Mesoporous*

Mate. **90**(1-3), 198–220 (Mar., 2006). ISSN 13871811.
[19] M. J. C. Noack, *Handbook of Porous Solids*, p. 2478. Wiley-VCH (2002).
[20] J. Caro, M. Noack, and P. Kölsch, Zeolite Membranes: From the Laboratory Scale to Technical Applications, *Adsorption.* **11**(3-4), 215–227 (July, 2005). ISSN 0929-5607
[21] J. Ramsay, S. Kallus, and N. Kanellopoulos, *Recent Advances in Gas Separation by Microporous Ceramic Membranes, Membrane Science and Technology*. Elsevier p. 373 (2000).
[22] A. Julbe, Zeolite membranes — A short overview, *Stud. Sur. Sci. Catal.* (157), 135 (2005).
[23] J. E. Lewis, G. R. Gavalas, and M. E. Davis, Permeation studies on oriented single-crystal ferrierite membranes, *AIChE Journal.* **43**(1), 83–90 (Jan., 1997). ISSN 0001-1541.
[24] Y. Wang, Y. Tang, A. Dong, X. Wang, N. Ren, W. Shan, and Z. Gao, Self-supporting porous zeolite membranes with sponge-like architecture and zeolitic microtubes, *Adv. Mater.* **14**(13-14), 994–997 (July, 2002). ISSN 0935-9648.
[25] J. Choi, H.-K. Jeong, M. A. Snyder, J. A. Stoeger, R. I. Masel, and M. Tsapatsis, Grain boundary defect elimination in a zeolite membrane by rapid thermal processing., *Science* (New York, N.Y.). **325**(5940), 590–3 (July, 2009). ISSN 1095-9203.
[26] G. Zhu, J. Wang, Y. Zhang, J. Lu, and J. Xiu, Preparation and permeability of ZSM-35 zeolite membranes on porous stainless steel tubes, *Frontiers of Chemical Engineering in China.* **1**(3), 217–220 (July, 2007). ISSN 1673-7369.
[27] A. Chiang and K.-j. Chao, Membranes and films of zeolite and zeolite-like materials, *J. Phys. Chem. Solids.* **62**(9-10), 1899–1910 (Sept., 2001). ISSN 00223697.
[28] M. P. Bernal, G. Xomeritakis, and M. Tsapatsis, Tubular MFI zeolite membranes made by secondary (seeded) growth, *Catal. Today.* **67**(1-3), 101–107 (May, 2001). ISSN 09205861.
[29] Y. Takata, T. Tsuru, T. Yoshioka, and M. Asaeda, Gas permeation properties of MFI zeolite membranes prepared by the secondary growth of colloidal silicalite and application to the methylation of toluene, *Microporous Mesoporous Mater.* **54**(3), 257–268 (July, 2002). ISSN 13871811.
[30] Z. Tang, S.-J. Kim, X. Gu, and J. Dong, Microwave synthesis of MFI-type zeolite membranes by seeded secondary growth without the use of organic structure directing agents, *Microporous Mesoporous Mater.* **118**(1-3), 224–231 (Feb., 2009). ISSN 13871811.
[31] Y. Li, M. Peratitus, G. Xiong, W. Yang, E. Landrivon, S. Miachon, and J. Dalmon, Nanocomposite MFI-alumina membranes via pore-plugging synthesis: Genesis of the zeolite material, *J. Membr. Sci.* **325**(2), 973–981 (Dec., 2008). ISSN 03767388.
[32] J. Hedlund, F. Jareman, A.-J. Bons, and M. Anthonis, A masking technique for high quality MFI membranes, *J. Membr. Sci.* **222**(1-2), 163–179 (Sept., 2003). ISSN 03767388.

[33] M. Yang, B. Crittenden, S. Perera, H. Moueddeb, and J.-A. Dalmon, The hindering effect of adsorbed components on the permeation of a non-adsorbing component through a microporous silicalite membrane: the potential barrier theory, *J. Membr. Sci.* **156**(1), 1–9 (Apr., 1999). ISSN 03767388.
[34] J. C. Poshusta, V. A. Tuan, E. A. Pape, R. D. Noble, and J. L. Falconer, Separation of light gas mixtures using SAPO-34 membranes, *AIChE Journal.* **46**(4), 779–789 (Apr., 2000). ISSN 00011541.
[35] X. Xu, X. Zhao, L. Sun, and X. Liu, Adsorption separation of carbon dioxide, methane, and nitrogen on Hβ and Na-exchanged β-zeolite, *J. Nat. Gas Chem.* **17**(4), 391–396 (Dec., 2008). ISSN 10039953.
[36] R. Krishna, A unified approach to the modelling of intraparticle diffusion in adsorption processes, *Gas Sep. Purifi.* **7**(2), 91–104 (Jan., 1993). ISSN 09504214.
[37] R. Krishna and L. van den Broeke, The Maxwell-Stefan description of mass transport across zeolite membranes, *Chem. Eng. J. Bioch. Eng. J.* **57**(2), 155–162 (Apr., 1995). ISSN 09230467.
[38] R. Krishna and J. Wesselingh, The Maxwell-Stefan approach to mass transfer, *Chem. Eng. Sci.* **52**(6), 861–911 (Mar., 1997). ISSN 00092509.
[39] R. Krishna, Diffusion of binary mixtures across zeolite membranes: entropy effects on permeation selectivity, *Int. Commun. Heat Mass Transf.* **28**(3), 337–346 (Apr., 2001). ISSN 07351933.
[40] T. Q. Gardner, A. I. Flores, R. D. Noble, and J. L. Falconer, Transient measurements of adsorption and diffusion in H-ZSM-5 membranes, *AIChE Journal.* **48**(6), 1155–1167 (June, 2002). ISSN 00011541.
[41] T. Q. Gardner, J. L. Falconer, and R. D. Noble, Adsorption and diffusion properties of zeolite membranes by transient permeation, *Desalination.* **149** (1-3), 435–440 (Sept., 2002). ISSN 00119164.
[42] T. Q. Gardner, J. B. Lee, R. D. Noble, and J. L. Falconer, Adsorption and diffusion properties of butanes in ZSM-5 zeolite membranes, *Ind. Eng. Chem. Res.* **41**, 4094–4105 (2002).
[43] T. Q. Gardner, J. L. Falconer, R. D. Noble, and M. M. P. Zieverink, Analysis of transient permeation fluxes into and out of membranes for adsorption measurements, *Chem. Eng. Sci.* **58**(10), 2103–2112 (May, 2003). ISSN 00092509. doi: 10.1016/S0009-2509(03)00016-2.
[44] T. Q. Gardner, J. L. Falconer, and R. D. Noble, Transient permeation of butanes through ZSM-5 and ZSM-11 zeolite membranes, *AIChE Journal.* **50**(11), 2816–2834 (Nov., 2004). ISSN 0001-1541. doi: 10.1002/aic.10217.
[45] T. Q. Gardner, J. G. Martinek, J. L. Falconer, and R. D. Noble, Enhanced flux through double-sided zeolite membranes, *J. Memb. Science.* **304**(1-2), 112–117 (Nov., 2007). ISSN 03767388.
[46] S. Li, J. G. Martinek, J. L. Falconer, R. D. Noble, and T. Q. Gardner, High-Pressure CO2 / CH 4 Separation Using SAPO-34 Membranes. *Ind. Eng. Chem. Res.* **44**(9) pp. 3220–3228 (2005).
[47] J. G. Martinek, T. Q. Gardner, R. D. Noble, and J. L. Falconer, Modeling transient permeation of binary mixtures through zeolite membranes, *Ind.*

Eng. Chem. Res. **45**(17), 6032–6043 (2006).

[48] B. Smit and R. Krishna, Molecular simulations in zeolitic process design, *Chem. Eng. Sci.* **58**(3-6), 557–568 (Feb., 2003). ISSN 00092509.

[49] G. Férey, Hybrid porous solids: past, present, future., *Chem. Soc. Rev.* **37** (1), 191–214 (Jan., 2008). ISSN 0306-0012.

[50] L. MacGillivray, *Metal Organic Framework: Design and Application*. John Wiley & Sons (2010).

[51] U. Mueller, M. Schubert, F. Teich, H. Puetter, K. Schierle-Arndt, and J. Pastré, Metal–organic frameworks—prospective industrial applications, *J. Mater. Chem.* **16**(7), 626 (Feb., 2006). ISSN 0959-9428.

[52] J. Gascon and F. Kapteijn, Metal-organic framework membranes–high potential, bright future?, *Angewandte Chemie* (International ed. in English). **49**(9), 1530–2 (Feb., 2010). ISSN 1521-3773.

[53] Y. Liu, Z. Ng, E. A. Khan, H.-K. Jeong, C.-b. Ching, and Z. Lai, Synthesis of continuous MOF-5 membranes on porous α-alumina substrates, *Microporous Mesoporous Mater.* **118**(1-3), 296–301 (Feb., 2009). ISSN 13871811.

[54] Y. Yoo, Z. Lai, and H.-K. Jeong, Fabrication of MOF-5 membranes using microwave-induced rapid seeding and solvothermal secondary growth, *Microporous Mesoporous Mater.* **123**(1-3), 100–106 (July, 2009). ISSN 13871811.

[55] H. Guo, G. Zhu, I. J. Hewitt, and S. Qiu, "Twin copper source" growth of metal-organic framework membrane: Cu(3)(BTC)(2) with high permeability and selectivity for recycling H(2)., *Journal of the Am. Chem. Soc.* **131** (5), 1646–7 (Feb., 2009). ISSN 1520-5126.

[56] R. Ranjan and M. Tsapatsis, Microporous metal organic framework membrane on porous support using the seeded growth method, *Chem. Mater.* **21**(20), 4920–4924 (Oct., 2009). ISSN 0897-4756.

[57] H. Bux, F. Liang, Y. Li, J. Cravillon, M. Wiebcke, and J. Caro, Zeolitic imidazolate framework membrane with molecular sieving properties by microwave-assisted solvothermal synthesis., *J. Am. Chem. Soc.* **131**(44), 16000–1 (Nov., 2009). ISSN 1520-5126.

[58] S. R. Venna and M. A. Carreon, Highly permeable zeolite imidazolate framework-8 membranes for CO2/CH4 separation., *J. Am. Chem. Soc.* **132** (1), 76–8 (Jan., 2010). ISSN 1520-5126.

[59] Y.-S. Li, F.-Y. Liang, H. Bux, A. Feldhoff, W.-S. Yang, and J. Caro, Molecular sieve membrane: supported metal-organic framework with high hydrogen selectivity., *Angewandte Chemie (International ed. in English).* **49**(3), 548–51 (Jan., 2010). ISSN 1521-3773.

[60] V. V. Guerrero, Y. Yoo, M. C. McCarthy, and H.-K. Jeong, HKUST-1 membranes on porous supports using secondary growth, *J. Mater. Chem.* **20**(19), 3938 (May, 2010). ISSN 0959-9428.

[61] Y. Liu, E. Hu, E. A. Khan, and Z. Lai, Synthesis and characterization of ZIF-69 membranes and separation for CO2/CO mixture, *J. Memb. Sci.* **353**(1-2), 36–40 (May, 2010). ISSN 03767388.

[62] M. C. McCarthy, V. Varela-Guerrero, G. V. Barnett, and H.-K. Jeong, Synthesis of zeolitic imidazolate framework films and membranes with con-

trolled microstructures., *Langmuir*. **26**(18), 14636–41 (Sept., 2010). ISSN 1520-5827.

[63] A. Huang, H. Bux, F. Steinbach, and J. Caro, Molecular-sieve membrane with hydrogen permselectivity: ZIF-22 in LTA topology prepared with 3-aminopropyltriethoxysilane as covalent linker., *Angewandte Chemie (International ed. in English)*. **49**(29), 4958–61 (July, 2010). ISSN 1521-3773.

[64] Y. Li, F. Liang, H. Bux, W. Yang, and J. Caro, Zeolitic imidazolate framework ZIF-7 based molecular sieve membrane for hydrogen separation, *Journal of Membrane Science*. **354**(1-2), 48–54 (May, 2010). ISSN 03767388.

[65] S. Takamizawa, Y. Takasaki, and R. Miyake, Single-crystal membrane for anisotropic and efficient gas permeation., *J. Am. Chem. Soc.* **132**(9), 2862–3 (Mar., 2010). ISSN 1520-5126.

[66] A. Huang, W. Dou, and J. Caro, Steam-stable zeolitic imidazolate framework ZIF-90 membrane with hydrogen selectivity through covalent functionalization., *J. Am. Chem. Soc.* **132**(44), 15562–4 (Nov., 2010). ISSN 1520-5126.

[67] A. Huang and J. Caro, Covalent post-functionalization of zeolitic imidazolate framework ZIF-90 membrane for enhanced hydrogen selectivity., *Angewandte Chemie* (International ed. in English). **50**(21), 4979–82 (May, 2011). ISSN 1521-3773.

[68] H. Bux, A. Feldhoff, J. Cravillon, M. Wiebcke, Y.-S. Li, and J. Caro, Oriented Zeolitic Imidazolate Framework-8 Membrane with Sharp H_2/C_3H_8 Molecular Sieve Separation, *Chem. of Mater.* **23**(8), 2262–2269 (Apr., 2011). ISSN 0897-4756.

[69] Y. Liu, G. Zeng, Y. Pan, and Z. Lai, Synthesis of highly c-oriented ZIF-69 membranes by secondary growth and their gas permeation properties, *J. Membr. Sci.* **379**(1-2), 46–51 (Sept., 2011). ISSN 03767388.

[70] Y. Pan and Z. Lai, Sharp separation of C2/C3 hydrocarbon mixtures by zeolitic imidazolate framework-8 (ZIF-8) membranes synthesized in aqueous solutions., *Chem. Comm. (Cambridge, England)*. **47**(37), 10275–7 (Oct., 2011). ISSN 1364-548X.

[71] J. A. Bohrman and M. A. Carreon, Synthesis and CO2/CH4 separation performance of Bio-MOF-1 membranes., *Chem. Comm. (Cambridge, England)*. **48**(42), 5130–2 (May, 2012). ISSN 1364-548X.

[72] X. Dong, K. Huang, S. Liu, R. Ren, W. Jin, and Y. S. Lin, Synthesis of zeolitic imidazolate framework-78 molecular-sieve membrane: defect formation and elimination, *J. Mater. Chem.* **22**(36), 19222 (Aug., 2012). ISSN 0959-9428.

[73] O. Tzialla, C. Veziri, X. Papatryfon, K. Beltsios, A. Labropoulos, B. Iliev, G. Adamova, T. Schubert, M. Kroon, M. Francisco, L. Zubeir, G. Romanos, and G. N. Karanikolos, Zeolite Imidazolate Framework–Ionic Liquid Hybrid Membranes for Highly Selective CO 2 Separation, *J. Phys. Chem. C*. **117** (36), 18434–18440 (Sept., 2013). ISSN 1932-7447.

[74] K. Huang, S. Liu, Q. Li, and W. Jin, Preparation of novel metal-carboxylate system MOF membrane for gas separation, *Sep. Purif. Technol.* **119**, 94–101 (Nov., 2013). ISSN 13835866.

[75] Z. Xie, T. Li, N. L. Rosi, and M. A. Carreon, Alumina-supported cobalt–adeninate MOF membranes for CO2/CH4 separation, *J. Mater. Chem. A.* **2**(5), 1239 (Dec., 2014). ISSN 2050-7488.

[76] H. Hayashi, A. P. Côté, H. Furukawa, M. O'Keeffe, and O. M. Yaghi, Zeolite A imidazolate frameworks., *Nature Materials.* **6**(7), 501–6 (July, 2007). ISSN 1476-1122.

[77] J. An, R. P. Fiorella, S. J. Geib, and N. L. Rosi, Synthesis, structure, assembly, and modulation of the CO2 adsorption properties of a zinc-adeninate macrocycle., *J. Am. Chem. Soc.* **131**(24), 8401–3 (June, 2009). ISSN 1520-5126.

[78] J. An, S. J. Geib, and N. L. Rosi, Cation-triggered drug release from a porous zinc-adeninate metal-organic framework., *J. Am. Chem. Soc.* **131** (24), 8376–7 (June, 2009). ISSN 1520-5126.

[79] J. An, S. J. Geib, and N. L. Rosi, High and selective CO2 uptake in a cobalt adeninate metal-organic framework exhibiting pyrimidine- and amino-decorated pores., *J. Am. Chem. Soc.* **132**(1), 38–9 (Jan., 2010). ISSN 1520-5126.

[80] J. An and N. L. Rosi, Tuning MOF CO2 adsorption properties via cation exchange., *J. Am. Chem. Soc.* **132**(16), 5578–9 (Apr., 2010). ISSN 1520-5126.

[81] J. An, C. M. Shade, D. A. Chengelis-Czegan, S. Petoud, and N. L. Rosi, Zinc-adeninate metal-organic framework for aqueous encapsulation and sensitization of near-infrared and visible emitting lanthanide cations., *J. Am. Chem. Soc.* **133**(5), 1220–3 (Mar., 2011). ISSN 1520-5126.

[82] T. Li, D.-L. Chen, J. E. Sullivan, M. T. Kozlowski, J. K. Johnson, and N. L. Rosi, Systematic modulation and enhancement of CO2:N2 selectivity and water stability in an isoreticular series of bio-MOF-11 analogues, *Chem. Sci.* **4**(4), 1746 (Mar., 2013). ISSN 2041-6520.

[83] J.-R. Li, Y. Ma, M. C. McCarthy, J. Sculley, J. Yu, H.-K. Jeong, P. B. Balbuena, and H.-C. Zhou, Carbon dioxide capture-related gas adsorption and separation in metal-organic frameworks, *Coord. Chem. Rev.* **255**(15-16), 1791–1823 (Aug., 2011). ISSN 00108545.

[84] R. Krishna and J. van Baten, Unified Maxwell-Stefan description of binary mixture diffusion in micro- and meso-porous materials, *Chem. Eng. Sci.* **64**, 3159–3178 (2009).

[85] R. Krishna and J. M. van Baten, Highlighting pitfalls in the Maxwell-Stefan modeling of water-alcohol mixture permeation across pervaporation membranes, *J. Memb. Sci.* **360**, 476–482 (2010).

[86] R. Krishna and J. M. van Baten, Investigating the potential of MgMOF-74 membranes for CO2 capture, *J. Memb. Sci.* **377**, 249–260 (2011).

[87] R. Krishna and J. M. van Baten, Maxwell-Stefan modeling of slowing-down effects in mixed gas permeation across porous membranes, *J. Memb. Sci.* **383**, 289–300 (2011).

[88] C. Sitprasert, F. Wang, V. Rudolph, and Z. Zhu, Ideal and mixture permeation selectivity of flexible prototypical zeolitic imidazolate framework – 8 Membranes, *Chem. Eng. Sci.* **108**, 23–32 (2014).

[89] M. Arjmandi and P. Majid, Mixed matrix membranes incorporated with cubic-MOF-5 for improved polyetherimide gas separation membranes: Theory and experiment, *J. Ind. Eng. Chem.* in press (2014).
[90] Existing plants – Emissions and capture program goals NETL (2009).
[91] E. Favre, Carbon dioxide recovery from post-combustion processes: Can gas permeation membranes compete with absorption?, *J. Memb. Sci.* **294** (1-2), 50–59 (May, 2007). ISSN 03767388.
[92] W. J. Koros and R. Mahajan, Pushing the limits on possibilities for large scale gas separation: Which strategies?, *J. Memb. Sci.* **175**(2), 181–196 (2000). ISSN 0376-7388.
[93] H. Lin and B. Freeman, Gas solubility, diffusivity and permeability in poly(ethylene oxide), *J. Memb. Sci.* **239**(1), 105–117 (Aug., 2004). ISSN 03767388.
[94] H. Okamoto, K.-I.; Umeo, N.; Okamyo, S.; Tanaka, K.; Kita, Selective permeation of carbon dioxide over nitrogen through polyethylene oxide containing polyimide membranes, *Chem. Lett.* **22**, 225–228 (1993).
[95] J. Huang, J. Zou, and W. S. W. Ho, Carbon Dioxide Capture Using a CO2 - Selective facilitated transport membrane, *Ind. Eng. Chem. Res.* **47**(4), 1261–1267 (Feb., 2008). ISSN 0888-5885.
[96] R. Bredesen, K. Jordal, and O. Bolland, High-temperature membranes in power generation with CO2 capture, *Chem. Eng. and Process.: Process Intensification.* **43**(9), 1129–1158 (Sept., 2004). ISSN 02552701.
[97] G. Zolandz, R.R.; Ho, W.S.W.; Sirkar, K.K., Eds.;, In Fleming Membrane Handbook, Chapman and Hall, New York, (1992).
[98] J. C. White, P. K. Dutta, K. Shqau, and H. Verweij, Synthesis of ultrathin zeolite Y membranes and their application for separation of carbon dioxide and nitrogen gases., *Langmuir* **26**(12), 10287–93 (June, 2010). ISSN 1520-5827.
[99] Y. Cui, H. Kita, and K.-i. Okamoto, Preparation and gas separation performance of zeolite T membrane, *J. Mater. Chem.* **14**(5), 924 (Feb., 2004). ISSN 0959-9428.
[100] D. W. Shin, S. H. Hyun, C. H. Cho, and M. H. Han, Synthesis and CO2/N2 gas permeation characteristics of ZSM-5 zeolite membranes, *Microporous Mesoporous Mater.* **85**(3), 313–323 (Nov., 2005). ISSN 13871811.
[101] J. van den Bergh, W. Zhu, J. Gascon, J. Moulijn, and F. Kapteijn, Separation and permeation characteristics of a DD3R zeolite membrane, *J. Memb. Sci.* **316**(1-2), 35–45 (May, 2008). ISSN 03767388.
[102] M. L. Carreon, S. Li, and M. A. Carreon, AlPO-18 membranes for CO2/CH4 separation., *Chem. Comm. (Cambridge, England).* **48**(17), 2310–2 (Mar., 2012). ISSN 1364-548X.
[103] Y.-S. Li, H. Bux, A. Feldhoff, G.-L. Li, W.-S. Yang, and J. Caro, Controllable synthesis of metal-organic frameworks: From MOF nanorods to oriented MOF membranes., *Advanced Materials (Deerfield Beach, Fla.).* **22** (30), 3322–6 (Aug., 2010). ISSN 1521-4095.
[104] Z. Zhao, X. Ma, A. Kasik, Z. Li, and Y. S. Lin, Gas Separation Properties of Metal Organic Framework (MOF-5) Membranes, *Ind. Eng. Chem. Res.*

52(3), 1102–1108 (Jan., 2013). ISSN 0888-5885.

[105] S. Li and C. Q. Fan, High-Flux SAPO-34 Membrane for CO 2 /N 2 Separation, *Ind. Eng. Chem. Res.* **49**(9), 4399–4404 (May, 2010). ISSN 0888-5885.

[106] L. M. Robeson, The upper bound revisited, *J. Memb. Sci.* **320**(1), 390–400 (2008).

[107] T. C. Merkel, H. Lin, X. Wei, and R. Baker, Power plant post-combustion carbon dioxide capture: An opportunity for membranes, *J. Memb. Sci.* **359** (1-2), 126–139 (Sept., 2010). ISSN 03767388.

[108] R. W. Baker and K. Lokhandwala, Natural gas processing with membranes: An Overview, *Ind. Eng. Chem. Res.* **47**(7), 2109–2121 (Apr., 2008). ISSN 0888-5885.

[109] R.W. Spillman, Economics of gas separation by membranes, *Chem. Eng. Progress.* **85**, 41 (1989).

[110] S. I. S. H Ohya , V V Kudryavsev, *Polyimide Membranes: Applications, Fabrications and Properties, CRC Press*(1997).

[111] R. Baker, I. Pinnau, Z. He, K. Amo, A. DaCosta, and R. Daniels. Nitrogen gas separation using organic vapor resistant membranes. Technical report, US Patent 6,572,680 (2003).

[112] S. R. Venna and M. A. Carreon, Amino-Functionalized SAPO-34 Membranes for CO2/CH4 and CO2/N2 Separation., *Langmuir* **27**(6), 2888–94 (Mar., 2011). ISSN 1520-5827.

[113] M. A. Carreon, S. Li, J. L. Falconer, and R. D. Noble, Alumina-supported SAPO-34 membranes for CO2/CH4 separation., *J. Am. Chem. Soc.* **130** (16), 5412–3 (Apr., 2008). ISSN 1520-5126.

[114] M. Carreon, S. Li, J. Falconer, and R. Noble, SAPO-34 Seeds and Membranes Prepared Using Multiple Structure Directing Agents, *Adv. Mater.* **20**(4), 729–732 (Feb., 2008). ISSN 09359648.

[115] R. Zhou, E. W. Ping, H. H. Funke, J. L. Falconer, and R. D. Noble, Improving SAPO-34 membrane synthesis, *J. Memb. Sci.* **444**, 384–393 (Oct., 2013). ISSN 03767388.

[116] E. W. Ping, R. Zhou, H. H. Funke, J. L. Falconer, and R. D. Noble, Seeded-gel synthesis of SAPO-34 single channel and monolith membranes, for CO2/CH4 separations, *J. Memb. Sci.* **415-416**, 770–775 (Oct., 2012). ISSN 03767388.

[117] H. H. Funke, M. Z. Chen, A. N. Prakash, J. L. Falconer, and R. D. Noble, Separating molecules by size in SAPO-34 membranes, *J. Memb. Sci.* **456**, 185–191 (Apr., 2014). ISSN 03767388.

[118] S. R. Venna, M. Zhu, S. Li, and M. A. Carreon, Knudsen diffusion through ZIF-8 membranes synthesized by secondary seeded growth, *J. Porous Materials.* **21**(2), 235–240 (Dec., 2014). ISSN 1380-2224.

[119] Y. Hu, X. Dong, J. Nan, W. Jin, X. Ren, N. Xu, and Y. M. Lee, Metal-organic framework membranes fabricated via reactive seeding., *Chem. Comm.* (Cambridge, England). **47**(2), 737–9 (Jan., 2011). ISSN 1364-548X.

[120] A. Bétard, H. Bux, S. Henke, D. Zacher, J. Caro, and R. A. Fischer, Fabrication of a CO2-selective membrane by stepwise liquid-phase deposition of an alkylether functionalized pillared-layered metal-organic framework

[Cu2L2P]n on a macroporous support, *Microporous Mesoporous Mater.* **150**, 76–82 (Mar., 2012). ISSN 13871811.

[121] S. Aguado, C.-H. Nicolas, V. Moizan-Baslé, C. Nieto, H. Amrouche, N. Bats, N. Audebrand, and D. Farrusseng, Facile synthesis of an ultramicroporous MOF tubular membrane with selectivity towards CO2, *New Journal of Chemistry.* **35**(1), 41 (Jan., 2011). ISSN 1144-0546.

[122] A. Huang, Y. Chen, N. Wang, Z. Hu, J. Jiang, and J. Caro, A highly permeable and selective zeolitic imidazolate framework ZIF-95 membrane for H2/CO2 separation., *Chem. Comm.* (Cambridge, England). **48**(89), 10981–3 (Nov., 2012). ISSN 1364-548X.

[123] A. J. Brown, J. R. Johnson, M. E. Lydon, W. J. Koros, C. W. Jones, and S. Nair, Continuous polycrystalline zeolitic imidazolate framework-90 membranes on polymeric hollow fibers., *Angewandte Chemie* (International ed. in English). **51**(42), 10615–8 (Oct., 2012). ISSN 1521-3773.

[124] A. Huang, Q. Liu, N. Wang, and J. Caro, Organosilica functionalized zeolitic imidazolate framework ZIF-90 membrane for CO2/CH4 separation, *Microporous Mesoporous Mater.* **192**, 18–22 (July, 2014). ISSN 13871811.

[125] T. C. Merkel, M. Zhou, and R. W. Baker, Carbon dioxide capture with membranes at an IGCC power plant, *J. Memb. Sci.* **389**, 441–450 (Feb., 2012). ISSN 03767388.

[126] Y. Ma, *Advanced Membrane Technology and Applications*, chapter Hydrogen separation membranes, pp. 671–684. John Wiley & Sons (2008).

[127] A. Y. Ku, P. Kulkarni, R. Shisler, and W. Wei, Membrane performance requirements for carbon dioxide capture using hydrogen-selective membranes in integrated gasification combined cycle (IGCC) power plants, *J. Memb. Sci.* **367**(1-2), 233–239 (Feb., 2011). ISSN 03767388.

[128] F. Guazzone, J. Catalano, I. P. Mardilovich, J. Kniep, S. Pande, T. Wu, R. C. Lambrecht, S. Datta, N. K. Kazantzis, and Y. H. Ma, Gas permeation field tests of composite Pd and Pd–Au membranes in actual coal derived syngas atmosphere, *Int. J. Hydrogen Energ.* **37**(19), 14557–14568 (Oct., 2012). ISSN 03603199.

[129] P. K. Liu, M. Sahimi, and T. T. Tsotsis, Process intensification in hydrogen production from coal and biomass via the use of membrane-based reactive separations, *Curr. Opin. Chem. Eng.* **1**(3), 342–351 (Aug., 2012). ISSN 22113398.

[130] K. Ramasubramanian, Y. Zhao, and W. Winston Ho, CO 2 capture and H 2 purification: Prospects for CO 2 -selective membrane processes, *AIChE Journal.* **59**(4), 1033–1045 (Apr., 2013). ISSN 00011541.

[131] V. I. Bondar, B. D. Freeman, and I. Pinnau, Gas transport properties of poly(ether-b-amide) segmented block copolymers, *J. Polym. Sci. Part B: Polym. Phys.* **38**(15), 2051–2062 (Aug., 2000). ISSN 0887-6266.

[132] W. Yave, A. Car, J. Wind, and K.-V. Peinemann, Nanometric thin film membranes manufactured on square meter scale: ultra-thin films for CO2 capture., *Nanotechnology.* **21**(39), 1–7 (Oct., 2010). ISSN 1361-6528.

[133] W. Yave, A. Car, and K.-V. Peinemann, Nanostructured membrane material designed for carbon dioxide separation, *J. Membr. Sci.* **350**(1-2),

124–129 (Mar., 2010). ISSN 03767388.

[134] S. R. Reijerkerk, K. Nijmeijer, C. P. Ribeiro, B. D. Freeman, and M. Wessling, On the effects of plasticization in CO2/light gas separation using polymeric solubility selective membranes, *J. Membr. Sci.* **367**(1-2), 33–44 (Feb., 2011). ISSN 03767388.

[135] H. Lin, Z. He, Z. Sun, J. Vu, A. Ng, M. Mohammed, J. Kniep, T. C. Merkel, T. Wu, and R. C. Lambrecht, CO2-selective membranes for hydrogen production and CO2 capture — Part I: Membrane development, *J. Membr. Sci.* **457**, 149–161 (May, 2014). ISSN 03767388.

[136] J. Lindmark and J. Hedlund, *From Zeolites to Porous MOF Materials - The 40th Anniversary of International Zeolite Conference, Proceedings of the 15th International Zeolite Conference.* vol. 170, *Studies in Surface Science and Catalysis*, Elsevier (2007).

[137] J. Lindmark, J. Hedlund, S. K. Wirawan, D. Creaser, M. Li, D. Zhang, and X. Zou, Impregnation of zeolite membranes for enhanced selectivity, *J. Membr. Sci.* **365**(1-2), 188–197 (Dec., 2010). ISSN 03767388.

[138] W. J. Bakker, F. Kapteijn, J. Poppe, and J. A. Moulijn, Permeation characteristics of a metal-supported silicalite-1 zeolite membrane, *J. Membr. Sci.* **117**(1-2), 57–78 (Aug., 1996). ISSN 03767388.

[139] S. Li, M. A. Carreon, Y. Zhang, H. H. Funke, R. D. Noble, and J. L. Falconer, Scale-up of SAPO-34 membranes for CO2/CH4 separation, *J. Membr. Sci.* **352**(1-2), 7–13 (Apr., 2010). ISSN 03767388.

[140] M. Li and M. Dincă, Selective formation of biphasic thin films of metal–organic frameworks by potential-controlled cathodic electrodeposition, *Chem. Sci.* **5**(1), 107 (Nov., 2014). ISSN 2041-6520.

[141] C. R. Wade, M. Li, and M. Dincă, Facile deposition of multicolored electrochromic metal-organic framework thin films., *Angewandte Chemie* (International ed. in English). **52**(50), 13377–81 (Dec., 2013). ISSN 1521-3773.

[142] C. J. Brinker, Y. Lu, A. Sellinger, and H. Fan, Evaporation-Induced Self-Assembly: Nanostructures Made Easy, *Adv. Mater.* **11**(7), 579–585 (May, 1999). ISSN 0935-9648.

[143] D. Grosso, F. Cagnol, G. Soler-Illia, E. Crepaldi, H. Amenitsch, A. Brunet-Bruneau, A. Bourgeois, and C. Sanchez, Fundamentals of Mesostructuring Through Evaporation-Induced Self-Assembly, *Adv. Funct. Mater.* **14**(4), 309–322 (Apr., 2004). ISSN 1616-301X.

[144] M. Carreon and V. Guliants, *Nanoporous Solids, Recent Advances and Prospects*, chapter Mesostructuring Metal Oxides through Evaporation Induced Self-Assembly: Fundamentals and Applications, pp. 407–432. Number 16. Elsevier (2008).

Chapter 21

Chemistry in Confined Pore Domains

Megan M. Otting, Yazhou Ji, Ryan M. Richards and Brian G. Trewyn
Department of Chemistry and Geochemistry
Colorado School of Mines, Golden, CO, 80401
rrichard@mines.edu, btrewyn@mines.edu

Recent years have seen advances in the synthesis and applications of mesoporous nanoparticle materials in molecular capture and separation, biomolecular delivery to animal and plant cells, and heterogeneous catalysis. Researchers at the Colorado School of Mines (CSM) are investigating mesoporous nanoparticles of varying composition and unique surface chemistries and have extended the synthesis methods to control morphology and surface functionality; advances that significantly increase the number of potential applications. Not only have we investigated the release from the mesopores but the uptake and sequestration of various high value and value-added biomolecules that are produced by microorganisms including microalgae, bacteria, yeast, and fungus into the pores. Additionally, multi- and single-site catalysis on porous supports is at the forefront of our research.

21.1. Introduction

Designing, developing, and synthesizing functional materials with well-defined pore structure, controlled morphology, and tunable surface chemistry is a burgeoning area of research undergoing a revolution in materials chemistry. We are currently focusing our research efforts on the synthesis, characterization, and application in catalysis, selective separations, and biomolecular delivery to plant and animal cells of mesoporous silica (MS) and ordered mesoporous carbon (OMC) materials.

Scientists led by Kresge at the Mobil Oil Corporation (MCM-41) and a research group led by Kuroda at Waseda University first discovered the synthesis of MS materials in the early 1990s [1, 2]. Both groups demonstrated a supramolecular synthetic mechanism involving both organic and

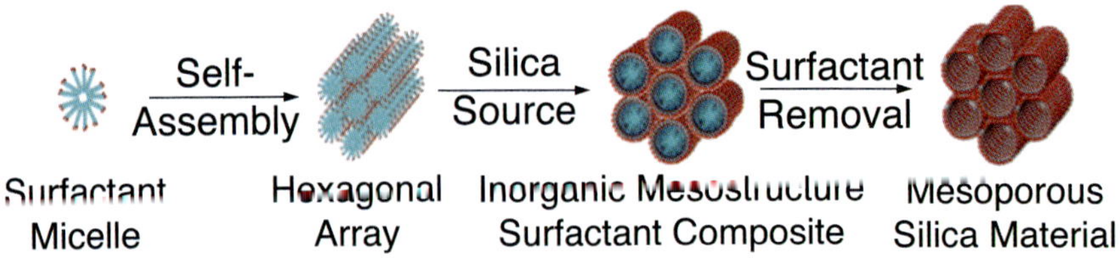

Fig. 21.1. Formation mechanism of mesoporous silica materials.

inorganic species in which the organization of the silica polymer was directed by cationic ammonium micelles (Fig. 21.1). Alternatively, the synthesis of OMC materials utilizes presynthesized MS particles as a "hard" template to form the mesopore framework along with thermal treatments of carbon precursors to fill in the pores, followed by HF or OH^- treatment to remove the silica. The unique structural features of these materials are a critical prerequisite for utilization in catalysis, separations, and bioactive molecule and gene delivery. These materials should have the following desirable characteristics to be fully functional:

(1) Chemical, thermal, and mechanical stability
(2) Ordered pores and controlled pore morphology
(3) Large surface area and tunable pore size
(4) Functionalizable interior and exterior surfaces

21.2. Synthesis and Properties

Utilizing structure-directing templates (cationic and non-ionic surfactants), the pore structure is formed in either an acidic or basic aqueous solution by adding silica precursor, which undergoes hydrolysis followed by condensation to polymerize around the templates. After a variable hydrothermal treatment time, the as-synthesized MS materials are filtered and washed. Post-synthetic removal of the organic pore-forming surfactants is accomplished by either high temperature calcination or acidified alcohol wash (ethanol/methanol) to achieve open pores (Fig. 21.1). The latter is critical when including covalently bound surface functional organic ligands, which will otherwise be pyrolyzed by the calcination step along with the surfactants.

A typical synthesis leads to surface areas in the range of 400–700 m^2 g^{-1} for nonionic surfactant templated MS materials with pore sizes ranging

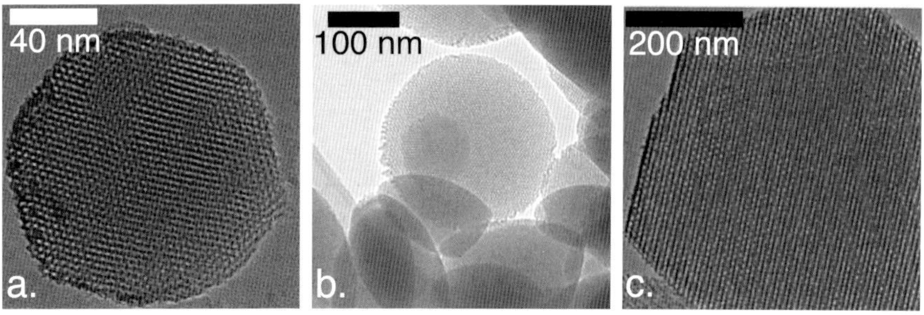

Fig. 21.2. Transmission electron micrographs of mesoporous silica particles with pore sizes tuned to (a) 3 nm, (b) 5 nm, and (c) 10 nm.

from 6 to 12 nm, and 900-1200 m^2 g^{-1} for cationic surfactant template MSN with pore sizes ranging from 2.5-4 nm. One of the important properties of MS materials is the uniquely small pore size distribution and uniform particle morphology from a single sample preparation. Product yields vary but typical research scale preparations yield anywhere from 2-4 g of MS particles, depending on the specific synthesis method. For each specific application, variation in the particle morphology and pore size is frequently desired. For example, we have procedures for loading biomolecules (proteins and peptides) into the pores for molecular delivery into cells and in these instances it is important to purposely synthesize pores that will be large enough to fit enzymes without denaturing but not too big that they readily diffuse in and out during loading procedures [3–5]. We have established methods to manipulate pore size including incorporating molecular swelling agents, which embed themselves in the organic region of the micelles that template the pores, and hydrothermal treatments, transmission electron microscopy (TEMs) of MS materials synthesized with different pore sizes can be observed in Fig. 21.2. Electron micrographs of MS templates with the corresponding OMC can be observed in Fig. 21.3. The MS templates (Figs. 21.3(a) and 21.3(c)) clearly dictate the morphology of the OMC materials (Figs. 21.3(b) and 21.3(d)). Chemical and elemental analysis has confirmed that silica is completely removed during the acid or base etching procedure. Details on the synthesis and applications of the OMC materials can be found in the following text.

To utilize these materials for the aforementioned applications, it is critical to have established methods for controlling the degree of organic functionalization. Though organic functionalization of MS and OMC materials can be either covalent or non-covalent, most applications prefer co-

Fig. 21.3. Scanning electron micrographs of MS materials (a, c) used as a template for OMC materials (b, d).

valent immobilization to circumvent undesired leaching of surface modifying molecules. There are two common techniques used to functionalize the surfaces of mesoporous silica with organic moieties for supporting organic functional groups. The first approach, post-synthesis grafting, is the more popular method due to the fewer synthetic variables that go into the preparation. As is shown in Fig. 21.4(a), either a modified homogeneous catalyst or a metal-free organic ligand is covalently immobilized on the surface of the presynthesized mesoporous support through a silylation reaction, typically in a moisture-free environment. The lack of moisture is critical to avoid self-condensation of organosilanes. While the porous structure remains intact, the placement of the ligands/catalysts may not be uniform, but rather focused on the external surface and near the pore openings. This phenomenon is typically dependent on the pore size and the functional groups mass transport freedom into the pores. The reactiv-

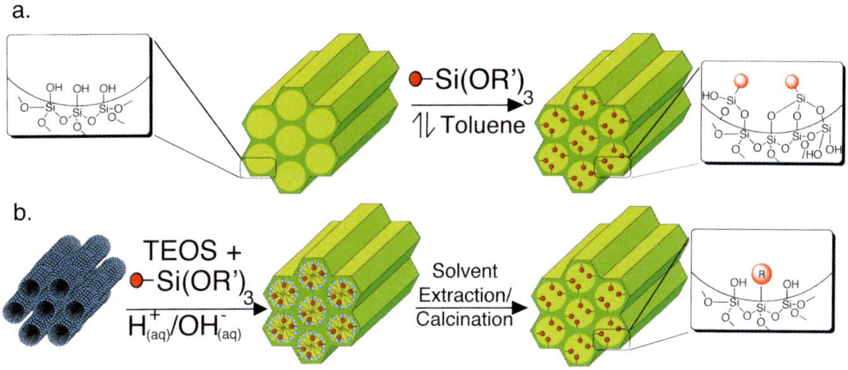

Fig. 21.4. Schematic illustration of the two most common techniques for surface functionalization of mesoporous materials. (a) describes the post-synthesis grafting method and (b) describes the in-situ co-condensation method.

ity of the surface silanol groups is a diffusion dependent characteristic and the kinetically most accessible regions are on the exterior surface and pore openings. The other approach, in situ co-condensation, is a direct synthesis method in which the organic functional group (organoalkoxysilane) is introduced to an aqueous solution (acidic or basic) of template-forming surfactant along with the silanol precursor (i.e. tetraethylorthosilicate, TEOS), illustrated in Fig. 21.4(b). The condensation of both the organoalkoxysilane and TEOS occurs simultaneously (co-condensation) leading to uniform distribution of the organic ligand on the surfaces of the mesoporous materials. In addition to a more homogeneous functionalization of the silica surface, the morphology of MCM-41 type silica materials can be altered with specific types and concentrations of organic ligands. For example, by co-condensing aminopropyl-trimethoxysilane (APTMS) with TEOS, mesoporous silica nanorods form (AP-MS). Synthesizing a series of these functionalized MSN materials demonstrated that co-condensation with the hydrophobic organoalkoxysilanes led to smaller particles than unfunctionalized MS materials whereas hydrophilic ligands selectively created larger particles (Fig. 21.5) [6, 7]. There are, however, some limitations with this method; the organic functional groups must be soluble in aqueous solutions and tolerate pH extremes since this synthesis is catalyzed by either a base or acid to activate the silicate hydrolysis. The incorporation of bulky organic functional groups is not always successful because the groups interfere with silica condensation. Finally, the amount of functional group incorporated in the mesoporous material this way cannot exceed 25% surface coverage

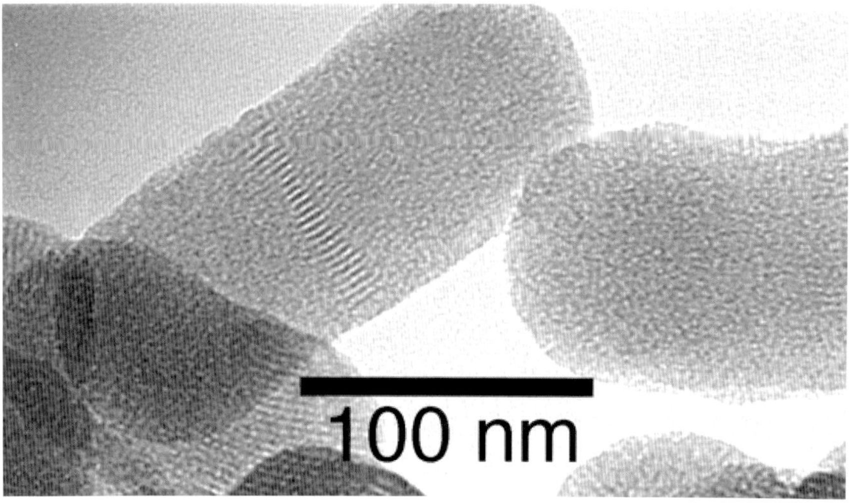

Fig. 21.5. TEM image of AP-MS nanorods synthesized via our co-condensation reaction conditions.

without having negative effects on the structural integrity.

Nanostructured catalysts offer the potential to tune properties based on their size, shape, and composition and have found use in a broad range of commercial processes. On the nanoscale (10^{-9} m, 6×10^4 times smaller than a human hair), metal particles demonstrate chemical and physical properties that are often not observed on the bulk or atomic scale. Since the initial report by Haruta demonstrating the high catalytic activity of nanoscale gold for oxidation of carbon monoxide there has been great interest in exploring the breadth of catalysis by gold [8–12]. As compared to Pt- and Pd- based catalysts, Au has the highest selectivity and is less prone to metal leaching, over-oxidation, and self-poisoning by strongly adsorbed byproducts for aqueous phase aerobic catalytic oxidation of alcohol [13]. It has been reported that the higher selectivity and absence of catalyst deactivation are likely due to a generally weaker adsorption of oxygen, hydrogen, reactants, and products on Au than other reactive metals [14, 15]. In addition, the adsorption and catalytic activity of Au strongly depends on its particle size which can be controlled by preparation methods [16]. However, as a soft metal, a major hindrance towards broad application of gold-based catalysts is that gold is very prone to aggregation and loss of activity. In an effort to minimize catalysts aggregation and sintering, dispersion of nanoscale particles on supporting materials is a promising method.

Incorporation of functional inorganic nanoparticle or multi-site reactive metal and metal oxide species is an important aspect of utilizing MS materials for heterogeneous catalysis. Similar to functionalizing with organic ligands, inorganic species can be incorporated via an in situ or post-synthetic method. We demonstrated the integration of noble metal nanoparticles, specifically gold, and earth-abundant elements in the form of metal oxides, sulfides, and phosphides incorporated into the silica without observing different phases through this in situ method [17, 18]. With the ambition of utilizing the thermal and chemical stability of porous silica to improve the durability, activity, and selectivity of nanostructured catalysts and gain a fundamental understanding of the relationships involved, we have pursued various strategies of interfacing these materials. The initial approach was to modify the surface of mesoporous silica SBA-15 and immobilize gold nanoparticles [19]. Although this approach yielded catalyst systems that demonstrated high conversions for the oxidation of cyclohexane, their durability was also limited. Gold nanoparticles confined in the walls of mesoporous silica for aerobic oxidation of alcohols was also reported previously, however, these catalysts showed irregular shapes and pores [20].

To obtain metal nanoparticle incorporated mesoporous silica either organic functional groups are decorated on the MS material via grafting or co-condensation to modify the surface chemistry and chelate metal for particle nucleation, or metals are directly chelated with silanol anions that are on the surface. In the case of gold nanoparticle (Au NP) functionalization, the interactions between the organic groups and the nanoparticles improved the immobilization of metals, however for Pd the organic ligand was not necessary, illustrating the uniqueness of metals. Combining synthesis methods (co-condensation and metal chelation), Richards and coworkers were able to intercalate gold in the mesoporous silica framework (GMS) by first synthesizing mesoporous silica surface decorated with thioethers (i.e., selectively incorporate the Au NPs in the walls of the pores [18]). This led to catalytic material that was more robust, but had equal catalytic activity as that of the surface impregnated Au NPs. Additionally, leaching becomes a major factor in the post-synthesis impregnation method and is substantially avoided by embedding the nanoparticles in the framework. Current research activities are investigating the thermal and hydrothermal stability of the silica support to optimize the conditions for complete condensation during synthesis. An example of a noble nanoparticle incorporated MS catalyst can be observed in Figs. 21.6 and 21.7. These synthetic methods

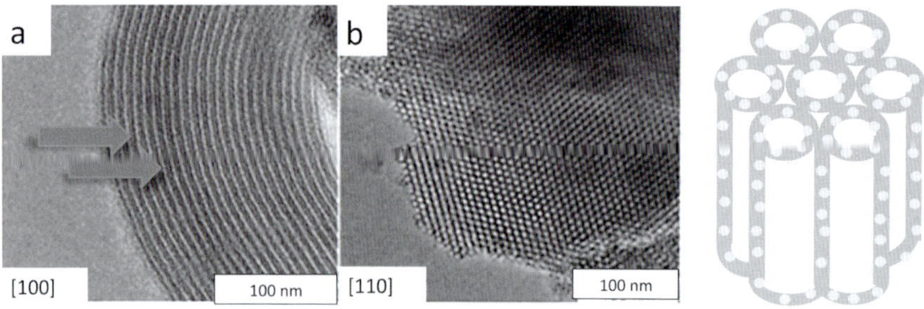

Fig. 21.6. HRTEM images of 2-2-GMS (a) and (b), the model of nanoreactor (right).

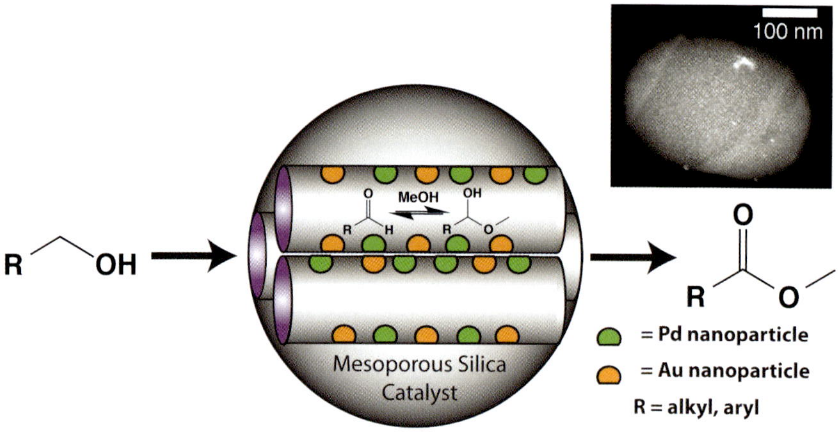

Fig. 21.7. Schematic representation of catalytic nanoparticle embedded MS materials performing as a heterogeneous oxidation catalyst converting alcohols to esters (a) and a TEM and STEM image of one of the nanoparticle embedded MS catalysts.

allow for control over the quantity of metal incorporated into the pores of the MS materials. By tuning the surface chemistry of the pores, either by adding more or less organic functional groups, or by modifying the pH to deprotonate more or less surface silanols, the pore environment around that active catalyst can be fine-tuned.

It remains desirable and challenging however, to locate nanoparticles in the framework of mesoporous materials while maintaining the periodic mesoporous structure. While a number of studies in this research area describe methods for anchoring nanoparticles on the walls of mesoporous silica, there are limited reports relating to intercalating nanoparticles in the walls of mesoporous materials. Garcia et al. incorporated γ-FeO nanoparticles in the walls of aluminosilicates [21]. In 2009, two reports involving gold nanoparticles intercalated into the walls of mesoporous silica through use of a thioether group were published [18, 22]. These reports, however, were primarily proof of concept studies and a number of questions regarding the role of the thioether agent on morphology, range of gold loading, and exact nature of the catalytic sites remain. Most importantly, the primary limitation to the selective location of nanoparticles is that there is a small amount of thioether group to attract gold while in contrast, the thioether group precursor (bis[3-(triethoxysilyl)propyl]-tetrasulfide, TESPTS) can form a microemulsion with the co-polymer template (P123) and thus could destroy or alter the periodic mesoporous structure of silica. This type of "enough is better than too much" has also been discussed in preparation of thioether-bridged mesoporous organosilicas utilizing TESPTS and tetramethoxysilane (TMOS) as the silica precursor [23].

As stated previously, MS materials are used as templates for ordered mesoporous carbon materials. Synthesis involves repeated hydrothermal treatments and thermal calcination of a carbon precursor (i.e., sucrose and furfural) followed by HF or base treatment to etch away the silica template [24]. Typical OMC surface areas synthesized by our method are in the 800 m^2 g^{-1} to 1100 m^2 g^{-1} range with pore sizes tunable between 3.5 nm to 7 nm and research-scale product yields approximately 2-4 g. Critical for many applications, the OMC materials are hydrothermally stable in high and low pH and, according to extensive Raman spectroscopy analyses, are graphitic and conductive. The unique synthetic method to produce OMC materials allows for morphology and particle size control, the morphology of the OMC particles shadows that of the MS templates exceedingly well as is clearly evident in Fig. 21.3. So well in fact that when hydrothermal treatments are employed to expand the pores in the templates at the expense of pore wall thickness and, since the formation of OMC is by reverse templating of MS particles, the pore size of OMC particles is controlled through synthetic control of MS templates.

21.3. Characterization

Heterogeneous catalytic systems and functionalized MS systems are complex due to the interactions involved between the support and the catalyst, the distribution of active sites, the uniformity of the catalytic surface, and the mechanisms involved in each reaction. In order to elucidate how the catalytic system works, including activity and selectivity, the use of multiple advanced techniques for qualitative and quantitative analysis are required. This section is intended to briefly outline the most common techniques used for the evaluation of surface, structural, pore, and functional group properties. To begin with, assessment of the surface, structural, and pore properties are determined by powder X-ray diffraction (XRD), nitrogen sorption analysis, and scanning and transmission electron microscopy (SEM and TEM, respectively). Typically, XRD diffraction peaks in the 2θ range of $1°$-$10°$ provide information regarding the arrangement of pores in mesoporous silica materials. For example, reflections corresponding to [100], [110], [200], and [210] planes can be observed for MCM-41 and SBA-15 type materials, consistent with a hexagonal arrangement of pores [2, 25]. Nitrogen sorption analysis gives valuable information on surface area, pore volume, pore size distribution, and pore surface properties of mesoporous materials. Generally, the IUPAC Type IV isotherm is observed for mesoporous materials with a hysteresis loop that can be used to predict the shape of the mesopores [26]. Along with the isotherm, the Brunauer-Emmett-Teller (BET) surface area based on multi-layer adsorption theory, and Barrett-Joyner-Halenda (BJH) pore volume based on the Kelvin equation are typically reported [27, 28]. Morphology, composition, and structural features are visualized using electron microscopy. Scanning electron microscopy is used to obtain information on overall particle morphology and uniformity in size and shape of mesoporous materials. In order to get detailed structural information, transmission electron microscopy is needed. TEM is able to surpass the power of XRD to determine pore structure and order in mesoporous materials [29, 30].

While elucidating structural properties is critically important, information on the chemistry and composition of mesoporous materials is often equally vital. Energy dispersive X-ray spectroscopy is frequently coupled with SEM and TEM to create a chemical map of the composition of materials. This is especially helpful when analyzing functionalized mesoporous materials. Another technique used to characterize heterogeneous catalysts is chemisorption, which probes the active sites. While a certain amount of catalyst may be present on the support, it is difficult to know how much of the catalyst is chemically accessible by the reactants. Various probe fluids can be introduced to the catalytic system and later removed in a controlled manner, thus showing the number of active sites, strength of these active sites, and oxidation and reduction profiles. Recent developments in solid-state nuclear magnetic resonance (ssNMR) including pulse techniques, cross polarization (CP), and magic angle spinning (MAS) have made it a very useful tool for studying heterogeneous catalysis including elucidating information on support surface-catalyst interactions [31]. Details about short-range order and structure of the catalyst can be obtained using ssNMR; for instance the structural details of a silica support can be observed using 29Si ssNMR [6, 32]. Often the catalyst material is studied using 13C and 1H (in the case of organometallic catalysts) or looking at the respective metal sites such as 195Pt [33, 34]. Structural stability, catalyst tethering, and inorganic-organic interfaces are just a few of the many characteristics of heterogeneous catalysts that can be studied with ssNMR. Ligand and catalyst loading onto support materials are often characterized using inductively coupled plasma (ICP) mass spectrometry or thermogravimetric analysis (TGA). ICP mass spectrometry is most useful for characterizing the amount of metal loaded onto the mesoporous support material. With high sensitivity (up to parts per trillion), ICP is also useful to determine metal leaching from the catalyst after a reaction. TGA is a technique often employed to determine ligand loading by monitoring weight loss during the thermal degradation of functional groups. It is especially beneficial with silica supports because the silica remains stable up to high temperatures, but organic functional groups degrade over the temperature range. Fourier transform infrared spectroscopy (FTIR) is used for qualitative analysis of the chemical composition of organic group modifiers, particularly of the reactive functional groups and the nature of bonding to the supports (i.e., electrostatic, covalent, etc.). Finally, X-ray photoelectron spectroscopy (XPS) is mainly utilized to determine the oxidation state of active metallic species on the surface of a catalytic material. Binding energies corresponding to specific states, such as Pt(II)/Pt(IV) or C-C or C=C, can be observed using this technique. Shifts in the binding energy help determine how the catalyst interacts with the support surface, and peak intensities can be used quantitatively to determine the amount of metal species present.

21.4. Applications

We have demonstrated that by modifying MS materials they can be used to deliver chemicals, proteins, and genomic materials to animal cells, plant cells and plant protoplasts [35, 36]. We designed a system that incorporated Au NP caps to cover the ends of the pores to entrap chemicals and proteins in the mesopores and non-covalently bind to genomic material on the external surface. As was mentioned above, the application of nanotechnology for intracellular delivery to plants is very limited. To investigate the potential application of MS nanoparticles (MSN) in this field a series of functionalized MSN materials were synthesized and tested in both protoplasts and plant cells. It was observed that while pristine MSNs cannot be internalized by tobacco mesophyll protoplasts, after decorating the surface with triethylene glycol (TEG), it was possible to internalize the nanoparticles. This was further corroborated by using the release of a plasmid DNA that encoded for the expression of green fluorescent protein (GFP) as a strategy to visualize the internalization of the vehicle. To address the internalization of plant cells a gold nanoparticle capped MSN system (Au-MSNs) was synthesized. The vehicle was introduced to the plant cells using a gene gun. As it will be described in more detail later, the release of both plasmid DNA and gene expression promoter was observed as proof of the efficient uptake of Au-MSNs in plant cells (Fig. 21.8) [37]. In a continuation of this work, we increased the pore size of the MSN and plated Au NPs on the surface through chemically initiated gold salt reduction [38]. These MSN materials maintained high surface area ($\approx$300 m^2 g^{-1}), large pore volume (0.88 ml g^{-1}), and pore sizes large enough to accommodate proteins even after gold was supported on the surfaces. Utilizing these materials we successfully demonstrated that proteins and DNA could be delivered to intact plant cells via gene bombardment. These experiments are seminal to plant science research because we demonstrated for the first time that porous particles can be modified to be used as vehicles for chemical, protein, and genomic material delivery to plant cells (Fig. 21.8).

Another advantage of surface functionalization of MSN is the ability to selectively sequester high-value molecules found in many biofuel feedstocks including microalgal oil and rapeseed oil. We have demonstrated that we can tune the physical properties (pore size, surface chemistry) of the MSN to selectively target free fatty acids, omega-3 fatty acids, tocopherol, and some sterols [39, 40]. These molecules are commercially interesting to pharmaceutical, neutraceutical, and cosmetic companies.

Types of MS and OMC supported catalysts can be divided into two major categories: single-site heterogeneous catalysis (SSHC) and multi-site heterogeneous catalysis (MSHC). While these are general categories, they are suitable for this review and both are developed in research laboratories at CSM. Multi-site heterogeneous catalysis, also called connected-site, is defined as closely packed atoms of reactive metal, metal oxides, alloys or, rarely, halides. Spatial independence is not a characteristic of this type of catalyst and strong interactions occur between active sites, making kinetic

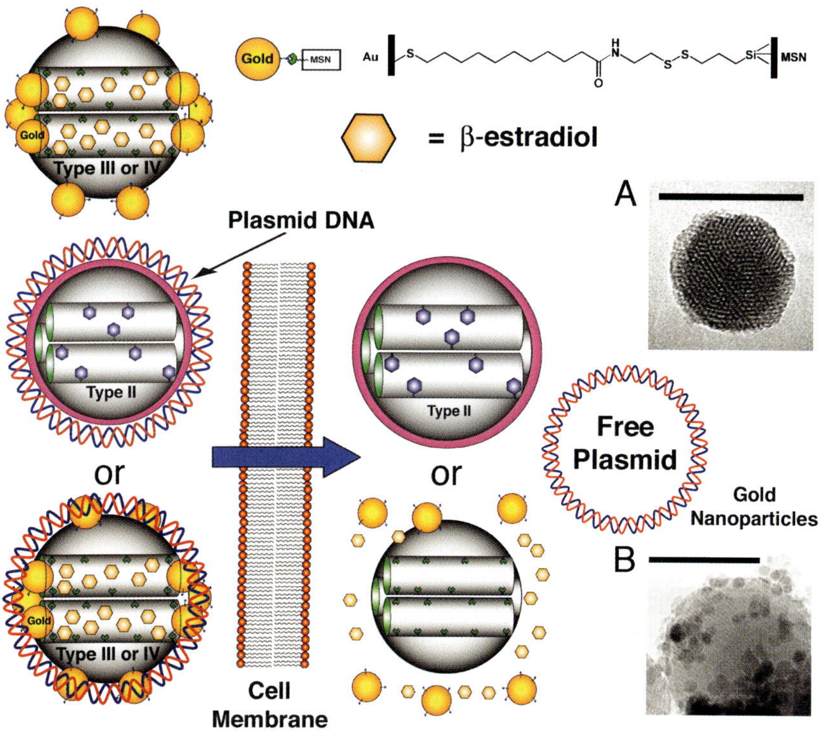

Fig. 21.8. Mesoporous silica nanoparticles for plant cell internalization with TEMs of Type II and Type IV MSN (a and b).

and thermodynamic analyses very challenging. Using the methods previously stated, we have demonstrated that active catalytic species can be incorporated into MS and OMC materials. First, by presynthesizing rhodium nanoparticles (2-3 nm diameters) we were able to incorporate these catalysts active for selective ethanol production from synthesis gas (CO and H2) and, since the NPs are in the silica framework and not on the surface, they are protected from high temperature and high-pressure sintering [41].

Recently, a novel method to synthesize a catalytic system was reported in which the gold nanoparticles are highly dispersed and all available evidence (nitrogen physisorption, X-ray diffraction, transmission electron microscopy) indicates that they are confined in the walls of mesoporous silica, not in the channels (Fig. 21.6) [18]. Thus, the channels will not be blocked and the gold nanoparticles are significantly more resistant to sintering under calcination (up to 650°C) or reaction conditions as compared to other

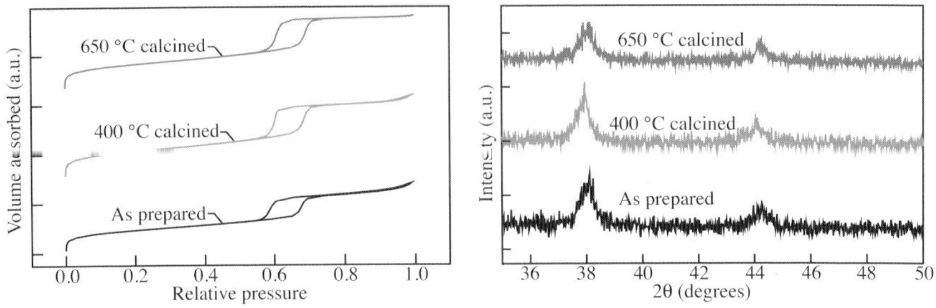

Fig. 21.9. Nitrogen sorption isotherms (above) and XRD patterns for GMS, a) as prepared, b) heated at 400°C for 3h, and c) after heating to 650°C for 3h.

silica supported gold systems. Further, these catalysts provided excellent catalytic performance in the oxidation of alcohols. The gold mesoporous silica (2-2-GMS) materials have similar nitrogen sorption isotherms to SBA-15 with very narrow pore-size distribution, which are typical for mesoporous materials. The surface area of 2-2-GMS is 800 m^2 g^{-1} with a pore size of 5.6 nm. Moreover, with a wall thickness of 4-8 nm, the materials are unusually stable to high temperature. To demonstrate the thermal stability of this system samples were heated to 400 and 650°C for 3 hours and characterized by nitrogen sorption and XRD.

As can be seen in Fig. 21.9, the integrity of both the porous network and the gold nanoparticles is unaffected by the thermal treatment inferring that the intercalation imparts thermal stability to the gold/mesoporous silica system. Examination of catalytic activity provides information regarding the accessibility of the metal as a function of the preparation procedure and resulting morphology. For the catalytic activity of the GMS material, the studies first focused on the oxidation of benzyl alcohol because the reaction is often employed as a model reaction for alcohol oxidation. The reaction pathway for the oxidation of benzyl alcohol has been suggested as (1) benzyl alcohol is oxidized into benzaldehyde; (2) benzaldehyde is further oxidized into benzoic acid; (3) benzyl alcohol and benzoic acid interact and produce benzyl benzoate. These supported catalysts performed as well or better than unsupported Au catalysts [42–46].

We also have extensive research activities in the development of catalysts supported on OMC materials. Since we use mesoporous silica as the template, there are a couple variations on the methods to get active catalysts, in nanoparticle form, supported on the OMC. For example, one method is to support the NPs on the surface of the MSN followed by repeated infusion of carbon precursor, hydrothermal, and high thermal treatments leading to active NP catalysts that are now embedded in the OMC. Embedding the NPs in either the silica for MS or carbon for OMC protects them and gives them more stability at pH and temperature extremes. This method, however, limits the exposed surface area of the active species. Interestingly, we have established another method to incorporate active metal species on OMC. To maximize the exposed surface area of catalytic NPs, we synthesize our MS template with the active species embedded in the silica, leading to MS with minimal exposed catalyst. By following this up with the preparation steps for OMC and removing the silica with either NaOH or HF we can synthesize OMC embedded metal nanoparticles with control over the amount of chemically accessible active catalyst.

In contrast, SSHC consist of spatially independent active sites and lack interactions between active sites making the interactions between each supported active site and reactant equivalent. This review will discuss SSHC supported on various mesoporous materials. We will focus, in detail, on several groundbreaking studies that have been recently published in the peer-reviewed literature. Two important structural features of SSHC are spatial separation and consistency in the structures of the active sites. This ensures constant energetic interaction between each active site and the reactant, thereby minimizing additional variables that frequently complicate catalytic studies. In many conditions, these catalysts are superior to MSHC where significant energetic interactions between the active sites lead to additional phenomena (i.e., oscillatory and chaotic behavior, making the chemistry tedious to interpret and comprehend even for simple chemical processes). While characterizing these catalysts remains challenging, computational and kinetic measurements of SSHC are less problematic than close packed heterogeneous catalysts. One of the remarkable features offered by these catalysts is spatial restriction in the pores of mesoporous materials, which has been utilized for asymmetric catalysis. For example, by tuning the surface chemistry of the pores through covalent modification, the pore environment can be hydrophobic, controlling which molecules are allowed to easily diffuse into and out of the porous network. Pairing this control over the pore environment with surface decorated catalytic species, the equilibrium of the esterification was shifted to the right because water, a byproduct in esterification, was too polar to stay in the hydrophobic pore. This inhibited the reverse reaction [47]. As the chemistry and synthetic control over particle and pore morphology advances, we have demonstrated that more complicated organometallic molecular catalysts can be

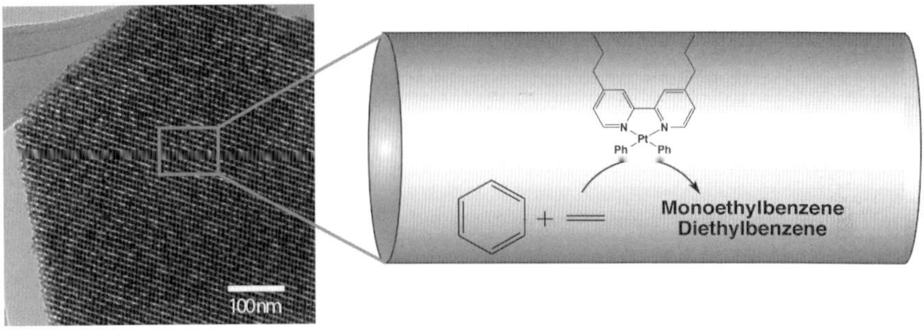

Fig. 21.10. Schematic representation of our C-H bond activating catalyst supported on MS materials.

functionalized on the surface of MS materials, which increases stability and changes product yield distribution dramatically. For example, one catalyst that we are using reacts ethylene and benzene to form ethylbenzene at 100°C (Fig. 21.10). By modifying and attaching the molecular catalyst to MS material, the catalyst stability increases and remains highly active at temperatures greater than 150°C and styrene, a major commodity to the polymer industry, becomes a key product at these elevated temperatures.

We have developed the first intracellular catalytic bioreactor supported on MS particles, utilizing the internal pores to entrap substrates and covalently bind enzymes to the external surface [48]. We also showed that catalysts can be protected from the reducing environment of cells when located inside the mesopores, yet remain active in viable cells.

Stack and coworkers have developed a novel two-step "click" chemistry synthesis for functionalizing the surface of mesoporous silica that is applicable to a wide range of organic functional groups [49]. By first co-condensing an azidopropyltriethoxysilane with SBA-15, a distribution of azide groups were loaded onto the surface with a tunable surface coverage of 2–50%. Following this, copper catalyzed azide alkyne cycloaddition ("click" chemistry) was used to immobilize various functional groups on the surface of the mesoporous support. By monitoring the fluorescence of immobilize pyrene groups, they were able to show that the distribution of the groups follows a random distribution rather than uniform surface distribution. This insight of the effects of co-condensation on organic functionalities on a mesoporous silica surface has led to a better understanding of catalyst/support interactions [49].

Utilizing novel chemistry, we have established methods to functionalize graphitic OMC with single-site catalysts [24]. This method involves

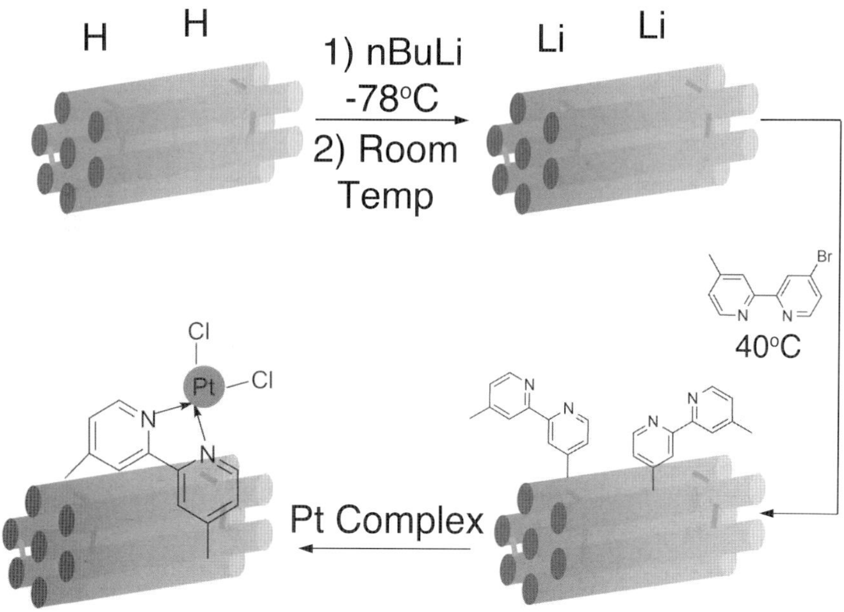

Fig. 21.11. Schematic representation illustrating our novel method for activating the surface of OMC for covalent ligand binding used for single-site catalysis.

lithiation of carbon-carbon bonds on the surface of OMC followed by covalent binding via a brominated ligand. The ligands are then metallated to achieve supported active catalysts (Fig. 21.11). The advantages of OMC are mechanical stability, hydrothermal and high and low pH stability, and conductivity. Though we are currently analyzing these catalysts for thermal catalysis, we believe they will be catalysts that will be active for electrocatalysis.

21.5. Conclusions and Outlook

In conclusion, we have spent a number of years advancing the knowledge and chemical capabilities of mesoporous silica and, more recently, ordered mesoporous carbon materials. We have demonstrated how these materials can be used for high-value molecule sequestration in the biofuel industry and how they can also be employed in the water purification industry to utilize high adsorption capacities and ability to functionalize to selectively retain and remove contaminants and pollutants. Additionally, these materials are used in plant science research using gene bombardment techniques to delivery chemicals, proteins, and genes to intact plant cells. This is a recent development in utilizing MSN in plant biology research. Finally, supporting and organizing catalytically active species, either multi- or single-site, on both MS and OMC materials has been extensively demonstrated. We have contributed considerably in discovering new synthetic methods to incorporate stable metal and metal oxide nanoparticles into the framework and on the surface of MS and OMC materials. Utilizing different synthetic techniques, we have established methods to incorporate site-isolated ligands covalently on the surfaces of MS and OMC materials to increase the stability and vary product yields of molecular organometallic catalysts.

Additionally, the research of robust nanoparticles in porous structures could be transformational for the field of catalysis because it could provide a paradigm whereby these materials could be employed in rigorous reaction conditions in which they would typically aggregate. Further, this approach offers a scalable approach to potentially independently alter the catalyst and porous environment. Since catalytic processes are of vital economic importance (approximately 35% of the gross world product), the benefits could be far reaching. An ideal "green" catalyst would be able to perform the oxidation using air as the oxidant in the absence of solvent or additional promoters. Finally, this material would be robust and recyclable. The ability to develop "green" catalysts would have a profound positive impact on the environment and improve the economic feasibility and resource efficiency of important industrial processes. As a proof of concept, studies with an emphasis on GMS have demonstrated the synthesis of a stable nanogold/mesoporous silica composite prototype with significantly improved catalytic activity for n-hexadecane and benzyl alcohol oxidation. Characterization by physisorption and transmission electron microscopy infers that the gold is intercalated into the walls of the porous silica. The catalyst systems are chemically and mechanically robust, can be separated easily, and are reusable, thus demonstrating attractive potential for practical applications.

Acknowledgements

Y.J. and R.M.R. acknowledge NSF (Grant CHE-1214068) and M.M.O and B.G.T. acknowledge the CSM Professional Development grant for financial support.

References

[1] S. Inagaki, Y. Fukushima, and K. Kuroda, Synthesis of highly ordered mesoporous materials from a layered polysilicate, *Chem. Commun.* **8**, 680–682 (1993).

[2] C. Kresge, M. Leonowicz, W. Roth, J. Vartuli, and J. Beck, Ordered mesoporous molecular sieves synthesized by a liquid-crystal template mechanism, *Nature.* **359**, 710–712 (1992).

[3] M. Joglekar, R. Roggers, Y. Zhao, and B. Trewyn, Interaction Effects of mesoporous silica nanoparticles with different morphologies on human red blood cells, *RSC Adv.* **3**, 2454–2461 (2013).

[4] I. Slowing, B. Trewyn, and V.-Y. Lin, Mesoporous silica nanoparticles for intracellular delivery of membrane-impermeable proteins, *J. Am. Chem. Soc.* **129**, 8845–8849 (2007).

[5] Y. Zhao, X. Sun, G. Zhang, B. Trewyn, I. Slowing, and V.-Y. Lin, Interaction of mesoporous silica nanoparticles with human red blood cell membranes: Size and Surface Effects, *ACS Nano.* **5**, 1366–1375 (2011).

[6] S. Huh, J. Wiench, J.-C. Yoo, M. Pruski, and V.-Y. Lin, Organic functionalization and morphology control of mesoporous silicas via a co-condensation synthesis method, *Chem. Mater.* **15**, 4247–4256 (2003).

[7] S. Huh, J. Wiench, B. Trewyn, S. Song, M. Pruski, and V.-Y. Lin, Tuning of particle morphology and pore properties in mesoporous silicas with multiple organic functional groups, *Chem. Commun.* pp. 2364–2365 (2003).

[8] M. Daniel and D. Astruc, Gold nanoparticles: assembly, supramolecular chemistry, quantum-size-related properties, and applications toward biology, catalysis, and nanotechnology, *Chem. Rev.* **104**, 293–346 (2004).

[9] M. Haruta, Gold as a novel catalyst in the 21st century: preparation, working mechanism and applications, *Gold Bull.* **37**, 27–36 (2004).

[10] M. Haruta, T. Kobayashi, H. Sano, and N. Yamada, Novel gold catalysts for the oxidation of carbon monoxide at a temperature far below 0°C, *Chem. Lett.* **16**, 405–408. (1987).

[11] A. Hashmi, K. Stephen, and G. Hutchings, Gold catalysis, *Angew. Chem. Int. Ed.* **45**, 7896–7936 (2006).

[12] C. Lui, Y. Guan, E. Hensen, J. Lee, and C. Yang, Nanocomposites as catalysts for direct propylene epoxidation with O and H mixtures, *J. Catal.* **282**, 94–102 (2011).

[13] C. Ma, B. Dou, J. Li, J. Cheng, Q. Hu, Z. Hao, and S. Qiao, Catalytic oxidation of benzyl alcohol on au or au-pd nanoparticles confined in mesoporous silica, *Appl. Catal., B.* **92**, 202–208 (2009).

[14] F. Boccuzzi, A. Chiorino, M. Manzoli, D. Andreeva, and T. Tabakova, FTIR Study of the low-temperature water-gas shift reaction on Au/Fe_2O_3 and Au/TiO_2 Catalysts, *J. Catal.* **188**, 176–185 (2003).

[15] M. Bron and R. Holze, Spectroelectrochemical Investigation of the Adsorption and oxidation of unsaturated C4-alcohols, *Surf. Sci.* **457**, 178–184 (2000).

[16] C. Baker, B. Shedd, R. Tseng, A. Martinez-Morales, C. Ozkan, M. Ozkan,

Y. Yang, and R. Kaner, Size control of gold nanoparticles grown on polyaniline nanofibers for bistable memory devices, *ACS Nano.* **5**, 3469–3474 (2011).

[17] S. Budhi, C. Peeraphatdit, S. Pylypenko, V. Nguyen, E. Smith, and B. Trewyn, enhanced metal loading in SBA-15-type catalysts facilitated by salt addition: synthesis, characterization and catalytic epoxide alcoholysis activity of molybdenum incorporated porous silica, *Appl. Catal., A.* **475**, 469–476 (2014).

[18] L. Chen, J. Hu, and R. Richards, Intercalation of aggregation-free and well-dispersed gold nanoparticles into the walls of mesoporous silica as a robust green catalyst for n-alkane oxidation, *J. Am. Chem. Soc.* **131**, 914–915 (2009).

[19] K. Zhu, J. Hu, and R. Richards, Aerobic oxidation of cyclohexane by gold nanoparticles immobilized upon mesoporous silica, *Cat. Lett.* **100**, 195–199 (2005).

[20] J. Hu, L. Chen, K. Zhu, A. Suchapar, and R. Richards, Aerobic oxidation of alcohols catalyzed by gold nanoparticles confined in the walls of mesoporous silica, *Cat. Today.* **122**, 277–283 (2007).

[21] C. Garcia, Y. Zhang, F. DiSalvo, and U. Wiesner, Mesoporous aluminosilicate materials with superparamagnetic particles embedded in the walls, *Angew. Chem. Int. Ed.* **42**, 1526–1530 (2003).

[22] E. Besson, A. Mehdi, C. Reye, and R. Corriu, Soft route for monodisperse gold nanoparticles confined within SH-functionalized walls of mesoporous silica, *J. Mater. Chem.* **19**, 4746–4752 (2009).

[23] J. Liu, Q. Yang, L. Zhang, D. Jiang, X. Shi, J. Yang, H. Zhong, and C. Li, Thioether-Bridged mesoporous organosilicas: mesophase transformations induced by the bridged organosilane precursor, *Adv. Funct. Mater.* **17**, 569–576 (2007).

[24] M. Joglekar, S. Pylypenko, M. Otting, J. Valenstein, and B. Trewyn, Universal and versatile route for selective covalent tethering of single-site catalysts and functional groups on the surface of ordered mesoporous carbons, *Chem. Mater.* **26**, 2873–2882 (2014).

[25] Z. Yuan, Q. Luo, J. Liu, T. Chen, J. Wang, and H. Li, Synthesis and characterization of boron-containing MCM-48 cubic mesoporous molecular sieves, *Microporous Mesoporous Mater.* **42**, 289–297 (2001).

[26] F. Rouquerol, J. Rouquerol, and K. Sing, *Adsorption by Powders and Porous Solids: Principles, Methodology and Applications.* Academic Press, San Diego, CA (1999).

[27] E. Barrett, L. Joyner, and P. Halenda, The determination of pore volume and area distributions in porous substances. I. Computations from nitrogen isotherms, *J. Am. Chem. Soc.* **73**, 373–380 (1951).

[28] S. Brunauer, P. Emmett, and E. Teller, Adsorption of gases in multimolecular layers, *J. Am. Chem. Soc.* **60**, 309–319 (1938).

[29] I. Daiz and A. Mayoral, TEM studies of zeolites and ordered mesoporous materials, *micron.* **42**, 512–527 (2011).

[30] O. Terasaki and T. Ohsuna. Structural study of microporous and mesoporous materials by transmission electron microscopy. In eds. S. M. Auerbach, K. A.

Carrado, and P. K. Dutta, *Handbook of Zeolite Science and Technology*. CRC Press (2003).

[31] C. Bonhomme, C. Coelho, N. Baccile, C. Gervais, T. Azais, and F. Boboneau, Advanced solid state NMR techniques for characterization of sol-gel-derived materials, *Acc. Chem. Res.* **40d**, 738–746 (2007).

[32] J. Tresboc, J. Wiench, S. Huh, V.-Y. Lin, and M. Pruski, Studies of organically functionalized mesoporous silicas using heteronuclear solid-state correlation NMR spectroscopy under fast magic angle spinning, *J. Am. Chem. Soc.* **127**, 7587–7593 (2005).

[33] N. Adachi Yoshia, Structural perameter of argonne coal samples. solid-state carbon-13 NMR spectroscopy, *Adv. Chem. Ser.* **229**, 269–279 (1993).

[34] S. Shylesh, A. Wagner, A. Siefert, S. Ernst, and W. Thiel, Cooperative acid-base effects with functionalized mesoporous silica nanoparticles: application in carbon-carbon bond formation reactions, *Chem.–Eur. J.* **15**, 7052–7062 (2009).

[35] B. Trewyn, S. Giri, I. Slowing, and V.-Y. Lin, Mesoporous silica nanoparticle based controlled release, drug delivery, and biosensor systems., *Chem. Commun.* pp. 3236–3245 (2007).

[36] B. Trewyn, I. Slowing, S. Giri, H.-T. Chen, and V.-Y. Lin, Synthesis and functionalization of a mesoporous silica nanoparticle based on the sol-gel process and applications in controlled release., *Acc. Chem. Res.* **40**, 846–853 (2007).

[37] F. Torney, B. Trewyn, V.-Y. Lin, and K. Wang, Mesoporous silica nanoparticles deliver dna and chemicals into plants, *Nat Nanotechnol.* **2**, 295–300 (2007).

[38] S. Martin-Ortigosa, J. Valenstein, V.-Y. Lin, B. Trewyn, and K. Wang, Gold functionalized mesoporous silica nanoparticle mediated protein and dna codelivery to plant cells via the biolistic method, *Adv. Funct. Mater.* **22**, 3576–3582 (2012).

[39] Y. Lee, R. Leverence, E. Smith, J. Valenstein, K. Kandel, and B. Trewyn, High-throughput analysis of algal crude oils using high resolution mass spectrometry., *Lipids.* **48**, 297–305 (2013).

[40] J. Valenstein, K. Kandel, F. Melcher, I. Slowing, V.-Y. Lin, and B. Trewyn, Functional mesoporous silica nanoparticles for the selective sequestration of free fatty acids from microalgal oil, *ACS Appl. Mater. Interf.* **4**, 1003–1009 (2012).

[41] Y. Huang, W. Deng, E. Guo, P. Chung, S. Chen, B. Trewyn, R. Brown, and V.-Y. Lin, Mesoporous silica nanoparticle-stabilized and manganese-modified rhodium nanoparticles as catalysts for highly selective synthesis of ethanol and acetaldehyde from syngas, *ChemCatChem.* **4**, 674–680 (2012).

[42] A. Abad, P. Concepcion, A. Corma, and H. Garcia, A collaborative effect between gold and a support induces the selective oxidation of alcohols, *Angew. Chem. Int. Ed.* **44**, 4066–4069 (2005).

[43] V. Choudhary, A. Dhar, P. Jana, R. Jha, and B. Uphade, A green process for chlorine-free benzaldehyde from the solvent-free oxidation of benzyl alcohol with molecular oxygen over a supported nano-size gold catalyst, *Green*

Chem. **7**, 768–770 (2005).

[44] D. Enache, J. Edwards, P. Landon, B. Solsona-Espriu, A. Carley, A. Herzing, M. Watanabe, C. Kiely, D. Knight, and G. Hutchings, Solvent-free oxidation of primary alcohols to aldehydes using Au-Pd/TiO_2 catalysts, *Science.* **311**, 362–365 (2006).

[45] D. Enache, D. Knight, and G. Hutchings, Solvent-free oxidation of primary alcohols to aldehydes using supported gold catalysts, *Cat. Lett.* **103**, 43–52 (2005).

[46] B. Guan, D. Xing, G. Cai, X. Wan, N. Yu, Z. Fang, L. Yang, and Z. Shi, Highly selective aerobic oxidation of alcohol catalyzed by a gold(I) complex with an anionic ligand, *J. Am. Chem. Soc.* **127** (2005).

[47] C. Tsai, H.-T. Chen, S. Althaus, K. Mao, T. Kobayashi, M. Pruski, and V. S.-Y. Lin, Rational catalyst design: A multifunctional mesoporous silica catalyst for shifting the reaction equilibrium by removal of byproduct, *ACS Catal.* **1**, 729–732 (2011).

[48] X. Sun, Y. Zhao, V.-Y. Lin, I. Slowing, and B. Trewyn, Luciferase and luciferin Co-immobilized mesoporous silica nanoparticle materials for intracellular biocatalysis., *J. Am. Chem. Soc.* **133**, 18554–18557 (2011).

[49] J. Nakazawa, B. Smith, and T. Stack, Discrete complexes immobilized onto click-sba-15 silica: Controllable loadings and the impact of surface coverage on catalysis, *J. Am. Chem. Soc.* **134**, 2750–2759 (2012).

Chapter 22

Fluid and Solute Transport in the Interstitial Space of Blood Clots

Adam R. Wufsus and Keith B. Neeves

Department of Chemical and Biological Engineering
Colorado School of Mines, Golden, CO 80401, USA
kneeves@mines.edu

Biological tissues can be described as porous media consisting of granular cells and a fibrous extracellular matrix. The transport of fluids and solutes through the interstitial space between cells regulates many physiological and pathological processes. In the case of blood clots, the ability of solutes to move into and out of the inner core of a clot dictates both growth and dissolution. Two physical properties that govern the rate of fluid and solute transport are the hydraulic permeability and diffusivity. Here, we report the experimental measurement of these two properties over a large range of clot compositions. Hydraulic permeability was calculated by measuring the interstitial fluid velocity through clots at a constant pressure gradient. The permeability changes by up to five orders-of-magnitude, spanning the range of 10^{-5}–$10^{-1}\mu m^2$, depending on the absolute and relative volume fractions of-cells (platelets) and a fibrous protein (fibrin). The permeability of clots can be predicted using solutions to Stokes flow in disordered fibrous media in the low cell volume fraction limit and by the Brinkman equation for clots with moderate to high cell volume fractions. The diffusivity of macromolecule tracers was measured using fluorescence recovery after photobleaching (FRAP) in fibrin gels. Diffusion of macromolecules becomes hindered even at low fibrin volume fractions (<0.05). These data are the first measurements of permeability and diffusivity over the physiological range of blood clot compositions, and thus will allow for better modeling and predictions of clot growth. These findings have important implications in the development of drug delivery strategies for disrupting clot formation and enhancing clot dissolution.

22.1. Introduction

Biological tissues are comprised of cells, blood and lymphatic vessels, and an extracellular space. This structure creates a porous medium with at least two length scales. One length scale is defined by the cells. Cells have a diameter of 1–25 μm and their volume fraction can range from 0.01–0.9. The space between cells can range from just few nanometers to hundreds of micrometers. Tissues such as cartilage are almost acellular, whereas the brain is typically >80% cells by volume. The second length scale is defined by the extracellular matrix (ECM), which is an array of fibrous proteins and sugars with diameters of 1–100 nm. The extracellular matrix is mostly water by volume (>90%) and extremely compressible. The space between fibers in the extracellular matrix can range from 10 nm to 100 μm. Taken together, these structures yield a mixed poroelastic medium where a set of roughly spherical objects (cells) are separated by a fibrous medium (ECM). Interstitial flows in many tissues are induced by the pressure gradient that exists between the blood vessels and the lymphatic vessels. These flows are typically small (0.1–1 cm s^{-1}), but are important for the exchange of nutrients and metabolic waste, the delivery and retention of drugs, and the development of nascent tissues and vessels [1].

Blood clots form at the site of vascular injury to stop the loss of blood out of the vessel in a process called hemostasis (Fig. 22.1). Excessive or deficient clot formation, thrombosis and bleeding, respectively, are the leading cause of death in the United States [2]. Blood vessels are lined with endothelial cells that maintain vessel patency by providing an inert surface that prevents blood cell adhesion. Endothelial cells also actively inhibit clot formation by secreting soluble chemicals (e.g. nitric oxide) that keep platelets, the primary blood cell responsible for clot formation, quiescent. At the site of a vascular injury, the endothelial cells lining is damaged or compromised, revealing the subendothelial matrix. This matrix promotes clot formation by providing an adhesive surface for platelet adhesion, primarily to collagens, and exposing procoagulant proteins (e.g. tissue factor) that initiate a series of biochemical reactions called coagulation. Coagulation is initiated by a protein called tissue factor that is presented on the surface of cells within the vessel wall and also circulates in the blood on the surface of white blood cells, cancer cells and cell fragments called microparticles. Coagulation is propagated on the surface of platelets by enzyme complexes that produce a burst in a protein called thrombin. Thrombin is both a platelet agonist and catalyst for the polymerization of a fibrous

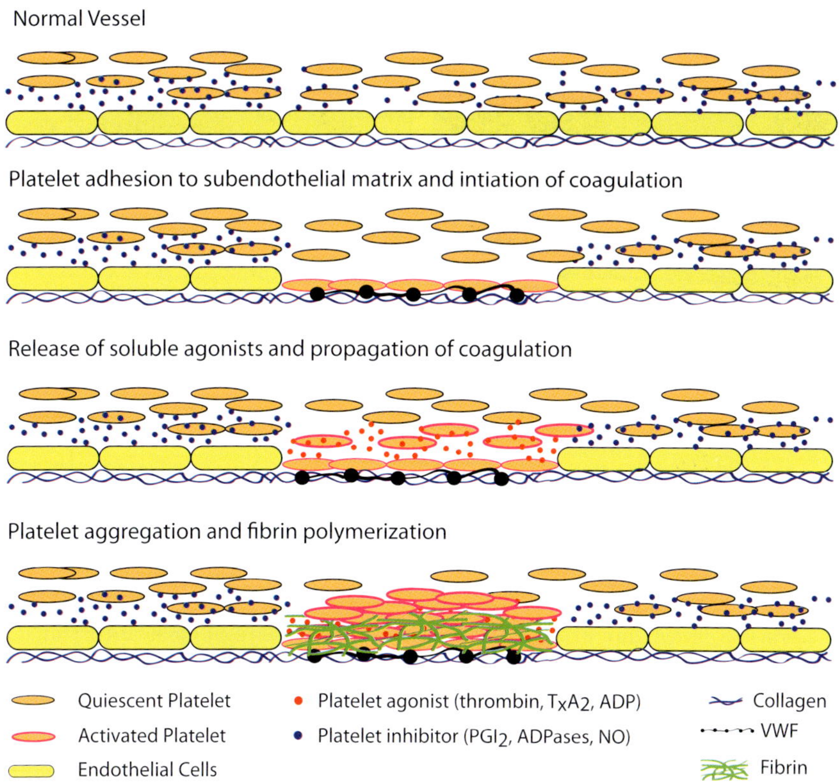

Fig. 22.1. Overview of clot formation. Endothelial cells secrete platelet inhibitors that keep platelets in a quiescent state. This include prostacyclin (PGI_2), ADPases and nitric oxide (NO) At the site of an injury the endothelial cells are removed and platelet adhere to and activate on the subendothelial matrix that contains various types of collagen and the adhesive protein von Willebrand factor (VWF). Once activated platelets secrete soluble agonists such as adenosine diphosphate (ADP) and thromboxane A_2 (TxA_2). These agonists activate new platelets which then aggregate with existing platelets in the clot. Platelets can also aggregate in the absence of these agonists at high shear stresses. Simultaneously the biochemical reaction pathway called coagulation takes place on procoagulant molecules in the subendothelial matrix and on the platelet surface. The end product of coagulation is an enzyme called thrombin that acts as a platelet agonist and a catalyst for the polymerization of the biopolymer fibrin.

polymer called fibrin.

The composition of a blood clot depends on the local blood flow velocity, the nature and extent of the injury and the vascular bed of the injury. In the large arteries where blood flow is fast, clots tend to be platelet rich be-

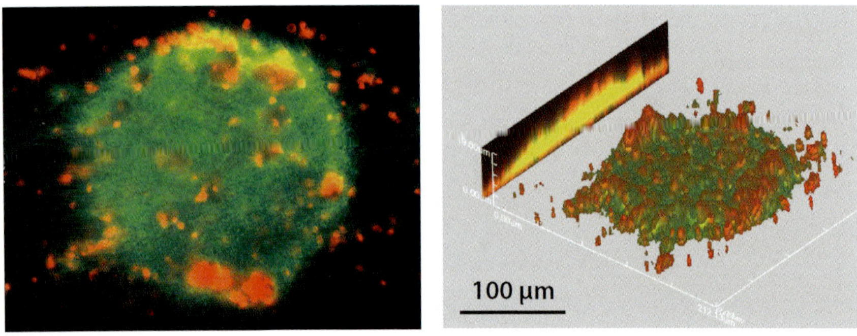

Fig. 22.2. Laser scanning confocal fluorescence images of a clot formed in an in vitro model of vascular injury. Human whole blood was perfused over a micropatterned spot containing collagen and the procoagulant protein tissue factor at a wall shear rate of 300 s^{-1}. Platelets are labeled red and fibrin(ogen) is labeled green. On the left is a top view of a clot and on the right a three-dimensional reconstruction. Note the heterogenous distribution of platelet and fibrin over a homogenous procoagulant surface.

cause the platelets can adhere and aggregate at high shear stresses whereas coagulation products tend to be diluted. These 'white clots' are characteristic of myocardial infarctions that form at the site of an atherosclerotic plaque rupture. In the veins of the extremities where blood flow is slow, clots tend to be fibrin and red blood cell rich because coagulation reactions and fibrin polymerization have a greater residence time in areas of recirculating flows such as in venous valve pockets. The fibrin entrains red blood cells to give what are referred to as 'red clots'. However, this white clot/red clot terminology oversimplifies the structure of the clot since both types of clots contain both platelets and fibrin, just in different relative densities. Clot composition is also highly heterogenous. Clots formed in large vessels have a laminate structure consisting of a series of platelet-rich and fibrin-rich layers known as the lines of Zahn. Even at a smaller length scale clots form regions of platelet-rich and fibrin-rich areas, although the mechanism of assembly is unclear (Fig. 22.2).

The deposition of fibrin and the aggregation of platelets results in a mixed porous medium structure similar to other biological tissues as described above. Scanning electron micrographs of focused ion beam milled clots show micrometer size pores in both platelet-rich and fibrin-rich regions of clots formed under flow in vitro (Fig. 22.3). Interstitial flows in these pores exist owing to the pressure gradient created by flowing blood along the periphery of a clot. There is growing evidence that fluid and solute transport within the pore space of clots plays a role in their growth,

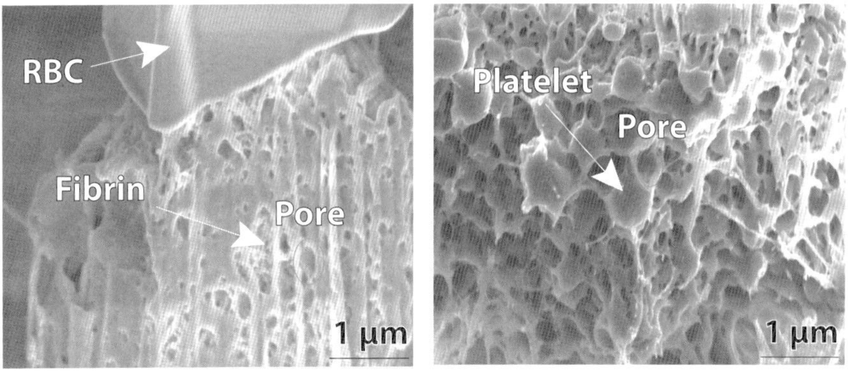

Fig. 22.3. Scanning electron micrographs of focused ion beam milling of blood clots formed in vitro on collagen-tissue factor substrates at 100 s^{-1}. Left: Interstitial pores in a fibrin-rich region of the clot. A red blood cell (RBC) drapes over the top of the clot. Right: Interstitial pores in a platelet-rich region of the clot.

arrest, and dissolution [3, 4]. However, there is a paucity of data on the physical properties of clots that dictate transport, including the hydraulic permeability and solute diffusivity, as well as a determination of the proper constitutive relationships. The data that does exist has been collected at fibrin and platelet concentrations that are orders-of-magnitude less than what has been reported in clots formed in vivo [5, 6].

In this chapter we use blood clots as a case study for the general features of porous media transport in tissues. In that, we describe methods for measuring fluid flow and solute transport at the microscopic and tissue length scales. The data from these experiments are used to derive constitutive relationships for the hydraulic permeability and diffusivity as a function of fibrin and platelet density. The implications of these findings are discussed in the context of the pathophysiology of bleeding and thrombosis and opportunities for drug delivery.

22.2. Theory

The permeability of a porous medium can be calculated from Darcy's law:

$$v = -\frac{k}{\mu}\nabla P \tag{22.1}$$

where v is the interstitial fluid velocity, k is the permeability, μ is the viscosity of the percolating fluid, and P is the pressure. The permeability is a function of the volume fraction of solids, pore size, and fiber or cell/cell aggregate size. The relative density of the two components of a clot—fibrin and platelets—dictates the interstitial fluid transport.

The diffusivity of a solute across a clot can be calculated from Fick's law:

$$j = -D\nabla c \tag{22.2}$$

where j is the flux, c is the concentration and D is the diffusivity. The diffusivity across a membrane is a function of solute size and the volume fraction of solids, pore size, and fiber or cell/cell aggregate size. Equations 22.1 and 22.2 were used to calculate permeability and diffusivity from experimental data. These experimental measurements were then compared to constitutive models of fluid and solute transport through porous media as described below.

Numerous analytical, computational and empirical relationships have been developed to predict the permeability of fibrous medium [7–14]. Most of these relationships have the general form of:

$$\frac{k_f}{a_f^2} = f(\phi_f) \tag{22.3}$$

where k_f is the permeability, a_f is the fiber radius, and ϕ_f is the fiber volume fraction. Two of the more successful models for predicting permeability of fibrous media are the Davies equation and the lattice Boltzmann (LB) simulations of Clague and colleagues [7, 15, 16]. The Davies equation is an empirical relationship based on air flow through fibrous media at low Reynolds numbers:

$$\frac{k_f}{a_f^2} = \left[16\phi_f^{1.5}(1 + 56\phi_f^3)\right]^{-1} \tag{22.4}$$

The Clague model is derived from LB simulation of flow through random fibrous media:

$$\frac{k_f}{a_f^2} = 0.51\left[\left(\frac{\pi}{4\phi_f}\right)^{0.5} - 1\right]^2 e^{-1.8\phi_f} \tag{22.5}$$

Both of these models require estimates of the fiber radius, a_f, and the fiber volume fraction, ϕ_f.

In platelet rich clots, viscous losses on the platelet surface will contribute to the overall hydrodynamic resistance through the clot in conjunction with losses on fibrin fibers. Platelets also increase the superficial velocity within the fibrin gel and increase the tortuosity. Platelet rich clots have elements of both granular (platelets) and fibrous (fibrin) media. Similar to the treatment of fibrin gels described above, two models are discussed that account for this type of mixed porous media. In the first model, the granular and fibrous media components are treated as independent hydraulic resistances that add in parallel:

$$\frac{1}{k_t} = \frac{1}{k_p} + \frac{1}{k_f} \tag{22.6}$$

where k_t is the overall permeability, k_p is the permeability of the platelets, and k_f is the permeability of the fibrin. We refer to Eq. 22.6 as the resistors-in-parallel model. This relationship is the most accurate method of averaging permeabilities in bimodal mixtures of fibers of different radii [11]. We measured k_f directly from platelet poor plasma. We estimated k_p using the Kozeny-Carmen equation for packed beds [17]:

$$k_p = \frac{\Psi^2(2a_p)^2(1-\phi_p)^3}{150\phi_p^3} \tag{22.7}$$

where Ψ is the sphericity, ϕ_p is the platelet volume fraction, and a_p is the platelet radius. The diameter of a platelet is approximately 1 μm. Sphericity was calculated by assuming platelets are flat cylinders with a diameter of 1 μm and a height of 100 nm.

The second model we considered was developed by Ethier and considers a two-component fibrous medium with highly dissimilar fiber radii [18]. This model is particularly relevant for blood clots because the radius of fibrin fibers (50-100 nm) is much smaller than the size of single platelet (1 μm). In the model, flow through the medium is described using the higher-order Brinkman equation rather than Darcy's law:

$$\mu\nabla^2 v - \frac{\mu}{k_f} - \nabla P = 0 \tag{22.8}$$

where μ is the fluid viscosity, v is the fluid velocity, k_f is the fibrin permeability and P is the pressure. In the Ethier model, the composite medium can be thought of as a matrix of fine fibers embedded with large solid inclusions. The no-slip boundary condition on these large inclusions is satisfied by the Laplacian term in Eq. 22.8. Experimental measurements of permeability of PRC were compared to the unit-cell matching in Fig. 3b of [18]. Note that Eq. 22.8 is only applicable at length scales where the pressure gradient is balanced by the Laplacian of the velocity. This length is often referred to as the Brinkman or effective screening length, and is equal to the square root of the permeability, $\sqrt{k_f}$.

Table 22.1. Permeability of fibrin gels. Fibrin gels were made from fibrinogen solutions using 10 nM thrombin and 2.5 mM $CaCl_2$. The hydrated fiber radius, a_f, was measured from SEM images. Flow was induced through the fibrin gels by pressure gradients that range from 10^4–10^6 Pa/m. The permeability is reported as a fit to Eq. 22.1 with uncertainty estimated by the standard error (SE) of the fit. The effective screening length is the square root of the measured permeability.

ϕ_f	c_{Fbg} [mg mL^{-1}]	a_f [nm]	k_f (SE) [$\times 10^3 \mu m^2$]	Effective Screening Length [nm]
0.02	3	60	120 (15)	346
0.03	7	58	52 (14)	228
0.04	10	57	22 (3)	148
0.08	20	55	7.0 (0.2)	84
0.10	24	55	3.4 (0.6)	58
0.18	48	53	0.73 (0.02)	27
0.25	67	52	0.57 (0.07)	24
0.27	72	52	0.51 (0.07)	23
0.32	89	52	0.44 (0.04)	21
0.35	96	52	0.33 (0.09)	18
0.43	121	51	0.27 (0.12)	16
0.54	156	51	0.15 (0.02)	12

The diffusivity of a solute will be hindered in both fibrin gels and platelet rich clots because of an increased effective viscosity near the pore wall and because of size exclusion where the solute diameter is larger than the pore diameter. One model used to describe the hindered diffusivity of solutes is the effective medium model [19], which claims that the effective diffusivity is a function of pore size and thus should be directly related to permeability. Phillips and colleagues derived an expression for an effective medium that relates the permeability of a porous medium to the hindered diffusion coefficient [20]:

$$\frac{D}{D_o} = \left(1 + \sqrt{\frac{d^2}{4k} + \frac{d^2}{12k}}\right)^{-1} \tag{22.9}$$

where D is the diffusion coefficient of the solute in the clot, D_o is the diffusion coefficient of the solute in free solution, k is the measured permeability of a clot and d is the diameter of the solute.

22.3. Experimental Methods

Clots were formed by adding 10 nM α-thrombin to solutions containing 3–156 mg ml^{-1} of fibrinogen and 0-5 $\times$ 10^7 platelets μL^{-1} according to previously developed protocols [21]. To measure permeability, both fibrin gels and platelet rich clots (PRC) were formed in 3 mL syringes, which served as permeation chambers. The flow rate of Tris buffer saline (TBS) through these clots at a constant pressure was used to calculate the permeability using Darcy's law. Fiber diameters ($2a_f$) were measured by scanning electron microscopy and were found to be relatively insensitive to fibrinogen concentration over the range of 7 mg ml^{-1} to 156 mg ml^{-1}. The measured values of fiber diameter were roughly between 50-60 nm for all fibrinogen concentrations. Platelet aggregate size was measured by confocal microscopy of platelets labeled with an anti-CD41 fluorescent antibody.

Fluorescence recovery after photobleaching (FRAP) is often used to measure diffusion coefficients in biological systems such as lipid bilayers or in intracellular compartments. The solute to be measured is labeled with a fluorophore that is known to easily photobleach with a suitable laser intensity. The diffusion of fluorescent molecules into the bleached region is measured over time and the transient recovery of fluorescence is used to calculate a diffusion coefficient. Here, we use FRAP to measure the diffusion of FITC-labeled dextrans of 12, 29, and 54 nm in diameter in fibrin gels. Diffusion coefficients were calculated from the transient recovery profiles using the Matlab code described by Jönsonn and colleagues [22]. Owing to light scattering by platelets we could not use FRAP for platelet rich clots, and instead used a custom diaphragm diffusion cell as previously described [23].

22.4. Results and Discussion

The permeability of fibrin gels spans three orders of magnitude over a fiber volume fraction range of 0.02 to 0.54 (Table 22.1). Permeability was most sensitive to volume fraction in the dilute to semi-dilute range of 0.02–0.18 (Fig. 22.4). Permeability was less sensitive to volume fraction range of 0.18–0.54, decreasing by less than one order of magnitude. The measured dimensionless permeability (k_f/a_f^2) of fibrin gels was compared to predictions from the two previously discussed models. Both the Davies equation and LB simulations were in good agreement with experimental values over the entire range of volume fractions with the more complex LB simulation

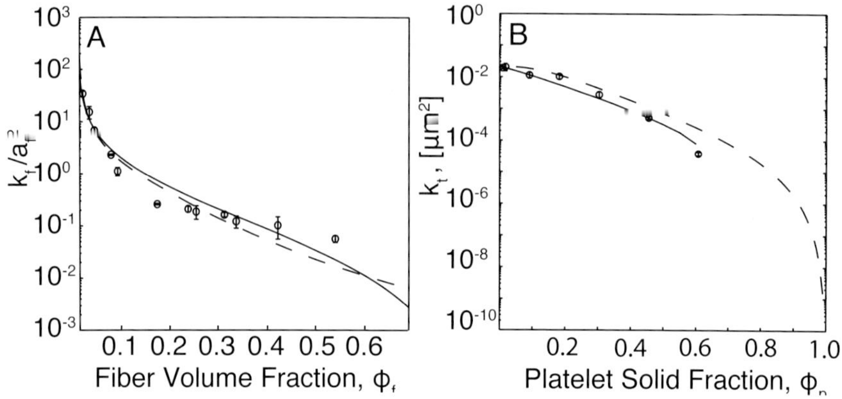

Fig. 22.4. A. Comparison of fibrin gel permeability to fibrous media models. The experimentally measured dimensionless permeability (k_f/a_f^2) of fibrin gels (○) compared with the Davies Equation, Eq. 22.4 (– –); lattice Boltzmann simulation results, Eq. 22.5 (—). B. Comparison of platelet rich clot permeability to mixed porous media models. The experimentally measured permeability, k_t, of platelet rich clots (○) compared with the resistor-in-parallel model, Eq. 22.6 (– –); Ethier's solution for $k_f/a^2_{\text{coarse}} = 1$ (—). Error bars represent standard deviations of n=3-4. Figure from Ref. [21].

being slightly more accurate (Fig. 22.4). Based on these comparisons, we conclude that fibrin gels over a fiber volume fraction of 0.02–0.54 are well described by models of disordered cylinders with uniform diameters. Gels made of 3–10 mg/mL fibrinogen consist of fibers spaced as far apart as approximately 10 μm within a given plane (Fig. 22.5). The effective hydrodynamic screening length for these gels, estimated as the square root of the permeability, is 148–346 nm (Table 22.1). Gels of 20–40 mg mL^{-1} fibrinogen consist of a dense mat of fibers with fibers spaced as far apart as approximately 1 μm within a given optical plane. These gels have an effective hydrodynamic screening length of 58–83 nm. Gels of 67 mg mL^{-1} and greater have no discernible porous structure at the resolution of the microscope and have an effective hydrodynamic screening length of 27 nm or less.

The permeability of PRC decreased by three orders-of-magnitude over a range of platelet volume fractions of 0.01 to 0.61 (Table 22.2). The platelet volume fraction range was calculated from laser scanning confocal microscopy images (Fig. 22.6). The permeability of bovine plasma clotted with 10 nM thrombin in the absence of any platelets was $1.2 \times 10^{-2} \pm 0.6 \times 10^{-2} \mu\text{m}^2$. We did not observe a contribution of the platelets

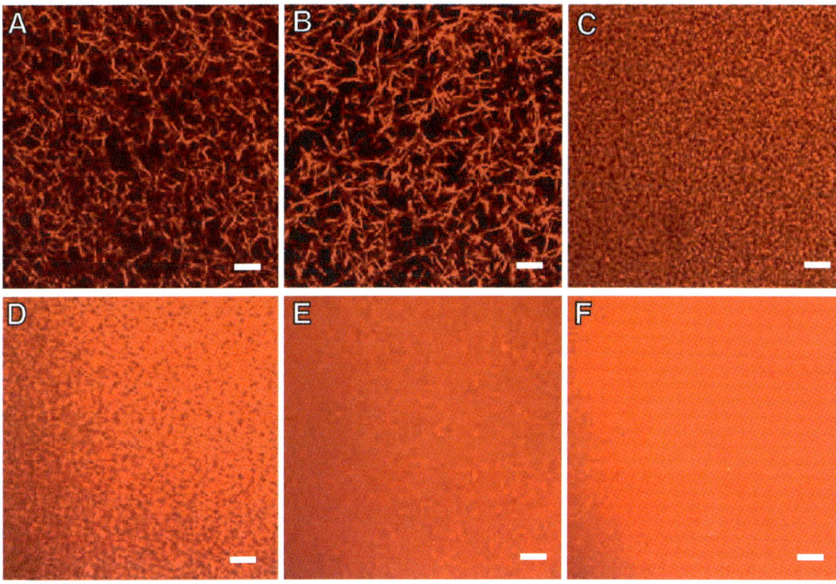

Fig. 22.5. Fibrin gel morphology as a function of fibrin density. Fibrin gels were made at various fibrinogen concentrations using 10 nM thrombin and imaged by confocal microscopy. Gels consisted of unlabeled fibrinogen and fluorescent labeled fibrinogen at a ratio 500:1 unlabeled:labeled. Final fibrinogen concentration of gels: (A) 3 mg mL^{-1}; (B) 10 mg mL^{-1}; (C) 20 mg mL^{-1}; (D) 40 mg mL^{-1}; (E) 60 mg mL^{-1}; (F) 100 mg mL^{-1}. Scale bar = 20 μm. Figure from Ref. [21].

to the permeability for platelet volume fractions of less than 0.19 (Table 22.2). At a platelet volume fraction of 0.19-0.31 we observed a modest contribution of the platelets to permeability with ~50% decrease compared to a pure fibrin gel. At higher platelet volume fractions (>0.31) the contribution of the platelets begins to dominate the overall hydraulic resistance. The permeability of a PRC with a platelet volume fraction of 0.37 was an order of magnitude less than that of a pure fibrin gel, decreasing from 1.5 $\times$ $10^{-2}\mu m^2$ to 1.1 $\times$ $10^{-3}\mu m^2$.

We compared the permeability of PRC to two mixed porous medium models; a resistors-in-parallel model, Eq. 22.6, and Ethier's numerical calculations of the Brinkman equation for fibrous medium with two dissimilar radii (Fig. 22.4). The resistors-in-parallel model treats the PRC as two independent and non-interacting media fibrous medium defined by fibrin and granular medium defined by platelets where the overall permeability is calculated by adding the resistances of each medium in parallel. The resistors-

Table 22.2. Permeability of platelet rich clots. Platelet rich clots (PRC) were made from suspension with different platelet densities, ρ_p, in platelet poor plasma and clotted with 10 nM thrombin in the presence of 2.5 mM $CaCl_2$. Platelet volume fractions, ϕ_p, were calculated from confocal microscopy images. Flow was induced through the clots by pressure gradients that range from 10^4–10^6 Pa m^{-1}. The radii of platelets and platelet aggregates, a_p, were estimated from confocal microscopy images of PRC. The permeability, k_t, is reported as a fit to Eq. 22.1 with uncertainty estimated by the standard error (SE) of the fit. The effective screening length is the square root of the measured permeability. Figure from [21]

ϕ_p	ρ_p [platelets μL^{-1}]	a_p [μm] (SD)	k_t (SE) [$\times 10^3 \mu m^2$]	Effective Screening Length [nm]
0.01	7×10^5	1 (0.2)	11 (1)	105
0.05	1×10^6	1 (0.3)	12 (1)	110
0.19	7×10^6	4 (4)	6.7 (0.8)	82
0.31	1×10^7	8 (3)	6.1 (0.7)	78
0.37	2×10^7	16 (5)	1.1 (0.5)	33
0.45	4×10^7	17 (6)	0.20 (0.02)	14
0.61	5×10^7	26 (6)	0.015 (0.002)	4

in-parallel model provides a good estimate of the permeability over most of the range of platelet volume fractions but begins to deviate as the platelet volume fraction increases beyond 0.3. Comparison of the data with the Ethier model shows that a PRC is well described as a Brinkman medium. This suggests that the Brinkman equation provides a better qualitative assessment of the interplay between fibrin and platelets as platelet volume fraction increases. Specifically, increases in platelet volume fraction result not only in viscous losses on the platelet surface, but also results in an increase of the tortuosity of the flow path and in an increase superficial velocity of the fluid through the fibrin gel.

The diffusion coefficients of solutes in fibrin gels and PRC were hindered compared to free solution. In 100 mg mL^{-1} fibrin gels we observed 60%, 66% and 85% reductions of diffusion coefficients for 12, 29, and 54 nm solutes (70 kDa, 500 kDa, 2000 kDa dextrans), respectively. Similar to the trend observed with the permeability, hindered diffusivity was most sensitive to volume fraction over the range 0.02–0.18 for all probes tested. This is expected as the permeability is an indirect measure of pore size. The effective medium model (Eq. 22.9) was used to predict the diffusion coefficient of the 12, 29, and 54 nm probes using the measured permeability as a model input (Fig. 22.7A). The effective medium model predicted the hindered diffusivity well for the 29 and 54 nm probes, but overestimated the

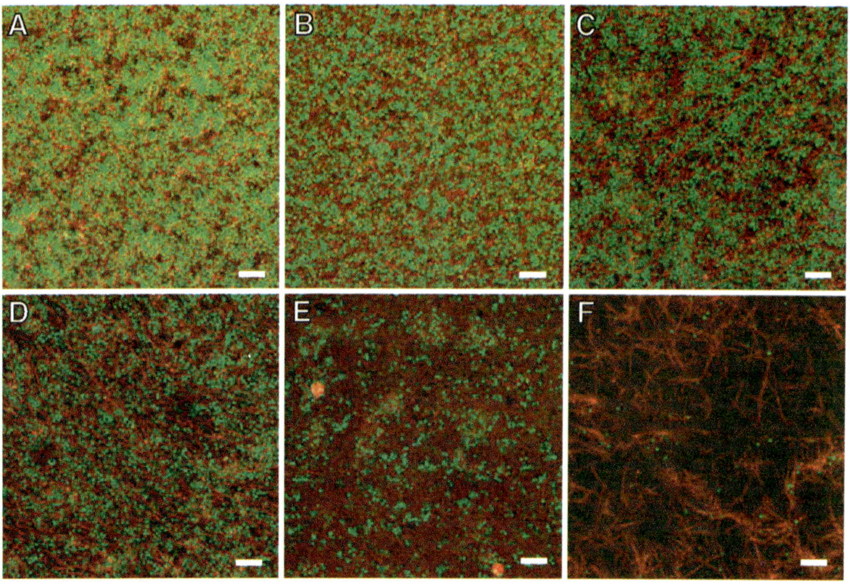

Fig. 22.6. Platelet rich clots morphology as function of platelet density. Platelet rich clots were made with platelets (green) diluted into plasma containing labeled fibrinogen (red) and imaged by confocal microscopy. Clotting was induced with 10 nM thrombin. The volume fraction of platelets, ϕ_p, for each panel is (A) 0.61; (B) 0.45; (C) 0.31; (D) 0.19; (E) 0.05; (F) 0.01. Scale Bar = 20 μm.

diffusivity of the 12 nm probe at higher fibrinogen concentrations. The diffusion coefficient of a 70 kDa dextran in a PRC decreased by as much as 93% compared to free solution as platelet volume fraction was increased to 0.61. The effective medium model was used to accurately predict the degree of hindrance using the measured permeability as a model input (Fig. 22.7(B)). Similar to the trend observed in permeability, the majority of the hindrance occurred above platelet volume fraction of 0.3.

22.5. Conclusions

The primary outcome of these studies is a catalog of physical properties that determine fluid and solute transport in clots as a function of their composition. These data provide bounds on how far zymogens and fibrinolytic agents can be expected to penetrate into clots with different compositions. For example, at pressures typical of large human arteries one would except fibrinolytic agents like tissue plasminogen activator (tPA) to penetrate no

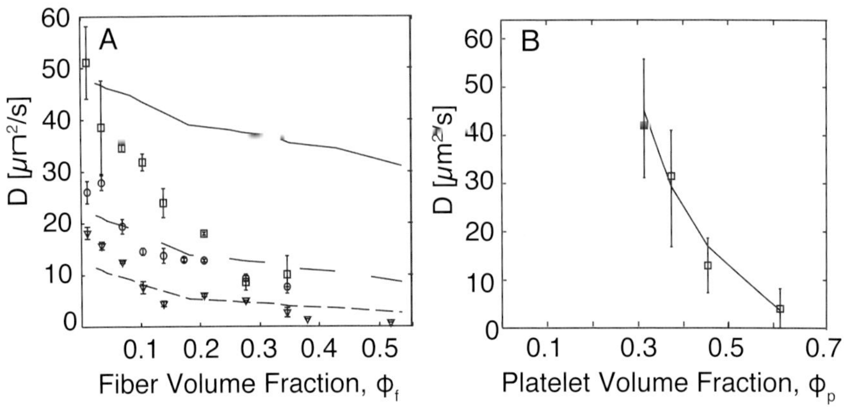

Fig. 22.7. Diffusion in fibrin gels and platelet rich clots. Measured diffusivity of 12 (□), 29 (○), 54 (▽) nm probes (70 kDa, 500 kDa, 2000 kDa dextrans) in (A) fibrin gels and (B) platelet rich clots compared the effective medium model, Eq. 22.9.

further than ∼100 μm over 1 hr into the interior platelet rich clot. This is too slow for most clinical applications and supports reports that fibrinolytic drugs have limited success on platelet rich clots [24]. In addition, the hindrance of transport of coagulation proteins and platelet agonists out of clots is a potential mechanism for the arrest of clots growth. Experiments from our group in both plasma and whole blood systems show evidence of self-limited growth of clots and fibrin gels that cannot be accounted for purely by inhibition of coagulation [25, 26]. For example, for the same 100 μm clot one would expect thrombin to be able to diffuse from the vessel wall to the lumen in ∼50 s; however, the hindrance caused by a dense platelet clot increases the time to nearly 12 min, by which time it would likely adsorb to fibrin or be inhibited by antithrombin III. Finally, physical properties presented here can be used to approximate the pore diameter, d_p, of a clot. This provides data to determine the size of a zymogen that is physically excluded from an interior region of a clot. For example, for a dense platelet clot $d_p = 4$ nm. This coincides with a recent study in an animal model of thrombosis that found a small probe (2.7 nm) was able to penetrate into a dense platelet core, but a large (6.4 nm) probe was limited [4]. These data will aid in the development of the growing number of computational models of clot formation, which, to date, have relied on semi-empirical correlations that are not valid for the wide range of solid fractions found in real clots [27].

Acknowledgments

This work was supported by a Grant–in–Aid award from the American Heart Association, National Science Foundation CAREER Award, the National Institutes of Health (R21NS082933 and R01HL120728), Bayer Hemophilia Awards Program, and US–Israel Binational Science Foundation.

References

[1] M. A. Swartz and M. E. Fleury, Interstitial Flow and Its Effects in Soft Tissues, *Ann. Rev. Biomed. Eng.* **9**(1), 229–256 (Aug., 2007).

[2] A. S. Go, D. Mozaffarian, V. L. Roger, E. J. Benjamin, J. D. Berry, W. B. Borden, D. M. Bravata, S. Dai, E. S. Ford, C. S. Fox, S. Franco, H. J. Fullerton, C. Gillespie, S. M. Hailpern, J. A. Heit, V. J. Howard, M. D. Huffman, B. M. Kissela, S. J. Kittner, D. T. Lackland, J. H. Lichtman, L. D. Lisabeth, D. Magid, G. M. Marcus, A. Marelli, D. B. Matchar, D. K. McGuire, E. R. Mohler, C. S. Moy, M. E. Mussolino, G. Nichol, N. P. Paynter, P. J. Schreiner, P. D. Sorlie, J. Stein, T. N. Turan, S. S. Virani, N. D. Wong, D. Woo, and M. B. Turner, The "Heart Disease and Stroke Statistics–2013 Update" and the need for a national cardiovascular surveillance system, *Circulation.* **127**(1), 21–23 (Jan., 2013).

[3] K. Leiderman and A. L. Fogelson, Grow with the flow: a spatial-temporal model of platelet deposition and blood coagulation under flow, *Math. Med. Biol.* **28**(1), 47–84 (Mar., 2011).

[4] T. J. Stalker, E. A. Traxler, J. Wu, K. M. Wannemahcer, S. L. Cermignano, R. Voronov, S. L. Diamond, and L. F. Brass, Hierarchical organization in the hemostatic response and its relationship to the platelet-signaling network, *Blood.* **121**(10), 1875–1885 (Mar., 2013).

[5] S. He, H. Cao, A. Antovic, and M. Blomback, Modifications of flow measurement to determine fibrin gel permeability and the preliminary use in research and clinical materials, *Blood Coagul. Fibrinolysis.* **16**(1), 61–67 (2005).

[6] R. C. Spero, R. K. Sircar, R. Schubert, R. M. Taylor, II, A. S. Wolberg, and R. Superfine, nanoparticle diffusion measures bulk clot permeability, *Biophys. J.* **101**(4), 943–950 (Aug., 2011).

[7] D. Clague, B. Kandhai, R. Zhang, and P. Sloot, Hydraulic permeability of (un) bounded fibrous media using the lattice Boltzmann method, *Phys. Rev. E.* **61**(1), 616 (2000).

[8] J. Drummond and M. Tahir, Laminar viscous flow through regular arrays of parallel solid cylinders, *Int. J. Multiph. Flow.* **10**(5), 515–540 (1984).

[9] J. Higdon and G. D. Ford, Permeability of three-dimensional models of fibrous porous media, *J. Fluid Mech.* **308**, 341–361 (1996).

[10] G. Jackson and D. F. James, The permeability of fibrous porous media, *The Canadian J. Chem. Eng.* **64**(3), 364–374 (1986).
[11] K. J. Mattern and W. M. Deen, "Mixing rules" for estimating the hydraulic permeability of fiber mixtures, *AIChE Journal.* **54**(1), 32–41 (2007).
[12] A. Sangani and A. Acrivos, Slow flow past periodic arrays of cylinders with application to heat transfer, *Int. J. Multiph. Flow.* **8**(4), 343–360 (1982).
[13] T. Stylianopoulos, A. Yeckel, J. J. Derby, X.-J. Luo, M. S. Shephard, E. A. Sander, and V. H. Barocas, Permeability calculations in three-dimensional isotropic and oriented fiber networks, *Phys. Fluids.* **20**(12), 123601 (2008).
[14] K. E. Thompson, Porescale modeling of fluid transport in disordered fibrous materials, *AIChE Journal.* **48**(7), 1369–1389 (2002).
[15] C. N. Davies, The Separation of Airborne Dust and Particles., *Arhiv za Higijenu Rada.* **1**(4), 393–427 (1950).
[16] D. S. Clague and R. J. Phillips, A numerical calculation of the hydraulic permeability of three-dimensional disordered fibrous media, *Phys. Fluids.* **9**, 1562–1572 (1997).
[17] W. L. McCabe, J. C. Smith, and P. Harriott, *Unit Operation in Chemical Engineering.* McGraw-Hill College (1993).
[18] C. Ethier, Flow through mixed fibrous porous materials, *AIChE Journal.* **37** (8), 1227–1236 (1991).
[19] D. L. Koch and J. F. Brady, The effective diffusivity of fibrous media, *AIChE Journal.* **32**(4), 575–591 (1986).
[20] R. J. Phillips, W. M. Deen, and J. F. Brady, Hindered transport of spherical macromolecules in fibrous membranes and gels, *AIChE Journal.* **35**(11), 1761–1769 (Nov., 1989).
[21] A. R. Wufsus, N. E. Macera, and K. B. Neeves, The hydraulic permeability of blood clots as a function of fibrin and platelet density, *Biophy. J.* **104**(8), 1812–1823 (Apr., 2013).
[22] P. Jönsson, M. P. Jonsson, J. O. Tegenfeldt, and F. Höök, A method improving the accuracy of fluorescence recovery after photobleaching analysis, *Biophy. J.* **95**(11), 5334–5348 (Dec., 2008).
[23] E. L. Cussler, *Diffusion: Mass Transfer in Fluid Systems*, 2nd edn. Cambridge University Press (1997).
[24] J. Vidmar, A. Blinc, E. Kralj, J. Balažic, F. Bajd, and I. Sersa, An MRI study of the differences in the rate of thrombolysis between red blood cell-rich and platelet-rich components of venous thrombi ex vivo, *J. Magn. Reson. Im.* **34**(5), 1184–1191 (Aug., 2011).
[25] A. A. Onasoga-Jarvis, T. J. Puls, S. K. O'Brien, L. Kuang, H. J. Liang, and K. B. Neeves, Thrombin generation and fibrin formation under flow on biomimetic tissue factor rich surfaces, *J. Thromb. Haemost.* **12**(3), 373–382 (2014).
[26] A. A. Onasoga-Jarvis, K. Leiderman, A. L. Fogelson, M. Wang, M. J. Manco-Johnson, J. A. Di Paola, and K. B. Neeves, The effect of factor VIII deficiencies and replacement and bypass therapies on thrombus formation under venous flow conditions in microfluidic and computational models, *PLoS ONE.* **8**(11), e78732 (Nov., 2013).

[27] A. L. Fogelson, Y. H. Hussain, and K. Leiderman, Blood clot formation under flow: The importance of factor XI depends strongly on platelet count, *Biophy. J.* **102**(1), 10–18 (Jan., 2012).

Author Index

Subject Index